ISBN 978-0-260-70873-1
PIBN 10023013

# AMERICAN
# Journal of Mathematics

EDITED BY

FRANK MORLEY

WITH THE COOPERATION OF

A. COHEN, CHARLOTTE A. SCOTT

AND OTHER MATHEMATICIANS

PUBLISHED UNDER THE AUSPICES OF THE JOHNS HOPKINS UNIVERSITY

*Πραγμάτων ἔλεγχος οὐ βλεπομένων*

VOLUME XXXVIII

BALTIMORE: THE JOHNS HOPKINS PRESS

LEMCKE & BUECHNER, *New York.*
G. E. STECHERT & CO., *New York.*
E. STEIGER & CO., *New York.*
ARTHUR F. BIRD, *London.*
WILLIAM WESLEY & SON, *London.*
A. HERMANN, *Paris.*
MAYER & MÜLLER, *Berlin.*

1916

The Lord Baltimore Press
BALTIMORE, MD., U. S. A.

# INDEX.

# AMERICAN
# ournal of Mathematics

EDITED BY

**FRANK MORLEY**

WITH THE COÖPERATION OF

A. COHEN, CHARLOTTE A. SCOTT

AND OTHER MATHEMATICIANS

PUBLISHED UNDER THE AUSPICES OF THE JOHNS HOPKINS UNIVERSITY

*Πραγμάτων ἔλεγχος οὐ βλεπομένων*

VOLUME XXXVIII, NUMBER 1

BALTIMORE: THE JOHNS HOPKINS PRESS

LEMCKE & BUECHNER, *New York.*
G. E. STECHERT & CO., *New York.*
E. STEIGER & CO., *New York.*
ARTHUR F. BIRD, *London.*
WILLIAM WESLEY & SON, *London.*
A. HERMANN, *Paris.*
MAYER & MÜLLER, *Berlin.*

JANUARY, 1916

Entered as Second-Class Matter at the Baltimore, Maryland, Postoffice.

# *The Oscillation of Functions of an Orthogonal Set.**

By O. D. Kellogg.

## 1. *Introductory.*

The sets of orthogonal functions which occur in mathematical physics have, in general, the property that each changes sign in the interior of the interval on which they are orthogonal once more than its predecessor. So universal is this property that such sets are frequently referred to as sets of "oscillating functions." The question arises, is this property of oscillation inherent in that of orthogonality? That it is not, is evidenced by a simple example. If the first function does not vanish, it is clear that the second must change signs, but the example shows that it does not follow that the third must change signs twice. Thus let $\phi_0(x)=1$, while $\phi_1(x)$ is defined on the interval $(0,1)$ as follows:

$$\begin{aligned}
\text{For } 0\leq x\leq \tfrac{1}{3}, &\quad \phi_1(x)=27x-8,\\
\tfrac{1}{3}\leq x\leq \tfrac{1}{2}, &\quad \phi_1(x)=18x-5,\\
\tfrac{1}{2}\leq x\leq \tfrac{2}{3}, &\quad \phi_1(x)=-18x+13,\\
\tfrac{2}{3}\leq x\leq 1, &\quad \phi_1(x)=1.
\end{aligned}$$

Let $\phi_2(x)=\phi_1(1-x)$. Then the three functions are orthogonal on the interval $(0,1)$, while the second two change signs but once each.

If $\phi_2(x)$ changes signs but once, say at $x=a$, the function $\phi_0(a)\phi_1(x)-\phi_1(a)\phi_0(x)$, which is orthogonal to $\phi_2(x)$, can not have $x=a$, where it vanishes, as the only point where it changes signs, since two orthogonal functions must be of the same sign in part of the interval and of opposite signs in part. Hence, if we make the supposition $\phi_0(x_0)\phi_1(x_1)-\phi_0(x_1)\phi_1(x_0)>0$ for $0<x_0<x_1<1$, $\phi_2(x)$ must change signs twice in the interior of the interval.

These considerations suggest the determinant condition of the next paragraph. It is interesting to note that it is essentially the condition that a function $c_0\phi_0(x)+c_1\phi_1(x)+\ldots\ldots+c_n\phi_n(x)$ can be found which will coincide with a given function $f(x)$ at any $n+1$ interior points of the interval, the difference being merely the substitution of a definite sign for the being different from

* Read before the American Mathematical Society, November 29, 1913.

zero. Also it should be noted that it is a condition satisfied by the more common orthogonal sets, like the sines, $\sin \pi x$, $\sin 2\pi x$, $\sin 3\pi x$, .... ; the Legendre polynomials; and the Bessel functions, $\sqrt{x}J_0(a_1x)$, $\sqrt{x}J_0(a_2x)$, $\sqrt{x}J_0(a_3x)$, ...., where $a_1, a_2, a_3, \ldots$ are the roots of $J_0(x)=0$. The subsequent paragraphs will be devoted to deriving some oscillation and other properties from the suggested conditions.

## 2. *Oscillation Properties.*

Let $\phi_0(x), \phi_1(x), \ldots, \phi_n(x), \ldots$ be a set of real continuous functions, normed and orthogonal, on the interval (0, 1). Let the determinants

$$D(x_0, x_1, x_2, \ldots, x_n) = \begin{vmatrix} \phi_0(x_0), & \phi_1(x_0), & \ldots, & \phi_n(x_0), \\ \phi_0(x_1), & \phi_1(x_1), & \ldots, & \phi_n(x_1), \\ \cdots & \cdots & \cdots & \cdots \\ \phi_0(x_n), & \phi_1(x_n), & \ldots, & \phi_n(x_n), \end{vmatrix}$$

be positive for any $x_0, x_1, \ldots, x_n$ in the interior of the interval (0, 1) and in ascending order of magnitude, this condition to hold for $n=0, 1, 2, \ldots$, $D_0(x_0)$ being understood as $\phi_0(x_0)$. Let $\Phi_{m,n}(x)$ denote the function $c_m\phi_m(x)+c_{m+1}\phi_{m+1}(x)+\ldots+c_n\phi_n(x)$.

The following theorems are either evident or are made so by the brief proofs indicated:

(1) *Given $n+1$ distinct points $x_0, x_1, \ldots, x_n$ in the interior of* (0, 1), *$c_0, c_1, \ldots, c_n$ may be so chosen that $\Phi_{0,n}(x)$ will take on any given values at these points.*

(2) *$\Phi_{0,n}(x)$ can not vanish at $n+1$ distinct points in the interior of* (0, 1) *without vanishing identically.*

I. *If $\Phi_{0,n}(x)$ vanishes at $n$ distinct points, it changes sign at each.* For it may be written in the form

$$\Phi_{0,n}(x) = kD(x, x_0, x_1, \ldots, x_{n-1}),$$

where $k$ is a constant other than zero, and $D(x, x_0, x_1, \ldots, x_{n-1})$ is positive or negative according as the number of inversions in the sequence $x, x_0, x_1, \ldots, x_{n-1}$ is even or odd.

(3) *$\phi_n(x)$ can not vanish more than $n$ times.* For otherwise $D(x_0, x_1, \ldots, x_n)$ could vanish, the arguments being all different.

II. *Every continuous function $\psi(x)$ orthogonal to $\phi_0(x), \phi_1(x), \ldots, \phi_n(x)$ on the interval* (0, 1) *changes signs at least $n+1$ times.* For, if $\psi(x)$ changed signs at $x_0, x_1, \ldots, x_k$ only, $k$ being less than $n$, we could determine a function $\Phi_{0,k+1}(x)$ which would vanish at these points, and which at one other point, $x_{k+1}$, would take on the value sgn $\psi(x_{k+1})$, by (1); moreover $\Phi_{0,k+1}(x)$ would be orthogonal to $\psi(x)$ since, $k+1$ being less than or equal to $n$, every term of $\Phi_{0,k+1}(x)$ would be orthogonal to $\psi(x)$. But this is impossible, since $\psi(x)\Phi_{0,k+1}(x)$ is continuous, not identically zero, and never negative, by I. Another aspect of this theorem is this: *The best approximation curve* $y=\Phi_{0,n}(x)$ *to a continuous curve* $y=f(x)$, *in the least square sense* (*i. e.*, $\int_0^1[f(x)-\Phi_{0,n}(x)]^2dx$, *a function of* $c_0, c_1, \ldots c_n$, *is a minimum*), *crosses this curve at least* $n+1$ *times.* For $f(x)-\Phi_{0,n}(x)$ is orthogonal to $\phi_0(x), \phi_1(x), \ldots, \phi_n(x)$.

III. From (3) and II, it follows that $\phi_n(x)$ *vanishes exactly n times, and changes sign at each zero.*

IV. *The system* $\phi_0(x), \phi_1(x), \ldots, \phi_n(x), \ldots$ *is closed with respect to all continuous functions having a finite number of sign changes;* i. e., there is no continuous function, other than 0, with a finite number of sign changes orthogonal to all the functions of the set, of which we of course suppose there are an unlimited number. More especially, the system is closed with respect to all continuous functions having only a finite number of maxima and minima.

V. *The function* $\Phi_{m,n}(x)=c_m\phi_m(x)+c_{m+1}\phi_{m+1}(x)+\ldots+c_n\phi_n(x)$, *for any fixed values of* $c_m, c_{m+1}, \ldots, c_n$, *not all zero, changes sign at least m times and at most n times.* This follows from (2), since $\Phi_{m,n}(x)$ is a special case of $\Phi_{0,n}(x)$, and from II.

## 3. *Separation Properties.*

VI. *No two successive functions* $\phi_{n-1}(x)$ *and* $\phi_n(x)$ *have a common root in the interior of* (0, 1). This may be proved by a *reductio ad absurdum.* Suppose $\phi_{n-1}(a)=\phi_n(a)=0$, $0<a<1$. We form $n-1$ functions $\Phi_{0,n-2}(x)$,

$$\begin{aligned} F_0(x) &= a_{00}\phi_0(x)+a_{01}\phi_1(x)+\ldots+a_{0,n-2}\phi_{n-2}(x),\\ F_1(x) &= a_{10}\phi_0(x)+a_{11}\phi_1(x)+\ldots+a_{1,n-2}\phi_{n-2}(x),\\ &\ldots\ldots\ldots\ldots\ldots\ldots\ldots\ldots\ldots\ldots,\\ F_{n-2}(x) &= a_{n-2,0}\phi_0(x)+a_{n-2,1}\phi_1(x)+\ldots+a_{n-2,n-2}\phi_{n-2}(x), \end{aligned}$$

subject to the conditions $F_0(a)\neq 0$, $F_1(a)=0$, $F_2(a)=0, \ldots, F_{n-2}(a)=0$ and that the functions $F_0(x), F_1(x), \ldots, F_{n-2}(x)$ be orthogonal on the interval (0, 1). If these functions are normed, the quantities $a_{ij}$ will be the direction cosines of $n-1$ perpendicular directions in hyperspace. The remaining conditions are then simply that one of these directions shall be given by the direction ratios $\phi_0(a), \phi_1(a), \ldots, \phi_{n-2}(a)$, where $\phi_0(a)\neq 0$, so that the conditions are certainly compatible.

Now $F_0(x)$ changes signs, at most, at $n-2$ points in the interior of (0, 1). In any event there are $n-2$ points, $a_1, a_2, \ldots, a_{n-2}$, dividing the interval (0, 1) into $n-1$ parts in each of which $F_0(x)$ does not change signs. As $F_0(a) \neq 0$, the point $a$ need not be one of the set $a_1, a_2, \ldots, a_{n-2}$. Suppose it comes between $a_i$ and $a_{i+1}$. The function $\Phi(x) = c_1 F_1(x) + c_2 F_2(x) + \ldots + c_{n-2} F_{n-2}(x) + c_{n-1}\phi_{n-1}(x) + c_n \phi_n(x)$ will be orthogonal to $F_0(x)$ for every choice of $c_1, c_2, \ldots, c_n$. We chose these constants so that

$$\int_0^{a_1} F_0(x)\Phi(x)\,dx = \int_{a_1}^{a_2} F_0(x)\Phi(x)\,dx = \ldots = \int_{a_i}^{a} F_0(x)\Phi(x)\,dx$$
$$= \int_a^{a_{i+1}} F_0(x)\Phi(x)\,dx = \ldots = \int_{a_{n-3}}^{a_{n-2}} F_0(x)\Phi(x)\,dx = 0.$$

This gives $n-1$ homogeneous linear equations in the $n$ constants $c_i$, so that the conditions are compatible, the $c_i$ not all vanishing. But as $\int_0^1 F_0(x)\Phi(x)\,dx = 0$, it follows also that $\int_{a_{n-2}}^1 F_0(x)\Phi(x)\,dx = 0$. But $F_0(x)$ keeps its sign in each interval of integration; hence $\Phi(x)$ must change signs in the interior of each, or $n$ times. But it also vanishes at one end point $a$, since each term vanishes there. Thus we are led to the conclusion that $\Phi(x)$, which is a function $\Phi_{0,n}(x)$, vanishes $n+1$ times, and this contradicts theorem V. Hence, $\phi_{n-1}(x)$ and $\phi_n(x)$ can not have a common root.

VII. *$\phi_{n-1}(x)$ changes sign between two successive roots of $\phi_n(x)$.* Suppose this were not the case. Let $a$ and $b$ be two successive roots of $\phi_n(x)$. We may then, without loss of generality, suppose $\phi_n(x)$ positive in the open interval $(a, b)$, and $\phi_{n-1}(x)$ positive in the closed interval $(a, b)$, for it will not vanish at $a$ or $b$, by VI. Consider the function $\Phi(x, t) = \phi_{n-1}(x) - t\phi_n(x)$. For $t=0$ it is positive in the closed interval $a \leq x \leq b$, and the same is true for sufficiently small positive values of $t$. For some finite positive value of $t$, say $T$, $\Phi(x, T)$ will have some negative values in the same interval. There is therefore a value of $t$, say $\tau$, such that for every positive $\varepsilon$ less than $\tau$ $\Phi(x, \tau-\varepsilon) > 0$ throughout the interval $a \leq x \leq b$, whereas $\Phi(x, \tau+\varepsilon)$ has negative values in this interval. The function $\Phi(x, \tau)$ can not be negative in the interval $(a, b)$, since $\Phi(x, t)$ is continuous. For the same reason its minimum can not be positive. Hence in the closed interval $(a, b)$, $\Phi(x, \tau)$ is never negative, but vanishes at one or more points, which, by VI, must be in the interior. But as $\Phi(x, \tau)$ is a function $\Phi_{n-1, n}(x)$, it must change signs at least $n-1$ times, by V. These sign changes must occur outside $(a, b)$, so that altogether $\Phi(x, \tau)$ vanishes $n$ times, without, however, changing signs at all $n$ points. This contradicts theorem I, so that $\phi_{n-1}(x)$ can not keep its sign between two successive roots of $\phi_n(x)$.

(4) *The roots of $\phi_n(x)$ and $\phi_{n-1}(x)$ occur alternately.* This is an immediate corollary of VII, since $\phi_n(x)$ has exactly $n$ roots and $\phi_{n-1}(x)$ exactly $n-1$.

## 4. *Remarks.*

Essentially the conditions on a set of functions which have been employed are (1) orthogonality on some interval $I$, and (2), $D(x_0, x_1, \ldots, x_n) > 0$ when $x_0, x_1, \ldots, x_n$ are in the interior of the interval $I$ and in ascending order of magnitude, for all $n$. The functions 1, $\sin x$, $\cos x$, $\sin 2x$, $\cos 2x$, .... do not satisify the second condition for every $n$, the interval $I$ being $(0, 2\pi)$. They do, however, for every even $n$. This suggests a modification of the condition in which we merely assume an infinite sequence of integers $n_1, n_2, n_3, \ldots$ for which the $D(x_0, x_1, \ldots, x_n) > 0$. Theorems of the same type as those obtained follow, with appropriate modification. Thus, in place of V we would have

V′. *If* $n_i < m$ *and* $n \leq n_j$, *then for any fixed* $c_m, c_{m+1}, \ldots, c_n$, *not all zero,* $\Phi_{m,n}(x) = c_m\phi_m(x) + c_{m+1}\phi_{m+1}(x) + \ldots + c_n\phi_n(x)$ *changes sign at least* $n_i + 1$ *times and at most* $n_j$ *times.*

The closure property of theorem IV is weaker than is desirable. All we can conclude from it, in the matter of the determination of a function by its generalized Fourier constants, is that the function is determined but for an additive function which is zero at an infinite number of points.

It is further desirable to make the connection of the above theory with that of integral equations. The chief matter of interest would be to find a condition on the kernel of the integral equation which corresponds to the determinant condition on its characteristic functions. I hope to have some results along these lines in the near future.

COLUMBIA, MO., *May* 11, 1915.

# *On Some Properties of the Medians of Closed Continuous Curves Formed by Analytic Arcs.*

By Arnold Emch.

## § 1. *Introduction.*

In two recent articles* of this journal I have investigated the properties of the medians of closed convex curves with continuous tangents and of closed convex curves formed by a finite number of ordinary analytic arcs, including convex rectilinear polygons. By means of continuous sets of medians corresponding to a continuous set of orthogonal directions it was possible to prove the theorem that to every such curve at least one square may be inscribed in such a manner that the vertices of the square lie on the curve.

In what follows I shall generalize the result for any closed continuous curve composed of a finite number of analytic arcs with a finite number of inflexions and other singularities.

The following two propositions of analysis situs will be used in this demonstration:

I. Consider a singly connected region $R$ of a plane, bounded by a continuous curve $B$ formed by analytic arcs without multiple points, and assume that the points of the boundary belong to $R$. Any continuous curve $C$ formed by analytic arcs without multiple points, that connects any two distinct points of the boundary and has no segment in common with the boundary, and whose points belong to $R$, separates this region into two distinct sub-regions $R_1$ and $R_2$. All points of $R_1$ and $R_2$, except those of $C$, shall belong to $R$.

Choose now any two points, $P_1$ in $R_1$ and $P_2$ in $R_2$, and join them by a continuous curve $K$, composed of analytic arcs, that has no segments or singular points in common with $C$, and lies entirely within $R$.

*Such a curve $K$ cuts $C$ in an odd number of points. When $P_1$ and $P_2$ are both in $R_1$ or in $R_2$, then $K$ cuts $C$ in an even number of points.*

II. Let $P_1, P_2, P_3, P_4$ be four distinct points of an ordinary closed continuous curve $B$ without multiple points, which follow each other in the given order when the curve is described uniformly in the same sense, and for which

$$P_1P_2 = P_2P_3 = P_3P_4 = P_4P_1.$$

The figure formed by four such points I call an *inscribed rhomb of the first kind.*

* Vol. XXXV, pp. 407–412; Vol. XXXVII, pp. 19–28.

Let $P_1'P_2'P_3'P_4'$ be another rhomb inscribed to the same curve $B$, such that the vertices follow each other in the given order when the curve is again described continuously in the same sense, and

$$P_1'P_4' = P_4'P_2' = P_2'P_3' = P_3'P_1'.$$

This figure shall be called an *inscribed rhomb of the second kind.*

*A continuous set of rhombs of the first kind and a continuous set of rhombs of the second kind attached to the same curve B can have no rhomb in common. In other words, it is impossible to deform a rhomb of the first kind continuously into a rhomb of the second kind.*

Indeed, when $P_1P_2P_3P_4$ is a rhomb of the first kind, to pass continuously in the same sense along the curve from $P_1$ to $P_3$, either the point $P_2$ or the point $P_4$ is met. Hence, a continuous change of the rhomb $P_1P_2P_3P_4$ in which we pass directly along the curve from $P_1$ to $P_3$ could only occur after $P_2$ and $P_4$ had crossed $P_3$ into the two parts of the curve $B$ determined by $P_1$ and $P_2$ as extremities. Hence, in the continuous deformation of the rhomb of the first kind we would have rhombs with two coincident consecutive vertices. This, however, is impossible on a curve without double points.

## § 2. *Medians or Mid-point-Curves of Systems of Parallel Chords.*

1. *Continuous Sets of Parallel Chords.* Consider first a closed continuous curve* with a continuous tangent without multiple points and with no other singularities than inflexions, and the system of all secants and tangents of this curve parallel to a given direction $\tau$. From the definition of such a curve it follows that there are two extreme tangents between which the curve is located. In case of an oval, there are just two and no other tangents to the curve for every direction. In case of a general closed ordinary curve we shall first assume that there are no multiple or inflectional tangents for a particular given direction $\tau$. Let $A_\tau$ and $Z_\tau$ be the points of tangency of the extreme tangents, and let the secant move parallel to $\tau$ from $A_\tau$ to $Z_\tau$. In the neighborhood of $A_\tau$ every secant cuts the curve in two points, $E_1$ and $E_2$ (Fig. 1). In case of more than two (an even number) tangents, the secant, in moving from $A_\tau$ to $Z_\tau$, becomes for the second time tangent to the curve at some point $T_1$. As this point counts for two points of intersection, there are on this secant an even number of intersections, beside this tangency. Evidently the system of parallel chords between $A_\tau$ and $T_1$ thus obtained forms a continuous set $A_\tau$, $E_1E_2$, $E_1'E_2'$, ...., $T_1S_1$. From this last position move the secant such that $T_1$ continues its motion along the curve in the same sense as from $A_\tau$ to $T_1$. As there is no tangency between $A_\tau$ and $S_1$, the $T_1$ must necessarily

* We restrict this expression throughout to curves as defined in the introduction.

move into a new point of tangency $T_2$, while the other extremity $S_1$ of the tangent chord moves to $S_2$. Again, the system of parallel chords $T_1S_1$, $B_2S'_1$, ...., $T_2S_2$ forms a continuous set. Continuing the motion of the secant such that $T_2$ continues to move on the curve in the same sense as from $T_1$ to $T_2$, and if there are no other tangents between $T_2S_2$ and $Z_\tau$, the chords parallel to $\tau$ between these two positions form a continuous set. In case that there are

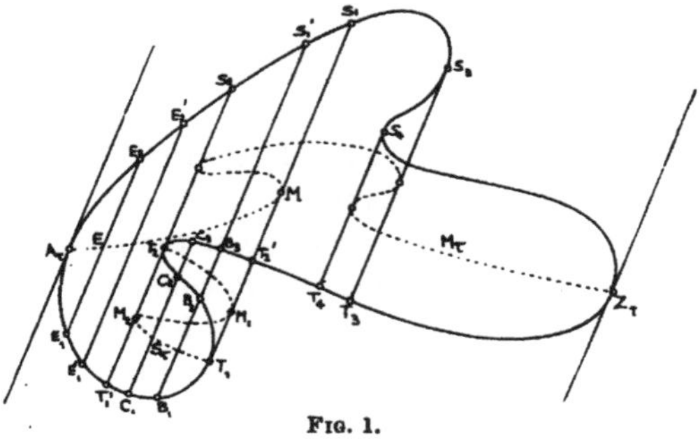

FIG. 1.

other tangents, either $T_2$ or $S_2$ will move into the next point of tangency. Suppose that $S_2$ move into such a point $S_3$, and $T_2$ into $T_3$; then the chords between $T_2S_2$ and $T_3S_3$ form also a continuous set. Continuing this process, all points of tangency, and finally $Z_\tau$, are reached. In this manner we have formed *a continuous set of chords parallel to a given direction between the two extreme tangents.*

Since $S_1T_1$ was the first tangent, we see that in the piece $A_\tau S_1$ of the curve there is no tangency. Hence, there lies an S-shaped portion of the curve between the tangents at $T_1$ and $T_2$. The chords determined by this curve, $T'_1T_1T_2T'_2$, form a continuous set $T_1$, $B_1B_2$, $C_1C_2$, $T'_1T_2$, $C_1C_3$, $B_1B_3$

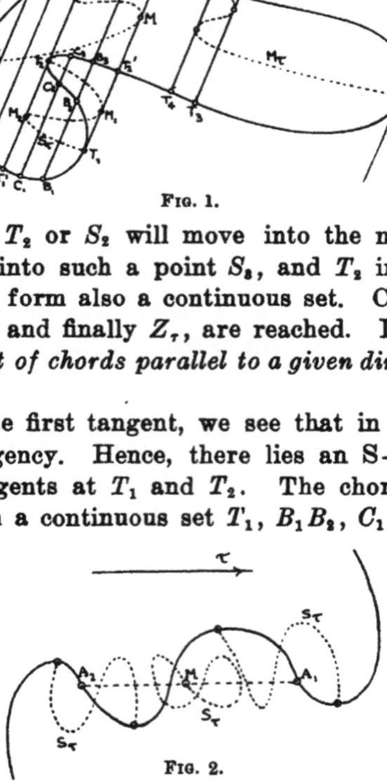

FIG. 2.

$T_1T'_2$, $B_2B_3$, $C_2C_3$, $T_2$. Such a set we may call a *secondary set.* Another example of a secondary set is shown in Fig. 2, consisting of all chords forming a continuous set with the chord $A_1A_2$ as one of the elements.

2. *Medians.* The locus of the mid-points of the chords of the continuous set of parallel chords between $A_\tau$ and $Z_\tau$ is a continuous curve with the extremities $A_\tau$ and $Z_\tau$. As before, this locus may be called the *median* $M_\tau$ of the

curve associated with the direction $\tau$. The loci of mid-points of the chords of secondary sets we call *secondary medians* $S_\tau$. Let $y=mx+b$ be the equation of a secant cutting the curve at $E_1$ and $E_2$. In the neighborhood of these points the coordinates of the points of the curve may be represented as uniform continuous differentiable functions of the parameter $b$:

$$x_1=f(b), \quad y_1=g(b), \quad (E_1),$$
$$x_2=h(b), \quad y_2=j(b), \quad (E_2).$$

For the coordinates of the mid-point $M$ of the chord $E_1E_2$ we get

$$\xi=\tfrac{1}{2}[f(b)+h(b)],$$
$$\eta=\tfrac{1}{2}[g(b)+j(b)].$$

When $b$ varies, these are the parametric equations of the median. Designating by $x$ and $y$ current coordinates, and by ($'$) derivatives with respect to $b$, the equations of the tangents to the curve at $E_1$ and $E_2$, and to the median at $M$, are

$$y_1'x-x_1'y+x_1'y_1-y_1'x_1=0,$$
$$y_2'x-x_2'y+x_2'y_2-y_2'x_2=0,$$
$$(y_1'+y_2')x-(x_1'+x_2')y+\tfrac{1}{2}(x_1'y_1+x_2'y_2-y_1'x_1-y_2'x_2+x_2'y_1+x_1'y_2-x_2y_1'+x_1y_2')=0.$$

But from $y=mx+b$ and $m=\dfrac{y_1-y_2}{x_1-x_2}$ we find by differentiation

$$y_1'=\frac{y_1-y_2}{x_1-x_2}x_1'+1, \quad y_2'=\frac{y_1-y_2}{x_1-x_2}x_2'+1,$$

and from this

$$x_1'y_1+x_2'y_2-y_1'x_1-y_2'x_2=x_2'y_1+x_1'y_2-x_2y_1'-x_1y_2'.$$

The equation of the tangent to the median assumes therefore the form

$$(y_1'+y_2')x-(x_1'+x_2')y+(x_1'y_1+x_2'y_2-y_1'x_1-y_2'x_2)=0,$$

and appears as the sum of the left-hand members of the equations of the tangents at $E_1$ and $E_2$. The three tangents are therefore concurrent.

This fact may be stated as

Theorem 1. *If $E_1$ and $E_2$ are the extremities, and $M$ the mid-point, of a chord in a continuous system of parallel chords of an ordinary curve, then the tangents to the curve at $E_1$ and $E_2$, and to the median at $M$, are concurrent.*

From this we gain immediately the

Corollary 1. *For a tangent chord like $T_1S_1$, the three tangents at $T_1$, $S_1$ and $M$ are concurrent at $S_1$, so that the median touches the tangent chord at $M$. When $T_1$ is a point of inflexion, then $M$ also is a point of inflexion of the median.*

Taking for $\tau$ a general direction (determined by slope $m$ as before), and excluding multiple tangents, the preceding results may be summed up in

Theorem 2. *The median of an ordinary closed curve without multiple points associated with a given direction is an ordinary continuous curve without*

***multiple points or cusps. Also the secondary medians are continuous and are always located between tangents that lie between $A_\tau$ and $Z_\tau$. All medians are tangent to the corresponding tangent chords.***

When $T_1$ is an inflexion, the secondary median in this neighborhood shrinks to the point $T_1$.

3. *Case of Double Tangent.* When there is a double tangent $d_2$ parallel to the given direction $\tau$, such that $d_2$ besides the points of tangency has no other points in common with the closed curve, choose slightly different from $\tau$ a direction $\tau_a$ and construct in the neighborhood of $d_2$ the medians $M_{\tau_a}$ and $S_{\tau_a}$, indicated by dash lines in Fig. 3. Now change the direction $\tau_a$ continuously

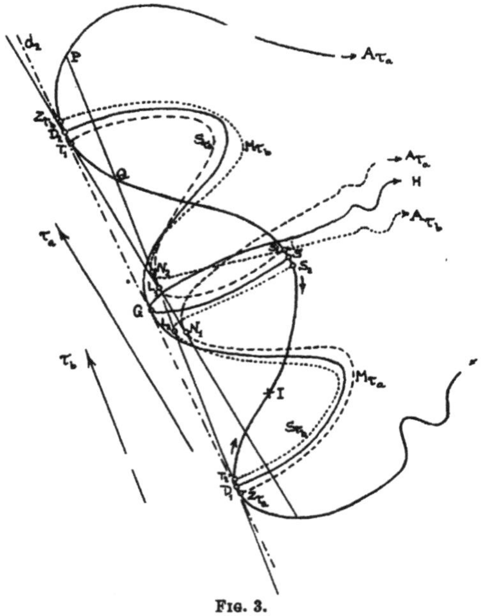


FIG. 3.

(in the figure clockwise) till it coincides with that of $\tau$, and continue the change in the same sense to some direction $\tau_b$ also near that of $\tau$. As $\tau_a$ approaches the direction of $d_2$, the points $N_1$ and $L_1$ both approach the mid-point $G$ between the points of tangency $D_1$ and $D_2$ of $d_2$; and when $\tau_a$ coincides with the direction of $d_2$, then, according to Theorem 1, the curves $M_\tau$ and $S_\tau$ will both be tangent to $d_2$ at $G$. In the figure they are drawn as heavy lines: $M_\tau$, from $D_1$ over $G$ to $H$, results by continuous deformation from $M_{\tau_a}$; $S_\tau$, from $D_2$ over $G$ to $S$, results continuously from $S_{\tau_a}$. As $\tau_a$ continues to change over $\tau$ to $\tau_b$, the characters of the curves $M_{\tau_a}$ and $S_{\tau_a}$ are partly interchanged. The part of $M_{\tau_a}$ from $Z_{\tau_a}$ to $N_1$ changes to the curve between $D_1$ and $G$, and from this to the part $T_2L_2$ of the curve $S_{\tau_b}$. At the same time the portion $L_1S_1$ of $S_{\tau_a}$ changes continuously into $L_2S_2$ of $S_{\tau_b}$. The remaining portions $N_1A_{\tau_a}$ of $M_{\tau_a}$ and $L_1T_1$ of $S_{\tau_a}$ change continuously into $M_{\tau_b}$. Both $M_{\tau_b}$ and $S_{\tau_b}$ are indicated by dotted curves. From this it is seen that in the change from $\tau_a$ to $\tau_b$ *in the neighborhood of a double tangent as assumed, there is a continuous change in the system of medians $M_\tau$ and $S_\tau$.*

In this change, $M_{\tau_a}$, in passing over into $M_{\tau_b}$, loses the portion $Z_{\tau_a}N_1$, while $N_1A_{\tau_a}$ will remain. $Z_{\tau_a}N_1$ acquires $L_1S_1$ to form after deformation $S_{\tau_b}$. $N_1A_{\tau_a}$, on the other hand, acquires $L_1T_1$ to form $M_{\tau_b}$.

4. *Continued Deformation of $S_{\tau_b}$.* When $\tau$ continues to change from $\tau_b$ in the same sense as before, then the point of tangency $T_2$ of the corresponding tangent will move along the curve as indicated by the arrow, as long as there

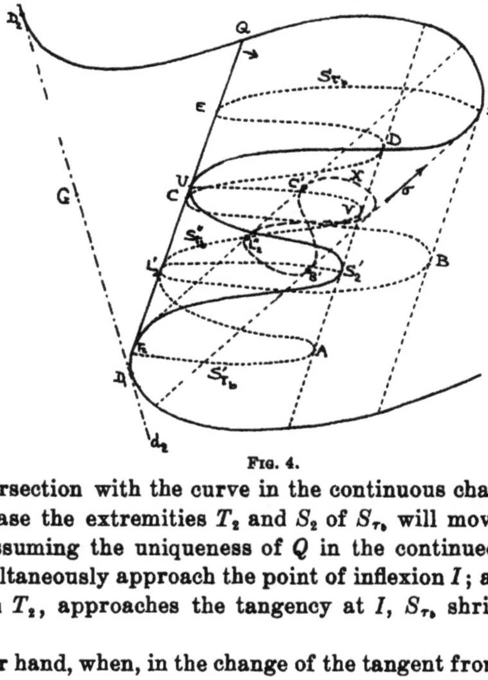


FIG. 4.

is no other intersection with the curve in the continuous change from $D_1D_2$ to $T_2Q$. In this case the extremities $T_2$ and $S_2$ of $S_{\tau_b}$ will move as indicated by the arrows. Assuming the uniqueness of $Q$ in the continued deformation, $S_2$ and $T_2$ will simultaneously approach the point of inflexion $I$; and as the tangent, proceeding from $T_2$, approaches the tangency at $I$, $S_{\tau_b}$ shrinks to nothing at the inflexion $I$.

On the other hand, when, in the change of the tangent from $T_2$, other points of intersection than $Q$ enter, this must occur the first time by a tangency, say at $U$ (Fig. 4). The point of tangency $L_2$ of $S_{\tau_b}$ with $T_2Q$ moves from $G$ continuously to the position $C$ in Fig. 4. The median $S_{\tau_b}$ consists of the arcs $T_2A$, $AL_2'$, $L_2'B$, $BC$, $CD$, $DE$, $ES_2$. To the portion $T_2L_2$ of $S_{\tau_b}$ in Fig. 3, corresponds the part of the curve, in Fig. 4, from $T_2$ over $A$, $L_2'$ and $B$ to $C$. Now, as $T_2Q$ continues to move as a tangent to the closed curve in the neighborhood of $T_2$ in the same sense, as indicated by the arrow, $S_{\tau_1}$ breaks up as in case of the double tangent $d_2$ in Fig. 3. After a slight deformation there will be a curve $S'_{\tau_1}$ arising from the arcs $T_2A$, $AL'_2$ of $S_{\tau_1}$ (Fig. 4), and a new arc $L'_2S'_2$. Another curve $S''_{\tau_1}$ arises from the remaining arcs $L'_2B$, $BC$, $CD$, $DE$, $ES_2$ of $S_{\tau_1}$ and the new arcs $UV$ and $VL'_2$. In Fig. 3 we see that when the piece $Z_{\tau_0}N_1$ drops from $M_{\tau_0}$ in the deformation to $M_{\tau_1}$, it is tangent to $d_2$ at $D_1$ and $G$. Likewise, at the moment when the piece $T_1L_1$ is added to $M_{\tau_1}$, it is tangent to $d_2$ at $D_2$ and $G$. Similarly in Fig. 4. When $T_2AL'_2$ drops from $S_{\tau_1}$, it is tangent to $T_2Q$ at $T_2$ and $L'_2$. The remaining piece $L'_2BC$ of the deformed part of $M_{\tau_0}$ that dropped in the position $D_1G$ connected with the double tangent $d_2$, is also tangent to $T_2Q$ at $L'_2$ and $C$. As to the further deformation of $S'_{\tau_1}$, either the case of its shrinkage to the point of inflexion between $T_2$ and $S'_2$ may occur, or the same process as described in connection with Fig. 4 may repeat itself. In the continued deformation of $S''_{\tau_1}$ similar conditions arise. We need only pay attention to the piece $L'_2BC$. Portions may drop off or be added as explained above, or the whole section may remain as is the case in Fig. 4. As the direction of $T_2Q$ moves to $\sigma$, $L'_2BC$ is deformed into $L''_2B'C'$, where the points of tangency $L''_2$ and $C'$ with the deformed tangent of $T_2Q$ are preserved. Owing to a break-up in the deformation caused by the double tangent in the neighborhood of $S_2$ and $S'_2$, another arc joins $L''_2B'C'$, so as to form an 8-shaped curve $X$, which finally shrinks to a point. To sum up, we may state

THEOREM 3. *When a median $M_\tau$ breaks up on account of an external double tangent $d_2$, it loses a section that touches $d_2$ in two points. On the other hand, it gains a section that also touches $d_2$ in two points, of which one coincides with one of the points of tangency of the other section. A secondary median $S_\tau$ is either deformed in such a manner that it ultimately vanishes at a point of inflexion, or, owing to further double tangents, is broken up into sub-medians $S'_\tau$, $S''_\tau$, . . . ., such that the lost and remaining sections, like $T_2AL'_2$ and $L'_2BC$, always touch the corresponding tangent $T_2Q$ in two points each.*

## § 3. *Case of Closed Curve Formed by Rectilinear Segments, generally by Analytic Arcs.*

A curve of this kind may contain reentrant angles, but shall have no double points, *i. e.*, points where two segments cross each other. Exactly as in the cases considered above, and in case of a convex polygon, it is found that with every direction $\tau$ is associated a median $M_\tau$ that extends from $A_\tau$ and $Z_\tau$, such that the whole curve lies between the straight lines through $A_\tau$ and $Z_\tau$ parallel to $\tau$. The extremities $A_\tau$ and $Z_\tau$ are either vertices of the polygon (curve) or mid-points of rectilinear segments. As in case of a convex polygon, a domain $(P)$ exists entirely within the domain $P$ enclosed by the curve, such that among the entire system of medians $M_\tau$, associated with all possible directions, no two medians intersect in the region within $P$ and outside of $(P)$.

In the construction of the median $M_\tau$ we must explain how to construct the segment of $M_\tau$ associated with a segment $BC$ of a reentrant angle (Fig. 5) when $BC$ is parallel to the direction $\tau$, and when $ABCD$ and $LK$ form parts of the contour. Draw $LL'' \parallel BC$, prolong $BC$ to the intersection $C''$ (or $B''$) with $KL$, and let $L'$ be the mid-point of $LL''$, $B'$ that of $BB''$. Then, $B'L'$ is a portion of the median, and its prolongation passes through the intersection $S$ of $BA$ and $KL$ produced. Also draw $KK'' \parallel BC$, and let $K'$ be the mid-point of $KK''$ and $C'$ that of $CC''$. Then, $B'C'$ and $C'K'$ are also portions of the median, and $B'C' = \frac{1}{2}BC$. Again, $DC$, $K'C'$ and $KL$ produced are concurrent at a point $T$.

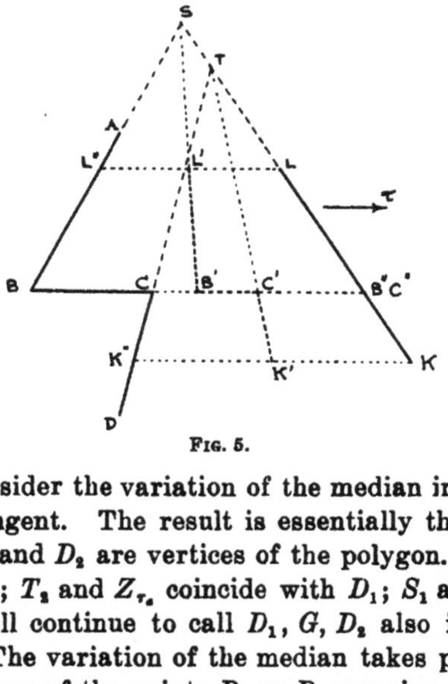


Fig. 5.

Next we must consider the variation of the median in the neighborhood of an external double tangent. The result is essentially the same as in case of Fig. 3, except that $D_1$ and $D_2$ are vertices of the polygon. $D_1S$ and $D_2S$ form now a reentrant angle; $T_2$ and $Z_{\tau_a}$ coincide with $D_1$; $S_1$ and $S_2$ with $S$; $T_1$ and $Z_{\tau_b}$ with $D_2$. We shall continue to call $D_1$, $G$, $D_2$ also in case of a polygon points of tangency. The variation of the median takes place according to the same principle, when one of the points $D_1$ or $D_2$ remains an ordinary tangency while the other is an angular point. Also the variation of secondary medians $S_\tau$ occurs according to the principles stated in connection with Fig. 5.

In Fig. 6 the change of $M_{\tau_a}$, $S_{\tau_a}$ into $M_{\tau_b}$, $S_{\tau_b}$ is shown for a special polygonal line $AD_1SBCD$, with the reentrant angle $D_1SB$ and with the prolongation of $CB$ or $d_2$ passing through $D_1$. Again, in the change from $\tau_a$ to $\tau_b$, passing over $d_2$, $M_{\tau_a}$ loses the portion with its extremities at $D_1$ and $G$ on $d_2$. (The notation and its meaning is the same as in Fig. 3.) On the other hand, the remainder of $M_{\tau_a}$ to pass into $M_{\tau_b}$ simply has to be joined to the piece resulting from the portion of $S_{\tau_a}$ that, at the moment when $\tau$ coincided with the direction of $d_2$, had its extremities at $D_2$ and $G$ in $d_2$.

When a closed curve without multiple points (where branches of the curve cross each other) formed by ordinary analytic arcs is given, we can inscribe to it a rectilinear polygon, as in case of a convex curve, and consider the given curve as the limit of such a polygon. As in the former case, the relations between $M_\sigma$ and $S_\sigma$ and their changes in the neighborhood of a line $d_2$ are not destroyed by the limit-process.

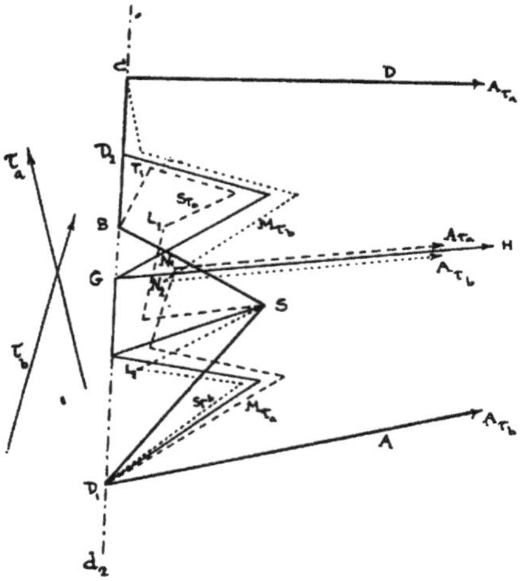

FIG. 6.

Calling a line $d_2$ that touches a curve in at least two distinct points, and with all points of the curve on one and the same side of $d_2$, a *base-line* of the curve, we can now state

THEOREM 4. *When $\tau$ changes continuously and in the same sense from $\tau_a$ to $\tau_b$, two directions close to that of a base-line, and enclosing it, then, in the change of $M_{\tau_a}$ to $M_{\tau_b}$ associated with those directions and a closed curve formed by ordinary analytic arcs without multiple points, $M_{\tau_a}$ drops the portion that has its extremities in $d_2$, and becomes $M_{\tau_b}$ by joining to its remainder the portion of $S_{\tau_a}$ that has its extremities also in $d_2$.*

*In the further deformation of the secondary median $S_{\tau_b}$, due to a continuous change of $\tau$ from $\tau_b$ in the same direction, the same process repeats itself: the parts dropped off and those that are added have their extremities in the corresponding internal base-line ($D_1 Q$ in Fig. 4) resulting continuously from $d_2$.*

§ 4. *Intersection of Medians Associated with a Pair of Orthogonal Directions.*

As has been shown in the previous papers, two medians $M_\sigma$ and $M_\tau$ associated with two orthogonal directions $\sigma$ and $\tau$ intersect in an odd number of points within the fundamental domain $(R)$. It is evident that no extremity $(A_\sigma, Z_\sigma)$ of one median can coincide with an extremity $(A_\tau, Z_\tau)$ of the other, or lie in the other, and be the center of an inscribed rhomb. Hence, when, in a continuous change $(\Sigma, T)$ of the pair $(\sigma, \tau)$ in which neither $\Sigma$ nor T encloses a base-line, points of intersection of $M_\sigma$ and $M_\tau$ disappear, they can not disappear separately. The only possibility is that two points first move to coincidence and then vanish. In other words, two distinct consecutive intersections of $M_\sigma$ and $M_\tau$ must first coincide by $M_\sigma$ and $M_\tau$ becoming tangent to each other. After the tangency $M_\sigma$ and $M_\tau$ separate. Suppose next that $\sigma$ approach the direction $\delta$ of an external base-line $d_2$. The points of $M_\tau$ in common with those of $M_\sigma$, when $\sigma=\delta$, if they exist, are the mid-points of chords, with the direction $\tau$, that have one of their ends in the portion of the closed curve immediately above the base-line ($D_1 S D_2$ in Fig. 3).

Clearly within the domain $(R)$* $M_\tau$ can have no points in common with $d_2$. When $\sigma=\delta$, let $M_a$ be the part which $M_\sigma$ drops, and $M_b$ the remainder. According to I, § 1, $M_a$ and $M_b$ cut $M_\tau$ respectively in an even and an odd number of points. Denoting by $S_a$ the portion of $S_\sigma$ with its extremities in $d_2$, then also $S_a$ cuts $M_\tau$ in an even number of points. Hence, *in passing from the intersections of $M_{\sigma_a}$ and $M_{\tau_a}$ to those of $M_{\sigma_b}$ and $M_{\tau_b}$ in the neighborhood of a base-line, an even number of points are lost and an even number gained.* The same is true when a change of $\tau$ encloses the direction of a base-line, and when in a special case $\sigma$ and $\tau$ simultaneously reach the directions of base-lines.

After $\sigma$ continues to change from $\delta$ in the same sense, the even number of points of intersection of $M_a$ and $M_\tau$, which belongs to the original system, begins to decrease as $M_a$ breaks up. The portion of $M_\tau$ that contains these points can have no points in common with the changed position of $d_2$, like $D_1 Q$ in Fig. 4. The points common to $M_a$ and $M_\tau$ are the mid-points of chords having one of their extremities again on the portion of the closed curve directly supported by the interior base-line $D_1 Q$. The same argument applies to the cases where $M_\tau$ breaks up. Hence also the points of intersection of $M_a$ and $M_\tau$ and possible further disintegrations, as long as they result from the original system of intersection by continuous change, when they exist, must disappear in pairs through previous tangencies of $M_a$ and $M_\tau$, or of parts resulting from them. We have therefore the

* When the curve has no other singular points than inflexions, this restriction is not necessary.

THEOREM 5. *In the change of the system of points of intersection of two medians associated with two continuously and uniformly changing orthogonal directions, points can disappear only after previous coincidence.*

*As no point can disappear singly, when $\sigma$ changes through a right angle, in general, these points in groups describe closed curves.*

### § 5. *Classification of Points of Intersection of Medians, and Mode of Description of Closed Curves by these Points.*

A point of intersection of two medians $M_\sigma$ and $M_\tau$ is either a center $A_i$ of an inscribed rhomb of the first kind of the closed curve $C$, or a center $B_k$ of an inscribed rhomb of the second kind. According to II, § 1, no point $A_i$ can ever coincide with a point $B_k$ through a tangency of $M_\sigma$ and $M_\tau$. The points $A_i$ and the points $B_k$, if they exist simultaneously, form therefore two separate systems. Now the number of $A_i$'s and $B_k$'s is odd, so that the number of $A_i$'s and the number of $B_k$'s can not both be either even or odd; *i. e.*, when one is even, the other is odd.

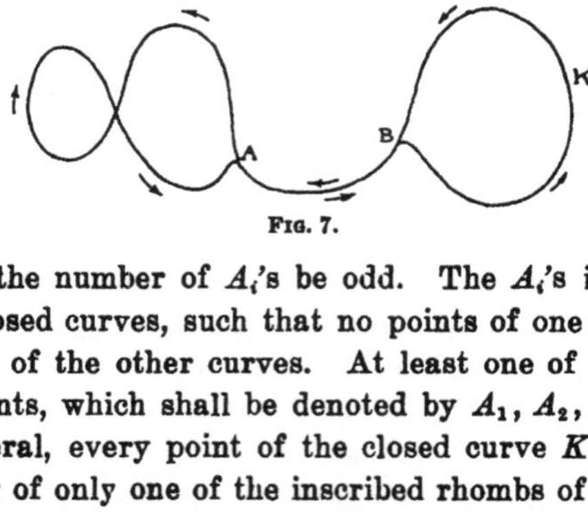


FIG. 7.

Suppose that the number of $A_i$'s be odd. The $A_i$'s in groups describe a finite number of closed curves, such that no points of one curve are permuted with points of any of the other curves. At least one of these groups has an odd number of points, which shall be denoted by $A_1, A_2, A_3, \ldots, A_n$ ($n$ odd and $< i$). In general, every point of the closed curve $K$ described by these points is the center of only one of the inscribed rhombs of the first kind. The curve $K$ may have double points and double arcs as shown in Fig. 7. Every double point and point of a double arc is the center of two inscribed rhombs. Higher multiplicities yield accordingly corresponding numbers of rhombs with the same center.

Consider now a quarter of a full revolution of the direction $\sigma$. Obviously, since the couple $(\sigma, \tau)$ leads to the same set of points $A_1, A_2, \ldots, A_n$ as the couple $(\tau, \sigma)$, after a quarter-revolution of $\sigma$ the same set of points appears. The points $A$ are merely permuted according to a definite law that depends upon the shape of the curve. During a quarter-revolution the same two consecutive points $A_\lambda$ and $A_{\lambda+1}$ can not vanish and reappear again through contact of $M_\sigma$ and $M_\tau$ without colliding with adjacent points. If they could, $A_\lambda$ and $A_{\lambda+1}$ would then describe a closed curve which necessarily would be identical with the curve $K$; and $A_\lambda$ and $A_{\lambda+1}$ would move into coincidence with other points of the set $(A)$, since these describe parts of the same curve. Supposing that $A_\lambda$ and $A_{\lambda+1}$ were describing only a part of $K$, and that $T_1$ were the point of contact of $M_\sigma$ and $M_\tau$ where $A_\lambda$ and $A_{\lambda+1}$ enter, and $T_3$ the point where they disappear through contact, then $A_\lambda$, to reach $T_3$ (Fig. 8), would have to pass over an infinite number of positions which $A_{\lambda+1}$ occupied before. There would therefore be an infinite number of points on $K$ as centers of more than one inscribed rhomb. $A_\lambda$ and $A_{\lambda+1}$ would describe a closed curve consisting of two coincident branches and forming a segment $\overrightarrow{XY}$ of the curve $K$. Denoting the remaining arc of $K$ by $\leftarrow XY \rightarrow$, this part would have two free ends $X$ and $Y$, which is not possible.

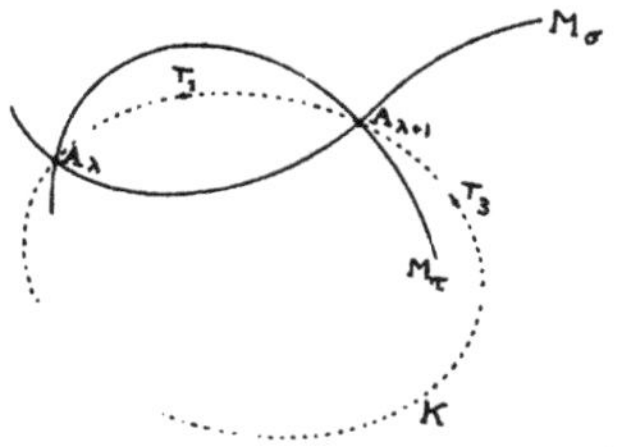


FIG. 8.

Any pair of orthogonal directions is contained in the set of all directions $\sigma$ within a quarter-revolution and the corresponding orthogonal directions $\tau$. Consequently, the set of points $(A)$ describes the curve $K$ just once when $\sigma$ turns through a right angle.

In the continuous change from $\sigma$ to $\tau$ an even number of points $(E)$ of the set $(A)$ vanish and reappear through contact of $M_\sigma$ and $M_\tau$. Consequently, in the same set $(A)$ an odd number of points $(F)$ exists, such that each of these points moves to a neighboring point of the set $(A)$ during the change from $\sigma$ to $\tau$. Exactly the same results are obtained with respect to a group of an odd number of $B_k$'s generating a closed curve.*

* In the deformation of secondary medians that result from the breaking up an $M_\sigma$, parts $S'$ may enter whose continuous deformations never will form a part of an $M_\sigma$. In the deformation of $S'$, its free end on $K$ may meet $M_\tau$ in a point $L$. In the further deformation the intersection of $S'$ and $M_\tau$, if it exists, describes a curve with two free ends $L$ and $L_1$ on $C$. The points of such a curve are also centers of inscribed rhombs, but there is no continuous connection between these points and the points of a curve $K$. The points $L$ and $L_1$ are centers of degenerate rhombs in which a pair of opposite vertices coincides with $L$ in one case and with $L_1$ in the other. Points of such a curve may even be centers of inscribed squares, although this is not in general true. As cases of this kind do not affect the set $(A)$ and the curves described by its points, no further attention will be given them.

## § 6. *System of Inscribed Rhombs Associated with a Curve K.*

Consider again the curve $K$ described by the set $(A)$. Denote the rhomb associated with the center $A_\iota$ by $R_1^\iota, R_2^\iota, R_3^\iota, R_4^\iota$. Let $A_\lambda, A_{\lambda+1}$ be two points of the type $(E)$ associated with the initial direction $\sigma$. To the change from $\sigma$ to $\tau$ corresponds a set of rhombs connecting continuously $R_1^\lambda R_2^\lambda R_3^\lambda R_4^\lambda$ and $R_1^{\lambda+1} R_2^{\lambda+1} R_3^{\lambda+1} R_4^{\lambda+1}$ in the given order. We say that in this case the two rhombs are directly connected, and that the continuous set connecting them is a *direct set*. Taking two points $A_\mu, A_{\mu+1}$ of the type $(F)$, then, in the change from $\sigma$ to $\tau$, as $A_\mu$ moves to $A_{\mu+1}$, $R_1^\mu R_2^\mu R_3^\mu R_4^\mu$ changes continuously into $R_1^{\mu+1} R_2^{\mu+1} R_3^{\mu+1} R_4^{\mu+1}$, such that after $\sigma$ has reached $\tau$, $R_1^\mu$ coincides with $R_2^{\mu+1}$, $R_2^\mu$ with $R_3^{\mu+1}$, $R_3^\mu$ with $R_4^{\mu+1}$, $R_4^\mu$ with $R_1^{\mu+1}$ (or, as another possibility: $R_1^\mu$ with $R_4^{\mu+1}$, $R_2^\mu$ with $R_1^{\mu+1}$, $R_3^\mu$ with $R_2^{\mu+1}$, $R_4^\mu$ with $R_3^{\mu+1}$). The two rhombs are connected by a continuous set of rhombs that permutes the original order of the diagonals. The two rhombs are said to be indirectly connected, and the set connecting them is an *indirect set.* Now, according to § 5, the number of direct sets is even, that of indirect sets odd. The effect of two indirect sets is evidently the same as that of a direct set. The effect of all direct and indirect sets is therefore that of an indirect set. Consequently, if we start from the rhomb $R_1 R_2 R_3 R_4$ with the center $A_1$, we can in succession pass through a continuous set of rhombs to those with centers $A_2, A_3, \ldots, A_n$, and finally back to the original rhomb in the order $R_2 R_3 R_4 R_1$ (or $R_4 R_1 R_2 R_3$). In other words, $R_1R_2R_3R_4$ can be continuously deformed into itself such that $R_1 R_2 R_3 R_4$ will respectively move to $R_2 R_3 R_4 R_1$.

The same is true with respect to a set $(B)$ of the second kind and of the same type.

We can therefore conclude as before, that *among such a system of rhombs there must be at least one square.*

But it is established that at least one such system, either of the first or of the second kind, always exists. There may, of course, be several systems; for example, one group of an odd number of $A$'s and two groups each of an odd number of $B$'s. In this case there would be at least one inscribed square of the first and two inscribed squares of the second kind.

*The theorem on the inscribed square of a closed curve is therefore proved for any closed curve formed by a finite number of analytic arcs without double points and with a finite number of inflexions and other singularities.*

The restriction on double points can, however, immediately be removed. From a closed curve with multiple points we can always in a number of ways detach closed curves without multiple points, to which, according to the foregoing result, a square may be inscribed. Hence the

THEOREM 6. *It is always possible to inscribe at least one square in a closed curve formed by a finite number of analytic arcs with a finite number of inflexions and other singularities.*

# *Theorems on the Groups of Isomorphisms of Certain Groups.*

By Louis Clark Mathewson.

## *Introduction.*

The properties of the group of isomorphisms of a group are such that the importance of these groups is being widely recognized. The theory of the group of isomorphisms includes the theory of primitive roots,* Fermat's and Wilson's theorem,† the determination of the various groups of which a given group is an invariant subgroup,‡ and the solvability of equations by means of radicals. Because of the numerous uses, and since only a limited amount of theory has been developed, this paper has for its object the presentation of several theorems bearing on the groups of isomorphisms of certain familiar groups.

The theory of an isomorphism between two different groups, which is indispensably connected with the complete determination of the intransitive groups of a given degree,§ seems to have been studied first. In general, two groups are said to be isomorphic, if to each operator of the first there corresponds one or more operators of the second, and *vice versa*, in such a way that to the product of any two operators of the first corresponds the product of the corresponding operators of the second. If the correspondence is $1:1$, the isomorphism is termed *simple*; if $a:1$, it is called *multiple*. The importance of the idea of simple isomorphism between two groups lies in the fact that abstractly two such groups have the same properties; that is, both are simple or both composite, abelian or non-abelian, etc. In 1868 Jordan introduced and defined an $a:1$-isomorphism,‖ and ten years later Capelli gave the generalized idea of an $a:b$-isomorphism.¶ Others who have studied these isomorphisms are Netto, Maillet, Dyck, Weber, Miller, Burnside, Dickson, etc.

---

* Miller, *Bulletin of American Mathematical Society* (2), Vol. VII (1900–01), p. 350.

† Miller, *Philosophical Magazine*, Vol. CCXXXI (1908), p. 224; *Transactions of American Mathematical Society*, Vol. IV (1903), p. 158; *Annals of Mathematics*, Vol. IV (1903), p. 188.

‡ Burnside, "Theory of Groups," 1897, § 165.

§ For a detailed discussion see Bolza, American Journal of Mathematics, Vol. XI (1889), pp. 195–214.

‖ *Comptes Rendus Acad. Sc. Paris*, Vol. LXVI (1868), p. 836.

¶ *Giornali di Matematiche*, Vol. XVI (1878), p. 33.

The theory of the group of isomorphisms of a group with itself had its genesis, like most other extensive theories of science, in the discovery and study of isolated cases, and some of these cases seem to have been worked out before it was realized that every finite group has a single group of isomorphisms which also is of finite order. Betti credits Galois with having given without proof an expression for the group of isomorphisms of a group of order $p^m$,* and Dickson makes a comment upon Betti's remark† and refers to a fragment of a posthumous paper, "Des équations primitives qui sont soluble par radicaux." ‡ In his own paper Betti obtains his "massimo moltiplicatore" of the cyclic group of order $p^m$. Jordan took the next step by starting with the abelian transitive linear group defined by $x_i' \equiv x_i + a_i$ $(i=1, \ldots, n)$, and found the non-homogeneous linear group as the largest group of linear substitutions under which the given group is invariant.§ Other special cases were observed by Gierster‖ and Klein.¶ As applied to finite abstract groups, the conception of the group of isomorphisms was developed by Hölder** and by Moore†† independently of each other.‡‡

If a group $G$ be written twice with 1:1-correspondence between its operators such that to the product of two operators of the first arrangement corresponds the product of the two corresponding operators in the second arrangement, the group is said to be *simply isomorphic with itself.*§§ Frobenius has called this an *automorphism,*‖‖ while Miller has employed the term *holomorphism.*¶¶ If $G$ be written as a regular substitution group, then operators on the same letters exist which transform it according to any of these simple isomorphisms.*** If any isomorphism is effected by an operator of $G$ itself, the isomorphism is called *cogredient* or *inner*; all others are called *contragredient* or *outer.* In his article of 1893 Hölder first proved that "the totality of the different isomorphisms of a group with itself form a group $I$," ††† and later states that the cogredient isomorphisms always form an invariant sub-

* *Annali di Scienze Matematiche e Fisiche,* Vol. VI (1855), p. 34.
† *Transactions of American Mathematical Society,* Vol. I (1900), p. 30, foot-note.
‡ "Oeuvres Mathématiques D'Evariste Galois," 1897, p. 58.
§ Traité des Substitutions, 1870, §§ 118, 119.
‖ *Mathematische Annalen,* Vol. XVIII (1881), p. 354.
¶ "Vorlesungen über das Ikosaeder," 1884, p. 232.
** *Mathematische Annalen,* Vol. XLIII (1893), pp. 313 *et seq.*
†† *Bulletin of American Mathematical Society,* Vol. I (1894–95), p. 61.
‡‡ *Bulletin of American Mathematical Society,* Vol. II (1895–96), p. 33, foot-note.
§§ Cf. Burnside, "Theory of Groups," 1911, p. 81.
‖‖ *Sizungsberichte der Berliner Akademie,* 1901, p. 1324.
¶¶ *Bulletin of American Mathematical Society* (2), Vol. IX (1902), p. 112.
*** Frobenius, *Berliner Sizungsberichte,* 1895, pp. 184, 185.
††† *Mathematische Annalen,* Vol. XLIII (1893), p. 314.

group of this group of isomorphisms *I.** It is readily seen that the group of cogredient isomorphisms is simply isomorphic with the quotient of the given group with respect to its central,† and if the given group $G$ has no invariant operators besides the identity, the group of cogredient isomorphisms is simply isomorphic with $G$ itself. If $G$ has no invariant operator besides the identity and admits of no contragredient isomorphism, Hölder has called it a *complete* group, and has proved explicitly that the symmetrical group of degree $n$ $(n>2, n\neq 6)$ is a complete group.* Frobenius defines as *characteristic* any operator or subgroup of a given group $G$ which is transformed into itself by all the operators of the group of isomorphisms of $G$; ‡ and Burnside has introduced the term *holomorph*, to mean the group composed of all the substitutions on the letters of a regular substitution group $G$ which transform $G$ into itself.§

Following the developments made by the two pioneers in the theory of the group of isomorphisms of a group, more general work began and applications of the theory were made. Burnside's "Theory of Groups" (1897) was the first text-book to elaborate the theory; Miller determined the groups of isomorphisms of all the substitution groups of degree less than 8, in connection with which he proved several far-reaching theorems.|| In all, upwards of forty articles have appeared in mathematical literature on isomorphisms of a group or on properties of the group of isomorphisms of a group. Many of these have been written by Miller, most of the others being by Burnside, Dickson and Moore, who have been mentioned previously, and by Young (J. W.) and Ranum.

## I. *Theorems on Certain Groups Involving a Characteristic Set.*

DEFINITION. *A characteristic set in a group $G$ consists of one or more complete sets of conjugates under the group of isomorphisms of $G$.*

This definition includes the notion of a characteristic set of operators of $G$ as well as that of a characteristic set of subgroups of $G$. The distinguishing property in a characteristic set, either of operators or of subgroups, is thus seen to be that the set contains a definite number of operators which as an aggregate correspond among themselves in all the isomorphisms of $G$. If a characteristic set consists of but two conjugates, it may be called a character-

* *Mathematische Annalen*, Vol. XLVI (1895), pp. 325 *et seq.*

† Given implicitly first in *Mathematische Annalen*, Vol. XLIII (1893), pp. 329, 330.

‡ Frobenius, *Berliner Sitzungsberichte*, 1895, pp. 184, 185.

§ "Theory of Groups," 1897, p. 228.

|| *Philosophical Magazine*, Vol. CCXXXI (1908), pp. 223 *et seq.*

istic pair; in case the set consists of a single operator or subgroup, this is characteristic as already noted.

It results immediately from the fundamental property of a characteristic set of subgroups that *if $G$ is the direct product of a characteristic set of subgroups, the $I$ of $G$ is the product of the groups of isomorphisms of these subgroups extended by operators corresponding to whatever isomorphisms are possible among these subgroups themselves.*

In the present discussion, unless the contrary is stated, the definition of a characteristic set of subgroups of $G$ will be restricted to mean a single complete set of conjugate subgroups under the group of isomorphisms of $G$. A few special products will be considered, the different cases arising from different kinds of groups in the characteristic set.

THEOREM I. *If $G$ is the $n$-th power of an indivisible group $H$ whose central is the identity, then the $I$ of $G$ is the $n$-th power of the group of isomorphisms of $H$ extended by operators which permute these $n$ groups of isomorphisms according to the symmetric group of degree $n$. If $i$ is the order of the group of isomorphisms of $H$, then the order of the $I$ of $G$ is $n!\,i^n$.**

For the proof of this theorem, as well as for the proof of a later theorem, it will be useful to have some auxiliary observations established. Accordingly, it will be shown first that *if $G$ is the direct product of two indivisible groups $G_1$ and $G_2$, neither of which contains an invariant operator besides the identity, then $G_1$ and $G_2$ form a characteristic pair (according to the unrestricted definition).*†

In proving that $G_1$ and $G_2$ form a characteristic pair (or set), it is necessary and sufficient to show that neither $G_1$ nor $G_2$ can correspond to any subgroup containing operators outside this pair. Now $G_1$ and $G_2$ have the following properties: neither is divisible and neither contains an invariant operator besides the identity; every operator of one is commutative with each operator of the other; and their direct product is $G$. Since they have similar properties, it will suffice to show that $G_1$ could not, in any of the isomorphisms of $G$ with itself, correspond to a subgroup outside this original pair. Suppose, if possible, that $G'$ were a subgroup neither $G_1$ nor $G_2$ which could correspond to $G_1$ in some isomorphism of $G$. Since $G_1$ is not a direct product, $G'$ could not be a direct product. Hence, this $G'$ must be constructed by some isomorphism between

---

* By the "$n$-th power" is meant the direct product of $n$ identical groups. An *indivisible* group is one which is not a direct product. *Divisible* is first used in English by Easton, "Constructive Development of Group Theory," 1902, p. 47, being a translation of "zerfallend," which was introduced by Dyck, *Mathematische Annalen*, Vol. XVII (1880), p. 482.

† Cf. Remak, *Crelle's Journal*, Vol. CXXXIX (1910–11), pp. 293–308.

the original factor groups, or one of them and a subgroup of the other, or between invariant subgroups from both of them. Supposing that $G'$ were formed by an $a:b$-isomorphism, it will be shown that it could not be simply isomorphic with $G_1$.

If $G'$ were formed by an $a:b$-isomorpnism, then $G_2$ could not correspond to itself (because not all of its operators would be commutative with its constituents in $G'$, since $G_2$ contains no invariant operator besides the identity), but must likewise have a $G''$ correspond to it in this particular isomorphism of $G$, and $G''$ would likewise be the result of some $a':b'$-isomorphism. Since every operator of $G_1$ is commutative with each operator of $G_2$, every operator of $G'$ must be commutative with each operator of $G''$, and accordingly the constituents of $G_1$ found in the operators of $G'$ must be commutative individually with the constituents from $G_1$ found in the operators of $G''$ (the same would be true respecting the constituents of $G_2$ as found in $G'$ and in $G''$). From this fact it will now be shown that $[G', G'']$ would not contain $G_1$. From the conditions just stated about the commutability of constituents, if any constituent of $G_1$ in $G'$ also appeared in $G''$, then $G_1$, if generated at all, would contain at least one invariant operator besides the identity; and if the constituents of $G_1$ in $G'$ were all different from those in $G''$, then $G_1$ would be a direct product. But both results are contrary to the original hypothesis that $G_1$ contain no invariant operator besides the identity and be indivisible. Hence, $[G', G'']$ does not contain $G_1$, or, in other words, $G$ is not generated by $G'$ and $G''$. Therefore, in no isomorphism of $G$ with itself can $G_1$ or $G_2$ correspond to a subgroup outside this generating pair, or $G_1$ and $G_2$ constitute a characteristic pair (or set) of subgroups of $G$.*

The development of this proof further shows that $G$ can not be the direct product of another pair of indivisible groups neither of which contains an invariant operator besides the identity. Moreover, no factor of $G$ could contain an invariant operator besides the identity, for then the central of $G$ would not be the identity. Accordingly, if $G$ is the direct product of two indivisible groups neither of which contains an invariant operator besides the identity, then if $G$ can be represented as the direct product of two other factors, at least one of these must be divisible. If $G$ were the product of several factors, the same line of proof would establish the generalization of this statement; viz.: *If $G$ is the direct product of $n$ indivisible groups $G_1, \ldots, G_n$ no one of which*

* If, in addition to the conditions already imposed upon $G_1$ and $G_2$, it is given that $g_1 \not\cong g_2$, then evidently $G_1$ and $G_2$ are themselves characteristic subgroups, and the $I$ of $G$ is the direct product of the groups of isomorphisms of these subgroups. Cf. Miller, *Transactions of American Mathematical Society*, Vol. I (1900), p. 396.

*contains an invariant operator besides the identity, and if $G$ can be represented as the direct product of other factors not all of which are identical with the former $n$, then at least one of this second set of factors is divisible.* From this it is evident that if $G$ were the product of $n$ such indivisible groups, no outside subgroup could correspond to any of these in any automorphism of $G$; and accordingly the preliminary observation becomes: *If $G$ is the direct product of $n$ indivisible groups no one of which contains an invariant operator besides the identity, then these $n$ groups form a characteristic set (according to the unrestricted definition).**

To proceed with the proof of Theorem I, let $H_1, H_2, \ldots, H_n$ be the $n$ similar groups each simply isomorphic with $H$, and let $I_1, I_2, \ldots, I_n$ be their respective groups of isomorphisms. These will also be simply isomorphic among themselves and have the same order $i$. Since $G = [H_1, \ldots, H_n]$, $G$ is invariant under each of these $I$'s and hence under their direct product. Moreover, in each of the $i^n$ different isomorphisms effected by $[I_1, \ldots, I_n]$ each of these $n$ generating $H$'s corresponds to itself. Now $G$ is the direct product of the $n$ indivisible groups no one of which contains an invariant operator besides the identity; hence, from the preceding observation no outside subgroup of $G$ can correspond to any of the $H$'s in any of the isomorphisms of $G$, or these $n$ $H$'s form a characteristic set.

From the fact that the original $n$ generating $H$'s constitute a characteristic set, the remaining isomorphisms of $G$ are those arising when the $n$ $H$'s correspond among themselves; and this is possible, since they all are simply isomorphic and distinct. Now if an operator $s_i$ of one of the $H$'s, say $H_i$, corresponds to an operator $s_j$ of $H_j$, then $H_i$ and $H_j$ must correspond in simple isomorphism throughout, because, since no $H$ is a direct product nor contains an invariant operator besides the identity, and since no operator from outside the groups of the generating set can correspond to an operator of the generating set, evidently the operators of $H_i$ are the only ones in the generating set which, with $s_i$, form a subgroup $H$ which could be simply isomorphic with $H_j$. Hence, the correspondences among the $n$ generating $H$'s of the characteristic set can be effected in exactly $n!$ ways, corresponding exactly to the symmetric group of degree $n$.

The $I$ of $G$, therefore, is of order $n!i^n$ and is simply isomorphic with $[I_1, I_2, \ldots, I_n]$ extended by operators which permute the $I$'s according to the symmetric group of order $n!$.

---

* Cf. Remak, *loc. cit.*, p. 304.

*Example.* Suppose that $G \equiv (abcd)$ pos $(efgh)$ pos $(ijkl)$ pos. Then the $I$ of $G$ is $[(abcd)$ all $(efgh)$ all $(ijkl)$ all$] \cdot (eai \cdot bfj \cdot cgk \cdot dhl) \cdot (ae \cdot bf \cdot cg \cdot dh)$.

By means of this theorem it is easy to establish the following one on a system of groups of isomorphisms which are complete groups:

THEOREM II. *If $H$ is the group of isomorphisms of a simple group $S$ of composite order, then the $I$ of the $n$-th power of $H$ is a complete group of order $n!h^n$ which is the $n$-th power of $H$ extended by operators which permute the $n$ $H$'s according to the symmetric group of degree $n$.*

In proving this theorem, use will be made of a proposition due to Burnside, "If $G$ is a simple group of composite order, or if it is the direct product of a number of isomorphic simple groups of composite order, the group of isomorphisms $L$ of $G$ is a complete group."*

Since $S$ is a simple group, it is indivisible and contains no invariant operator besides the identity. Hence, from Theorem I, the group of isomorphisms of the $n$-th power of $S$ is $L \equiv [H_1, \ldots, H_n]$ extended by operators which permute the $H$'s according to the symmetric group of degree $n$, and from Burnside's theorem, $L$ is a complete group.

The quoted theorem states that $H$ also is a complete group. It accordingly contains no invariant operator. Moreover, it is indivisible. To show this it will first be proved that *if a group which is the direct product of indivisible factors no one of which contains an invariant operator besides the identity, contains an invariant simple group, then that simple group is contained wholly in one of the factors.* For this simple group could not be the direct product of invariant subgroups from different factors, nor could it be the result of any isomorphism, not 1:1, of such subgroups from different factors, because in both cases the group formed would be composite. Since no factor contains an invariant operator besides the identity, the required group could not be the result of a 1:1-isomorphism, for in this case the group formed would not be invariant under the entire group. Hence, if an invariant simple subgroup occurs, it must be wholly in one of the factors.

Now to complete the proof that $H$ is indivisible, suppose it were divisible. $S$ is its own group of cogredient isomorphisms, and hence it is contained in $H$ invariantly. Then from the statement just established, if $H$ were divisible, $S$ would necessarily be contained wholly in one of the factors. But this would lead to an absurdity, for then $S$ would be transformed in the same way by more than one operator in its group of isomorphisms, since every operator of the other factors would transform it in exactly the same way as the identity.

---

* "Theory of Groups," 1911, p. 96.

Accordingly, $H$ is an indivisible group. Then since $H$ contains no invariant operator besides the identity and is indivisible, the preceding Theorem I can be applied to the $n$-th power of $H$, and the group of isomorphisms is seen to be $I\equiv[H_1, \ldots, H_n]$ extended by operators which permute the $H$'s according to the symmetric group of degree $n$. Furthermore, since $I$ is seen to be the same as the complete group $L$, $I$ is likewise a complete group.

*Example.* The group of isomorphisms of the simple group of order 168 and degree 7 is known to be simply isomorphic with the group of order 336 and degree 8. Hence, the $I$ of the $n$-th power of $(abcdefgh)_{336}$ is a complete group, which may as a substitution group be obtained according to the theorem.

COROLLARY I. *The $I$ of the $n$-th power of a symmetric group whose degree is $>4$ and $\neq 6$ is a complete group.*

Because these symmetric groups are the groups of isomorphisms of their respective alternating groups which are simple and of composite order.*

COROLLARY II. *If $H$ is the group of isomorphisms of a simple group $S$ of composite order, then the $I$ of the square of $H$ or of $S$ is a complete group which is the double holomorph of $H$.*

From the given theorem this $I$ is seen to be the square of $H$ extended by an operator of order 2 which interchanges the two systems. Since $H$ is a complete group, the resulting $I$ is the double holomorph of $H$.†

Again, in connection with the groups of isomorphisms of the squares of certain groups, another theorem which is closely related to the preceding two will be given.

THEOREM III. *If $G$ is the square of an indivisible group $H$ whose central is the identity and which is simply isomorphic with a characteristic subgroup of its group of isomorphisms, then the $I$ of $G$ is simply isomorphic with the double holomorph of the group of isomorphisms of $H$.*

In proving this theorem it may be supposed that $H$ is written as a regular substitution group. Then the operators transforming it according to its possible automorphisms can be written as substitutions on the same letters.‡ Furthermore, since $H$ contains no invariant operator besides the identity, and is thus its own group of inner isomorphisms, it is identical with the characteristic subgroup of its group of isomorphisms $(I_H)$ with which it was known to be simply isomorphic. Since no two operators of the group of isomorphisms of a group transform that group in the same way, no operator besides the

---

* See Burnside, "Theory of Groups," 1911, §§ 162, 139.

† This term was introduced by Miller. See *Transactions of American Mathematical Society*, Vol. IV (1903), p. 154.

‡ Frobenius, *Berliner Sitzungsberichte*, 1895, pp. 184, 185.

identity in $I_H$ can be commutative with all the operators of $H$. Hence, $I_H$ contains no invariant operator besides the identity, or is its own group of inner isomorphisms.

Now $H$ is a characteristic subgroup of $I_H$, and accordingly the possible automorphisms of $I_H$ with the operators of $H$ in some fixed correspondence form an invariant subgroup of the group of isomorphisms of $I_H$. But since no two operators in $I_H$ and outside $H$ transform the operators of $H$ in the same way, the total automorphism of $I_H$ is fixed if the automorphism of $H$ is fixed. Hence, the $I$ of the group of isomorphisms of $H$ is not of greater order than is $I_H$ itself. But its group of inner isomorphisms has been seen to be of that order. Accordingly, since $I_H$ contains no invariant operator besides the identity and admits of no outer isomorphism, it is a complete group.*

Finally, from Theorem I the $I$ of $G$ is seen to be the square of a complete group extended by an operator of order 2 which interchanges the two systems, and therefore this $G$ is the double holomorph of the group of isomorphisms of $H$.

*Example.* If $G \equiv (abcd)$ pos $(efgh)$ pos, then $I \equiv [(abcd)$ all $(efgh)$ all$]$ . $(ae \cdot bf \cdot cg \cdot dh)$, which is the double holomorph of the symmetric group of degree 4.

This part of the discussion is concluded with the following theorem and corollaries on conjoints:†

THEOREM IV. *If an indivisible group $H$ and its conjoint $H'$ generate $G$, where the order of the central of $H$ is not greater than 2, then if $G$ contains no other pair of subgroups simply isomorphic with $H$ and $H'$, the $I$ of $G$ is simply isomorphic with the square of the group of isomorphisms of $H$ extended by an operator of order 2 which simply interchanges the two systems.*

Suppose that $C$ is the central of $H$. From the nature of the construction of the conjoints, $C$ is also the central of $H'$, and the cross-cut of the two groups; accordingly, it is the central of $G$. By hypothesis $C$ is either the identity or the identity and a characteristic operator of order 2.

It will be proved that these two conjoints form a characteristic set or pair. Let the operators of $H$ outside $C$ be represented by $s$'s, those of $H'$ by $t$'s; thus,

$$H \equiv C, \ s_2, s_3, \ldots, s_h;$$
$$H' \equiv C, \ t_2, t_3, \ldots, t_h.$$

Since by hypothesis there are no other pairs of subgroups which could correspond to $H$ and $H'$, it is sufficient to show that no subgroup exists in $G$

* Burnside gives a different proof (*loc. cit.*, p. 95).

† The term *conjoint* is due to Jordan. See "Traité des Substitutions," 1870, p. 60.

which could correspond to one of these conjoints and with the other form a generating pair in some automorphism of $G$. Any such subgroup would have to contain an operator or some operators from outside $H$ and $H'$. But none of the operators outside $H$ and $H'$ is commutative with all the operators of $H$ or with all those of $H'$. This is because all such operators are of the form $s_i t_j$ $(i, j = 2, \ldots, h)$, where neither factor is from the central of $C$; so that the product can not be commutative with all of $s_2, \ldots, s_h$ nor with all of $t_2, \ldots, t_h$. Accordingly, no subgroup exists which could replace either conjoint in an automorphism of $G$; so that, under the hypotheses, they constitute a characteristic pair of subgroups.

Since the operators of each conjoint are commutative with each operator of the other, and each conjoint subgroup is independent of the other, evidently each can correspond exactly according to its own group of isomorphisms while the other remains fixed in identical correspondence. Furthermore, the conjoints themselves can correspond, and they correspond throughout if some operator outside the central of one is in correspondence with an operator of the other. That is, if some $s$ corresponds to some $t$, then $H$ and $H'$ correspond entirely. Otherwise, it would be possible to form from the operators of $H$ and $H'$ two groups each simply isomorphic with $H$, each containing both $s$'s and $t$'s, and they would possess the property that every operator of one was commutative with every operator of the other. Since $H$ is not a direct product, this would necessitate that some one or more of the $s$'s (or $t$'s) be commutative with all the other $s$'s (or $t$'s), making the order of the central greater than that of $C$, which would be contrary to hypothesis. Hence, if one operator (outside $C$) of $H$ corresponds to an operator of $H'$, all of $H$ corresponds to $H'$.

Hence, $G$ has the isomorphisms of $H$, combined with all those of $H'$, and finally those additional ones resulting from making $H$ and $H'$ correspond, and it has no more. The $I$ of $G$ is thus representable as a substitution group by extending the square of the group of isomorphisms of $H$ by an operator which simply interchanges the two systems.

Should $H$ contain no invariant operator besides the identity, from the auxiliary facts established in the proof of Theorem I, $H$ and its conjoint would constitute a characteristic pair, and the conditions would be a special case of Theorem I with $n=2$; viz.:

COROLLARY I. *If $G$ is generated by an indivisible group $H$ and its conjoint $H'$, where $H$ contains no invariant operator besides the identity, then the $I$ of $G$ is simply isomorphic with the square of the group of isomorphisms of $H$ extended by an operator of order 2 which simply interchanges the two systems.*

If, in this theorem or the corollary just stated, the group of isomorphisms of $H$ should be a complete group, the $I$ of $G$ would be simply isomorphic with the double holomorph of the group of isomorphisms of $H$.

If two conjoints such as those of the theorem generate a group $G$ which contains other pairs of simply isomorphic subgroups (which could correspond to $H$ and $H'$ in some automorphism of $G$), then from the preceding proof, where it was shown that no one subgroup could be in two different pairs of conjoints, the following statement can be made regarding the order of the group of isomorphisms of $G$:

COROLLARY II. *If $G$ is generated by an indivisible group $H$ and its conjoint $H'$, where the order of the central of $H$ is not greater than 2, then the order of the $I$ of $G$ is twice the product of the order of the group of isomorphisms of $H$ by the number of pairs of subgroups which can correspond to $H$ and $H'$ in the automorphisms of $G$.*

*Example.* The group $G = [(abcd)_8\ (efgh)_8]_{2,2}\ (ae \cdot bf \cdot cg \cdot dh)$, which is of degree 8, well illustrates most of the theory brought out in connection with this theorem. This group is generated by two conjoint quaternion groups having a common central (the identity and a characteristic operator of order 2) and nothing else in common; furthermore, it contains no other quaternion subgroups. Now the symmetric group of degree 4 is the group of isomorphisms of the quaternion group,* so that the $I$ of $G$ is simply isomorphic with the square of the symmetric group of degree 4 extended by an operator of order 2 which simply interchanges the two systems; or since the symmetric group of degree 4 is a complete group, this $I$ is the double holomorph of this symmetric group, its order being 1152.

Again, it may be noted that this $G$ is generated by a pair of conjoint octic groups. If there were but one pair, $I$ would be of order 128; but there are nine distinct pairs of octic groups, each pair having the properties of conjoints and being generators of $G$; so that the $I$ of $G$ is of order $9 \times 128 = 1152$.

## II. *Theorems on the Direct Product of Particular Dissimilar Groups.*

In the theorems of this section the group $G$ is the direct product of two subgroups, one of which ($A$) is characteristic, and the other ($B$) may have conjugates under the group of isomorphisms ($I_G$). $A$ and $B$, therefore, have the following properties: every operator of $B$ is commutative with each operator of $A$; the identity is their only common operator; $A$ and $B$ generate $G$; $A$ always corresponds to itself; and $B$ may correspond to some other subgroup.

* Cf. Miller, *Bulletin of American Mathematical Society* (2), Vol. V (1898–99), p. 294.

Since $A$ always corresponds to itself, any group $B'$ which corresponds to $B$ in any automorphism of $G$ can have only the identity in common with $A$; hence, $B'$ could not be the direct product of any subgroup from $A$ and any subgroup of $B$. Accordingly, if $B'$ exists, it must be formed by some isomorphism. Every operator of $B'$ must be commutative with each operator of $A$; hence, any operators of $A$ appearing as constituents in $B'$ must be from the central of $A$. $B'$ could not be formed by an isomorphism of any subgroup from $B$ of index $\gtreqless 2$ with a subgroup from the central of $A$, for the group generated by such a group and $A$ would not contain $B$, hence would not be $G$. Hence, since $B'$ is to be of the same order as $B$, every possible $B'$ must be formed by a $d:1$-isomorphism of $B$ with some subgroup from the central of $A$. Such a group and $A$ would evidently generate $B$ and accordingly $G$; moreover, it would be simply isomorphic with $G$ and would have only the identity in common with $A$. With this introduction the following auxiliary proposition will be established for immediate use in the following four theorems:

THEOREM V. *If the group $G$ is the direct product of two groups $A$ and $B$ such that* (a) *$A$ is characteristic in $G$;* (b) *$B$ and all its conjugates contain a common subgroup $D$ of index $p$ ($p$ being any prime);* (c) *the operators of order $p$ in the central of $A$ form a characteristic subgroup which admits of $\alpha$-holomorphisms,* $\alpha=1, \ldots, (p-1)$, under the $I$ of $A$: then the $I$ of $G$ is simply isomorphic with the direct product of the group of isomorphisms of $B$ ($I_B$) and a group $I'$.*

From the statement of the conditions it is obvious that all the conjugates of $B$ are formed by a $d:1$-isomorphism between $B$ and cyclic groups of order $p$ from the central of $A$, all of these conjugates having the same head $D$, which is thus a characteristic subgroup of $G$. It will first be proved that the $I$ of $G$ is simply isomorphic with a group generated by $I_A$, $I_B$, and operators representing certain transformations of $B$ into its conjugates. Ultimately it will be shown also that $I'$ *is generated by the group of isomorphisms of $A$ ($I_A$) and operators corresponding to the transformation of $B$ into its conjugates in which the order of the $B$-constituents is the same as the order of the operators of $B$ itself.*

In the automorphisms of $G$, $A$ can correspond according to all its possible isomorphisms while $B$ remains in identical correspondence, and *vice versa.* But each such set of transformations forms a group, $I_A$ and $I_B$, respectively, which have nothing more than the identity in common. Hence, the $I$ of $G$ has

* Defined by Young (J. W.), *Transactions of American Mathematical Society*, Vol. III (1902), p. 186.

subgroups simply isomorphic with $I_A$ and with $I_B$ (all the operators of one subgroup being commutative with each of those of the other) and a subgroup simply isomorphic with their direct product. When referring to these simply isomorphic subgroups, they will be called $I_A$, $I_B$, etc., rather than "the subgroup simply isomorphic with $I_A$," etc. With the characteristic subgroup $A$ in identical correspondence, $B$ can correspond to itself according to $I_B$ and can correspond to each of its conjugates. Thus, the group generated by $I_B$ and operators corresponding to the transformation of $B$ into its conjugates is an invariant subgroup of the $I$ of $G$, and since the quotient group is the $I_A$, and the $I$ of $G$ is known to contain $I_A$, none of whose operators (excepting the identity) is in the invariant subgroup, therefore the $I$ of $G$ is generated by $I_A$, $I_B$, and operators representing the transformation of $B$ into its conjugates.

It will now be proved that the $I$ of $G$ contains a subgroup, simply isomorphic with $I_B$, every operator of which is commutative with the operators effecting the automorphisms of $A$ and the transformation of $B$ into its conjugates leaving the order of the operators of $B$ unchanged as they become constituents of a $B'$.

Suppose that $m-1$ subgroups can be formed to correspond to $B$ (besides itself). $D$, the characteristic head of all these conjugates, and any one of the $m$ tails form a group simply isomorphic with $B$. Since $p$ is a prime, the co-sets with respect to $D$ in $B$ (or in a $B'$) are transformed among themselves cyclically if they are permuted at all; and if the co-sets of one conjugate are permuted, the corresponding co-sets of all the conjugates permute in the same order. Consider an operator which transforms $B$ into itself according to some one of its automorphisms. If each one of the following $m-1$ tails does not go into itself under this transformation of the operators of $B$, a different transforming operator can be found which will effect the same automorphism of $B$ and simultaneously transform each tail into itself. Such an operator can be found by taking the product of the original transforming operator by an operator from $I_A$ which effects the same $n$-holomorphism among the operators of order $p$ in $A$'s central as is effected by the original transforming operator among the co-sets of the tail of $B$. If $R$ is such an operator of the $I$ of $G$, and any operator of $B$, as $t_1$ (which for convenience may be supposed in the first co-set after $D$), is transformed into $t'_n$ in the $n$-th co-set (where $n=1, \ldots, \overline{p-1}$, supposing that the order of the co-sets is that of successive powers), and if $c$ is an operator of order $p$ from the central of $A$, then

$$R^{-1} t_1 r = t'_n \qquad (R^{-1} cR = c^n). \tag{1}$$

Now let $V$ be any operator of the $I$ of $G$ which transforms $B$ into a $B'$ leaving the order of the $B$-constituents the same as that of the corresponding operators of $B$. Then

$$V^{-1} t_1 V = t_1 c^i \qquad (V^{-1} t'_n V = t'_n c^{ni}). \tag{2}$$

Transforming (1) by $V$ and (2) by $R$ gives

$$V^{-1} R^{-1} t_1 RV = V^{-1} t'_n V = t'_n c^{ni},$$
$$R^{-1} V^{-1} t_1 VR = R^{-1} t_1 c^i R = t'_n c^{ni}.$$

Hence, $VR = RV$. But the totality of the $R$'s constitute a subgroup of the $I$ of $G$ simply isomorphic with $I_B$. Accordingly, each operator of this $R$-subgroup is commutative with each operator transforming $B$ into a $B'$, under the conditions stated.

It must also be shown that every $R$ is commutative with the operators effecting the automorphisms of $A$. The only automorphisms of $A$ which need be examined in this relation are those affecting the $c$'s in some way. If W is an operator effecting some such automorphism of $A$, it is commutative with $t$; hence, $WR$ and $RW$ transform $B$ in exactly the same way. Suppose that

$$W^{-1} cW = c' \qquad (W^{-1} tW = t); \tag{3}$$

then, from (1) and (3),

$$R^{-1} W^{-1} cWR = c'^n,$$
$$W^{-1} R^{-1} cRW = c'^n.$$

Hence, the $R$'s are individually commutative with the operators of $I_A$; and, accordingly, there is in the $I$ of $G$ a factor which is simply isomorphic with $I_B$.

Since the $I_B$ is a factor, the other factor ($I'$) depends only upon the isomorphisms of $A$ and those of the tails. Hence, in determining $I'$, the operators of $B$ can be used in some arbitrary order. If letters be placed before each of the $m(p-1)$ co-sets of the $m$ tails with respect to $D$ (a letter before each such co-set), and letters before such operators of $A$ outside its largest characteristic subgroup (none are needed if all the operators are commutative and of order $p$) as are necessary to determine its group of isomorphisms, then the permutations of these letters during the transformation of $B$ into each $B'$ ($B$-constituents remaining in the same order as the corresponding operators of $B$ itself), and during all the automorphisms of $A$, will determine $I'$.

In the following theorems $A$ and $B$ are of such forms that $I'$ is readily obtained.

NOTES. If in this proposition $p=2$, then the proof is much simpler, since there is but one co-set in each tail and hence no cyclical permutation of the co-sets in any tail.

If, in addition to the given hypotheses, the central of $A$ is known to be the identity, then $B$ also is characteristic and the $I$ of $G$ immediately determined. Or, if $B$ contains no invariant subgroup whose quotient group is simply isomorphic with a subgroup of the central of $A$, then $B$ also is characteristic and the $I$ of $G$ determined as before.*

* Miller, *Transactions of American Mathematical Society*, Vol. I (1900), p. 396.

*Example.* If $G \equiv (abcd)$ all $(efgh)_8$, then $A$ is the octic group, $B$ is the symmetric group $(abcd)$ all, and there is but one $B'$, which is formed by dimidiating $B$ with the central of $A$. The operator of order 2 which transforms $B$ into $B'$ is commutative not only with all the operators of $I_B$ but also with those of $I_A$ (because the operator of order 2 from $A$ used in the dimidiation is characteristic in $A$ and invariant under $I_A$). Therefore, the $I$ of $G$ is simply isomorphic with $(abcd)$ all $(efgh)_8(ij)$.

The following theorem is stated because it represents the common starting point of Theorems VII, VIII, IX and X, which follow; and a proof is here given to introduce the method of demonstration to be employed throughout. The latter four theorems are extensions in different directions of the one under immediate consideration.

THEOREM VI. *If a group $G$ is the direct product of the group of order $p$, $p$ being any prime, and a group $H$ which contains a characteristic subgroup $D$ of index $p$, neither $D$ nor the central of $H$ containing an invariant subgroup of index $p$, then the $I$ of $G$ is simply isomorphic with the direct product of the group of isomorphisms of $H$ and the holomorph of the group of order $p$.*

The cyclic group of order $p$ is characteristic, since it is the only subgroup of order $p$ in the central of $G$; and since $D$ is characteristic in $H$ and contains no invariant subgroup of index $p$, it too is characteristic in $G$. Each of these subgroups must correspond to itself in every isomorphisms of $G$; furthermore, $G$ must contain the product of these two subgroups as a characteristic subgroup of index $p$. After this characteristic subgroup of index $p$ is fixed in any isomorphism, there can correspond to any one of the remaining co-sets, with respect to $D$, any one of $p$ such co-sets and no more, and this correspondence then establishes the isomorphism of the entire group. The $I$ of $G$ thus contains an invariant cyclic subgroup of order $p$. For convenience, it may be supposed to be of degree $p$, a letter corresponding to each of the $p$ interchangeable co-sets. Now, if the operators of the characteristic subgroup $D$ are retained in the same isomorphism as before, there are exactly $p-1$ possible correspondences of the succeeding co-sets in the characteristic subgroup of index $p$, since any one of the $p-1$ co-sets can stand first and the order of the remaining ones is then determined. $I$ thus contains an invariant subgroup of order $p(p-1)$ which, moreover, contains the cyclic subgroup of order $p$ invariantly. Since this subgroup can be written on $p$ letters, it is the holomorph of the cyclic group of order $p$.

The only other isomorphisms of $G$ are those effected by the group of isomorphisms of $H$, and accordingly, from Theorem V, the $I$ of $G$ is simply isomorphic with the direct product of the group of isomorphisms of $H$ and the holomorph of the cyclic group of order $p$.

*Example.* If $G = (abcd)$ pos $(efg)$ cyc, then $H$ is the alternating group $(abcd)$ pos which contains a four-group as a characteristic subgroup $D$ of index 3, neither $D$ nor the central of $H$ containing an invariant subgroup of index $p$. Since the group of isomorphisms of this $H$ and the holomorph of the cyclic group of order 3 are known, the $I$ of $G$ is simply isomorphic with $(abcd)$ all $(efg)$ all.*

THEOREM VII. *If a group $G$ is the direct product of an abelian group $A$ of order $p^m$, type $(1, 1, 1, \ldots .)$, $p$ being any prime, and a group $H$ which contains a characteristic subgroup $D$ of index $p$, neither $D$ nor the central of $H$ containing an invariant subgroup of index $p$, then the $I$ of $G$ is simply isomorphic with the product of the group of isomorphisms of $H$ and the holomorph of $A$.*

The abelian group $A$ is characteristic in $G$, since it is the largest subgroup in the central of $G$ that contains operators of order $p$ only; and since $D$ is characteristic in $H$ and contains no invariant subgroup of index $p$, it too is characteristic in $G$. Each of these subgroups must correspond to itself in every automorphism of $G$; furthermore, $G$ must contain the product $B$ of these two subgroups as a characteristic subgroup of index $p$. The rest of $G$ consists of the product of the abelian group $A$ into the tail of $H$ (which is $H$ minus $D$).

Suppose, while determining the isomorphisms of the various subgroups which can be formed by isomorphism, that the operators of $H$ remain in some fixed order. Let the characteristic subgroup $B$ be in some identical correspondence; then any one of the first $p^m$ co-sets in the tail of $G$ can be taken to correspond with the one originally first. When this is selected, it with $B$ determines the entire isomorphism of $G$. There are, accordingly, exactly $p^m$ such isomorphisms, and they are commutative and each of order $p$. Hence, so far as the isomorphisms of these subgroups of $G$ are concerned, there is an invariant abelian subgroup of order $p^m$, type $(1, 1, 1, \ldots .)$. For convenience, this subgroup may be supposed to be written on $p^m$ letters, one corresponding to each of the first $p^m$ co-sets formed with respect to $D$, since the $p^m$ co-sets of any of the $p-1$ systems of similar co-sets correspond among themselves and only among themselves, excepting in isomorphisms of $H$ itself.

Next, consider the possible isomorphisms of the co-sets in $B$ (the operators of $H$ being in the same original order, and $D$ thereby in identical corre-

* Cf. Miller, *Philosophical Magazine*, Vol. CCXXXI (1908), pp. 223 *et seq.*

spondence). The arrangement of these co-sets is unrestricted by the order of the co-sets in the tail of $G$ ($= G$ minus $B$). Hence, since these $p^m-1$ co-sets differ from one another only by operators from $A$ by which $D$ is multiplied to give them, they may be made isomorphic exactly according to the group of isomorphisms of $A$.

This then effects all the possible isomorphisms of $A$ and of the $p^m$ subgroups simply isomorphic with $H$ (found by a $d:1$-isomorphism of $H$ with each of the $\frac{p^m-1}{p-1}$ subgroups of $A$,* each of order $p$ and each being here employable in $p-1$ ways, since any one of its operators of order $p$ can be taken as the generator). The number of these isomorphisms is $p^m$ times the order of the group of isomorphisms of $A$. Furthermore, the group of these isomorphisms can be written on the $p^m$ letters previously introduced, and it contains as an invariant subgroup, as was seen, an abelian group of order $p^m$, type $(1, 1, 1, \ldots.)$. But $p^m$ times the order of the group of isomorphisms of $A$ is the order of the holomorph of $A$, which is the maximum group on $p^m$ letters containing $A$ invariantly. Hence, this group is the holomorph of $A$.

Besides these isomorphisms already determined, the only other isomorphisms of $G$ are those effected by the group of isomorphisms of $H$, and from Theorem V this group is a factor in the $I$ of $G$. Therefore, the $I$ of $G$ is simply isomorphic with the direct product of the group of isomorphisms of $H$ and the holomorph of $A$.

*Example.* If $G \equiv (abcd)$ all $(ef)(gh)$, then $H$ is the symmetric group of degree 4 and $A$ is the four-group. Since $H$ is then a complete group and the holomorph of the four-group is the symmetric group of order 24, the $I$ of $G$ is simply isomorphic with $(abcd)$ all $(efgh)$ all.

The groups of isomorphisms of abelian groups of order $p^m$, type $(1, 1, 1, \ldots.)$, have been studied by Moore,† so that the holomorph of such a group is a definitely determined group. When $p=2$, it is interesting to note that, if $m=2$ ($A$ is then the four-group), the holomorph is the symmetric group of order 24; if $m=3$, the holomorph of $A$ is the group of order 1344 and degree 8;‡ if $m=4$, the holomorph of $A$ is a primitive substitution group of degree 16, order 8!8.§

* Burnside, "Theory of Groups," 1911, p. 110.

† Moore, *Bulletin of American Mathematical Society* (2), Vol. II (1895), pp. 33–43. Cf. Burnside, "Theory of Groups," 1911, §§ 89, 90.

‡ For more information and references on this interesting group see Miller, AMERICAN JOURNAL OF MATHEMATICS, Vol. XXI (1899), p. 337.

§ Miller, AMERICAN JOURNAL OF MATHEMATICS, Vol. XX (1898), p. 233.

COROLLARY. *If $G$ is the direct product of a symmetric group of degree $n$, $n \neq 2$ nor 6, and an abelian group of order $2^m$, type $(1, 1, \ldots)$, then the $I$ of $G$ is simply isomorphic with the direct product of that symmetric group and the holomorph of the given abelian group.*

*If $n=2$, $G$ is simply an abelian group of order $2^{m+1}$, type $(1, 1, \ldots)$. If $n=6$, the $I$ of $G$ is the direct product of the holomorph of the given abelian group and an imprimitive group of order 1440 and degree 12, this being the group of isomorphisms of the symmetric group of order 720.**

THEOREM VIII. *If a group $G$ is the direct product of a cyclic group of order $p^m$, $p$ being an odd prime, and a group $H$ which contains a characteristic subgroup $D$ of index $p$, neither $D$ nor the central of $H$ containing an invariant subgroup of index $p$, then the $I$ of $G$ is simply isomorphic with the direct product of the cyclic group of order $p^{m-1}$, the holomorph of the group of order $p$, and the group of isomorphisms of $H$.*

From the hypotheses, $D$ is characteristic in $G$, as is also the cyclic group $(E)$ of order $p^m$ and the single subgroup of order $p$ contained in it, these latter two being the only subgroups of their respective orders in the central of $G$. From Theorem V, the $I$ of $G$ will be the direct product of the group of isomorphisms of $H$ and a group $I'$ which is generated by the group of isomorphisms of the cyclic group and operators corresponding to the subgroups to which $H$ can correspond. It remains to determine $I'$.

Now $G$ is generated by the characteristic cyclic group of order $p^m$ and $H$ or any group simply isomorphic with $H$ which can be formed by an isomorphism of $H$ with subgroups from $E$. The only subgroup from $E$ which can be thus used is the single one of order $p$, since $D$ is the subgroup which in $H$ must be employed as the invariant subgroup, and any subgroup of order greater than $p$ from $E$ would then result in a group of order greater than $H$, which then could not be put into simple isomorphism with it. Since this subgroup of order $p$ can be generated by any one of its $p-1$ operators of order $p$, $H$ and the $p-1$ different groups which can be formed simply isomorphic constitute a characteristic set of $p$ subgroups; and with $E$ in fixed isomorphism, the resulting invariant subgroup in $I'$ is a cyclic group of order $p$.

The group of isomorphisms of $E$ is a cyclic group of order $\phi(p^m) = p^{m-1}(p-1)$.† In this the operators of order $p^n$, $n=1, \ldots, m-1$, leave the

* Miller, *Bulletin of American Mathematical Society* (2), Vol. I (1895), p. 258; and Hölder, *Mathematische Annalen*, Vol. XLVI (1895), p. 345.

† Burnside, "Theory of Groups," 1911, § 88; also Miller, *Transactions of American Mathematical Society*, Vol. IV (1903), pp. 153–160.

operators of order $p$ in $E$ invariant, simply transforming among themselves in the possible ways the operators of order $p^i$, $i>1$, in $E$ whose first power in the subgroup of order $p$ is the same operator. Since these operators of order $p$ are invariant under this cyclic group of order $p^{m-1}$, the subgroup of order $p$ in $I'$ will have each of its operators commutative with each of the operators of this cyclic group. But the cyclic subgroup of order $p-1$ in the group of isomorphisms of $E$ transforms the cyclic subgroup of order $p$ in $E$ exactly according to its possible isomorphisms, or the cyclic subgroup of order $p$ in $I'$ is transformed according to its own group of isomorphisms. Hence, $I'$ contains a cyclic group of order $p^{m-1}$ and a cyclic group of order $p$ extended by operators transforming it into all its possible isomorphisms, these operators being commutative individually with those of the cyclic group of order $p^{m-1}$. Therefore, $I'$ is the direct product of the cyclic group of order $p^{m-1}$ and the holomorph of the group of order $p$.

If the cyclic group $E$ were of order $2^m$, then its group of isomorphisms would be an abelian group of order $2^{m-1}$, type $(m-2, 1)$.* The operator of order 2 used from $E$ in forming the only conjugate of $H$ would be characteristic, hence unaffected by any operator of the group of isomorphisms of $E$. Accordingly, $I'$ would be an abelian group of order $2^m$, type $(m-2, 1, 1)$, and the theorem would be:

THEOREM IX. *If a group $G$ is the direct product of a cyclic group of order $2^m$ and a group $H$ which contains a characteristic subgroup $D$ of index 2, neither $D$ nor the central of $H$ containing an invariant subgroup of index 2, then the $I$ of $G$ is simply isomorphic with the direct product of an abelian group of order $2^m$, type $(m-2, 1, 1)$ and the group of isomorphisms of $H$.*

*Example.* If $G = (abcd)$ all $(efgh)$ cyc, then $H = (abcd)$ all contains the tetrahedral group as a characteristic subgroup $D$ of index 2, neither $D$ nor the central of $H$ having an invariant subgroup of index 2. The $I$ of $G$, accordingly, is simply isomorphic with $(abcd)$ all $(ef)$ $(gh)$.

NOTE. It may be noted that in the preceding Theorems VI, VII, VIII and IX the fact that $I_H$ is a factor in the $I$ of $G$ could be established in each case independently of the Theorem V. The problem is simple if $p=2$. If $p$ is an odd prime, then in each case there is a holomorph of an abelian group of odd order $p^m$, type $(1, 1, \ldots.)$, which holomorph is a complete group.† By showing that this holomorph is an invariant subgroup of the $I$ of $G$, it is known to be a factor.‡

THEOREM X. *If $G$ is the direct product of the group of order 2 and a group $H$ which contains in a characteristic series* § *$n$ successive subgroups*

---

* See preceding foot-note references.

† Burnside, "Theory of Groups," 1897, p. 239.

‡ Hölder, *Mathematische Annalen*, Vol. XLVI (1895), p. 325.

§ Defined by Frobenius, *Berliner Sitzungsberichte*, 1895, p. 1027.

*each of index* 2 *in the preceding, neither the last,* $H_n$ *of index* $2^n$ *in* $H$, *nor the central of* $H$ *containing an invariant subgroup of index* 2, *then the* $I$ *of* $G$ *is simply isomorphic with the direct product of the group of isomorphisms of* $H$ *and an abelian group of order* $2^n$, *type* $(1, 1, \ldots .)$.

While this theorem does not come directly under the problem considered in Theorem V, still it is easy to show that $I_H$ is here a factor in the $I$ of $G$. As suggested in a note after Theorem V, since the prime is 2, there is but one co-set outside the characteristic subgroup of index $p$ in any subgroup of the characteristic series. Hence, there is no cyclic permutation among the co-sets of a tail, and in transforming $H$ into any of its conjugates an operator $(t)$ either goes into some operator $(t')$ of the same co-set (or head) or goes into such an operator $(t')$ multiplied by the operator of order 2. If one operator of a co-set is multiplied by one of the operators from the group of order 2, all of the operators of that co-set are multiplied by that same operator; say by $c$. Let $R$ be an operator transforming $H$ according to any one of its automorphisms; then

$$R^{-1}tR=t', \qquad (R^{-1}cR=c). \tag{1}$$

If $V$ be an operator transforming $H$ into any one of its conjugates without altering the order of the $H$-constituents from that of the order of the corresponding operators of $H$, then

$$V^{-1}tV=tc, \qquad (V^{-1}t'V=t'c). \tag{2}$$

Transforming (1) by $V$ and (2) by $R$ gives, respectively,

$$V^{-1}R^{-1}tRV=t'c,$$
$$R^{-1}V^{-1}tVR=t'c.$$

Hence, $VR=RV$, and $I_H$ is accordingly a factor in the $I$ of $G$. Since the group of isomorphisms of the group of order 2 is the identity, it remains only to determine the possible isomorphisms of $H$ with its conjugates when the order of the $H$-constituents is the same as the order of the corresponding operators of $H$ itself.

The operator of order 2 in the group of order 2 is characteristic in $G$. $H_n$ is a characteristic subgroup of $G$. Let the operators of $H$ be in some fixed order and let $H_n$ remain in identical correspondence while the subgroup $I'$ of the $I$ of $G$ is determined. Each isomorphism of $H$ with an $H'$ is fixed by $n$ correspondences, one for each of $H, H_1, \ldots, H_{n-1}$, and there are two options in each case (since there are two conjugates in each case). These $n$ correspondences are independent of one another, so that there are in all $2^n$ subgroups in this set of conjugates (including $H$ itself). Moreover, the operator transforming $H$ into any of these $H'$'s is of order 2, and all such transformations are commutative. Hence, $I'$ is an abelian group of order $2^n$ containing no operator of order greater than 2. Therefore, the $I$ of $G$ is simply isomorphic with the direct product of an abelian group of order $2^n$, type $(1, 1, \ldots.)$, and the group of isomorphisms of $H$.

*Example.* If $G \equiv (abc)$ all $(def)$ all $(gh)$, then $H \equiv (abc)$ all $(def)$ all, in which $H_1 \equiv [(abc)$ all $(def)$ all$]$ pos, which is a characteristic subgroup of index 2; $H_2 \equiv (abc)$ cyc $(def)$ cyc, which is a characteristic subgroup of index $2^2$ in $H$, and neither $H_2$ nor the central of $H$ contains an invariant subgroup of index 2. Since the group of isomorphisms of $H$ is the double holomorph of the symmetric group of order 6, or is $(abcdef)_{72}$, therefore the $I$ of $G$ is seen to be simply isomorphic with $(abcdef)_{72}\,(ghij)_4$.

THEOREM XI. *If $G$ is the direct product of a metabelian group of order $p(p-1)$ and a cyclic group of order $p-1$, then the $I$ of $G$ is simply isomorphic with the direct product of the holomorphs of the cyclic groups of order $p$ and $p-1$.*

The problem is insignificant if $p$ is the even prime; hence, $p$ will be supposed in the following proof to be an odd prime.

$G$ contains the cyclic group $(C)$ of order $p-1$ as a characteristic subgroup, since it is the central of the direct product. The metacyclic group $(M)$ of order $p(p-1)$ is a complete group, being the holomorph of the cyclic group of order $p$,* which cyclic group is a characteristic subgroup of $M$. In the automorphisms of $G$, $M$ corresponds to itself and to conjugates formed by multiple isomorphism of $M$ with $C$ and with all its various subgroups.

The conditions are not exactly those named in Theorem V, since here the characteristic subgroup of order $p$ in $M$ is not usually (only when $p=3$) of prime index. It must be noted, however, that this same characteristic subgroup of index $p-1$ occurs in all the conjugates of $M$ and that the transformations of the co-sets of $M$ are always cyclic with respect to this subgroup of index $p-1$. Accordingly, by means of equations like those employed in proving Theorem V, it can be shown immediately that the $I$ of $G$ contains the group of isomorphisms of $M$ as a factor, the other factor $(I')$ being the group of isomorphisms of the cyclic group of order $p-1$ extended by operators corresponding to the transformation of $M$ into its conjugates without changing the order of the operators of $M$ as they become constituents in the conjugates. Since $M$ is its own group of automorphisms, it remains to show that $I'$ is the holomorph of the cyclic group of order $p-1$.

---

* Burnside, "Theory of Groups," 1897, p. 239.

Now every subgroup of a cyclic group is cyclic and characteristic. If the order of any subgroup (including the identity and the given cyclic group itself) is $d$, it has exactly $\phi(d)$ automorphisms, so that the number of conjugates of $M$ is $p-1$; because, from number theory, "If $d_1, d_2, \ldots, d_r$ be the different divisors of $m$,

$$\sum_{j=1}^{r} \phi(d_j) = |m|."*$$

If $M$ is transformed into the conjugate formed by a multiple isomorphism between $M$ and the cyclic group of order $p-1$ itself, the transforming operator is seen to be of order $p-1$; and, furthermore, if this transformation be repeated, it gives the entire set of $p-1$ conjugates. Accordingly, the subgroup in $I'$ which corresponds to the transformation of $M$ into all its conjugates (order of $M$-constituents being the same as the order of the operators of $M$) is a cyclic group of order $p-1$, which can be written as a transitive substitution group on $p-1$ letters, a letter corresponding to each conjugate. If the operators of $M$ remain in some fixed order and the characteristic cyclic group of order $p-1$ in $G$ is transformed according to its group of isomorphisms, the $p-1$ conjugate subgroups (in the set with $M$) are permuted according to this group of isomorphisms. Hence, the $p-1$ subgroups are permuted according to a group generated by a cyclic group of order $p-1$ and the group of isomorphisms of a cyclic group of order $p-1$. This group can be represented on $p-1$ letters and contains the cyclic group of order $p-1$ invariantly. Accordingly, $I'$ is the holomorph of the cyclic group of order $p-1$, and the $I$ of $G$ is simply isomorphic with the direct product of the holomorph of order $p(p-1)$ and the holomorph of the cyclic group of order $p-1$.

*Example.* Let $G \equiv (abcde)_{20}\,(fghi)$ cyc. Then the metacyclic group is $M \equiv (abcde)_{20}$, which is the holomorph of the group of order 5. The holomorph of $(fghi)$ cyc is an octic group. Hence, the $I$ of $G$ is simply isomorphic with $(abcde)_{20}\,(fghi)_8$.

## III. *Theorems on Some Extended Groups.*

THEOREM XII. *Suppose the group $H$ written as a regular substitution group and its group of isomorphisms $I_H$ written on the same letters. If $H$ be extended by an invariant substitution of order 2 from $I_H$ such that it transforms every operator of the central $C$ of $H$ into its inverse*† *and leaves $H$ characteristic in the newly formed group $G$, then the $I$ of $G$ is simply isomorphic with the substitution group generated by $C$ and $I_H$.*

*The order of the $I$ of $G$ equals the product of the order of $C$ by the order of $I_H$.*

---

* Lucas, "Théorie des Nombres," 1891, p. 400.

† Miller, *Transactions of American Mathematical Society*, Vol. X (1909), pp. 471–478.

Since $H$ is characteristic in $G$, when $H$ is fixed in identical correspondence, the possible isomorphisms of the latter half of $G$ determine an invariant subgroup of $I$. If $s$ is the operator of order 2 by which $H$ is extended, the products of $s$ by each of the operators of $C$ transform $H$ exactly as $s$ itself does, and all these products are of order 2; moreover, these are the only ones having these properties. Furthermore, the automorphisms which are effected if they in turn stand in correspondence with $s$, exactly correspond to a group (invariant in $I$) simply isomorphic with $C$.

The other possible isomorphisms arise when the subgroup $H$ corresponds to itself in all the ways it may as a subgroup of $G$, which evidently could not be more ways than if it were an independent group by itself. But here it can correspond exactly according to its own group of isomorphisms, since its group of isomorphisms always transforms it into itself, and at the same time, by hypothesis, transforms $s$ into itself. Hence, $G$ is invariant under the group of isomorphisms $I_H$, and the order of the $I$ of $G$ equals the product of the order of $C$ by the order of $I_H$, since the quotient group of $I$ with respect to a group simply isomorphic with $C$ has been seen to be $I_H$. Now, in the isomorphisms of $G$ the operator $s$ and its $c-1$ conjugates were first permuted according to an invariant subgroup (of $I$) simply isomorphic with $C$, and then exactly as the operators of $C$ itself were permuted by the substitutions of $I_H$; moreover, these and their products are the only permutations of these $c$ conjugates. Hence, the invariant subgroup (in $I$) simply isomorphic with $C$ is transformed by the rest of the operators of $I$ in exactly the same way as $C$ is transformed by $I_H$. Now the first power of any operator of $I_H$ appearing in $C$ is the identity; otherwise some non-identity operator of $I_H$ would transform $H$ in the same way as the identity itself does, which is impossible. Moreover, $C$ is invariant under $I_H$, since it is a characteristic subgroup of $H$. Hence, if $C$ were extended by the operators of $I_H$, the resulting group would be of order equal to the order of $I$. With respect to the invariant subgroup $C$, its quotient group would be simply isomorphic with $I_H$. In this group, $C$ is transformed exactly as the simply isomorphic invariant subgroup in $I$ is transformed; also every operator of this group represents a different isomorphism of $G$. Hence, the group formed would be simply isomorphic with the $I$ of $G$.*

NOTES. If, in the preceding theorem, $H$ were the direct product of the group $C$ (abelian) and a group (which could contain no invariant operator besides the identity), then the $I$ of the extended group ($G$) would be simply isomorphic with the direct product of the holomorph of $C$ and the group of isomorphisms of the other factor.

If the central $C$ of $H$ contained no operator of even order, then the substitutions of $C$ would actually transform the $c$ conjugates (of $s$) according to the group said to be simply isomorphic with $C$, and simultaneously leave $H$ fixed in identical correspondence. In this case $[C, I_H]$ is a substitution group simply isomorphic with $I$ and on the same letters as $G$, so that its operators actually transform $G$ according to all its automorphisms.

* Miller, *Bulletin of American Mathematical Society* (2), Vol. III (1896–97), p. 218, Th. II.

The fact that $I$ is here written as a regular group makes the method for obtaining the $I$ of $G$ seem formidable; still, since the central $C$ frequently is of low order, a group simply isomorphic with the $I$ of $G$ can often be easily constructed.* That is, *if $G$ is formed by extending a group $H$ by an operator of order 2 which transforms each operator of the central ($C$) of $H$ into its inverse and leaves $H$ characteristic in $G$ and such that it can still correspond to itself according to its own group of isomorphisms ($I_H$), then the $I$ of $G$ is simply isomorphic with the group obtained by extending a group $C'$, simply isomorphic with $C$, by the operators of a group simply isomorphic with $I_H$, which operators transform $C'$ exactly as the corresponding operators of $I_H$ transform $C$, their first powers appearing in $C'$ being the identity.*

Thus, if $H \equiv (abc)$ cyc $(defg)_8$, it is of order 24, its central $C \equiv (abc)$ cyc $(df \cdot eg)$ being a cyclic group of order 6, and its group of isomorphisms simply isomorphic with the direct product of an octic group and a group of order 2.† Suppose $H$ is extended by an operator of order 2, say $ab$, which transforms each operator of $C$ into its inverse; then $H$ remains characteristic in the newly formed group $G$ (since $H$ is generated by the operators of $G$ whose orders are divisible by 3). Hence, the $I$ of $G$ is simply isomorphic with a group generated by a cyclic group of order 6 and a group simply isomorphic with $I_H$ (under which that cyclic group is invariant) which transforms the operators of that cyclic group into their inverses and is such that the first power of any of its operators appearing in that cyclic group is the identity. These conditions are fulfilled by $[(abc)(de)][(ab)(fghi)_8]$. This is the group $(abc)$ all $(de)(fghi)_8$, or $I$ is simply isomorphic with the direct product of $G$ and a group of order 2.†

Now if $A$ is any abelian group of even order, it may be extended by an operator of order 4 which has its square in $A$ and which transforms each operator of $A$ into its inverse; all of the operators in the extension will be of order 4, and all have a common square in $A$. In connection with the preceding theorem, accordingly, the following proposition is obvious, since its proof is precisely as that just made:

*If $G$ is formed by extending a group $H$, whose central $C$ is of even order, by an operator of order 4 whose square is in $C$ and which transforms each operator of $C$ into its inverse and leaves $H$ characteristic in $G$ and such that it can correspond according to its own group of isomorphisms ($I_H$), then the $I$ of $G$ is simply isomorphic with the group obtained by extending a group $C'$ simply isomorphic with $C$, by the operators of a group simply isomorphic with*

---

* Miller, *Bulletin of American Mathematical Society* (2), Vol. III (1896–97), p. 218, Th. II.

† Miller, *Philosophical Magazine*, Vol. CCXXXI (1908), pp. 231, 232.

*$I_H$, which operators transform $C'$ exactly as the corresponding operators of $I_H$ transform $C$, their first powers appearing in $C'$ being the identity.*

If $H$ is an abelian group, it is its own central, and the extended group $G$ is in one case the general dihedral group and in the other the general dicyclic group. Theorems regarding the groups of isomorphisms in these special cases have been established by Miller; thus: *

1. *If an abelian group $H$ which involves operators whose orders exceed 2 is extended by means of an operator of order 2 which transforms each operator of $H$ into its inverse, then the group of isomorphisms of this extended group is the holomorph of $H$.*

2. *The group of isomorphisms of the general dicyclic group as regards an abelian group which is not both of order $2^m$ and type $(2, 1, 1, \ldots.)$, is the holomorph of this abelian group.*

The following theorem and its corollary cover cases of certain extensions of groups which contain no invariant operators besides the identity:

THEOREM XIII. *If a group $H$, whose central is the identity, is a characteristic subgroup of a group $G$ which is simply isomorphic with an invariant subgroup of the group of isomorphisms of $H$ $(I_H)$, then the $I$ of $G$ is simply isomorphic with $I_H$.*

In proving this theorem it may be supposed that $H$ is written as a regular substitution group and that $I_H$ is written on the same letters. Since $H$ is its own group of inner isomorphisms, it will be an invariant subgroup of the substitution group $I_H$, and will be a characteristic subgroup of an invariant subgroup $G'$ of $I_H$, where $G'$ is simply isomorphic with $G$. Abstractly, $G$ and $G'$ have the same group of automorphisms.

Since $H$ is characteristic in $G'$, the $I$ of $G$ will contain an invariant subgroup corresponding to the possible isomorphisms of $G'$ when the operators of $H$ remain in fixed correspondence. But $H$ is extended by operators of its own group of isomorphisms to form $G'$, and as these operators all transform $H$ differently, this invariant subgroup is just the identity. Hence, the $I$ of $G'$ can not be greater than the group of isomorphisms of $H$. But $G'$ is invariant under the group of isomorphisms of $H$ and is transformed differently by each operator of this group. Therefore, the $I$ of $G$ is simply isomorphic with the group of isomorphisms of $H$.

*Example.* If $H = (abcd)$ pos $(efgh)$ pos, then its group of isomorphisms is the double holomorph of the symmetric group of degree 4; viz., $(abcd)$ all

* Miller, *Philosophical Magazine*, Vol. CCXXXI (1908), pp. 224, 225.

$(efgh)$ all $(ae \cdot bf \cdot cg \cdot dh)$. $H$ is a characteristic subgroup of $G = [(abcd)$ all $(efgh)$ all$]$ pos, which is invariant in the double holomorph of the symmetric group of degree 4. Hence, the $I$ of $G$ is the double holomorph $(abcd)$ all $(efgh)$ all $(ae \cdot bf \cdot cg \cdot dh)$.

COROLLARY. *If the group of inner isomorphisms of a group $H$, whose central is the identity, is a characteristic subgroup of the group of isomorphisms of $H$ ($I_H$), then the $I$ of any group $G$ which contains $H$, and which is simply isomorphic with an invariant subgroup of $I_H$, is simply isomorphic with $I_H$.*

If $H$ is written as a regular group, $I_H$ can be written on the same letters and will contain $H$ as a characteristic subgroup. Hence, $I_H$ is a complete group (see proof of Theorem V), and accordingly contains no other subgroup simply isomorphic with $H$. For this reason $H$ is characteristic in any subgroup of $I_H$, or is characteristic in $G$. The rest of the proof follows from the preceding theorem.

*Example.* If $H$ is the dihedral group $[(abcde)_{10}\,(fgh)$ all$]$ dim, order 30, its group of isomorphisms is the double holomorph of the cyclic group of order 15, or $I_H = (abcde)_{20}\,(fgh)$ all, order 120. Now $H$ contains no invariant operator besides the identity and is characteristic in $I_H$, because it is formed by the group $G_{15}$ composed of all the operators of orders 3, 5 and 15 and every operator of order 2 (in $I_H$) which transforms each operator of $G_{15}$ into its inverse. Hence, since any subgroup of index 2 in $I_H$ is an invariant subgroup, $(abcde)_{10}\,(fgh)$ all, which is of order 60 and contains $H$, has $I_H$ for its group of isomorphisms.

# *Self-Projective Rational Sextics.**

By R. M. Winger.

## *Introduction. General Considerations.*

1. Elsewhere† the various types of self-projective rational quartic and quintic curves have been tabulated. In this connection two curves are said to be of different *type* if they are invariant under different groups, or if, when invariant under the same group, they are projectively distinct, and one is not a special case of the other, in the sense that one can not be obtained by continuous variation of the arbitrary constants in the parametric equations of the other. In the present paper a complete classification of self-projective ternary rational sextics‡ is undertaken, and some of the more immediate properties are inferred.

The study is interesting, not only as illustrating a fruitful method of investigation, but also because it parallels the theories of binary and ternary collineation groups. In particular, several classical configurations associated with the curves are brought to light.

Since we are dealing with parallel theories some conventions of language are convenient. Thus, we restrict *involution* to the binary domain, *i. e.*, the parameter. We may, however, use "point" in place of the awkward "parameter of a point" when the meaning is clear. Binary and ternary groups of order $n$ are designated respectively by $g_n$ and $G_n$, and the corresponding curve by $\rho_n^6$. An involutory ternary collineation is called invariably a *reflexion*.

2. It is essential to bear in mind the relation between the binary and ternary groups. Thus, any rational curve is invariant under the general binary linear transformation of the parameter which is a continuous three-parameter

---

* Read before the American Mathematical Society, September 9, 1913.

† American Journal of Mathematics, Vol. XXXVI, January, 1914. This paper is referred to in the sequel as "D." It is believed that all types of quartics are obtained there, but the list of quintics would have to be supplemented to conform to the present criterion. Thus the curve (foot-note, p. 68) should be included as a distinct type. There is also a second $\rho_3^5$, $x_0 = t^5 + t^2$, $x_1 = t^3$, $x_2 = 1$.

‡ This implies a classification of self-projective rational sextics in $S_3$, since the ternary sextic projectively defines a quaternary sextic, the line sections of the one and the plane sections of the other being apolar binary forms. The problem for the general plane sextic is solved by A. H. Tappan, American Journal of Mathematics, Vol. XXXVII, July, 1915.

group, but the effect may be regarded merely as a change of the *naming* of the points. Hence, in general, the curve is not invariant under the ternary group so generated. *If, however, a binary $g_n$ permutes the $\infty^2$ line sections, the curve admits a ternary $G_n$ and is self-projective.*

Again, *if a rational curve admits a ternary $G_n$, this $G_n$ on the points is effected by a binary group on the parameter of order $n$ and not less,* else the curve would be a locus of fixed points under some elements of $G_n$, which is impossible, since there is only a finite number of such points.

*A cyclic $g_n$ has just two fixed points, and at these points the directions (tangents) are also fixed. The corresponding $G_n$ has the same fixed points and directions, but it may have other fixed points on the curve, which, therefore, must be multiple points of order $n$.*

There is a single type of cyclic $g_n$, viz.,

$$t' = \varepsilon t, \quad \varepsilon = e^{\frac{2\pi i}{n}}.$$

But the ternary cyclic groups separate broadly into two classes: (1) the homologies with multipliers 1, 1, $\varepsilon^a$ with a center and axis, the one a point of fixed lines and the other a line of fixed points; and (2) those with a fixed triangle and multipliers 1, $\varepsilon$, $\varepsilon^b$ $(a=1, \ldots, n-1;\ b=2, \ldots, n-1)$. These latter divide into types projectively distinct as $n$ becomes larger.

The following principles are very useful in studying rational curves admitting a reflexion. A fixed parameter of an involution can lie only at the center or on the axis of reflexion. All other intersections of the axis or of lines on the center must be conjugate pairs of the involution. Hence, *the intersections of the axis are either contacts, simple or of higher order, of tangents from the center, or multiple points. And the tangents from the center either have their contacts on the axis or are multiple tangents.*

3. The accompanying table exhibits in canonical form the parametric equations of the various types of self-projective rational sextics, together with the generating transformations, binary and ternary, of their characteristic groups. For the dihedral groups only transformations for the corresponding invariant cyclic subgroups are given. They are to be combined in every case with $t'=1/t$, $x_0'=x_1$, $x_1'=x_0$, $x_2'=x_2$. Cyclic, dihedral, tetrahedral, octahedral and icosahedral groups are designated respectively by $C, D, T, O$ and $I$; $u$ stands for the substitution $t=\frac{t'+i}{t'-i}$, while $v$ denotes the transformation $T$ of Klein;* $\omega^3=i^4=\varepsilon^5=1$. For the tetrahedral sextic,

$$r=\frac{16a}{4-a^2}, \quad s=\frac{3a^2+20}{a^2-4}.$$

* "Ikosaeder," p. 41.

Following the table, the trilinear equations of some of the curves are given.

| | *Transformations.* | | | | *Equations.* | | | |
|---|---|---|---|---|---|---|---|---|
| | $t'$ | $x_0'$ | $x_1'$ | $x_2'$ | $x_0$ | $x_1$ | $x_2$ | |
| $C_2$ | $-t$ | $x_0$ | $x_1$ | $-x_2$ | $t^6+at^4+bt^2$ | $ct^4+dt^2+1$ | $t^5+et^3+t$ | (1) |
| $C_2$ | $-t$ | $-x_0$ | $-x_1$ | $x_2$ | $t^5+at^4$ | $bt^3+t$ | $t^6+ct^4+dt^2+1$ | (2) |
| $C_3$ | $\omega t$ | $\omega x_0$ | $\omega^2 x_1$ | $x_2$ | $t^4+at$ | $bt^5+t^2$ | $t^6+ct^3+1$ | (3) |
| $C_3$ | $\omega t$ | $x_0$ | $x_1$ | $\omega x_2$ | $t^6+at^3$ | $bt^3+1$ | $t^4+t$ | (4) |
| $C_3$ | $\omega t$ | $\omega x_0$ | $\omega x_1$ | $x_2$ | $t$ | $t^4$ | $t^6+at^3+1$ | (5) |
| $C_4$ | $it$ | $-x_0$ | $x_1$ | $ix_2$ | $t^6+at^2$ | $bt^4+1$ | $t^5+t$ | (6) |
| $C_4$ | $it$ | $-x_0$ | $x_1$ | $-ix_2$ | $t^6$ | $t^4+1$ | $t^3$ | (7) |
| $C_5$ | $\varepsilon t$ | $x_0$ | $\varepsilon x_1$ | $\varepsilon^4 x_2$ | $t^5+1$ | $t^6+at$ | $t^4$ | (8) |
| $C_5$ | $\varepsilon t$ | $\varepsilon^3 x_0$ | $\varepsilon x_1$ | $x_2$ | $t^3$ | $t^6$ | $t^5+1$ | (9) |
| $C_5$ | $\varepsilon t$ | $\varepsilon x_0$ | $\varepsilon x_1$ | $x_2$ | $t$ | $t^6$ | $t^5+1$ | (10) |
| $C_6$ | $-\omega^2 t$ | $-\omega^2 x_0$ | $-x_1$ | $x_2$ | $t$ | $t^3$ | $t^6+1$ | (11) |
| $C_6$ | $-\omega^2 t$ | $\omega x_0$ | $-x_1$ | $x_2$ | $t^2$ | $t^3$ | $t^6+1$ | (12) |
| $C_6$ | $-\omega^2 t$ | $-\omega^2 x_0$ | $\omega^2 x_1$ | $x_2$ | $t$ | $t^4$ | $t^6+1$ | (13) |
| $C_6$ | $-\omega^2 t$ | $-\omega^2 x_0$ | $\omega x_1$ | $x_2$ | $t$ | $t^2$ | $t^6+1$ | (14) |
| $C_\infty$ | $\alpha t$ | $\alpha^6 x_0$ | $\alpha x_1$ | $x_2$ | $t^6$ | $t$ | $1$ | (15) |
| $D_4$ | $-t$ | $x_0$ | $x_1$ | $-x_2$ | $t^6+at^4+bt^2$ | $bt^4+at^2+1$ | $t^5+et^3+t$ | (16) |
| $D_4$ | $-t$ | $-x_0$ | $-x_1$ | $x_2$ | $t^5+at^3$ | $at^3+t$ | $t^6+bt^4+bt^2+1$ | (17) |
| $D_6$ | $\omega t$ | $\omega x_0$ | $\omega^2 x_1$ | $x_2$ | $t^4+at$ | $at^5+t^2$ | $t^6+bt^3+1$ | (18) |
| $D_8$ | $it$ | $-x_0$ | $x_1$ | $ix_2$ | $t^6+at^2$ | $at^4+1$ | $t^5+t$ | (19) |
| $D_8$ | $it$ | $-x_0$ | $x_1$ | $-ix_2$ | $t^6+at^2$ | $at^4+1$ | $t^3$ | (20) |
| $D_{10}$ | $\varepsilon t$ | $\varepsilon x_0$ | $x_1$ | $\varepsilon^3 x_2$ | $t^6+at$ | $at^5+1$ | $t^3$ | (21) |
| $D_{12}$ | $-\omega^2 t$ | $-\omega^2 x_0$ | $\omega x_1$ | $x_2$ | $t$ | $t^5$ | $t^6+1$ | (22) |
| $T_{12}$ | $d_4, u$ | $D_4$ (16), | | $U$ | $t^6+rt^4+st^2$ | $st^4+rt^2+1$ | $t^5+at^3+t$ | (23) |
| $O_{24}$ | $d_8, u$ | $D_8$ (19), | | $U$ | $t^6-5t^2$ | $-5t^4+1$ | $t^5+t$ | (24) |
| $O_{24}$ | $d_8, u$ | $D_8$ (20), | | $U$ | $t^6+3t^2$ | $3t^4+1$ | $t^3$ | (25) |
| $I_{60}$ | $d_{10}, v$ | $D_{10}$, | | $V$ | $t^6+3t$ | $3t^5-1$ | $-5t^3$ | (26) |

(7) $(x_2^2-x_0x_1)^3+x_0^4x_2^2=0$, (9) $(x_0^2-x_1x_2)^3+x_0x_1^5=0$,

(11) $(x_0^3+x_1^3)^2-x_0^3x_1x_2^2=0$, (12) $x_0^6+x_1^6-x_0^3x_1^2x_2=0$,

(13) $(x_0^3+x_1^3)^3-x_0^3x_1x_2^3=0$, (14) $x_0^6+x_1^6-x_0^4x_1x_2=0$,

(22) $(x_0^3-x_1^3)^2+x_0x_1x_2^2(4x_0x_1-x_2^2)=0$.

## I. *The Cyclic Sextics.*

4. The curves invariant under the cyclic groups are referred uniformly to a fixed triangle, when the equations appear in the canonical form.

Curve (1): $x_2$ is the axis of reflexion and cuts out the fixed points of the involution and two double points. Four double lines meet at the center $\xi_2$.*

* The vertex of the triangle of reference opposite $x_i$ will be denoted by $\xi_i$.

5. Curve (2): There is a biflecnode at $\xi_2$, the center of reflexion, whose parameters are the fixed points of the involution. $x_2$ is the axis and cuts out three double points. The intersections of the axis being completely accounted for, two double lines meet at the center.

The remaining six double points must lie in pairs on lines through the center, *i. e.*, *six of the double points are harmonically perspective from a seventh.*

6. Curve (3) has a double point at $\xi_2$ whose parameters are fixed under the binary $g_3$. $x_0$ and $x_1$ are the nodal tangents.

7. Curve (4): $\xi_2$ is the center and $x_2$ the axis of the homology. $x_2$ cuts out a point of inflexion and a cusp, whose parameters are the fixed points of the $g_3$. The remaining intersections of $x_2$ being fixed points of $G_3$, in consequence of a theorem of § 2, must constitute a triple point.

8. Curve (5): The center of the homology is a multiple point arising from the junction of a cusp of higher order with an undulation, equivalent to one cusp, three double points, six flexes and nine double lines. The parameters are the fixed points of $g_3$. The intersections of the axis $x_2$ are fixed points of $G_3$ and must constitute, therefore, two triple points. The remaining six flexes lie in threes on lines through the center.

9. The cyclic $G_5$: Since the curve is of order 6 and the group of order 5, one line section must be of the form (1) $t^6+at$. The only other binomial line section is (2) $t^5+b$. The remaining possibilities are (3) $t^2$, $t^3$, $t^4$. Inasmuch as sections (1) and (3) contain a common factor, (2) must always be used. We have then three types of cyclic $\rho_5^6$, the first two belonging to the same group, the third to an homology.

10. Curve (8) has a "touching undulation tangent,"* $x_2$, which consumes two flexes and four double lines, and whose contacts are the fixed points of the binary group.

The six flexes of curve (9) lie on a conic whose remaining intersections fall at the triple point $\xi_2$.

11. Curve (10) is the sextic case of the rational $\rho_{n-1}^n$, $x_0=t$, $x_1=t^n$, $x_2=t^{n-1}+1$, or $x_0(x_0+x_1)^{n-1}-x_1 x_2^{n-1}=0$. $\xi_2$ is the center and $x_2$ the axis of the homology. The curve has an $(n-1)$-fold point on $x_2$. $x_0$ and $x_1$ are tangents with $(n-1)$-point and $n$-point contacts, respectively, consuming $(2n-5)$ flexes and $(n-3)^2$ double lines. There remain $(n-1)$ flexes and $(n-1)(n-3)$ double lines. The flexes lie on a line through the center.

12. The four varieties of $\rho_6^6$ are found at once from the formula, "D.," § 10.†

* *i. e.*, an undulation tangent with an additional (simple) contact.

† $x_0=t^r$, $x_1=t^s$, $x_2=t^6+1$, $r \neq s$, $1 \leqq \left\{ \begin{matrix} r \\ s \end{matrix} \right\} \leqq 5$.

The binary group contains as subgroups a $g_3$ and a $g_2$; *i. e.*, the curves admit a reflexion. Note, however, that there are two types of ternary $G_6$, the first three curves belonging to one, and the fourth to the other. These types may be characterized, the one by the fact that its subgroup of order 3 is an homology, the other as possessing a pencil of proper fixed conics.

13. Curve (11) has a multiple point at $\xi_2$ which is equivalent to two cusps, five double points, one double line and two flexes, consuming in all six flexes and twelve double lines. There are, however, but two distinct parameters, the fixed points of the binary group.

Sets of conjugate points are cut out by the line pairs $x_2^2+\lambda x_1^2=0$. For example, for particular $\lambda$ there are the following pairs: (a) one on three other double points, $\lambda=0$; (b) one on six other flexes, $5\lambda=16$; (c) one which is a pair of triple tangents, $\lambda=-4$.

14. Curve (12) has a fivefold point at $\xi_2$, due to a cusp falling at a special triple point. The multiple point is equivalent to three cusps and seven double points, and consumes therefore six flexes and fifteen double lines. The two parameters are fixed under the binary group. The cuspidal tangent, $x_1$, is the axis of reflexion, the complete intersections falling at $\xi_2$. $x_0$ counts as a simple tangent from $\xi_1$, the center of reflexion. The other six tangents, the curve being of class 7, must consist of three double lines.

As before, sets of conjugate points are cut out by the line pairs $x_2^2+\lambda x_1^2=0$. The following pairs are of interest: (a) one on the six flexes, $2\lambda=1$; (b) one pair of triple lines, $\lambda=-4$; (c) one on six contacts of the three double lines, $2\lambda=-9$. All singularities of the curve are now accounted for, the three triple tangents counting for six double lines.

15. Curve (13): This is obtained from (5), by placing $a=0$, which amounts to imposing a reflexion. The curve has the same singularities as before. $x_0$, the cuspidal tangent, is the axis of reflexion, all intersections falling at $\xi_2$. $x_1$ counts three times as a tangent from $\xi_0$, the center of reflexion, the other six tangents forming double lines.

Sets of conjugate points are cut out by the line pairs $x_0^2+\lambda x_1^2=0$, six of the intersections having been taken up at $\xi_2$. For varying $\lambda$ the following special pairs are to be noted, one on (a) two triple points, $\lambda=1$; (b) contacts of three double lines, $2\lambda=-1$; (c) six flexes, $2\lambda=5$; (d) six contacts of tangents from $\xi_1$, $5\lambda=-1$.

16. Curve (14) has a fivefold point at $\xi_2$, equivalent to three cusps and seven double points, caused, however, by a simple branch passing through a special fourfold point whose four parameters coincide. $\xi_0$ is the center and

$x_0$ the axis of reflexion. The six intersections of the axis lie at the multiple point. $x_1$ counts once as a tangent from $\xi_0$, the other six tangents being three double lines.

Sets of six conjugate points are cut out by the pencil of double-contact conics $x_0^2+\lambda x_1 x_2=0$, six of the intersections falling at the multiple point. Of these conics may be mentioned: (a) one on the six flexes; (b) three each on six contacts of double lines.

17. If $\rho^6$ admits a cyclic $G_7$, it admits an infinite group.* There is just one type, (15). There is a hyperosculation point at 0 and the dual singularity at $\infty$, the two consuming all of the singularities of the curve.

## II. *The Dihedral Sextics.*

18. Closely associated with dihedral groups is the question of symmetry.† The group method is particularly effective in discussing the invariant curves, the properties of which are of unusual interest.

Curve (16): The fixed parameters of the three involutions are cut out by the axes of reflexion and name contacts of tangents from the centers. The other intersections of the axes must be interchanged under the involutions; that is, *two double points lie on each axis.* The remaining four double points are a set of four under the ternary group. Since the intersections of the axes are completely accounted for, the other tangents from the centers must be double tangents. Hence, *four double lines meet at each center of reflexion.*

19. Curve (17): A second vertex of the triangle of centers is a biflecnode and the properties of $\rho_2^6$, (2), are accordingly repeated. The third center, however, is off the curve, and the fixed parameters of the corresponding involution are cut out by the third axis, naming contacts of tangents from the center. Hence, *four double lines meet at this center and the fixed points of all three involutions lie on its axis.*

The remaining double points are a set of four under $G_4$. We may say, then: It happens twice that six double points are harmonically perspective from a seventh, while from the third center run two lines each carrying two double points and two lines carrying three each.

20. Curve (18): The three axes of reflexion meet at a double point, whose parameters, we saw (§ 3), are fixed under the cyclic $g_8$. The fixed points of the three involutions lie in pairs on the axes and are contacts of tangents from the centers. As above, it follows that *a second double point lies on each axis. The other six double points are a general set and lie on one of the double-contact conics* $x_0 x_1+\lambda x_2^2=0$, which in general cut out two sets of conjugate points.

* "D.", § 10.

† "D.," § 4.

All intersections of the axes are accounted for. Hence, *four double lines meet at each center.*

The flexes must constitute two sets of six, which lie on two of the invariant conics.

21. Curve (19): The only special sets of parameters are given by $t$, $t^4+1$ and $t^4-1$, the first two being cut out by $x_2$. The two pairs $t^4-1$ lie on the axes $x_0^2-x_1^2=0$, being contacts of tangents from the corresponding centers. The residual intersections of these axes are pairs of double points. Since there are no other fixed parameters, all the intersections of the axes $x_0^2+x_1^2=0$ must be interchanged by their respective involutions. That is, *three double points lie on each of these axes.*

Now the cyclic $G_4$ contains a reflexion, whose center is the intersection, $\xi_2$, of the four axes above and whose axis, $x_2$, is their line of centers. The fixed parameters of the involution are 0 and $\infty$, while $t^4+1$ names two double points. The flex form* is found to be $(t^4+1)\{at^8+(3a^2-16a+15)t^4+a\}$. Hence, *the two double points on $x_2$ are biflecnodes.*

The biflecnodes are centers of reflexion, the parameters being fixed points of the corresponding involutions. The four tangents from each must make up two double lines, since all of the intersections of the axes are double points; while four double lines meet at each of the other three centers. This accounts for sixteen of the twenty-four double lines.

General conjugate sets of eight points are cut out by the system of invariant conics, all of which have contact with $\rho^6$ at 0 and $\infty$. Of these may be mentioned: (a) one on eight flexes; (b) one on eight flex tangents; (c) one with four additional contacts; (d) two each on four double points; (e) two each on eight tangents at double points; (f) three each on four double lines; (g) three each on eight contacts of double lines. (c) is a perspective conic.

Many interesting special cases arise for particular values of $a$. Among these are roots of the discriminant $(a-1)(a-3)(a-5)(3a-5)$ of the flex form, exclusive of $t^4+1$, whose vanishing, owing to the symmetry, means four pairs of equal roots. To these values of $a$ correspond the following curves: (1) $a=1$, degenerate; (2) $a=3$, cusps at $t^4-1$; (3) $a=5$, two double points, each with two five-point contact tangents; (4) $3a=5$, undulations at $t^4-1$. (2) is self-dual, having four double points, four double lines, two double-cusp tangents and two biflecnodes.

22. Curve (20): The line $x_2$ is a double-cusp tangent, which counts for a double line, besides consuming eleven double lines and four flexes. Four double lines meet at the center $\xi_2$.

---

* J. I. Tracey, *Johns Hopkins University Circular*, July, 1913, has developed the flex form for the general rational plane curve.

On each of the other four axes lie two double points and two contacts of tangents from the centers, the fixed points of the corresponding involutions; and at each of the four centers two double lines meet, accounting for the remaining double lines.

All other sets of conjugate points are sets of eight, and are cut out by the pencil of invariant conics, which pass through the cusps. Of these conics we note: (1) one on the eight flexes; (2) one on eight flex tangents; (3) two each with four contacts besides the cusps; (4) two each on four double points, *i. e.*, on six double points; (5) two each on eight tangents at double points; (6) two each on four double lines; (7) two each on eight contacts of double lines.

The flex form is $3at^8+(a^2-27)t^4+3a$. Corresponding to the factors of the discriminant are two pairs of projectively equivalent curves, characterized as follows: (1) $a=3$, cusps at $t^4-1$; (1′) $a=-3$, cusps at $t^4+1$; (2) $a=9$, undulations at $t^4+1$; (2′) $a=-9$, undulations at $t^4-1$.

23. Curve (21): The line $x_2$ is a double-flex tangent, which counts for two flex tangents and four double lines, the contacts being fixed points of the cyclic $g_5$.

The fixed points of the five involutions lie in pairs on the five axes and are contacts of tangents from the five centers. The other intersections of the axes are double points, and the other tangents from the centers are double lines. That is, *two double points lie on each axis and two double lines meet at each center.*

The double-flex tangent is the common chord of the pencil of invariant conics, the contacts being at the flexes. The remaining intersections of the conics are sets of ten conjugate points. Noteworthy among these conics are the following: (1) one on ten, *i. e.*, on all twelve flexes; (2) one on ten flex lines; (3) two each on five double points; (4) two each on ten tangents at double points; (5) two each with five contacts; (6) two each on five double lines; (7) two each on ten contacts of double lines. The equation of (1) is

$$3x_0x_1+(5a^2-30)x_2^2=0.$$

The factors of the discriminant of the flex form, exclusive of 0 and $\infty$, and the significance of their vanishing are: (1) $2a=3$, cusps at $t^5-1$; (1′) $2a=-3$, cusps at $t^5+1$; (2) $4a=9$, undulations at $t^5+1$; (2′) $4a=-9$, undulations at $t^5-1$. Corresponding to these values of $a$ are two pairs of projectively equivalent curves.

24. Curve (22):* The vertex $\xi_2$ is a double point, the tangent at each branch of which has five-point contact, consuming six flexes and six double lines.

---

* See "D.," § 10.

The nodal parameters are fixed under the cyclic $g_6$, the node itself being a fixed point of the complete ternary $G_{12}$. $\xi_2$ is also a center of reflexion with $x_2$ as axis.

The other six axes of reflexion constitute two non-conjugate sets of three lines,* $x_0^3-x_1^3=0$, $x_0^3+x_1^3=0$, meeting at the double point $\xi_2$. Designate these respectively as lines $\alpha$ and $\beta$ and the corresponding centers by points $a$ and $b$.

Of the special sets of parameters, $t^6-1$ lie in pairs on axes $\alpha$ and name contacts of tangents from centers $a$. A double point completes the intersections of each of these three axes. Accordingly, the remaining tangents from centers $a$ are double tangents; in other words, *four double lines meet at each center $a$.*

The other special set of parameters, $t^6+1$, cut out by $x_2$, names three double points. Since all other parameters on the curve are sets of twelve, the six flexes must be one of these special sets of six. Hence, *the three double points on $x_2$ are biflecnodes,* one lying on each axis $\alpha$ and being a center $b$.

Again, all the intersections of each axis $\beta$ must be interchanged by an involution. Hence, *the residual six double points lie in pairs on the axes $\beta$.* That is, *six double points are perspective from a seventh,* the axis of perspective being $x_2$, the line of three biflecnodes. In addition *they are triply perspective, the centers and axes of perspective being the centers $a$ and the axes $\alpha$ of reflexion.*

*From each biflecnode run three double lines,* so that the aggregate of eighteen is now accounted for. Thus, on the line $x_2$ are three points, $a$, each carrying four double lines, and three biflecnodes, $b$, each carrying two double lines.

The invariant conics all touch the tangents of the special double point $\xi_2$ where they meet the line $x_2$, accounting thus for eight common lines. Of these conics, whose complete intersections are sets of conjugate points, we record: (1) one with six contacts, $t^6-1$; (2) one on six ordinary nodes and (3) one on twelve nodal tangents of the same; (4) one on the six flex tangents; (5) three each on six double lines; (6) three each on twelve contacts of double lines. (1) is perspective. All twenty common lines of each are accounted for except those of (2) and (6).

## III. *The Tetrahedral, Octahedral and Icosahedral Sextics.*

25. The equations of the tetrahedral sextic (23) are found from those of the dihedral $\rho_4^6$, (16), by requiring that the quadratics naming the fixed points of the involutions, *i. e.*, that the axes of reflexion, be permuted cyclically by the transformation $u$.

---

* Consult "D.," § 4.

The parameters of the curve in general occur in conjugate sets of twelve. But there are two special sets of four and one set of six, corresponding to the vertices of the two tetrahedra, $T$ and $T'$, and to the mid-points, $O$, of the edges.* The six points $O$ are the fixed points of the involutions and lie in pairs on the axes of reflexion. The axes cut out, besides, six double points whose parameters are a general set of twelve. The remaining four double points then give rise to a set of eight. Since there is no set of eight, these points must be the two sets $T$ and $T'$. Now, the points $T$ are interchanged in pairs by the three involutions, and the same is true of the points $T'$. Hence, *there is one point of each set at each of the four double points; that is to say, the parameters of each double point are the fixed points of a cyclic* $g_3$. The effect of each of the ternary $G_3$'s, therefore, is to leave one double point fixed, while it permutes cyclically the other three. Since all special sets of points are accounted for, the flexes are a general set of twelve. Four double lines meet at each center.

26. Curve (24): When $a=0$ or $\infty$ in the equations of the tetrahedral sextic, the curve admits the complete octahedral $G_{24}$. The binary group contains only three special sets of points fewer than twenty-four, viz., †

$$O,\ t(t^4-1)\,;\quad C,\ t^8+14t^4+1\,;\quad M,\ t^{12}-33t^8-33t^4+1.$$

Since the $G_{24}$ contains three conjugate dihedral $G_8$'s, anything true once for a $\rho_8^6$ will be true three times in general for the octahedral sextic, and the properties of the latter can be inferred readily from a consideration of the former, of which it is a special case. Thus, each $G_8$ possesses a pencil of invariant conics with contacts at the fixed points of an involution of the "four-group" $g_4$. *There is, however, a single fixed conic of* $G_{24}$. *Belonging to each system, it has double contact with each conic of the three conjugate systems and six contacts with the curve. These contacts, being a set of six, are the vertices of the octahedron, O.*

*The flexes are a set of twelve parameters, and coincide therefore with the points M*, as is verified at once from Tracey's equation. But we saw (§ 21) that four of the flexes of $\rho_8^6$, (19), lie at two double points. Hence, in this case, *all of the flexes fall at double points, giving rise to six biflecnodes. The remaining four double points are named by a set of eight parameters, which constitute therefore the cube vertices C.*

Now, each of the biflecnodes of $\rho_8^6$ lies on a line with a pair of other double points. And since the double points must be treated symmetrically, this will happen for every biflecnode. Hence, *the biflecnodes are the complete intersections of four lines, which form with the four ordinary double*

* Klein, "Ikosaeder," Chapter II, § 11.

† Klein, "Ikosaeder," p. 54.

*points a quadrangle-quadrilateral configuration,** a self-dual configuration such that each side of the quadrilateral is the polar of one vertex of the quadrangle with reference to the triangle of the other three.

The biflecnodes are centers of reflexion, from each of which run two double lines, while four double lines meet at the centers of reflexion of the four-group.

27. Curve (25): This is seen to be the *projective astroid,* the miscalled "hypocycloid of four cusps," since there are six. The group property might have been anticipated, since the curve is the exact dual of the projective emniscate, long known to admit the octahedral group.†

All of the flexes and twenty-one of the double lines are consumed in the formation of the cusps. There are three double-cusp tangents, which count as double lines. The only other singularities of the curve are four ordinary double points.

*The cusps, whose parameters are the octahedron vertices, lie on a conic, the invariant conic of $G_{24}$. The other double points, being a set of eight, are the cube vertices.* Each of the three conjugate sets of invariant conics has a double-cusp tangent as common chord with the contacts of the cusps. One conic of each set has four contacts with the curve. In other words, *there are three conics, each on two cusps, and having four contacts. These contacts constitute a set of twelve, and must therefore be the median points M.*

It may be remarked that the ternary octahedral group is the group of permutations on the four ordinary double points of both types of sextics invariant under it.

28. The icosahedral sextic is a curve of particular interest, not only on account of its striking and beautiful properties, but because it is the curve of lowest order other than a conic to admit the icosahedral group. It may be obtained from our $\rho_{10}^6$, (21), by imposing the further condition of invariance under the transformation $T$ of Klein (p. 41).

Many properties of $\rho_{60}^6$ can be deduced from the dihedral sextics, of which it is a special case. Thus, we saw that $\rho_{10}^6$ has a singular line counting for two flex tangents and four double lines. Hence, *the icosahedral sextic has six double-flex tangents consuming all the flex and double lines.* Their equations are

$$x_2=0, \quad \varepsilon^i x_0+\varepsilon^{-i}x_1+x_2=0, \quad \varepsilon^5=1, \quad i=0, \ldots, 4.$$

Again, the double-contact conics of a $G_{10}$ have as common chord a multiple tangent with the two flexes as points of contact. Of these conics there are: (1) two each on five double points; (2) two each on ten tangents at these

* Veblen and Young, "Projective Geometry," Vol. I, p. 44.

† Berzolari, *Istituto Lombardo Rendiconti,* Series II, Vol. XXXVII, pp. 277 and 304.

double points; (3) two each with five contacts; (4) one on five double-flex tangents, which make up the total of twenty common lines. (5) Among the invariant conics of a dihedral $G_6$ there is one on six double points. The dihedral groups being conjugate, theorems (1)–(4) are true six times, theorem (5) ten times.

*The flexes, being a set of twelve, are the icosahedron vertices I. The double points are a set of twenty parameters, which are therefore the dodecahedron vertices H. The fixed points of the fifteen involutions lie in pairs on the fifteen axes of reflexion, constituting a set of thirty, which are the mid-points of the edges, J.**

29. We saw ("D.," § 4) that every ternary dihedral group has a fixed line, the line of centers, and a fixed point, the intersection of the axes of reflexion. The dihedral $G_4$, however, containing only reflexions, has three such points and lines, forming a triangle. Thus, there are associated with the $G_{60}$, by means of the dihedral subgroups, six points, say $a$, ten points $b$ and fifteen points $c$, and the corresponding lines $\alpha, \beta, \gamma$. To these we now direct our attention.

In the first place, these points and lines are poles and polars with respect to the systems of double-contact conics set up by the groups ("D.," § 4). But *there is a single conic, N, invariant under the whole group, and belonging, therefore, to each system, namely the conic on the twelve flexes:* $x_0x_1+x_2^2=0$. Hence, the points and lines belong to the polar system of the conic $N$.

The points $c$ and lines $\gamma$ are the centers and axes of the fifteen reflexions. Lines $\gamma$ meet by fives at points $a$, by threes at points $b$, and in pairs at points $c$, which accounts for the total of 105 intersections. A dual statement holds for points $c$. Now, the line of centers of each dihedral $G_{10}$ is a double-flex tangent. That is, *the multiple lines are the lines α.* Again, we saw (§ 20) that the fixed point of the ternary dihedral $G_6$ is a double point of $\rho^6$. Hence, *the ten double points are the points b.* From each center of reflexion run two simple tangents, which have their contacts on the corresponding axis. There can be no other simple tangents and, the curve being of class 10, two double-flex tangents meet at each center. That is, *the fifteen centers of reflexion are the complete intersections of the six singular lines.*

To sum up, *the complete junctions of the six points a are the axes of reflexion γ which meet by threes in the ten double points b. The lines α are the double-flex tangents. Their fifteen intersections are the centers of reflexion c, which lie by threes on lines β. In other words, the points and lines define the dual configuration of a tenfold Brianchon six-point and a tenfold Pascal hexagram.*

THE UNIVERSITY OF OREGON, *January*, 1914.

* These invariant forms are given, Klein, p. 56.

# *On Linear Difference and Differential Equations.**

By Clyde E. Love.

## I. *Introduction.*

1. In two papers† appearing in 1899 and in 1900 respectively, Dini developed certain general theorems on linear differential equations, and applied them to the integration of such equations for large values of the independent variable. Some years later Ford‡ carried out the parallel investigation for linear difference equations. In the present paper an effort will be made to adapt the methods of Dini and Ford to the study of somewhat more general classes of equations than were considered in the above-mentioned investigations. It will therefore be well, for purposes of comparison, to state briefly the nature of these earlier results.

2. Dini's work may be summarized as follows:

In the differential equation

$$y^{(n)}+[a_1+\alpha_1(x)]y^{(n-1)}+\ldots+[a_n+\alpha_n(x)]y=0, \tag{1}$$

suppose that the roots $\mu_1, \mu_2, \ldots, \mu_n$ of the characteristic equation

$$\mu^n+a_1\mu^{n-1}+\ldots+a_n=0$$

are distinct. Let $\tau(x)$ § be a suitably chosen positive function of $x$ for which the integral

$$T(x)=\int_x^\infty \tau(t)\,dt$$

exists, and suppose that the functions $\alpha_1(x), \alpha_2(x), \ldots, \alpha_n(x)$, in addition to satisfying certain conditions as to continuity, etc., have the property that

$$|\alpha_r^{(s)}(x)| \leqq \tau(x), \qquad r=1, 2, \ldots, n;\ s=0, 1, \ldots, n.$$

Then, if the roots $\mu_1, \mu_2, \ldots, \mu_n$ all have the *same real part* $\bar{\mu}$, the general integral of (1) may be written, for large real and positive values of $x$, in the form

$$y=c_1 e^{\mu_1 x}+c_2 e^{\mu_2 x}+\ldots+c_n e^{\mu_n x}+e^{\bar{\mu} x}\varepsilon(x),$$

where $c_1, c_2, \ldots, c_n$ are arbitrary constants, while $\varepsilon(x)$ vanishes at infinity to at least as high an order as does $T(x)$.

---

* Read before the American Mathematical Society, April 2, 1915.

† *Annali di Matematica*, Ser. 3, Vol. II (1899), pp. 297–324; *ibid.*, Vol. III (1900), pp. 125–183.

‡ *Annali di Matematica*, Ser. 3, Vol. XIII (1907), pp. 263–328.

§ In Dini's notation, $\frac{\tau(x)}{x}$.

In case the roots are not all distinct, but all have the same real part, the behavior of the general solution is determined under suitably changed hypotheses regarding the functions $\alpha_1(x)$, $\alpha_2(x)$, ...., $\alpha_n(x)$.

Unfortunately the assumption that all the roots shall have the same real part diminishes greatly the range of applicability of these results. This assumption is essential to the method used by Dini, the method serving in the general case merely to determine the behavior of certain particular solutions.

3. Ford's results are in substance as follows:

In the difference equation

$$\mathfrak{a}_0(x)y(x+n)+\mathfrak{a}_1(x)y(x+n-1)+\dots+\mathfrak{a}_n(x)y(x)=0, \qquad (2)$$

let us write the coefficients in the form

$$\mathfrak{a}_r(x)=a_r+\alpha_r(x), \qquad r=0, 1, \dots, n,$$

where $a_0 \neq 0$, $a_n \neq 0$, and suppose that the roots of the characteristic equation

$$a_0\mu^n+a_1\mu^{n-1}+\dots+a_n=0$$

are distinct. Let $\tau(x)$* be a suitably chosen positive function of $x$ such that the series

$$T(x)=\sum_{t=x+1}^{\infty}\tau(t)$$

converges, and suppose that each of the functions $\alpha_0(x)$, $\alpha_1(x)$, ...., $\alpha_n(x)$ has the property, for sufficiently large values of $x$, that

$$|\alpha_r(x)|<\tau(x), \qquad r=0, 1, \dots, n.$$

Then, if the roots $\mu_1, \mu_2, \dots, \mu_n$ all have the *same modulus* $\bar{\mu}$, the general solution of (2) takes for sufficiently large positive integral values of $x$ the form

$$y(x)=c_1\mu_1^x+c_2\mu_2^x+\dots+c_n\mu_n^x+\bar{\mu}^x\epsilon(x),$$

where $c_1, c_2, \dots, c_n$ are arbitrary constants, while $\epsilon(x)$ is infinitesimal in $1/x$ of an order at least as high as that of $T(x)$.

For the case of multiple roots, all having the same modulus, the analogous result is also obtained.

When the roots do not all have the same modulus, the method gives only the behavior of certain particular solutions.

4. The first part of the present paper is devoted to linear difference equations, the problem treated being closely related to the work of Ford just described. The results obtained constitute an advance over those of Ford in three principal respects: the restriction that the roots of the characteristic equation shall have the same modulus is dispensed with; the behavior of the

* In Ford's notation, $\frac{\tau(x)}{x}$.

$n$ fundamental integrals is determined separately; and the results are applied to the integration of the non-homogeneous equation.

The second part of the paper, dealing with linear differential equations, is similarly related to the work of Dini. The results of this part likewise possess the three characteristics just mentioned. The behavior of the derivatives of the integrals is also discussed.

The results of both Ford and Dini for the case of distinct roots fall out as special cases of certain theorems which Horn* has obtained by a somewhat different method. We shall therefore be chiefly concerned in the following pages with the case of multiple roots.

## PART I. LINEAR DIFFERENCE EQUATIONS.

### II *The Fundamental Theorems.*

5. As a preliminary step we shall develop two general theorems on linear difference equations.†

In the difference equation

$$a_0(x)y(x+n)+a_1(x)y(x+n-1)+\ldots\ldots+a_n(x)y(x)=0, \qquad (3)$$

suppose that the coefficients are defined for all integral values of $x$ in the interval $x \geqq x_0$, and that $a_n(x)$ never vanishes in the same interval.

Let us choose $n$ *auxiliary functions* $z_1(x), z_2(x), \ldots, z_n(x)$, each defined in the interval $x \geqq x_0-n$, and having the property that the determinant

$$Q(x) = \begin{vmatrix} z_1(x) & z_1(x-1) & \ldots\ldots & z_1(x-n+1) \\ z_2(x) & z_2(x-1) & \ldots\ldots & z_2(x-n+1) \\ \ldots & \ldots & \ldots & \ldots \\ z_n(x) & z_n(x-1) & \ldots\ldots & z_n(x-n+1) \end{vmatrix}$$

never vanishes in that interval.

Let $A_r(x)$ be the minor of $Q(x)$ with respect to the element $z_r(x)$. Put

$$Z_r(x)=z_r(x)a_n(x)+z_r(x-1)a_{n-1}(x-1)+\ldots\ldots+z_r(x-n)a_0(x-n),$$
$$r=1, \ldots, n,$$

and denote by $q(x, t)$ the determinant formed from $Q(x)$ by replacing $z_r(x)$ by $Z_r(t)$, $r=1, \ldots, n$. Also make the following definitions:

$$f_r(x)=\frac{(-1)^n C_r A_r(x)}{a_n(x)\,Q(x)}, \qquad r=1, \ldots, n,$$

where $C_1, C_2, \ldots, C_n$ are arbitrary constants,

$$K(x, t)=\frac{(-1)^n q(x, t)}{a_n(x)\,Q(x)},$$

---

* *Journal für Mathematik*, Vol. CXXXVIII (1910), pp. 159–191.

† These theorems differ only slightly from certain theorems given by Ford (*loc. cit.*).

$$u_{r,0}(x) = v_{r,0}(x) = f_r(x), \tag{4}$$

$$u_{r,m}(x) = \sum_{t=x+1}^{\infty} K(x,t)\, u_{r,m-1}(t), \qquad m=1, 2, \ldots, \tag{5}$$

$$v_{r,m}(x) = \sum_{t=x_0}^{x} K(x,t)\, v_{r,m-1}(t), \qquad m=1, 2, \ldots. \tag{6}$$

Then we have

THEOREM A. *If for all integral values of* $x \geqq x_0$

(a) *the series* $\sum_{m=0}^{\infty} |u_{i,m}(x)|$, *where* $i$ *is one of the numbers* $1, \ldots, n$, *converges to a limit* $U_i(x)$ *such that*

(b) *the series* $\sum_{t=x+1}^{\infty} |K(x,t)|\, U_i(t)$ *converges also;*

(c) *the series*

$$y_i(x) = \sum_{m=0}^{\infty} u_{i,m}(x) \tag{7}$$

*defines a function such that each of the series* $\sum_{t=x+1}^{\infty} y_i(t)\, Z_r(t)$, $r=1, \ldots, n$, *converges,*

*then* $y_i(x)$ *as given by* (7) *is a solution of* (3) *for all values of* $x \geqq x_0+n-1$.

THEOREM B. *If the series*

$$y_i(x) = \sum_{m=0}^{\infty} v_{i,m}(x),$$

*where* $i$ *is one of the numbers* $1, \ldots, n$, *is convergent, then* $y_i(x)$ *is a solution of* (3) *for* $x \geqq x_0+n-1$.

6. To prove Theorem A, we note first that (2) may be written in the form*

$$\alpha_0(x)\, \Delta^n y(x) + \alpha_1(x)\, \Delta^{n-1} y(x+1) + \ldots + \alpha_n(x)\, y(x+n) = 0,$$

where

$$\alpha_r(x) = \frac{(-1)^{n-r}}{(n-r)\,!} \sum_{i=0}^{r} \frac{(n-i)\,!}{(r-i)\,!} a_{n-i}(x), \qquad r=0, 1, \ldots, n.$$

Now we have at once

$$y_i(x) = f_i(x) + \sum_{m=0}^{\infty} \sum_{t=x+1}^{\infty} K(x,t)\, u_{i,m}(t).$$

In view of hypothesis (b), we may invert the order of summation, and write

$$y_i(x) = f_i(x) + \sum_{t=x+1}^{\infty} K(x,t) \sum_{m=0}^{\infty} u_{i,m}(t)$$

$$= f_i(x) + \sum_{t=x+1}^{\infty} K(x,t)\, y_i(t). \tag{8}$$

Next, let us use the symbols $z_r$, $\alpha_s$, $p_{r,m}$, $q_{r,m}$ to denote the functions $z_r(x-n+1)$, $\alpha_s(x-n+1)$, $p_{r,m}(x)$, $\Delta p_{r,m}(x)$ respectively, and place

---

* Cf. Ford, *loc. cit.*

$$p_{r,0} = z_r \alpha_0,\ p_{r,1} = z_r \alpha_1 - q_{r,0},\ \ldots,\ p_{r,n-1} = z_r \alpha_{n-1} - q_{r,n-2}, \qquad r=1,\ldots,n;$$

$$\Phi_r(x) = \sum_{t=x+1}^{\infty} Z_r(t)\, y_i(t), \qquad r=1,\ldots,i-1,\, i+1,\ldots,n;$$

$$\Phi_i(x) = \sum_{t=x+1}^{\infty} Z_i(t)\, y_i(t) + C_i;$$

$$P(x) = \begin{vmatrix} p_{1,n-1} & p_{1,n-2} & \ldots & p_{1,0} \\ p_{2,n-1} & p_{2,n-2} & \ldots & p_{2,0} \\ \ldots & \ldots & \ldots & \ldots \\ p_{n,n-1} & p_{n,n-2} & \ldots & p_{n,0} \end{vmatrix}, \quad P_i(x) = \begin{vmatrix} \Phi_1(x) & p_{1,n-2} & \ldots & p_{1,0} \\ \Phi_2(x) & p_{2,n-2} & \ldots & p_{2,0} \\ \ldots & \ldots & \ldots & \ldots \\ \Phi_n(x) & p_{n,n-2} & \ldots & p_{n,0} \end{vmatrix}.$$

Upon substituting the values of $f_i(x)$ and $K(x,t)$ in (8), we find after certain transformations that

$$y_i(x) = \frac{P_i(x)}{P(x)}.$$

Consider now the system of $n$ functions $\eta_0,\ \eta_1,\ \ldots,\ \eta_{n-1}$, each defined for all values of $x \geq x_0$ by the system of equations

$$p_{r,n-1}\,\eta_0 + p_{r,n-2}\,\eta_1 + \ldots + p_{r,0}\,\eta_{n-1} = \Phi_r(x), \qquad r=1,\ldots,n. \tag{9}$$

We have at once

$$\eta_0(x) = y_i(x).$$

Further, let us take the first difference of each member of (9) and write $\zeta_s$ for $\Delta\eta_s(x)$, thus:

$$p_{r,n-1}\zeta_0 + p_{r,n-2}\zeta_1 + \ldots + p_{r,0}\zeta_{n-1} + q_{r,n-1}\eta_0(x+1) + q_{r,n-2}\eta_1(x+1)$$
$$+ \ldots + q_{r,0}\,\eta_{n-1}(x+1) = -y_i(x+1)Z_r(x+1). \tag{10}$$

But $y_i(x+1) = \eta_0(x+1)$, and $Z_r(x+1) = z_r\alpha_n - q_{r,n-1}$. Equation (10) thus becomes

$$z_r\alpha_0\zeta_{n-1} + (z_r\alpha_1 - q_{r,0})\,\zeta_{n-2} + \ldots + (z_r\alpha_{n-1} - q_{r,n-2})\zeta_0$$
$$+ (z_r\alpha_n - q_{r,n-1})\eta_0(x+1) + q_{r,0}\,\eta_{n-1}(x+1) + q_{r,1}\,\eta_{n-2}(x+1)$$
$$+ \ldots + q_{r,n-1}\,\eta_0(x+1) = 0,$$

or

$$\theta z_r + \theta_1 q_{r,0} + \ldots + \theta_{n-1} q_{r,n-2} = 0, \qquad r=1,\ldots,n, \tag{11}$$

where

$$\theta = \alpha_0\zeta_{n-1} + \alpha_1\zeta_{n-2} + \ldots + \alpha_{n-1}\zeta_0 + \alpha_n\eta_0(x+1),$$

and

$$\theta_s = \eta_{n-s}(x+1) - \zeta_{n-s-1}, \qquad s=1,\ldots,n-1. \tag{12}$$

The system (11) is a set of $n$ homogeneous linear equations in $\theta, \theta_1, \ldots, \theta_{n-1}$. Since $q_{r,s-1} = z_r\alpha_s - p_{r,s}$, it appears that the discriminant of the system does not vanish. Hence

$$\theta \equiv \theta_1 \equiv \ldots \equiv \theta_{n-1} \equiv 0. \tag{13}$$

We shall now show that

$$\eta_s(x) = \Delta^s y_i(x-s). \tag{14}$$

Assuming that

$$\eta_{s-1}(x) = \Delta^{s-1} y_i(x-s+1), \tag{15}$$

we find

$$\Delta\eta_{s-1}(x) = \Delta^s y_i(x-s+1).$$

But since

$$\theta_{n-s} = \eta_s(x+1) - \Delta\eta_{s-1}(x) = 0,$$

we have

$$\eta_s(x+1) = \Delta^s y_i(x-s+1),$$

or

$$\eta_s(x) = \Delta^s y_i(x-s). \tag{16}$$

Since the assumption (15) is true for $s=1$, formula (14) is established; whence, taking the first difference of each member of (16), we obtain

$$\Delta\eta_s(x) = \Delta^{s+1} y(x-s). \tag{17}$$

Now upon substituting in $\theta$ the values of $\Delta\eta_{n-1}(x), \ldots, \Delta\eta_0(x)$ as obtained from (12), (13) and (17), we find

$$\alpha_0 \Delta^n y_i(x-n+1) + \alpha_1 \Delta^{n-1} y_i(x-n+2) + \ldots + \alpha_n y_i(x+1) = 0,$$

or, for values of $x \geqq x_0+n-1$,

$$\alpha_0(x)\, \Delta^n y_i(x) + \alpha_1(x)\, \Delta^{n-1} y_i(x+1) + \ldots + \alpha_n(x)\, y_i(x+n) \equiv 0,$$

which was to be proved.

The proof of Theorem B may be carried out similarly.

### III. *The Homogeneous Equation: Distinct Roots.**

7. In the difference equation

$$Y(x) = \mathfrak{a}_0(x) y(x+n) + \mathfrak{a}_1(x) y(x+n-1) + \ldots + \mathfrak{a}_n(x) y(x) = 0, \tag{2}$$

wherein

$$\mathfrak{a}_r(x) = a_r + \alpha_r(x), \qquad r=0, 1, \ldots, n,$$

let us assume that $a_0 \neq 0$, $a_n \neq 0$, and that the roots $\mu_1, \mu_2, \ldots, \mu_n$ of the characteristic equation

$$\phi(\mu) = a_0 \mu^n + a_1 \mu^{n-1} + \ldots + a_n = 0 \tag{18}$$

are distinct.

Suppose further that a positive function $\tau(x)$ exists such that

$$|\alpha_r(x)| \leqq \tau(x), \qquad r=0, 1, \ldots, n,$$

where $\tau(x)$ has the following properties:

(a) The series $T(x) = \sum_{t=x+1}^{\infty} \tau(t)$ is convergent;

(b) the ratio $\dfrac{x\tau(x)}{T(x)}$ remains finite for large values of $x$;

(c) the function $e^{ax}\tau(x)$ increases monotonically with $x$ for all values of $a>0$.

* As noted above, the results for this case are hardly new (cf. Horn, *loc. cit.*). However, it is believed that the treatment from the standpoint of Ford's theory may be of some interest. Further, having in mind the work of this section, we are enabled to omit many details in the discussion of the more complicated case following.

These properties are evidently possessed by each of the functions

$$\frac{1}{x^\nu}, \quad \frac{1}{x(\log x)^\nu}, \quad \frac{1}{x \log x (\log \log x)^\nu}, \ldots, \qquad \nu > 1.$$

For simplicity we shall assume that $\tau(x)$ may be taken as one of this set of functions.

8. Let

$$\delta = \begin{vmatrix} 1 & \mu_1 & \ldots & \mu_1^{n-1} \\ 1 & \mu_2 & \ldots & \mu_2^{n-1} \\ \ldots & \ldots & \ldots & \ldots \\ 1 & \mu_n & \ldots & \mu_n^{n-1} \end{vmatrix} = \prod_{r,s} (\mu_r - \mu_s), \qquad r > s,$$

and denote by $\delta_r$ the cofactor of the element $\mu_r^{n-1}$. Also put

$$\mu = \mu_1 \mu_2 \ldots \mu_n,$$
$$\sigma_{1,r} = \mu_r^n \alpha_0(x-n) + \mu_r^{n-1} \alpha_1(x-n+1) + \ldots + \alpha_n(x).$$

With this notation we find

$$Q(x) = \delta \mu^{-x},$$
$$f_r(x) = (-1)^n C_r \delta_r \mu_r^x / \delta \mathfrak{a}_n(x), \qquad r = 1, \ldots, n, \tag{19}$$
$$Z_r(x) = z_r(x) \sigma_{1,r}(x),$$
$$K(x,t) = \sum_{r=1}^{n} \delta_r \mu_r^{x-t} \sigma_{1,r}(t) / \delta \mathfrak{a}_n(x). \tag{20}$$

We note that a constant $M_1$ can be found so large that

$$|\sigma_{1,r}(x)| \leqq M_1 \tau(x), \qquad r = 1, \ldots, n. \tag{21}$$

Let our notation be such that

$$|\mu_1| \leqq |\mu_2| \leqq \ldots \leqq |\mu_n|, \tag{22}$$

and consider first the case in which these quantities $|\mu_r|$ are all distinct.

We have

$$|K(x,t) f_1(t)| \leqq \sum_{r=1}^{n} |\delta_1 \delta_r C_1 \mu_r^{x-t} \mu_1^t \sigma_{1,r}(t) / \delta^2 \mathfrak{a}_n(x) \mathfrak{a}_n(t)|$$
$$\leqq |f_1(x)| \sum_{r=1}^{n} \left| \delta_r \left(\frac{\mu_1}{\mu_r}\right)^{t-x} \sigma_{1,r}(t) \Big/ \delta \mathfrak{a}_n(t) \right|.$$

Now since $|\mu_1| \leqq |\mu_r|$, $r = 1, \ldots, n$, it is evident that the quantity

$$\left| \delta_r \left(\frac{\mu_1}{\mu_r}\right)^{t-x} \Big/ \delta \mathfrak{a}_n(t) \right|$$

remains less than some fixed constant for all values of $x$ and $t$ in question. Hence, by (21), a constant $M_2$ can be found such that

$$|K(x,t) f_1(t)| < M_2 |f_1(x)| \sum_{r=1}^{n} \frac{|\sigma_{1,r}(t)|}{M_1}$$
$$< n M_2 |f_1(x)| \tau(t). \tag{23}$$

9. We are now in position to show that the conditions of Theorem A are satisfied by the function

$$y_1(x) = \sum_{m=0}^{\infty} u_{1,m}(x), \tag{24}$$

where $u_{1,m}(x)$, $f_1(x)$ and $K(x, t)$ are given by (4) and (5), (19), and (20) respectively.

We shall first prove by mathematical induction that

$$|u_{1,m}(x)| \leq [M_2 n T(x)]^m |f_1(x)|, \qquad m = 0, 1, 2, \ldots . \tag{25}$$

Assume that this formula is true for $u_{1,m-1}(x)$. Then

$$|u_{1,m}(x)| \leq \sum_{t=x+1}^{\infty} |K(x,t) f_1(t)| [M_2 n T(t)]^{m-1}$$
$$\leq [M_2 n T(x)]^{m-1} \sum_{t=x+1}^{\infty} |K(x,t) f_1(t)|,$$

since $T(x)$ is a monotonically decreasing function. By (23), we have at once the relation (25). To complete the proof by induction, we need only point out that our assumption regarding $u_{m-1}(x)$ is justified, by (4), for $m=1$.

Let $x$ now be restricted to values so large that

$$M_2 n T(x) \leq \theta, \qquad 0 < \theta < 1,$$

where $\theta$ is a constant independent of $x$. Then, as a consequence of (25), condition (a) of Theorem A is satisfied, by Weierstrass's test; and further, we find

$$U_1(x) = \sum_{m=0}^{\infty} |u_{1,m}(x)| \leq \frac{|f_1(x)|}{1 - M_2 n T(x)} \leq \frac{|f_1(x)|}{1-\theta}. \tag{26}$$

As regards (b), we have

$$\sum_{t=x+1}^{\infty} |K(x,t)| U_1(t) \leq \frac{1}{1-\theta} \sum_{t=x+1}^{\infty} |K(x,t) f_1(t)|$$
$$\leq \frac{M_2 n}{1-\theta} |f_1(x)| T(x) \leq \frac{\theta}{1-\theta} |f_1(x)|,$$

so that this condition is satisfied also.

From (19), (21) and (26) it follows at once that (c) holds.

Thus $y_1(x)$ as given by (24) is a solution of (2) for large values of $x$. As a consequence of (25) we have

$$|y_1(x) - f_1(x)| \leq \sum_{m=1}^{\infty} |u_{1,m}(x)|$$
$$\leq \frac{n M_2 T(x) \, |f_1(x)|}{1 - M_2 n T(x)};$$

whence we may write

$$y_1(x) = \mu_1^x [1 + \epsilon_1(x)],$$

where $\epsilon_1(x)$ vanishes, when $x = \infty$, to at least as high an order as that of $T(x)$.

10. In the above argument, the convergence of the series defining $u_{1,m}(x)$ depends on the fact that $|\mu_1| \leq |\mu_r|$, $r=1, \ldots, n$. When we try to find a solution $y_2(x)$ corresponding to the root $\mu_2$, the process breaks down by reason of the fact that $|\mu_2| > |\mu_1|$. We may avoid the difficulty by the use of Theorem B, as will now be shown.

The difference-equation

$$Z(x) = a_n(x)z(x) + a_{n-1}(x-1)z(x-1) + \ldots + a_0(x-n)z(x-n) = 0 \quad (27)$$

evidently satisfies all the conditions that have been imposed upon equation (2). The roots of its characteristic equation

$$a_n\mu^n + a_{n-1}\mu^{n-1} + \ldots + a_0 = 0$$

are $\frac{1}{\mu_n}, \frac{1}{\mu_{n-1}}, \ldots, \frac{1}{\mu_1}$. Therefore, by the result of the preceding paragraph, equation (27) has a solution corresponding to the root $\frac{1}{\mu_n}$ of the form

$$z_n(x) = \mu_n^{-x}[1+\eta_n(x)], \quad (28)$$

where $\eta_n(x)$ has the property assigned above to $\varepsilon_1(x)$.

Let us now choose a set of auxiliary functions in which $z_1, z_2, \ldots, z_{n-1}$ are as before, while $z_n$ is given by (28). With this choice of $z_n$, we have

$$Z_n(x) \equiv 0.$$

The functions $f_r(x)$ and $K(x, t)$ are not materially changed, except that in $K(x, t)$ the summation now runs from $r=1$ to $r=n-1$.

Consider the function

$$y_n(x) = \sum_{m=0}^{\infty} v_{n,m}(x),$$

where $v_{n,m}(x)$ is given by (4) and (6). We find

$$K(x, t)\, f_n(t) = \sum_{r=1}^{n-1} C_n \mu_r^{x-t} \mu_n^t \rho_n(x)\, \sigma_{n,r}(t),$$

where $\rho_n(x)$ remains finite for large values of $x$, while a constant $M_3$ can be found such that

$$|\sigma_{n,r}(x)| \leq M_3 \tau(x), \qquad r=1, \ldots, n.$$

Therefore a constant $M_4$ can be chosen so large that

$$|K(x, t)\, f_n(t)| \leq |f_n(x)|\, M_4 \sum_{r=1}^{n-1} \left| \left(\frac{\mu_n}{\mu_r}\right)^{t-x} \right| \tau(t).$$

It can now be shown by mathematical induction that

$$|v_{n,m}(x)| \leq [M_4(n-1)\, x\tau(x)]^m\, |f_n(x)|. \quad (29)$$

For, this is true for $v_{n,0}(x)$, by definition (cf. (4)). Assume it to hold for $v_{n,m-1}(x)$. Then

$$|v_{n,m}(x)| \leq \sum_{t=x_0}^{x} |K(x,t)\, v_{n,m-1}(t)|$$
$$\leq [M_4(n-1)]^{m-1} \sum_{t=x_0}^{x} |K(x,t)\, f_n(t)|\, [t\tau(t)]^{m-1}$$
$$\leq M_4^m (n-1)^{m-1} |f_n(x)| \sum_{t=x_0}^{x} \left\{ t^{m-1}[\tau(t)]^m \sum_{r=1}^{n-1} \left| \left(\frac{\mu_n}{\mu_r}\right)^{t-x} \right| \right\}. \tag{30}$$

Now, each of the functions $t^{m-1}[\tau(t)]^m \left|\left(\frac{\mu_n}{\mu_r}\right)^t\right|$, $r=1, \ldots, n-1$, increases monotonically with $t$. Thus we may replace each of these functions in the right member of (30) by its largest value, viz., its value when $x=t$, which gives (29) at once.

As a consequence of (29), we have

$$|y_n(x)| \leq \sum_{m=0}^{\infty} |v_{n,m}(x)| \leq \frac{|f_n(x)|}{1-M(n-1)\, x\tau(x)} < \frac{|f_n(x)|}{1-\theta},$$

provided $x$ be taken so large that

$$M(n-1)\, x\tau(x) < \theta, \qquad 0<\theta<1.$$

Thus, for such values of $x$, Theorem B is satisfied, and $y_n(x)$ is a solution of (2). Further, by (29),

$$|y_n(x) - f_n(x)| \leq \sum_{m=1}^{\infty} |v_{n,m}(x)|$$
$$\leq \frac{M(n-1)\, x\tau(x)}{1-M(n-1)\, x\tau(x)},$$

so that

$$y_n(x) = \mu_n^x [1+\varepsilon_n(x)], \tag{31}$$

where $\varepsilon_n(x)$ is infinitesimal in $1/x$ to at least as high an order as that of $x\tau(x)$, and therefore, by hypothesis, to as high an order as that of $T(x)$.

11. To find a third particular solution of (2), we may proceed as follows. Equation (27) has, by (31), a solution

$$z_1(x) = \mu_1^{-x}[1+\eta_1(x)].$$

Introducing this in place of the original $z_1(x)$, thus making $Z_1(x) \equiv 0$, we find that, in $K(x,t)$, the summation runs only from $r=2$ to $r=n-1$. Proceeding exactly as in § 9, we find an integral $y_2(x)$ of (2) having the form

$$y_2(x) = \mu_2^x [1+\varepsilon_2(x)],$$

where $\varepsilon_2(x)$ has the usual property for large $x$.

Now if we introduce in place of $z_{n-1}(x)$ the solution

$$z_{n-1}(x) = \mu_{n-1}^{x}[1+\eta_{n-1}(x)]$$

of (27), we may obtain as in § 10 a solution of (2) of the form

$$y_{n-1}(x) = \mu_{n-1}^{x}[1+\varepsilon_{n-1}(x)].$$

The remainder of the process is obvious. Using Theorem A and Theorem B alternately, we obtain $n$ linearly independent integrals of (2) having, for sufficiently large values of $x$, the form

$$y_i = \mu_i^x[1+\varepsilon_i(x)], \qquad i=1,\ldots,n. \tag{32}$$

12. In case the quantities $|\mu_r|$, $r=1,\ldots,n$, are not all different, suppose they fall into $l$ groups containing $n_1, n_2, \ldots, n_l$ roots respectively ($n_1+n_2+\ldots+n_l=n$), in such a way that

$$|\mu_1|=|\mu_2|=\ldots=|\mu_{n_1}|<|\mu_{n_1+1}|=\ldots=|\mu_{n_1+n_2}|<\ldots<|\mu_{n-n_l+1}|=\ldots=|\mu_n|.$$

Then by using Theorem A as in § 9 we find $n_1$ independent integrals of (2) corresponding to that group of roots whose moduli are equal and least. With a properly revised system of auxiliary functions Theorem B gives $n_l$ solutions corresponding to the group of roots with greatest modulus. Going on in this way, we obtain again (32).

In particular, if the roots all have the same modulus, which is the case treated by Ford, the complete result can be written down by the method of § 8.

IV. *The Homogeneous Equation: Multiple Roots.*

13. Suppose that the characteristic equation (18) has $l$ distinct roots $\mu_1, \ldots, \mu_l$, occurring $n_1, \ldots, n_l$ times respectively: $n_1+\ldots+n_l=n$. Let $n'$ be the largest of the numbers $n_1, \ldots, n_l$. Further, let the coefficients in equation (2) be such that, when $x$ is large,

$$|a_r(x)|<\tau(x)/x^{2n'-2}, \qquad r=1,\ldots,n,$$

$\tau(x)$ having the same meaning as before.

Let us take $n$ auxiliary functions of the form*

$$z_{r,q}(x)=x^{n_r-q}\mu_r^{-x}, \qquad r=1,\ldots,l;\ q=1,\ldots,n_r. \tag{33}$$

14. Place

$$\delta=\prod_{r,s} 0!\,1!\ldots(n_r-1)!\,\mu_r^{\frac{1}{2}n_r(n_r-1)}(\mu_r-\mu_s)^{n_rn_s}, \qquad r=1,\ldots,l;\ r>s;$$

also

$$\mu=\mu_1^{n_1}\mu_2^{n_2}\ldots\mu_l^{n_l}.$$

Then we find

$$Q(x)=\delta\mu^{-x},$$

and

$$f_{r,q}(x)=x^{q-1}\mu_r^x\rho_{r,q}(x), \qquad r=1,\ldots,l;\ q=1,\ldots,n_r, \tag{34}$$

where $\rho_{r,q}(x)$ remains finite and different from 0 as $x=\infty$. Also

$$Z_{r,q}(x)=z_{r,q}(x)\,\sigma_{r,q}(x),$$

where a constant $M_5$ can be found such that

* For brevity we write $z_{n_1+\ldots+n_{r-1}+q}=z_{r,q}$, with a similar meaning for $f_{r,q}$, etc.

$$|\sigma_{r,q}(x)| < M_5 \tau(x)/x^{2n'-2}.$$

Thus

$$K(x,t) = \sum_{r=1}^{l} \sum_{q=1}^{n_r} x^{q-1} \mu_r^x \nu_{r,q}^t t^{n_r-q} \mu_r^{-t} \sigma_{r,q}(t), \tag{35}$$

where $\nu_{q,r}(x)$ remains finite as $x=\infty$.

From (34) and (35) we have

$$|K(x,t)\, f_{1,k}(t)| \leq \sum_{r=1}^{l} \sum_{q=1}^{n_r} x^{q-1} t^{n_r-q+k-1} |\mu_r^{x-t} \mu_1^t \nu_{r,q}(x)\, \sigma_{r,q}(t)\, \rho_{1,k}(t)|$$

$$\leq M_6 |f_{1,k}(x)| \sum_{r=1}^{l} \sum_{q=1}^{n_r} x^{q-k} t^{n_r-q+k-2n'+1} \left|\left(\frac{\mu_1}{\mu_r}\right)^{t-x}\right| \tau(t), \quad k=1,\ldots,n_1,$$

where $M_6$ is the maximum value, for all values of $x$ and $t$ in question, of the quantity $M_5 \left| \frac{\rho_{1,k}(t)}{\rho_{1,k}(x)} \nu_{r,q}(x) \right|$. This latter inequality may be written

$$|K(x,t) f_{1,k}(t)| \leq M_6 |f_{1,k}(x)| \sum_{r=1}^{l} \sum_{q=1}^{n_r} x^{1-k} \left(\frac{x}{t}\right)^{q-1} t^{n_r+k-2n'} \left|\left(\frac{\mu_1}{\mu_r}\right)^{t-x}\right| \tau(t).$$

Since $|\mu_1| \leq |\mu_r|$, $k \geq 1$, $t > x$, $q \geq 1$, $2n' \geq n_r + k$, we have

$$|K(x,t)\, f_{1,k}(t)| \leq M_6 n |f_{1,k}(t)|\, |\tau(t)|. \tag{36}$$

15. It can now be shown that each of the functions

$$y_{1,k}(x) = \sum_{m=0}^{\infty} u_{1,k,m}(x), \qquad k=1,\ldots,n_1, \tag{37}$$

as given by (7) is a solution of (2) for large values of $x$.

We shall first prove by mathematical induction that

$$|u_{1,k,m}(x)| \leq [M_6 n\, T(x)]^m |f_{1,k}(x)|, \tag{38}$$

where

$$T(x) = \sum_{t=x+1}^{\infty} \tau(t).$$

This is true of $u_{1,k,0}(x)$ by definition (cf. (4)). Assume it to hold for $u_{1,k,m-1}(x)$. Then

$$|u_{1,k,m}(x)| \leq \sum_{t=x+1}^{\infty} |K(x,t)\, u_{1,k,m-1}(t)|$$

$$\leq (M_6 n)^{m-1} \sum_{t=x+1}^{\infty} |K(x,t)\, f_{1,k}(t)|\, [T(t)]^{m-1}$$

$$\leq (M_6 n)^m |f_{1,k}(x)| \sum_{t=x+1}^{\infty} [T(t)]^{m-1} \tau(t),$$

by (36). Since $T(x)$ decreases monotonically with $x$, we may write

$$|u_{1,k,m}(x)| \leq (M_6 n)^m [T(x)]^{m-1} |f_{1,k}(x)| \sum_{t=x+1}^{\infty} \tau(t),$$

which is the desired relation.

Let us henceforth restrict ourselves to values of $x$ so large that

$$M_6 n\, T(x) \leq \theta, \qquad 0 < \theta < 1.$$

Then we have by (38)

$$U_{1,k}(x) = \sum_{m=0}^{\infty} |u_{1,k,m}(x)| \leq \frac{|f_{1,k}(x)|}{1-M_6 n T(x)} \leq \frac{|f_{1,k}(x)|}{1-\theta}, \tag{39}$$

so that condition (a) of Theorem A is satisfied.

Further, we have by (39) and (36)

$$\sum_{t=x+1}^{\infty} |K(x,t)|\, U_{1,k}(t) \leq \frac{1}{1-\theta} \sum_{t=x+1}^{\infty} |K(x,t)\, f_{1,k}(t)| \leq \frac{M_6 n}{1-\theta} |f_{1,k}(x)|\, T(x),$$

whence condition (b) is satisfied.

Finally, denoting by $M_7$ the maximum value of $M_5|\rho_{1,k}(x)|$, we find

$$|y_{1,k}(x)\, Z_{r,q}(x)| \leq \frac{1}{1-\theta} |f_{1,k}(x)|\, x^{n_r-q+2-2n'} |\mu_r^{-x} \sigma_{r,q}(x)| \leq \frac{M_7}{1-\theta} \left| \left(\frac{\mu_1}{\mu_r}\right)^x \right| x^{n_r-q+2-2n'} \tau(x).$$

Since $|\mu_1| \leq |\mu_r|$ and $2n' \geq n_r - q + 2$, this shows that

$$\sum_{t=x+1}^{\infty} |y_{1,k}(t)\, Z_{r,q}(t)| \leq \frac{M_7 T(x)}{1-\theta},$$

whence (c) is satisfied.

The proof that each of the functions $y_{1,k}(x)$ as given by (37) satisfies (2) for sufficiently large values of $x$ is now complete. Further, as a consequence of (38) we have

$$|y_{1,k}(x) - f_{1,k}(x)| \leq \sum_{m=1}^{\infty} |u_{1,k,m}(x)| \leq \frac{M_6 n T(x)}{1-M_6 n T(x)} |f_{1,k}(x)|,$$

whence we may write

$$y_{1,k}(x) = x^{k-1} \mu_1^x [1 + \varepsilon_{1,k}(x)], \qquad k = 1, \ldots, n_1, \tag{40}$$

where $\varepsilon_{1,k}(x)$ is infinitesimal in $1/x$ of at least as high an order as that of $T(x)$.

16. To obtain further particular solutions of (2), we shall proceed as in § 10. The equation $Z(x) = 0$ (cf. (27)) has, by (40), a group of $n_l$ solutions corresponding to the $n$-fold root $1/\mu_l$ of its characteristic equation

$$a_n \mu^n + \ldots + a_0 = 0,$$

which solutions may be expressed in the form

$$z_{l,q}(x) = x^{n_l-q} \mu_l^{-x} [1 + \eta_{l,k}(x)], \qquad q = 1, \ldots, n_l.$$

Let us introduce these functions into the auxiliary system (33) in place of the functions $z_{l,q}(x)$ as originally chosen, thus making

$$Z_{l,q}=0, \qquad q=1,\ldots,n_l.$$

Thus in $K(x,t)$ the summation with respect to $r$ runs from $r=1$ to $r=l-1$ only.

We may now show by Theorem B, by much the same argument as that used in § 10, that (2) has a group of integrals

$$y_{l,k}(x)=x^{k-1}\mu_l^x[1+\varepsilon_{l,k}(x)], \qquad k=1,\ldots,n_l.$$

The remainder of the argument for the case where the quantities $|\mu_r|$ are distinct, and also for that where these quantities are not distinct, may be easily filled in by referring to §§ 12–13.

17. The results of this section may be summarized in

THEOREM I. *Let $\tau(x)$ be one of the system of functions**

$$\frac{1}{x^\nu}, \quad \frac{1}{x(\log x)^\nu}, \quad \frac{1}{x\log x(\log\log x)^\nu}, \quad \ldots, \qquad \nu>1,$$

*and put*

$$T(x)=\sum_{t=x+1}^{\infty}\tau(t).$$

*In the difference equation*

$$Y(x)=\mathfrak{a}_0(x)y(x+n)+\mathfrak{a}_1(x)y(x+n-1)+\ldots+\mathfrak{a}_n(x)y(x)=0,$$

*where $\mathfrak{a}_r(x)=a_r+\alpha_r(x)$, $r=0,1,\ldots,n$, suppose that the characteristic equation*

$$\phi(\mu)=a_0\mu^n+a_1\mu^{n-1}+\ldots+a_n=0$$

*has $l$ different roots $\mu_1,\mu_2,\ldots,\mu_l$ occurring $n_1,n_2,\ldots,n_l$ times respectively $(n_1+n_2+\ldots+n_l=n)$, and let $n'$ be the largest of the numbers $n_1,n_2,\ldots n_l$.*

*If a function $\tau(x)$ exists such that, for sufficiently large values of $x$,*

$$|\alpha_r(x)|\leqq\tau(x)/x^{2n'-2}, \qquad r=0,1,\ldots,n,$$

*then for the same values of $x$ the equation $Y(x)=0$ has $n$ linearly independent solutions $y_{i,k}(x)$ expressible in the form*

$$y_{i,k}(x)=x^{k-1}\mu_i^x[1+\varepsilon_{i,k}(x), \qquad i=1,2,\ldots,l;\ k=1,2,\ldots,n_i,$$

*where $\varepsilon_{i,k}(x)$ vanishes at infinity to at least as high an order as that of $T(x)$.*

## V. *The Non-homogeneous Equation.*

18. With the results of the previous section before us, it is a simple matter to determine the behavior of the complete solution of the non-homogeneous equation

$$Y(x)=X(x), \tag{41}$$

with proper restrictions on the right-hand member. This subject we shall discuss briefly in the present section.

* The function $\tau(x)$ is chosen in this way for simplicity. Cf. § 7.

Let the homogeneous equation $Y(x)=0$ corresponding to (41) satisfy the conditions of Theorem I, and suppose that $X(x)$ has the form

$$X(x)=\nu^x\,[b+\beta(x)]=\nu^x h(x),$$

where $\nu$ and $b$ are constants different from 0 and $\beta(x)$ satisfies the conditions that are imposed by Theorem I upon the functions $a_r(x)$.

We begin by transforming equation (41) into a homogeneous equation of order $n+1$ by the usual process. Equation (41) thus takes the form

$$\begin{aligned}&\mathbf{a}_0(x+1)h(x)y(x+n+1)+[\mathbf{a}_1(x+1)h(x)-\nu\,\mathbf{a}_0(x)h(x+1)]y(x+n)\\&\quad+[\mathbf{a}_2(x+1)h(x)-\nu\,\mathbf{a}_1(x)h(x+1)]y(x+n-1)+\ldots.\\&\quad+[\mathbf{a}_n(x+1)h(x)-\nu\mathbf{a}_{n-1}(x)h(x+1)]y(x+1)-\nu\mathbf{a}_n(x)h(x+1)y(x)=0.\end{aligned}\tag{42}$$

Thus if we write (42) in the form

$$\begin{aligned}&[a_0 b+\gamma_0(x)]y(x+n+1)+[(a_1-a_0\nu)b+\gamma_1(x)]y(x+n)+\ldots.\\&\quad+[(a_n-a_{n-1})b+\gamma_n(x)]y(x+1)+[-a_n\nu\,b+\gamma_{n+1}(x)]y(x)=0,\end{aligned}\tag{43}$$

it appears that the functions $\gamma_0(x),\gamma_1(x),\ldots.,\gamma_{n+1}(x)$ satisfy all the conditions that have been imposed upon $a_0(x),\,a_1(x),\ldots.,\,a_n(x)$.

The characteristic equation

$$(\mu-\nu)\phi(\mu)=a_0\mu^{n+1}+(a_1-a_0\nu)\mu^n+\ldots.+(a_n-a_{n-1}\nu)\mu-a_n\nu=0$$

evidently has the roots $\mu_1,\ldots.,\mu_l,\nu$, where $\mu_r$ occurs $n_r$ times, $r=1,\ldots.,l$.

19. Suppose first that

$$\nu\neq\mu_r,\qquad r=1,\ldots.,l.$$

Then by Theorem I, equation (43) has a particular solution $\bar{y}(x)$ expressible for large values of $x$ in the form

$$\bar{y}(x)=c\,\nu^x[1+\bar{\epsilon}(x)],$$

where $c$ is an arbitrary constant and $\bar{\epsilon}(x)$ vanishes to at least as high an order as that of $T(x)$ when $x=\infty$. Further, since $\bar{y}(x)$ is evidently not a solution of the equation $Y(x)=0$, it must with proper choice of $c$ satisfy equation (41). Substituting in (41), we find that $\bar{y}(x)$ will be a solution of that equation also, provided we take

$$c=\frac{b}{\phi(\nu)}.$$

20. Next, suppose $\nu$ is a root of the equation $\phi(\mu)=0$, say $\nu=\mu_h$. Two cases must be distinguished.

If $n_h<n'$, then equation (43) has by Theorem I a particular integral

$$\bar{y}(x)=c\,x^{n_h}\nu^x[1+\bar{\epsilon}(x)].$$

Substituting in (41), we find that $\bar{y}(x)$ satisfies that equation provided

$$c=\frac{b}{\phi^{(n_h)}(\nu)}.$$

On the other hand, if $n_h = n'$, we must assume that for large $x$

$$|\gamma_r(x)| < \tau(x)/x^{2n'}, \qquad r=0, 1, \ldots, n+1.$$

In particular, this will happen if the functions $\alpha_0(x), \alpha_1(x), \ldots, \alpha_n(x), \beta(x)$ all satisfy this same condition. With this assumption the present case reduces to the one immediately preceding.

The results of this section, when combined with Theorem I, evidently give the behavior of the complete solution of (41).

## PART II. LINEAR DIFFERENTIAL EQUATIONS.

### VI. *The Fundamental Theorems.*

21. We shall make use of two general theorems on linear differential equations, which will now be stated.*

In the differential equation

$$y^{(n)} + a_1(x)\, y^{(n-1)} + \ldots + a_n(x)\, y = 0, \tag{44}$$

suppose that the coefficients $a_1(x), \ldots, a_n(x)$, together with their first $n$ derivatives, are continuous for all sufficiently large positive values of $x$. Let $z_1(x), \ldots, z_n(x)$ denote $n$ auxiliary functions of $x$, which, together with their first $n$ derivatives, are likewise continuous when $x$ is large, and are such that the determinant

$$Q(x) = \begin{vmatrix} z_1(x) & z_1'(x) \ldots & z_1^{(n-1)}(x) \\ z_2(x) & z_2'(x) \ldots & z_2^{(n-1)}(x) \\ \ldots & \ldots & \ldots \\ z_n(x) & z_n'(x) \ldots & z_n^{(n-1)}(x) \end{vmatrix}$$

never vanishes. Let $A_r(x)$ be the minor of $Q(x)$ with respect to the element $z_r^{(n-1)}(x)$. Also put

$$Z_r(x) = z_r(x)\, a_n(x) - [z_r(x)\, a_{n-1}(x)]' + \ldots$$
$$+ (-1)^{n-1}[z_r(x)\, a_1(x)]^{(n-1)} + (-1)^n z_r^{(n)}(x),$$

and denote by $q(x, t)$ the determinant formed from $Q(x)$ by replacing the elements $z_r^{(n-1)}(x)$ by $Z_r(t)$, $r=1, \ldots, n$.

Place

$$K(x,t) = (-1)^{n-1} q(x,t)/Q(x),$$
$$f_r(x) = u_{r,0}(x) = v_{r,0}(x) = (-1)^{n-1} C_r A_r(x)/Q(x), \qquad r=1, \ldots, n,$$

where $C_r$ is an arbitrary constant,

$$u_{r,m}(x) = \int_x^\infty K(x,t)\, u_{r,m-1}(t)\, dt, \qquad m=1, 2, \ldots,$$
$$v_{r,m}(x) = \int_{x_0}^x K(x,t)\, v_{r,m-1}(t)\, dt, \qquad m=1, 2, \ldots.$$

* These theorems, which differ only slightly from those of Dini (*loc. cit.*), have been stated and proved in a previous paper by the present writer (AMERICAN JOURNAL OF MATHEMATICS, Vol. XXXVI (1914), pp. 151–166. They are restated here for convenience of reference.

Then we have

THEOREM C. *If for all values of* $x \geqq x_0$

(a) *the series*

$$y_i(x) = \sum_{m=0}^{\infty} u_{i,m}(x) \tag{45}$$

*converges;*

(b) *the series for* $y_i(t)$, *when multiplied by* $K(x,t)$, *may be integrated term by term with respect to* $t$ *from* $x$ *to* $\infty$;

(c) *each of the integrals*

$$\int_x^{\infty} y_i(t)\, Z_r(t)\, dt, \qquad r=1, \ldots, n,$$

*has a meaning,*

*then for such values of* $x$ *the function* $y_i(x)$ *is an integral of* (44).

THEOREM D. *If for all values of* $x \geqq x_0$

(a) *the series*

$$y_i(x) = \sum_{m=0}^{\infty} v_{i,m}(x)$$

*converges;*

(*b*) *the series for* $y_i(t)$, *when multiplied by* $K(x,t)$, *may be integrated term by term with respect to* $t$ *from* $x_0$ *to* $x$,

*then for such values of* $x$ *the function* $y_i(x)$ *is an integral of* (44).

## VII. *The Homogeneous Equation: Multiple Roots.*

22. Since the case of distinct roots is included in that of multiple roots, we pass at once to the general case.

Consider the differential equation

$$Y(x) = y^{(n)} + [a_1 + \alpha_1(x)]\, y^{(n-1)} + \ldots + [a_n + \alpha_n(x)]\, y = 0, \tag{1}$$

where the functions $\alpha_1(x), \ldots, \alpha_n(x)$, together with their first $n$ derivatives, are continuous for large values of $x$.

Let $\tau(x)$ denote one of the set of functions

$$\frac{1}{x^{\nu}}, \quad \frac{1}{x\,(\log x)^{\nu}}, \quad \frac{1}{x \log x\,(\log\log x)^{\nu}}, \ldots, \qquad \nu > 1,$$

and set

$$T(x) = \int_x^{\infty} \tau(t)\, dt.$$

Suppose that the characteristic equation

$$\phi(\mu) = \mu^n + a_1 \mu^{n-1} + \ldots + a_n = 0 \tag{46}$$

corresponding to (1) has $l$ distinct roots $\mu_1, \mu_2, \ldots, \mu_l$ occurring $n_1, \ldots, n_l$ times respectively ($n_1 + \ldots + n_l = n$). Assume further that, for large values of $x$, a function $\tau(x)$ can be chosen such that

$$|\alpha_r^{(s)}(x)| \leqq \tau(x)/x^{2n'-2}, \qquad r=1, \ldots, n;\ s=0, 1, \ldots, 2n-1,$$

where $n'$ is the largest of the numbers $n_1, \ldots, n_l$.

Choose $n$ auxiliary functions of $x$ of the form *

$$z_{r,q}(x) = x^{n_r-q}e^{-\mu_r x}, \qquad r=1, \ldots, l;\ q=1, \ldots, n_r.$$

Placing

$$\delta = \prod_{r,s} 0!1!\ldots(n_r-1)!(\mu_r+\mu_s)^{n_r n_s}, \qquad r=1, \ldots, l;\ r>s,$$

we find

$$Q(x) = \delta e^{-\mu x},$$

where

$$\mu = n_1\mu_1 + n_2\mu_2 + \ldots + n_l\mu_l,$$

so that the hypothesis $Q(x) \neq 0$ is satisfied. Also

$$Z_{r,q}(x) = x^{n_r-q}e^{-\mu_r x}\rho_{r,q}(x),$$

where a constant $M_1$ can be found such that, for sufficiently large $x$,

$$|\rho_{r,q}(x)| \leqq M_1\tau(x)/x^{2n'-2}.$$

Further, with proper choice of $C_{r,q}$, we have

$$f_{r,q}(x) = x^{q-1}e^{\mu_r x};$$

also

$$K(x,t) = \sum_{r=1}^{l}\sum_{q=1}^{n_r} x^{q-1}e^{\mu_r(x-t)}t^{n_r-q}\sigma_{r,q}(t),$$

where $\sigma_{r,q}(t)$ is a constant multiple of $\rho_{r,q}(t)$.

Let the roots $\mu_1, \ldots, \mu_l$ of (46) be so arranged that

$$R[\mu_1] \leqq R[\mu_2] \leqq \ldots \leqq R[\mu_l],$$

where $R[\mu]$ means the real part of $\mu$, and suppose in the first place that these real parts are all distinct.

23. Let us form as trial solutions of (1) a group of functions $y_{1,k}(x)$, $k=1, \ldots, n_1$, as given by (45).

We proceed to prove by mathematical induction that

$$|u_{1,k,m}(x)| \leqq [M_1 n T(x)]^m |f_{1,k}(x)|, \qquad k=1, \ldots, n_1. \tag{47}$$

This relation is true for $u_{1,k,0}(x)$ by definition. Assume it to hold for $u_{1,k,m-1}(x)$. Then

$$\begin{aligned}|u_{1,k,m}(x)| &\leqq \int_x^\infty |K(x,t)\,u_{1,k,m-1}(t)|\,dt\\ &\leqq (M_1 n)^{m-1}\int_x^\infty [T(t)]^{m-1}|K(x,t)\,f_{1,k}(t)|\,dt.\end{aligned} \tag{48}$$

* Cf. foot-note, p. 67.

But

$$|K(x,t)f_{1,k}(t)| \leq |f_{1,k}(x)| \sum_{r=1}^{l}\sum_{q=1}^{n_r} x^{q-k}|e^{(\mu_r-\mu_1)(x-t)}|\, t^{n_r-q+k-1}|\sigma_{r,q}(t)|$$
$$\leq |f_{1,k}(x)|\, M_1\tau(t)\sum_{r=1}^{l}\sum_{q=1}^{n_r}|e^{(\mu_r-\mu_1)(x-t)}|\, x^{q-k}\, t^{n_r-q+k-2n'+1}$$
$$\leq |f_{1,k}(x)|\, M_1\tau(t)\sum_{r=1}^{l}\sum_{q=1}^{n_r}|e^{(\mu_r-\mu_1)(x-t)}|\, x^{1-k}\left(\frac{x}{t}\right)^{q-1} t^{n_r+k-2n'}$$
$$\leq |f_{1,k}(x)|\, M_1 n\tau(t), \tag{49}$$

since

$$R[\mu_r-\mu_1]\geq 0,\quad x\leq t,\quad k\geq 1,\quad q\geq 1,\quad 2n'\geq n_r+k.$$

Substituting (49) in (48), we find

$$|u_{1,k,m}(x)| \leq (M_1 n)^m\, |f_{1,k}(x)| \int_x^\infty [T(t)]^{m-1}\tau(t)\,dt. \tag{50}$$

Now the function $T(x)$ decreases monotonically as $x$ increases, so that in the right member of (50) we may replace $T(t)$ by its greatest value $T(x)$. Whence

$$|u_{1,k,m}(x)| \leq (M_1 n)^m\, [T(x)]^{m-1}\, |f_{1,k}(x)| \int_x^\infty \tau(t)\,dt,$$

and formula (47) is established.

Let $x$ now be taken so large that

$$M_1 n\, T(x)\leq\theta,\qquad 0<\theta<1, \tag{51}$$

where $\theta$ is a constant independent of $x$. Then by (47) we have

$$\sum_{m=0}^{\infty}|u_{1,k,m}(x)| \leq \frac{|f_{1,k}(x)|}{1-M_1 n\, T(x)} \leq \frac{|f_{1,k}(x)|}{1-\theta}. \tag{52}$$

Thus condition (a) of Theorem C is satisfied.

To show that (b) is satisfied we make use of the following theorem:*

"If $\Sigma f_m(x)$ converges uniformly in any fixed interval $a\leq x\leq b$, where $b$ is arbitrary, and if $\phi(x)$ is continuous for all finite values of $x$, then

$$\int_a^\infty \phi(x)\,\Sigma f_m(x)\,dx = \Sigma\int_a^\infty \phi(x)\, f_m(x)\,dx,$$

provided that the integral $\int_a^\infty |\phi(x)|\,\Sigma\,|f_m(x)|\,dx$ is convergent."

Let us take $a=x$, $x=t$, $\phi(t)=K(x,t)$, $f_m(t)=u_{1,k,m}(t)$. The uniform convergence of the series $\sum_{m=0}^{\infty} u_{1,k,m}(t)$ in any interval $x\leq t\leq b$ follows at once from (47) and (51), by Weierstrass's test. The continuity of the functions $\sigma_{r,q}(t)$, and hence that of $K(x,t)$, follows from our assumptions regarding the continuity of the coefficients $a_1(x),\ldots, a_n(x)$ and their derivatives. And the convergence of the integral $\int_x^\infty |K(x,t)|\sum_{m=0}^{\infty}|u_{1,k,m}(t)|\,dt$ is a consequence of (52) and (49).

* Bromwich, "Infinite Series" (1908), p. 453.

Finally, we note that

$$|y_{1,k}(x)\,Z_{r,q}(x)| \leq |Z_{r,q}(x)| \sum_{m=0}^{\infty} |u_{1,k,m}(x)|$$
$$\leq \frac{1}{1-\theta} M_1 x^{k+1+n_r-q-2n'} |e^{(\mu_1-\mu_r)x}| \tau(x)$$
$$\leq \frac{M_1 \tau(x)}{1-\theta}.$$

Thus (c) holds, and the conditions of Theorem C are satisfied by the functions $y_{1,k}(x)$, $k=1, \ldots, n_1$.

We have now proved that these functions are integrals of (1) for values of $x$ so large that (51) holds. Further, by (47), we have

$$|y_{1,k}(x)-f_{1,k}(x)| \leq \sum_{m=1}^{\infty} |u_{1,k,m}(x)|$$
$$\leq \frac{M_1 n T(x)\,|f_{1,k}(x)|}{1-\theta}.$$

Thus we may write

$$y_{1,k}(x) = x^{k-1} e^{\mu_1 x} [1+\epsilon_{1,k}(x)], \qquad k=1, \ldots, n_1, \tag{53}$$

where $\epsilon_{1,k}(x)$ vanishes with increasing $x$ to at least as high an order as that of $T(x)$.

24.* We shall prove next that

$$\lim_{x=\infty} y_{1,k}^{(s)}(x)/y_{1,k}(x) = \mu_1^s, \qquad s=1, \ldots, n;$$

or in other words, that we may write

$$y_{1,k}^{(s)}(x) = x^{k-1} e^{\mu_1 x}[\mu_1^s+\zeta_{1,k,s}(x)], \tag{54}$$

where

$$\lim_{x=\infty} \zeta_{1,k,s}(x) = 0.$$

To do this, let

$$K_g(x,t) = \frac{\partial^g}{\partial x^g} K(x,t),$$
$$K_g^{(h)}(x,x) = \frac{d^h}{dx^h} \{[K_g(x,t)]_{t=x}\}.$$

It can be shown by elementary calculus, as a result of our hypotheses concerning $\alpha_1(x), \ldots, \alpha_n(x)$ and their derivatives, that, if $h \leq n-1$,

$$|K_g^{(h)}(x,x)| \leq M_2 \tau(x)/x^{n'-1}, \tag{55}$$

where $M_2$ is a sufficiently large constant.

We note that, as a result of the conditions of Theorem C, the function $y_{1,k}(x)$ satisfies the singular Volterra equation

* The present section, being relatively unimportant, will be condensed as much as possible. The missing details are readily supplied.

$$y_{1,k}(x) = f_{1,k}(x) + \int_x^\infty K(x,t)\, y_{1,k}(t)\, dt. \tag{56}$$

Differentiating (56)* $s-1$ times in succession ($s \leq n-1$), we find

$$y_{1,k}^{(s-1)}(x) = f_{1,k}^{(s-1)}(x) + S_1 y_{1,k}^{(s-2)}(x) + S_2 y_{1,k}^{(s-3)}(x) + \ldots + S_{s-1} y_{1,k}(x) + \int_x^\infty K_{s-1}(x,t) y_{1,k}(t)\, dt, \tag{57}$$

where $S_1, S_2, \ldots, S_{s-1}$ are linear and homogeneous in the derivatives $K_g^{(h)}(x, x)$, so that by virtue of (55) constants $M_3$ and $M_4$ can be found such that

$$|S_i(x)| \leq M_3 \tau(x)/x^{n'-1}, \qquad i=1, \ldots, s-1, \tag{58}$$

$$|S_i'(x)| \leq M_4 \tau(x)/x^{n'-1}, \qquad i=1, \ldots, s-1. \tag{59}$$

We are now able to prove (54) by induction. The formula is known to hold for $y_{1,k}(x)$. Let it be supposed true for

$$y_{1,k}'(x),\ y_{1,k}''(x),\ \ldots,\ y_{1,k}^{(s-1)}(x), \qquad s \leq n-1.$$

Differentiating (57), we find

$$y_{1,k}^{(s)}(x) = f_{1,k}^{(s)}(x) + S_1 y_{1,k}^{(s-1)}(x) + (S_1' + S_2) y_{1,k}^{(s-2)}(x) + \ldots + (S_{s-2}' + S_{s-1})\, y_{1,k}'(x) + S_{s-1}' y_{1,k}(x) - K_{s-1}(x,x)\, y_{1,k}(x) + \int_x^\infty K_s(x,t)\, y_{1,k}(t)\, dt. \tag{60}$$

Denoting by $x^{k-1}|e^{\mu_1 x}|\, W(x)$ the sum of the absolute values of all the terms in the right member of (60) except the first and last, we have by virtue of (55), (58) and (59)

$$\lim_{x=\infty} W(x) = 0.$$

Evidently

$$\lim_{x=\infty} x^{1-k} e^{-\mu_1 x} f_{1,k}^{(s)}(x) = \mu_1^s.$$

As regards the last term of (60), we find by inspecting the form of $K_s(x,t)$ that

$$\left|\int_x^\infty K_s(x,t)\, y_{1,k}(t)\, dt\right| \leq \int_x^\infty |K_s(x,t)\, y_{1,k}(t)|\, dt \leq M_5 \int_x^\infty |K(x,t) f_{1,k}(t)|\, dt,$$

where $M_5$ is a certain constant. But by (49) this becomes

$$\left|\int_x^\infty K_s(x,t)\, y_{1,k}(t)\, dt\right| \leq M_1 M_5 n\, x^{k-1} |e^{\mu_1 x}|\, T(x).$$

We have thus proved that

$$\lim_{x=\infty} x^{1-k} e^{-\mu_1 x} y_{1,k}^{(s)}(x) = \mu_1^s, \qquad s=1, \ldots, n-1.$$

For the case $s=n$, we have directly from (1) that

$$\lim_{x=\infty} x^{1-k} e^{-\mu_1 x} y_{1,k}^{(n)}(x) = -a_1 \mu_1^{n-1} - a_2 \mu_2^{n-2} - \ldots - a_n = \mu_1^n.$$

Thus (54) is established.

25. If we try to apply the process of § 23 to find a group of integrals $y_{2,k}$ of (1) corresponding to the root $\mu_2$, we find that some of the improper integrals occurring diverge because of the fact that $R[\mu_2 - \mu_1] > 0$. We shall therefore employ Theorem D, first making a change in our system of auxiliary functions.

* Cf. Goursat-Hedrick, "Mathematical Analysis," p. 370.

The differential equation

$$Z(x)=[a_n+\alpha_n(x)]\,z(x)-\frac{d}{dx}\{[a_{n-1}+\alpha_{n-1}(x)]\,z(x)\}+\dots$$
$$+(-1)^{n-1}\frac{d^{n-1}}{dx^{n-1}}\{[a_1+\alpha_1(x)]\,z(x)\}+(-1)^n\frac{d^n}{dx^n}z(x)=0$$

evidently satisfies all the conditions that have been imposed upon equation (1). It therefore has a group of solutions, given by (53), corresponding to that root of its characteristic equation having the least real part. The roots are evidently those of (46) with signs changed, so that the solutions in question may be put in the form

$$z_{l,q}(x)=x^{n_l-q}e^{-\mu_l x}[1+\eta_{l,q}(x)], \qquad q=1,\dots,n_l. \tag{61}$$

Upon introducing the functions $z_{l,q}(x)$ as given by (61) in our auxiliary system in place of $z_{l,q}(x)$ as originally chosen, we find that, in $K(x,t)$, $r$ takes only the values $1,\dots,l-1$. We are now able to show by Theorem D that (1) is satisfied by each of the functions

$$y_{l,k}(x)=\sum_{m=0}^{\infty}v_{l,k,m}(x), \qquad k=1,\dots,n_l,$$

where, in $v_{l,k,m}(x)$, the functions $f_{l,k}(x)$ and $K(x,t)$ must of course be formed in terms of our revised auxiliary system.

First, we have

$$f_{r,q}(x)=x^{q-1}e^{\mu_r x}\lambda_{r,q}(x),$$

where it follows from (54) that $\lambda_{r,q}(x)$ remains less than some fixed constant for large values of $x$. Hence,

$$|K(x,t)\,f_{l,k}(t)|\leq\sum_{r=1}^{l-1}\sum_{q=1}^{n_r}x^{q-1}\,|e^{\mu_r(x-t)+\mu_l t}|\,t^{n_r-q+k-1}|\nu_{r,q}(t)|,$$

where a constant $M_6$ can be found such that

$$|\nu_{r,q}(x)|\leq M_6\tau(x)/x^{2n'-2}.$$

Thus

$$|K(x,t)\,f_{l,k}(t)|\leq M_6\tau(t)\,|f_{l,q}(x)|\sum_{r=1}^{l-1}\sum_{q=1}^{n_r}|e^{(\mu_r-\mu_l)(x-t)}|\,x^{q-k}\,t^{n_r-q+k-2n'+1}.$$

We shall now prove by induction that

$$|v_{l,k,m}(x)|\leq[M_6(n-1)x\tau(x)]^m|f_{l,k}(x)|.$$

This is evidently true of $v_{l,k,0}(x)$. Suppose it to hold for $v_{l,k,m-1}(x)$. Then

$$|v_{l,k,m}(x)|\leq\int_{x_0}^{x}|K(x,t)\,v_{l,k,m-1}(t)|\,dt$$
$$\leq[M_6(n-1)]^{m-1}\int_{x_0}^{x}[t\tau(t)]^{m-1}|K(x,t)\,f_{l,k}(t)|\,dt$$
$$\leq M_6^m(n-1)^{m-1}|f_{l,k}(x)|\int_{x_0}^{x}[t\tau(t)]^{m-1}\tau(t)\sum_{r=1}^{l}\sum_{q=1}^{n_r}|e^{(\mu_l-\mu_r)(t-x)}|\,x^{q-k}\,t^{n_r-q+k-2n'+1}\,dt$$
$$\leq M_6^m(n-1)^{m-1}|f_{l,k}(x)|\int_{x_0}^{x}[t\tau(t)]^{m-1}\tau(t)\sum_{r=1}^{l-1}\sum_{q=1}^{n_r}|e^{(\mu_l-\mu_r)(t-x)}|\,x^{q-k}\,t^{k-q}\,dt,$$

since $2n' \geqq n_r + 1$. Now it results from the form of $\tau(x)$ that each of the functions $e^{R[\mu_l - \mu_r]t}\, t^{m-1-q+k}\,[\tau(t)]^m$, $r=1, \ldots, l-1$, increases monotonically with $t$. We may therefore replace each function by its maximum value, *i. e.*, its value when $t=x$, in the right member of the inequality last written. Thus

$$|v_{l,k,m}(x)| \leqq [M_6(n-1)\,\tau(x)]^m x^{m-1}\, |f_{l,k}(x)| \int_{x_0}^{x} dx$$
$$< [M_6(n-1)\, x\tau(x)]^m\, |f_{l,k}(x)|,$$

which is the required result.

It can now be shown without difficulty that for sufficiently large values of $x$ the functions $y_{l,k}(x)$ satisfy the conditions of Theorem D, and therefore satisfy (1); also that we may write

$$y_{l,k}(x) = x^{k-1} e^{\mu_l x}[1+\varepsilon_{l,k}(x)], \qquad k=1, \ldots, n_l.$$

The result for $y_{l,k}^{(s)}(x)$ analogous to (54) is also readily established.

26. The remainder of the process is precisely like that used at the corresponding stage of the work on difference equations (§ 11). We find ultimately $n$ linearly independent integrals of (1), of the form indicated in the theorem below.

The modification required in case the real parts of the roots $\mu_1, \mu_2, \ldots, \mu_l$ are not all distinct is also obvious. In particular the case treated by Dini, in which the roots all have the same real part, yields to Theorem A at once.

27. We summarize in

THEOREM II. *Let $\tau(x)$ be one of the system of functions**

$$\frac{1}{x^\nu}, \quad \frac{1}{x(\log x)^\nu}, \quad \frac{1}{x \log x (\log\log x)^\nu}, \quad \ldots, \qquad \nu>1,$$

*and put*

$$T(x) = \int_x^\infty \tau(t)\, dt.$$

*In the differential equation*

$$Y(x) = y^{(n)} + [a_1+\alpha_1(x)]\, y^{(n-1)} + \ldots + [a_n+\alpha_n(x)]\, y = 0,$$

*suppose that, when $x$ is large, the functions $\alpha_1(x), \alpha_2(x), \ldots, \alpha_n(x)$ are continuous and possess $2n-1$ continuous derivatives. Suppose also that the characteristic equation*

$$\phi(\mu) = \mu^n + a_1\mu^{n-1} + \ldots + a_n = 0$$

*has $l$ different roots $\mu_1, \mu_2, \ldots, \mu_l$, occurring $n_1, n_2, \ldots, n_l$ times respectively $(n_1+n_2+\ldots+n_l=n)$, and let $n'$ be the largest of the numbers $n_1, n_2, \ldots, n_l$.*

*If a function $\tau(x)$ exists such that for sufficiently large values of $x$*

$$|\alpha_r^{(s)}(x)| \leqq \tau(x)/x^{2n'-2}, \qquad r=0,1,\ldots,n;\ s=0,1,\ldots,2n-1,$$

* The function $\tau(x)$ is chosen in this way merely for simplicity. Cf. § 7.

*then for the same values of $x$ the equation $Y(x)=0$ has $n$ linearly independent solutions $y_{i,k}(x)$ expressible in the form*

$$y_{i,k}(x)=x^{k-1}e^{\mu_i x}[1+\varepsilon_{i,k}(x)], \qquad i=1,2,\ldots,l;\ k=1,2,\ldots,n_i,$$

*where $\varepsilon_{i,k}(x)$ vanishes at infinity to at least as high an order as that of $T(x)$, while further*

$$y_{i,k}^{(s)}(x)=x^{k-1}e^{\mu_i x}[\mu_i^s+\zeta_{i,k,s}(x)], \qquad s=1,2,\ldots,n,$$

*where*

$$\lim_{x=\infty}\zeta_{i,k,s}(x)=0.$$

## VIII. *The Non-homogeneous Equation.*

28. In the equation

$$Y(x)=X(x), \tag{62}$$

let the right-hand member have the form

$$X(x)=e^{\nu x}[b+\beta(x)],$$

where $\nu$ and $b$ are constants and $b\neq 0$. Let the functions $\alpha_1(x)$, $\alpha_2(x)$, ...., $\alpha_n(x)$, $\beta(x)$ be continuous and possess $2n+1$ continuous derivatives, and suppose a function $\tau(x)$ exists such that

$$|\alpha_r^{(s)}(x)|\leqq\tau(x)/x^{2n'-2}, \qquad r=1,\ldots,n;\ s=0,1,\ldots,2n+1,$$
$$|\beta^{(s)}(x)|\leqq\tau(x)/x^{2n'-2}, \qquad s=0,1,\ldots,2n+1.$$

The work of determining by Theorem II the behavior of a particular solution of (62) is so similar to that of §§ 18–20 that we give only the results.

If $\nu\neq\mu_r$, $r=1,\ldots,l$, equation (62) has a particular integral of the form

$$\bar{y}=\frac{b}{\phi(\nu)}e^{\nu x}[1+\bar{\varepsilon}(x)],$$

where $\bar{\varepsilon}(x)$ vanishes at infinity to as high an order as that of $T(x)$.

If $\nu=\mu_h$, where $n_h<n'$, equation (62) has a particular integral

$$\bar{y}=\frac{b}{\phi^{(n_h)}(\nu)}x^{n_h}e^{\nu x}[1+\bar{\varepsilon}(x)]. \tag{63}$$

If $\nu=\mu_h$, where $n_h=n'$, equation (62) has still an integral of the form (63), provided we assume that

$$|\alpha_r^{(s)}(x)|\leqq\tau(x)/x^{2n'}, \qquad r=1,\ldots,n;\ s=0,1,\ldots,2n+1,$$
$$|\beta^{(s)}(x)|\leqq\tau(x)/x^{2n'}, \qquad s=0,1,\ldots,2n+1.$$

The results of this section, together with Theorem II, determine the behavior of the complete integral of (62).

UNIVERSITY OF MICHIGAN.

# *The Uniform Motion of a Sphere through a Viscous Liquid.*

By R. W. Burgess.

1. The problem of finding at any point in a viscous liquid the velocity due to the uniform rectilinear motion of a sphere was first attacked by Stokes in 1851 and an approximate solution obtained. From this was derived the well-known Stokes' law, that a force $6\pi\mu Wa$ is required to maintain a uniform velocity $W$ of a sphere of radius $a$ through a liquid whose coefficient of viscosity is $\mu$. This solution was proposed merely as a limiting form to which the distribution of velocities may be supposed to tend as $W$ approaches the limit zero, but has nevertheless proved itself well in accord with experimental facts, at least as a close first approximation. The serious mathematical incompleteness of the solution has been recognized, however, and several attempts have been made to improve on it. Whitehead, in 1888, attempted to modify this solution by taking into account terms depending on squares of velocities, but found it impossible by his method to satisfy the boundary conditions at infinity. In 1893, Rayleigh showed that Stokes' solution would be accurate if certain additional forces were introduced, the magnitude of the required forces furnishing some indication of the error of the solution. In 1911, Oseen found a solution which is satisfactory at infinity; it also gives a good approximation near the sphere if the velocity or the radius of the sphere is small. Shortly afterwards Lamb, using a different method, presented Oseen's results in a simpler form, but still subject to the same limitations. In 1913, F. Noether attempted to take into consideration the terms neglected by Stokes, but merely derived Whitehead's correction over again. A few months later Oseen took up the same problem and tried to apply the principle of his previous solution, but fell into the same error that he had carefully pointed out in Whitehead's work. In the following paper Oseen's results are obtained by a simpler method than that of either Oseen or Lamb. The method is generalized, so as to give a solution that satisfies the boundary conditions at the sphere to any desired degree of approximation. A correction is determined which takes account of all the terms quadratic in velocities, and differs from the corrections of Whitehead, Noether and Oseen in that it satisfies the boundary conditions at infinity.

2. Using cylindrical coordinates $(r, z)$, let $u$, $w$ denote the components of the velocity, $p$ the pressure, $d$ the density, $\nu\left(=\frac{\mu}{d}\right)$ the kinematic coefficient of viscosity, and $V$ the potential of the impressed forces. The differential equations for the motion of a viscous liquid, in the case when the $Z$-axis is an axis of symmetry, are*

$$\left.\begin{aligned}\frac{\partial u}{\partial t}+u\frac{\partial u}{\partial r}+w\frac{\partial u}{\partial z}&=\frac{\partial Q}{\partial r}+\nu\left(\nabla^2 u-\frac{u}{r^2}\right),\\ \frac{\partial w}{\partial t}+u\frac{\partial w}{\partial r}+w\frac{\partial w}{\partial z}&=\frac{\partial Q}{\partial z}+\nu\nabla^2 w,\end{aligned}\right\} \qquad (1)$$

where

$$\nabla^2=\frac{\partial^2}{\partial r^2}+\frac{1}{r}\frac{\partial}{\partial r}+\frac{\partial^2}{\partial z^2},$$

$$Q=-V-p/d.$$

Eliminating $Q$ from these equations, we obtain

$$\frac{\partial}{\partial z}\left(\frac{\partial u}{\partial t}+u\frac{\partial u}{\partial r}+w\frac{\partial w}{\partial z}\right)-\frac{\partial}{\partial r}\left(\frac{\partial w}{\partial t}+u\frac{\partial w}{\partial r}+w\frac{\partial w}{\partial z}\right)$$
$$=\nu\left[\frac{\partial}{\partial z}\left(\nabla^2 u-\frac{u}{r^2}\right)-\frac{\partial}{\partial r}\nabla^2 w\right]. \qquad (2)$$

This equation is simplified by making use of the equation of continuity†

$$\frac{\partial u}{\partial r}+\frac{u}{r}+\frac{\partial w}{\partial z}=0.$$

If we put

$$u=-\frac{1}{r}\frac{\partial \psi}{\partial z}, \quad w=\frac{1}{r}\frac{\partial \psi}{\partial r}, \qquad (3)$$

this relation is satisfied; hence the velocities may be found by differentiation from this function $\psi$, known as Stokes' current function, and the problem is reduced to the determination of $\psi$. The curves of intersection of planes through the $z$-axis with the surface $\psi=$ a constant, will be the lines of flow or stream lines; that is, the lines indicating the direction of the resultant velocity at any point. This system of curves provides a convenient method of representing graphically the motion given by any function $\psi$.

* Bassett, "Hydrodynamics," Vol. II, p. 244. † Bassett, "Hydrodynamics," Vol. I, pp. 8, 12.

Substituting in (2) the values of $u$ and $w$ given in (3), we have*

$$\left(\frac{\partial}{\partial t}+u\frac{\partial}{\partial r}+w\frac{\partial}{\partial z}-\frac{2u}{r}-\nu D\right)D\psi=0, \tag{4}$$

where

$$D\equiv\frac{\partial^2}{\partial r^2}-\frac{1}{r}\frac{\partial}{\partial r}+\frac{\partial^2}{\partial z^2}.$$

The preceding equation applies to any case in which the motion of the liquid is symmetrical with respect to the $z$-azis. We desire to discuss in this paper the motion due to a sphere which has been moving for an infinite time with uniform velocity $W$ along the $z$-axis, and whose center is at $(0, Wt)$ at time $t$. Since the conditions at a point depend only on its position relative to the sphere, $\psi$ is evidently a function of $r$ and $z-Wt$ only. If $z'=z-Wt$, we have

$$\frac{\partial\psi}{\partial z}=\frac{\partial\psi}{\partial z'}\frac{\partial z'}{\partial z}=\frac{\partial\psi}{\partial z'},\quad \frac{\partial\psi}{\partial t}=\frac{\partial\psi}{\partial z'}\frac{\partial z'}{\partial t}=-W\frac{\partial\psi}{\partial z'}.$$

Equation (4) then becomes

$$\left(u\frac{\partial}{\partial r}+(w-W)\frac{\partial}{\partial z'}-\frac{2u}{r}-\nu D\right)D\psi=0, \tag{5}$$

a form on which we shall base our further discussion.

3. The problem we shall consider is the determination of the function $\psi$ of $r$ and $z'$ which satisfies (5) and is such that the velocity of the liquid at the surface of the sphere is the same as the velocity of the sphere, and that the velocity of the liquid at an infinite distance from the sphere vanishes. This problem was first attacked by Stokes† in 1851 and solved, approximately, for the case in which the liquid is very viscous; that is, $\nu$ large, and the velocities sufficiently small so that the terms involving the products of velocities and accelerations are comparatively unimportant. On these assumptions the approximate form of (5) is

$$\nu D\cdot D\psi=0, \tag{6}$$

for which we may easily obtain Stokes' solution

$$\psi=\tfrac{1}{4}Wa^2\left(\frac{3\rho}{a}-\frac{a}{\rho}\right)\sin^2\theta,$$

where $a$ is the radius of the sphere, $\rho=\sqrt{r^2+z^2}$ and

$$\tan\theta=\frac{r}{z}. \tag{7}$$

* Cf. Bassett, "Hydrodynamics," Vol. II, p. 259.

† *Cambridge Transactions*, Vol. IX (1851); *Sci. Papers*, Vol. III, p. 55.

From this is derived Stokes' well-known formula $F=6\pi\mu Wa$ for the force required to maintain the motion. This formula has proved so accurate for small velocities that the approximations by which it was obtained have not always been fully recognized. Stokes himself, however, proposed the solution (7) only as a limiting form to which the velocities may be supposed to tend as $W$ approaches zero.

In 1888, Whitehead* attempted to modify this solution to take account of the neglected quadratic terms. The attempt was successful in finding correction terms satisfactory as far as the differential equation (5) and the boundary conditions at the sphere were concerned, but these terms failed, as Whitehead himself noticed, to satisfy the condition that the velocity at an infinite distance vanishes. The correction terms for the velocities along and perpendicular to the radius vector were found to be

$$\left.\begin{aligned} R_1&=-\frac{3}{16}\frac{W^2a}{\nu}\left(\frac{a}{\rho}-1\right)^2\left(\frac{a^2}{\rho^2}+\frac{a}{\rho}+2\right)P_2(\cos\theta),\\ \Theta_1&=-\frac{3}{16}\frac{W^2a}{\nu}\left(\frac{a}{\rho}-1\right)\left(\frac{a^3}{\rho^3}+\frac{a^2}{2\rho^2}+\frac{a}{2\rho}+2\right)\sin\theta\cos\theta.\end{aligned}\right\} \tag{8}$$

It is obvious that when $\rho$ is infinite these terms do not vanish as required.

In 1893, Rayleigh† showed that Stokes' solution would be accurate if certain additional forces were introduced; the ratio of the forces thus required to the viscous forces actually considered gives a means of estimating the accuracy of the approximation. This ratio can easily be shown to be $\frac{W\rho}{\nu}$, which increases indefinitely with $\rho$; Stokes' solution is therefore not valid at infinity.

4. In 1911, Oseen‡ introduced a new approximate solution, valid at infinity. Consider the term $(w-W)\frac{\partial}{\partial z}$ in equation (5). Near the surface of the sphere it is true that $w-W$ is very small, and the neglect of this term is justified, but at any considerable distance it is not; in fact, $w-W$ is very nearly $-W$, not zero. Oseen shows that the ratio of the neglected terms to the terms giving the viscous forces is $\frac{Wz}{\nu}$, which becomes infinite with $z$, no matter how small we take $W$. On account of these considerations at a large distance from the sphere, we take as an approximate form of (5)

$$\left(-W\frac{\partial}{\partial z'}-\nu D\right)D\psi=0. \tag{9}$$

* *Quarterly Journal of Mathematics*, Vol. XXIII (1888), pp. 143–152.
† *Philosophical Magazine*, Ser. 4, Vol. XXXVI (1893), p. 354.
‡ *Arkiv för Matematik, Astronomi, och Fysik*, Bd. 6 (1911), No. 29.

Oseen's method is not founded on a consideration of equation (5) or (9), nor does he use the current function; but the above remarks embody his argument. He obtains a solution which, expressed as a current function, satisfies (9) and the boundary conditions at infinity, but satisfies the boundary conditions at the surface of the sphere only to the first power of the quantity $\frac{Wa}{\nu}$, which is assumed to be small.

5. The essential points of this work of Oseen were restated by Lamb* shortly afterwards, with the use of a less intricate method of analysis. Lamb points out especially the significance of the result, showing that the equation (9), or rather the similar equations used by Oseen, represents more accurately than does Stokes' equation (6) the nature of the motion at a large distance from the sphere. Lamb shows, moreover, that Oseen's solution more nearly represents the wake which follows a moving object, in that the velocity preceding the sphere is markedly smaller than that following.

Two years later F. Noether,† referring in his work to that of Oseen, determined as a correction to Stokes' current function (7)

$$\psi_1 = \frac{3}{32}\frac{W^2a}{\nu}\sin^2\theta\cos\theta\left(\frac{\rho-a}{\rho}\right)^2(2\rho^2+a\rho+a^2). \tag{10}$$

Corresponding to this, the corrections for the velocities are

$$R_1 = -\frac{3}{16}\frac{W^2a}{\nu}P_2(\cos\theta)\left(\frac{\rho-a}{\rho}\right)^2\left(2+\frac{a}{\rho}+\frac{a^2}{\rho^2}\right),$$

$$\Theta_1 = \frac{3}{16}\frac{W^2a}{\nu}\sin\theta\cos\theta\left(1-\frac{a}{\rho}\right)\left(2+\frac{a}{2\rho}+\frac{a^2}{2\rho^2}+\frac{a^3}{\rho^3}\right),$$

which are exactly (8), as found by Whitehead.

In the same year Oseen‡ again published an article on the subject criticizing certain portions of Noether's work, and giving as the value of Stokes' current function

$$\psi_2 = \frac{W}{4}\sin^2\theta\left(1-\frac{a}{\rho}\right)^2\left[\left(1+\frac{3}{8}\left|\frac{Wa}{\nu}\right|\right)\rho(2\rho+a) + \frac{3}{8}\frac{Wa}{\nu}\cos\theta\,(2\rho^2+a\rho+a^2)\right] - \frac{W}{2}\rho^2\sin^2\theta. \tag{11}$$

But this function yields values of $R$ and $\Theta$ which fail to vanish at infinity,

---

* *Philosophical Magazine*, Ser. 6, Vol. XXI (1911), pp. 112–121.

† *Zeitschrift für Mathematik und Physik*, Bd. 62.

‡ *Arkiv för Matematik, Astronomi, och Fysik*, Bd. 9, No. 16.

exactly as Whitehead's resolution did, and therefore, as Oseen pointed out in his earlier paper, is not admissible.

6. We shall now consider in detail equation (9),

$$\left(-W\frac{\partial}{\partial z'}-\nu D\right)D\psi=0. \tag{9}$$

As these operations are commutative, a solution will be given by

$$\psi=\psi'+\psi'', \tag{12}$$

where

$$D\psi'=0 \tag{13}$$

and

$$\left(\nu D+W\frac{\partial}{\partial z'}\right)\psi''=0. \tag{14}$$

We shall obtain the solution of these separately. Equation (13) has been treated very fully by Sampson,* but for convenience of reference we shall here give in detail the solution so far as is necessary for our purposes. Transforming to polar coordinates

$$\rho=\sqrt{r^2+z'^2},\quad \theta=\cos^{-1}\frac{z'}{\sqrt{r^2+z'^2}}=\sin^{-1}\frac{r}{\sqrt{r^2+z'^2}},$$

our operation $D$ becomes

$$\frac{\partial^2}{\partial\rho^2}+\frac{\sin\theta}{\rho^2}\frac{\partial}{\partial\theta}\left(\frac{1}{\sin\theta}\frac{\partial}{\partial\theta}\right).$$

We desire, then, the solution of the equation

$$\frac{\partial^2\psi}{\partial\rho^2}+\frac{\sin\theta}{\rho^2}\frac{\partial}{\partial\theta}\left(\frac{1}{\sin\theta}\frac{\partial\psi}{\partial\theta}\right)=0. \tag{13}$$

A particular solution is

$$\psi_n=c_n\frac{\sin\theta}{\rho^n}\frac{\partial P_n}{\partial\theta}, \tag{13A}$$

where $P_n$ is the zonal harmonic of degree $n$, and therefore satisfies the differential equation

$$n(n+1)P_n\sin\theta+\frac{\partial}{\partial\theta}\left(\sin\theta\frac{\partial P_n}{\partial\theta}\right)=0. \tag{15}$$

For substituting in (13) the given value of $\psi_n$, we have for the left-hand side

$$\frac{n(n+1)c_n}{\rho^{n+2}}\sin\theta\frac{\partial P_n}{\partial\theta}+\frac{c_n\sin\theta}{\rho^{n+2}}\frac{\partial}{\partial\theta}\left\{\frac{1}{\sin\theta}\frac{\partial}{\partial\theta}\left(\sin\theta\frac{\partial P_n}{\partial\theta}\right)\right\};$$

* *Philosophical Transactions*, Vol. CLXXXII (1891), p. 449.

that is, using (15),

$$\frac{c_n \sin\theta}{\rho^{n+2}}\left[n(n+1)\frac{\partial P_n}{\partial\theta}-\frac{\partial}{\partial\theta}\{n(n+1)P_n\}\right],$$

which is identically zero. It is well known that

$$P_n=A_n\left\{\cos^n\theta-\frac{n(n-1)}{2(2n-1)}\cos^{n-2}\theta+\frac{n(n-1)(n-2)(n-3)}{2\cdot 4(2n-1)(2n-3)}\cos^{n-4}\theta-\ldots\ldots\right\}. \quad (16)$$

Hence, by (14),

$$\psi_n=\frac{c_n\sin^2\theta}{\rho^n}\left[\cos^{n-1}\theta-\frac{(n-1)(n-2)}{2(2n-1)}\cos^{n-3}\theta+\ldots\ldots\right.$$
$$\left.+(-1)^r\frac{(n-1)(n-2)\ldots\ldots(n-2r)}{2^r\,\underline{|r}\,(2n-1)\ldots\ldots(2n-2r+1)}\cos^{n-2r+1}\theta+\ldots\ldots\right] \quad (13\text{B})$$

for all positive integral values of $n$. For $n=0$, we have $\psi$ a function of $\theta$ only; then

$$\frac{1}{\sin\theta}\frac{\partial\psi_0}{\partial\theta}=c_0, \quad \psi_0=c_0\cos\theta.$$

Placing $n$ successively 1, 2, 3, .... in (13B), we have

$$\left.\begin{aligned}
\psi_1&=c_1\frac{\sin^2\theta}{\rho},\\
\psi_2&=c_2\frac{\sin^2\theta\cos\theta}{\rho^2},\\
\psi_3&=c_3\frac{\sin^2\theta\,(5\cos^2\theta-1)}{\rho^3},\\
\psi_4&=c_4\frac{\sin^2\theta\cos\theta\,(7\cos^2\theta-3)}{\rho^4}.
\end{aligned}\right\} \quad (17)$$

7. We desire also the solution of equation (14), which is, if $2k=\frac{W}{\nu}$,

$$\left(\frac{\partial^2}{\partial r^2}-\frac{1}{r}\frac{\partial}{\partial r}+\frac{\partial^2}{\partial z'^2}+2k\frac{\partial}{\partial z'}\right)\psi''=0. \quad (14\text{A})$$

Letting $\psi''=e^{-kz'}\phi(r,z')$, we have at once

$$\left(\frac{\partial^2}{\partial r^2}-\frac{1}{r}\frac{\partial}{\partial r}+\frac{\partial^2}{\partial z'^2}-k^2\right)\phi=0,$$

or, in polar coordinates,

$$\frac{\partial^2\phi}{\partial\rho^2}+\frac{\sin\theta}{\rho^2}\frac{\partial}{\partial\theta}\left(\frac{1}{\sin\theta}\frac{\partial\phi}{\partial\theta}\right)-k^2\phi=0. \quad (18)$$

Assume $\phi=R\Theta$, where $R$ and $\Theta$ are respectively functions of $\rho$ and $\theta$ only.

$$\Theta\frac{\partial^2 R}{\partial\rho^2}+R\frac{\sin\theta}{\rho^2}\frac{\partial}{\partial\theta}\left(\frac{1}{\sin\theta}\frac{\partial\Theta}{\partial\theta}\right)-k^2R\Theta=0,$$

$$\frac{\rho^2}{R}\frac{\partial^2 R}{\partial\rho^2}-k^2\rho^2=-\frac{\sin\theta}{\Theta}\frac{\partial}{\partial\theta}\left(\frac{1}{\sin\theta}\frac{\partial\Theta}{\partial\theta}\right)=n(n+1),\text{ say.}$$

We have, then, to determine $R$ and $\Theta$, the two equations:

$$\frac{\partial^2 R}{\partial\rho^2}-k^2R-\frac{n(n+1)}{\rho^2}R=0, \tag{19}$$

$$n(n+1)\Theta+\sin\theta\frac{\partial}{\partial\theta}\left(\frac{1}{\sin\theta}\frac{\partial\Theta}{\partial\theta}\right)=0. \tag{20}$$

In this form, for $n$ any positive integer, $R$ is a solution in finite form of Riccati's equation; that is,

$$R_n=\rho^{n+1}\left(\frac{1}{\rho}\frac{\partial}{\partial\rho}\right)^{n+1}(Ae^{k\rho}+Be^{-k\rho}), \tag{21}$$

and $\Theta$ is $\sin\theta\frac{\partial P_n}{\partial\theta}$. We have, therefore,

$$\psi''=\sum_{n=0}^{n=\infty}e^{-k\rho\cos\theta}\sin\theta\frac{\partial P_n}{\partial\theta}R_n. \tag{22}$$

We have, then, as a general function satisfying the differential equation (9)

$$\psi=e^{-k\rho(1+\cos\theta)}\left[C+D\cos\theta+b_1\sin^2\theta\left(k+\frac{1}{\rho}\right)\right.$$
$$\left.+b_2\sin^2\theta\cos\theta\left(k^2+\frac{3k}{\rho}+\frac{3}{\rho^2}\right)+\dots\right]$$
$$+L\cos\theta+\frac{M\sin^2\theta}{\rho}+\frac{N\sin^2\theta\cos\theta}{\rho^2}+\frac{P\sin^2\theta}{\rho^3}(5\cos^2\theta-1)+\dots \tag{23}$$

8. To adapt this solution to the problem of the sphere moving with uniform velocity $W$, we must have, if

$$\left.\begin{aligned}R&=\frac{1}{\rho^2\sin\theta}\frac{\partial\psi}{\partial\theta}\text{ is the radial velocity}\\ \text{and}\quad\Theta&=-\frac{1}{\rho\sin\theta}\frac{\partial\psi}{\partial\rho}\text{ is the velocity perpendicular to this,}\end{aligned}\right\} \tag{24}$$

(1) $R$ and $\Theta$ always finite,
(2) $R=0$ and $\Theta=0$ at infinity,
and (3) $R=W\cos\theta$, $\Theta=-W\sin\theta$ on the surface of the sphere, *i. e.*, when $\rho=a$. (25)

The terms we have chosen are all such that $R$ and $\Theta$ derived from them vanish at an infinite distance, and, if we take $C+D\cos\theta=b_0(1-\cos\theta)$, are everywhere finite, since $\frac{\partial\psi}{\partial\theta}$ and $\frac{\partial\psi}{\partial\rho}$ are both divisible by $\sin\theta$. The last condition [25(3)] remains to be satisfied by a proper combination of the functions and determination of the constants.

Oseen's solution is equivalent to taking the terms in (23) whose coefficients are $b_0$, $L$, and $M$; that is to say,

$$\psi=b_0e^{-k\rho(1+\cos\theta)}(1-\cos\theta)+L\cos\theta+\frac{M\sin^2\theta}{\rho}. \tag{26}$$

Using the conditions [25(3)] at the surface of the sphere to determine the constants, we have

$$\left.\begin{aligned} W\cos\theta&=-\frac{L}{a^2}+\frac{2M\cos\theta}{a^3}+\frac{b_0e^{-ka(1+\cos\theta)}}{a^2}\{1+ka(1-\cos\theta)\},\\ -W\sin\theta&=\frac{M\sin\theta}{a^3}+\frac{b_0k}{a}e^{-ka(1+\cos\theta)}\sin\theta. \end{aligned}\right\} \tag{27}$$

Equating the constant terms in each of these two equations and the coefficients of $\cos\theta$ on each side of the first, we have

$$\frac{L}{a^2}=\frac{b_0}{a^2},\quad W=\frac{2M}{a^3}-\frac{2b_0k}{a},\quad -W=\frac{M}{a^3}+\frac{b_0k}{a}.$$

It is to be noticed that we have neglected all squares and higher powers of the quantity $ka$, which must therefore be assumed to be small. Hence we derive

$$L=b_0=-\frac{3a\nu}{2},\quad M=-\frac{Wa^3}{4}.$$

Substituting these values in (26), we have

$$\psi=-\frac{3a\nu}{2}[e^{-k\rho(1+\cos\theta)}(1-\cos\theta)+\cos\theta]-\frac{Wa^3\sin^2\theta}{4\rho}. \tag{28}$$

If we transform this value of $\psi$ into cylindrical coordinates, find $u$ and $w$ from equations (3), and resolve $u$ in the directions of the $x$ and $y$ axes, we get expressions for the velocities which, except for notation, are exactly those derived by Oseen. As explained in section 4, this solution differs from Stokes' solution (7) fundamentally in that it satisfies the differential equation (9), which for large values of $\rho$ is a close approximation to the complete equation (5). We can see its significance by throwing it into several different forms. In the first place, since a constant makes no difference in a function used solely in the form of its derivatives, we may write it

$$\psi=\frac{Wa^2\sin^2\theta}{4}\left[3\,\frac{e^{-k\rho(1+\cos\theta)}-1}{-ka(1+\cos\theta)}-\frac{a}{\rho}\right]. \tag{28A}$$

Expanding the exponential term, this is

$$\psi=\frac{Wa^2}{4}\left(\frac{3\rho}{a}-\frac{a}{\rho}\right)\sin^2\theta-\frac{3kaW}{8}\rho^2\sin^2\theta(1+\cos\theta)+\ldots.. \tag{28B}$$

The first term of this expansion is Stokes' solution; the following terms contain powers of $\frac{Wd\rho}{\mu}$ and are therefore less important than the first when $\rho$ is small (*i. e.*, near the sphere) and $W$ is small (*i. e.*, for slow motion). As Oseen says, then, in the neighborhood of the sphere the solution approaches that of Stokes as a limit when $W$ approaches zero. It is important to notice, however, that the character of the solution is essentially different from that of Stokes. If we plot the stream-lines for Stokes' solution, we notice that we have symmetry with regard to the plane $z'=0$; for Oseen's, however, the stream-lines are more crowded following the sphere. If, for instance, we set the first two terms of (28) equal to a constant; that is, $e^{-k\rho(1+\cos\theta)}(1-\cos\theta)+\cos\theta=c$, the curves thus resulting, more conveniently expressed in the form

$$\rho=\frac{\log\dfrac{1-\cos\theta}{c-\cos\theta}}{k(1+\cos\theta)},$$

are asymptotic to the lines $\theta=\cos^{-1}c$ before the sphere and run off in a parabolic form in the wake of the sphere. The simplest way of seeing the difference between the conditions before and after the sphere, is to notice that the velocities will each have an exponential factor $-k\rho(1+\cos\theta)$; when $\theta$ is in the first quadrant, this term will be smaller than when $\theta$ is in the second quadrant, for the same values of $\rho$. Since this corresponds with our observations of fluid motion, the solution derived by Oseen appears worthy of further consideration and improvement.

Expanding the boundary conditions as given in (27), we have

$$\left.\begin{aligned} W\cos\theta&=-\frac{L}{a^2}+\frac{2M\cos\theta}{a^3}+\frac{b_0}{a^2}\left\{1-ka(1+\cos\theta)+\frac{k^2a^2}{\underline{|2}}(1+\cos\theta)^2-\ldots.\right\}\\ &\qquad\times\{1+ka(1-\cos\theta)\},\\ -W\sin\theta&=\frac{M\sin\theta}{a^3}+\frac{b_0k\sin\theta}{a}\left\{1-ka(1+\cos\theta)+\frac{k^2a^2}{\underline{|2}}(1+\cos\theta)^2-\ldots.\right\}.\end{aligned}\right\} \tag{27A}$$

From these equations it is clear that on the surface of the sphere the velocity components found in (28) differ from the velocity components of the sphere by terms whose ratio to the terms $W\cos\theta$ and $-W\sin\theta$, is $ka$. We proceed to improve this approximation by using two more terms from (20), and by this means making the above-mentioned ratio equal to $k^2a^2$.

9. Let

$$\psi=\sin^2\theta\left\{b_0\frac{e^{-k\rho(1+\cos\theta)}-1}{1+\cos\theta}+b_1\left(k+\frac{1}{\rho}\right)e^{-k\rho(1+\cos\theta)}+\frac{M}{\rho}+\frac{N\cos\theta}{\rho^2}\right\} \tag{29}$$

$$\begin{aligned}=\sin^2\theta\Big\{&-b_0k\rho+b_1\left(k+\frac{1}{\rho}\right)+\frac{M}{\rho}-\frac{N}{\rho^2}\left[\frac{b_0k^2\rho^2}{2}-b_1k(1+k\rho)+\frac{N}{\rho^2}\right](1+\cos\theta)\\&+\left[-\frac{b_0k^3\rho^3}{\lfloor 3}+\frac{b_1k^2\rho^2}{\lfloor 2}\left(k+\frac{1}{\rho}\right)\right](1+\cos\theta)^2+\ldots\\&+\left[b_0\frac{(-k\rho)^{n+1}}{\lfloor n+1}+\frac{b_1(-k\rho)^n}{\lfloor n}\left(k+\frac{1}{\rho}\right)\right](1+\cos\theta)^n+\ldots\Big\}\end{aligned} \tag{29A}$$

$$=\sin^2\theta[B_0+B_1(1+\cos\theta)+B_2(1+\cos\theta)^2+\ldots+B_n(1+\cos\theta)^n+\ldots]. \tag{29B}$$

The boundary conditions (21) and (22) at the surface of the sphere give

$$R=\frac{1}{\rho^2\sin\theta}\frac{\partial\psi}{\partial\theta}=W\cos\theta,$$

$$\Theta=-\frac{1}{\rho\sin\theta}\frac{\partial\psi}{\partial\rho}=-W\sin\theta,\text{ when }\rho=a.$$

As the first of these is to be true, for all values of $\theta$, we may integrate, giving $\psi=\frac{1}{2}Wa^2\sin^2\theta$, when $\rho=a$. We have, therefore,

$$\frac{1}{2}Wa^2\sin^2\theta=\sin^2\theta[B_0+B_1(1+\cos\theta)+\ldots],$$

$$Wa\ \sin^2\theta=\sin^2\theta\left[\frac{\partial B_0}{\partial\rho}+\frac{\partial B_1}{\partial\rho}(1+\cos\theta)+\ldots\right].$$

Dividing by $\sin^2\theta$ and equating like powers of $(1+\cos\theta)$, we obtain

$$\left.\begin{aligned}\frac{1}{2}Wa^2=B_0\Big]_{\rho=a},\qquad & Wa=\frac{\partial B_0}{\partial\rho}\Big]_{\rho=a},\\ 0=B_n]_{\rho=a},\qquad & 0=\frac{\partial B_n}{\partial\rho}\Big]_{\rho=a}.\end{aligned}\right\} \tag{30}$$

Using the values of $B_0$, $B_1$, etc., given above in equation (29A), we find that we can satisfy the four equations for $B_0$, $B_1$, $\frac{\partial B_0}{\partial\rho}$ and $\frac{\partial B_1}{\partial\rho}$, neglecting squares and higher powers of $ka$, with the following values for the constants:

$$\left.\begin{aligned}b_0=-\frac{3}{4}Wa^2\left(\frac{1}{ka}+\frac{3}{4}\right),\qquad & M=\frac{Wa^3}{2}\left(1-\frac{3}{2}ka\right),\\ b_1=-\frac{3}{4}Wa^3\left(1-\frac{3}{4}ka\right),\qquad & N=-\frac{3}{8}Wa^4\cdot ka.\end{aligned}\right\} \tag{31}$$

To satisfy the boundary conditions (30) exactly, we should have $B_n$ and $\frac{\partial B_n}{\partial \rho}=0$ when $\rho=a$; it is obvious that, while the given values of the constants do not effect this, the terms which should but do not vanish all contain at least the square of $ka$ as a factor, and our error in approximately satisfying the boundary conditions at the sphere is proportional to the square instead of the first power of the small quantity $ka$. We see, therefore, that while solution (28) holds as valid in the limit when $ka$ approaches zero, the solution just found is a better approximation when $ka$ is sufficiently small so that its square, but not its first power, may be neglected. If a closer approximation is desired, two more terms may be added, namely:

$$b_2 e^{-k\rho(1+\cos\theta)} \sin^2\theta \cos\theta\left(k^2+\frac{3k}{\rho}+\frac{3}{\rho^2}\right)+\frac{P\sin^2\theta(5\cos^2\theta-1)}{\rho^3}, \tag{32}$$

and we shall then be able to satisfy exactly the boundary conditions $B_2=0$, $\frac{\partial B_2}{\partial \rho}=0$ when $\rho=a$, making our error proportional to $k^3a^3W$. Similarly, we can add any number of pairs of functions and make our error proportional to as high a power of $ka$ as we desire. On account also of the $\lfloor n$ in the denominator of the $n$-th term of (29), this error is more and more negligible as we go on. We have shown, then, how to obtain as accurate a solution of equation (9) as may be desired.

10. In this differential equation (9) which we have just discussed, we have neglected certain terms which were included in our fundamental equation (5). We shall now include these and proceed to obtain a solution of the complete equation (5) accurate as far as the terms involving $k^2a^2$ or $k^2\rho^2$ as a factor. When transformed to polar coordinates, equation (5) becomes

$$\begin{aligned}\Bigg[\frac{1}{\rho^2\sin\theta}\left(\frac{\partial\psi}{\partial\theta}\frac{\partial}{\partial\rho}-\frac{\partial\psi}{\partial\rho}\frac{\partial}{\partial\theta}\right)+\frac{2}{\rho^2\sin^2\theta}\left(\cos\theta\frac{\partial\psi}{\partial\rho}-\frac{\sin\theta}{\rho}\frac{\partial\psi}{\partial\theta}\right)\\ -\nu\left\{\frac{\partial^2}{\partial\rho^2}+\frac{\sin\theta}{\rho^2}\frac{\partial}{\partial\theta}\left(\frac{1}{\sin\theta}\frac{\partial}{\partial\theta}\right)\right\}\\ -W\left(\cos\theta\frac{\partial}{\partial\rho}-\frac{\sin\theta}{\rho}\frac{\partial}{\partial\theta}\right)\Bigg]\left[\frac{\partial^2\psi}{\partial\rho^2}+\frac{\sin\theta}{\rho^2}\frac{\partial}{\partial\theta}\left(\frac{\partial\psi}{\sin\theta\,\partial\theta}\right)\right]=0.\end{aligned} \tag{5A}$$

Let us assume that the desired approximate solution of this equation is of the form

$$\psi=\sin^2\theta[A_0+A_1\cos\theta], \tag{33}$$

where $A_0$ and $A_1$ are functions of $\rho$, and $A_1$ contains $ka$ as a factor. This assumption appears appropriate in view of the fact that, to the desired degree of approximation, each of the solutions (7), (28) and (29) already found is of this type. Substituting this value of $\psi$ in the differential equation (5A), omitting all terms which contain $k^2a^2$ as a factor, we have, after some simplification,

$$\left(\frac{2A_0}{\rho^2}-W\right)\left\{\frac{\partial}{\partial\rho}\left(\frac{\partial^2 A_0}{\partial\rho^2}-\frac{2A_0}{\rho^2}\right)-\frac{2}{\rho}\left(\frac{\partial^2 A_0}{\partial\rho^2}-\frac{2A_0}{\rho^2}\right)\right\}\cos\theta$$
$$-\nu\left[\frac{\partial^4 A_0}{\partial\rho^4}-\frac{4}{\rho^2}\frac{\partial^2 A_0}{\partial\rho^2}+\frac{8}{\rho^3}\frac{\partial A_0}{\partial\rho}-\frac{8A_0}{\rho^4}\right.$$
$$\left.+\left(\frac{\partial^4 A_1}{\partial\rho^4}-\frac{12}{\rho^2}\frac{\partial^2 A_1}{\partial\rho^2}+\frac{24}{\rho^3}\frac{\partial A_1}{\partial\rho}\right)\cos\theta\right]=0. \quad (34)$$

In order to satisfy this relation, we must in the first place have

$$\frac{\partial^4 A_0}{\partial\rho^4}-\frac{4}{\rho^2}\frac{\partial^2 A_0}{\partial\rho^2}+\frac{8}{\rho^3}\frac{\partial A_0}{\partial\rho}-\frac{8A_0}{\rho^4}=0. \quad (35)$$

That is to say, any value assumed for $A_0$ must be such that $A_0\sin^2\theta$ satisfies $D^4\psi=0$, since (35) results from this part of (5A), which was used by Stokes. The complete solution of this equation can easily be shown to be

$$A_0=\frac{a_0}{\rho}+a_1\rho+a_2\rho^2+a_3\rho^4.$$

In order to satisfy (34), we must also have

$$\frac{1}{\nu}\left(\frac{2A_0}{\rho^2}-W\right)\left(\frac{\partial}{\partial\rho}-\frac{2}{\rho^2}\right)\left(\frac{\partial^2 A_0}{\partial\rho^2}-\frac{2A_0}{\rho^2}\right)=\frac{\partial^4 A_1}{\partial\rho^4}-\frac{12}{\rho^2}\frac{\partial^2 A_1}{\partial\rho^2}+\frac{24}{\rho}\frac{\partial A_1}{\partial\rho}. \quad (36)$$

Hence we may easily determine the value of $A_1$ corresponding to any value of $A_0$.

11. The first obvious assumption is to take $A_0$ from Stokes' solution (7),

$$A_0=b_0\left(\frac{3\rho}{a}-\frac{a}{\rho}\right).$$

Then

$$\frac{\partial^2 A_0}{\partial\rho^2}-\frac{2A_0}{\rho^2}=-\frac{6b_0}{a\rho}, \quad (37)$$

and (34) becomes

$$\rho^4\frac{\partial^4 A_1}{\partial\rho^4}-12\rho^2\frac{\partial^2 A_1}{\partial\rho^2}+24\rho\frac{\partial A_1}{\partial\rho}=\frac{18b_0}{\nu a}\left(\frac{6b_0\rho}{a}-\frac{2b_0a}{\rho}-W\rho^2\right). \quad (38)$$

Assume a solution,

$$A_1=d_1\rho^2+d_2\rho+d_3+\frac{d_4}{\rho}+\frac{d_5}{\rho^2}, \quad (39)$$

in which the third and fifth terms are parts of the complementary function, the fifth term being essentially $\psi_2$ of equation (17). We could now easily obtain the value of $A_1$; our work is simplified, however, by first determining $b_0$. As our solution must satisfy the boundary conditions at the surface of the sphere, we have, as in (25),

$$\frac{1}{a^2 \sin\theta}\frac{\partial\psi}{\partial\theta}\Big]_{\rho=a} = W\cos\theta, \quad \frac{1}{a\sin\theta}\frac{\partial\psi}{\partial\rho}\Big]_{\rho=a} = W\sin\theta. \tag{40}$$

Since $\psi = \sin^2\theta(A_0 + A_1\cos\theta)$,

$$Wa^2\cos\theta = 2A_0\cos\theta + A_1[-1+3\cos^2\theta],$$

$$Wa = \frac{\partial A_0}{\partial\rho} + \frac{\partial A_1}{\partial\rho}\cos\theta.$$

We must therefore have

$$\left.\begin{aligned} \frac{Wa^2}{2} &= A_0\Big]_{\rho=a}, & Wa &= \frac{\partial A_0}{\partial\rho}\Big]_{\rho=a}, \\ A_1\Big]_{\rho=a} &= 0, & \frac{\partial A_1}{\partial\rho}\Big]_{\rho=a} &= 0. \end{aligned}\right\} \tag{41}$$

From these conditions on $A_0$, we find $b_0 = \frac{1}{4}Wa^2$. Substituting this value for $b_0$, (38) becomes

$$\rho^4\frac{\partial^4 A_1}{\partial\rho^4} - 12\rho^2\frac{\partial^2 A_1}{\partial\rho^2} + 24\rho\frac{\partial A_1}{\partial\rho} = \frac{9W^2a}{4\nu\rho}[-2\rho^3 + 3a\rho^2 - a^3]. \tag{42}$$

Hence we easily obtain

$$A_1 = -\frac{3}{32}\frac{W^2a}{\nu\rho^2}[2\rho^4 - 3a\rho^3 + d_3\rho^2a^2 - a^3\rho + d_5]. \tag{43}$$

Using (41), we have

$$d_3 = d_5 = 1. \tag{44}$$

$$\therefore \quad A_1 = -\frac{3}{32}\frac{W^2a}{\nu}\left(1-\frac{a}{\rho}\right)^2(2\rho^2 + a\rho + a^2).$$

The solution given by this assumption of the value of $A_0$ is therefore

$$\psi = \frac{1}{4}Wa\sin^2\theta\left\{\left(3\rho - \frac{a^2}{\rho}\right) - \frac{3W}{8\nu}\left(1-\frac{a}{\rho}\right)^2(2\rho^2 + a\rho + a^2)\cos\theta\right\}, \tag{45}$$

which is exactly the correction found by Noether, as stated in (10); the velocities derived from this are given in (8), and were first found by Whitehead. The objection which Whitehead himself raised against this approximation still stands, however; that is, the velocities do not vanish at infinity as required by the conditions of the problem.

12. In order to avoid this difficulty at infinity, let us take $A_0$ from the solution of an equation which is, as Stokes' is not, a good approximation at a

large distance from the sphere. The simplest one of this type is (28). We shall consider this as our basis and determine a correction, accurate as far as terms involving $k^2a^2$ or $k^2\rho^2$ as a factor, which shall take account of the quadratic terms. Accordingly we take

$$A_0=b_0\left(\frac{3\rho}{a}-\frac{a}{\rho}\right)-\frac{3}{8}kaW\rho^2. \tag{46}$$

Determining $b_0$ at once from equations (41), we have

$$\frac{Wa^2}{2}=2b_0-\frac{3}{8}kaWa^2,$$

$$Wa=\frac{4b_0}{a}-\frac{3}{4}kaWa;$$

that is

$$b_0=\frac{Wa^2}{4}\left(1+\frac{3}{4}ka\right).$$

Substituting the value of $A_0$ from (46) in (36), and neglecting, as we are doing throughout, all terms involving the square of $ka$, we again obtain (42). The approximate solution $\psi=\sin^2\theta(A_0+A_1\cos\theta)$ derived by this method is then

$$\psi=\frac{1}{4}W\sin^2\theta\left[\left(1+\frac{3}{4}ka\right)\left(3\rho a-\frac{a^3}{\rho}\right)-\frac{3ka}{2}\rho^2 -\frac{3}{4}ka\cos\theta\left(1-\frac{a}{\rho}\right)^2(2\rho^2+a\rho+a^2)\right], \tag{47}$$

which is in substance exactly equation (11), as derived by Oseen. In this form, however, the solution is still subject to the objection that it gives velocities which do not vanish at infinity.

Solution (47) is therefore unsatisfactory; but if we notice that the correction taking into account the quadratic terms is based on solution (28), and must therefore be added to it to give the required result, this difficulty disappears. The term $-\frac{3}{8}Wka\rho^2\sin\theta\cos\theta$ of (47) is already included in the exponential term of (28); combining with (28) the terms of (47) not already included, we obtain

$$\psi=-\frac{3\nu a}{2}\left(1+\frac{3}{4}ka\right)\sin^2\theta\,\frac{e^{-k\rho(1+\cos\theta)}-1}{1+\cos\theta} -\frac{Wa^3}{4}\left(1+\frac{3}{4}ka\right)\frac{\sin^2\theta}{\rho}+\frac{3Wka}{16}\sin^2\theta\cos\theta\left(3a\rho-a^2+\frac{a^3}{\rho}-\frac{a^4}{\rho^2}\right). \tag{48}$$

Hence, as usual,

$$R=\frac{1}{\rho^2\sin\theta}\frac{\partial\psi}{\partial\theta}=-\frac{3\nu a}{2\rho^2}\left(1+\frac{3}{4}ka\right)[e^{-k\rho(1+\cos\theta)}\{1+k\rho(1-\cos\theta)\}-1]$$

$$-\frac{Wa^3}{2}\left(1+\frac{3}{4}ka\right)\frac{\cos\theta}{\rho^3}+\frac{3Wka}{16\rho^2}(-1+3\cos^2\theta)\left(3a\rho-a^2+\frac{a^3}{\rho}-\frac{a^4}{\rho^2}\right), \quad (49)$$

$$\Theta=-\frac{1}{\rho\sin\theta}\frac{\partial\psi}{\partial\rho}=-\frac{3Wa}{4}\left(1+\frac{3}{4}ka\right)\sin\theta\, e^{-k\rho(1+\cos\theta)}$$

$$-\frac{Wa^3}{4}\left(1+\frac{3}{4}ka\right)\frac{\sin\theta}{\rho^3}-\frac{3Wka}{16\rho}\sin\theta\cos\theta\left[3a-\frac{a^3}{\rho^2}+\frac{2a^4}{\rho^3}\right]. \quad (50)$$

Using these velocities, we find with Oseen that Stokes' law should be

$$F=6\pi\mu Wa\left(1+\frac{3Wda}{8\mu}\right).$$

As is obvious, these velocities, (49) and (50), both vanish when $\rho$ is infinite; when $\rho=a$, they are, except for terms involving the squares or higher powers of $ka$, equal to $W\cos\theta$ and $-W\sin\theta$, respectively; and since, omitting squares and higher powers of $ka$ and $k\rho$, (48) is the same as (47), to that approximation they satisfy the complete differential equation (5). The solution, then, is superior to any previously given. The difficulty first raised by Whitehead, that a certain term gives velocities which do not vanish at infinity, is met by showing that this term is included in an exponential expression.

CORNELL UNIVERSITY, *May*, 1915.

# *Note on the Theory of Optical Images.*

By George Steić.

L. Mandelstam* obtains the result, that for a given aperture of a diaphragm only those linear structures will be similarly reproduced—through a lens-system—which satisfy the integral equation

$$f(\xi)=\lambda \int_{-a}^{+a} f(x) K(x-\xi)\, dx,$$

in the case of an illuminated structure; and the integral equation

$$f^2(\xi)=\lambda \int_{-a}^{+a} f^2(x) K^2(x-\xi)\, dx,$$

in the case of a self-luminous structure. Here $f(x)$ denotes the distribution of amplitudes in the linear object, or — briefly — the structure itself. $K(x-\xi)=K(s)$ is a function which depends only on the aperture of the diaphragm; when the aperture is circular, we have $K(s)=\frac{J_1(c\cdot s)}{s}$.† $\lambda$ is a positive constant. $2a$ is the length of the linear object; the problem is more general, when the length of the structure is taken to be infinite.

Mandelstam assumes for the solution of the above equations

$$f(x)=\cos(bx+\phi),$$

which is permissible on certain suppositions. For the reproduced images we have the relation

$$\Phi(\xi)=\frac{1}{\lambda_b} f(x_b),$$

where $\frac{1}{\lambda_b}$ satisfies, for illuminated and self-luminous structures of infinite length, the equations

$$\frac{1}{\lambda_b}=2\int_0^\infty K(s)\cdot \cos bs\cdot ds$$

and

$$\frac{1}{\lambda_b}=2\int_0^\infty K^2(s)\cdot \cos bs\cdot ds$$

respectively.

---

* Weber's *Festschrift* (1912), "Über eine Anwend. d. Integralgleich.," p. 237.

† $J_1(c\cdot s)$ = Bessel's function; $c=\frac{2\pi r}{l\cdot f}$ = const.; $r$ = radius of the aperture; $l$ = wave length of light; $f$ = focus distance of the lens.

Mandelstam obtains explicit results for $\frac{1}{\lambda_b}$ in the case of a square aperture of the diaphragm, both for illuminated and for self-luminous structures; he obtains also an explicit result for a circular aperture, but only for illuminated structures. In the case of a circular aperture in the diaphragm and self-luminous structures of infinite length, he points out, with Lord Rayleigh,* that the solution of the problem depends on the evaluation of the integral

$$\int_0^\infty \left[\frac{J_1(c\cdot s)}{s}\right]^2 \cdot \cos bs \cdot ds, \tag{I}$$

which figures in the expression of $\frac{1}{\lambda_b}$ for self-luminous structures and circular aperture of the diaphragm. We easily recognize that such functions as

$$\left[\frac{J_1(c\,s)}{s}\right]^2 \cdot \cos b\,s \text{ or } \left[\frac{J_1(c\,s)}{s}\right]^2 \cdot \sin b\,s, \qquad J_1(c\,s) = \text{Bessel's function,}$$

are finite and continuous for all values of $s$.

The purpose of this note is the evaluation of the former integral for the case $b \lesseqqgtr 2c$.

We will start by taking into consideration the following integral:

$$\Phi(\varepsilon) = c\int_0^\varepsilon \int_0^\infty \left[\frac{J_1(z)}{z}\right]^2 e^{i2z\varepsilon}\, dz\, d\varepsilon, \tag{II}$$

where $\varepsilon = \frac{b}{2c}$ and $i = \sqrt{-1}$, $J_1(z) =$ Bessel's function. For the present it is sufficient to suppose $0 \leq |\varepsilon| < M =$ finite constant. Considering fixed, for the moment, the upper limit of the outer integral, we can write

$$\Phi(\varepsilon) = c\int_0^\infty \left[\frac{J_1(z)}{z}\right]^2 dz \int_0^\varepsilon e^{i2z\varepsilon}\, d\varepsilon. \tag{II'}$$

It is easy to see that

$$\text{(I)} = \text{real part of } \left[\frac{d\Phi(\varepsilon)}{d\varepsilon}\right], \tag{III}$$

because from (II′) we get

$$\frac{d\Phi(\varepsilon)}{d\varepsilon} = c\int_0^\infty \left[\frac{J_1(z)}{z}\right]^2 e^{i2z\varepsilon}\, dz,$$

the real part of which, after putting $z = cs$ and remembering $\varepsilon = \frac{b}{2c}$, gives (I).

---

* *Sc. Pap.*, Vol. IV, p. 260.

We are going to transform the integral (II), and for this reason we take the well-known formula* existing for Bessel's functions,

$$J_n(as)\cdot J_n(bs)=\frac{s^n\cdot a^n\cdot b^n}{2^n\sqrt{\pi}\,\Pi(n-\frac{1}{2})}\int_0^\pi\frac{J_n(s\sqrt{a^2+b^2-2ab\cos\phi})}{(\sqrt{a^2+b^2-2ab\cos\phi})^n}\sin^{2n}\phi\cdot d\phi.$$

Let us put here $n=1$, $a=b=c=\frac{2\pi r}{lf}$ ; further remembering that $\sqrt{1-\cos\phi}$ $=\sqrt{2}\sin\frac{\phi}{2}$, then putting $\frac{\phi}{2}=\psi$, and $\sin\psi=y$, we have

$$\left[\frac{J_1(cs)}{s}\right]^2=\frac{2c}{\sqrt{\pi}\cdot\Pi(\frac{1}{2})}\cdot\int_0^1\frac{J_1(s\cdot 2cy)}{s}y\cdot\sqrt{1-y^2}\,dy.$$

Multiplying this equation by $e^{ib\epsilon}$ and integrating from 0 to $\infty$ with respect to $s$, we have

$$\int_0^\infty\left[\frac{J_1(cs)}{s}\right]^2 e^{ibs}\,ds=\frac{2c}{\sqrt{\pi}\,\Pi(\frac{1}{2})}\int_0^1 y\cdot\sqrt{1-y^2}\,dy\int_0^\infty\frac{J_1(s\cdot 2cy)}{s}\cdot e^{ibs}\,ds.$$

Putting in the integral on the left $cs=z$ and remembering that $\frac{b}{c}=2\epsilon$, then integrating on both sides from 0 to $\epsilon$, we get

$$\begin{aligned}\Phi(\epsilon)&=c\int_0^\epsilon\int_0^\infty\left[\frac{J_1(z)}{z}\right]^2 e^{i2\epsilon z}\,dz\,d\epsilon\\&=\frac{2c}{\sqrt{\pi}\,\Pi(\frac{1}{2})}\int_0^\epsilon d\epsilon\int_0^1 y\sqrt{1-y^2}\,dy\int_0^\infty\frac{J_1(2cys)}{s}e^{ibs}\,ds.\end{aligned}\qquad\text{(II'')}$$

Let us consider

$$I_1=\int_0^\infty\frac{J_1(2cys)}{s}e^{ibs}\,ds.$$

Putting here $2cy\cdot s=t$ and considering $y$ as constant during the integration, we get

$$I_1=\int_0^\infty\frac{J_1(t)}{t}e^{i\cdot\frac{\epsilon}{y}\cdot t}\,dt.$$

But†

$$\int_0^\infty\frac{J_1(t)}{t}\cdot e^{ixt}\cdot dt=e^{i\arcsin x}\text{ for }0\leq x\leq 1.$$

So we will have

$$I_1=e^{i\arcsin\left(\frac{\epsilon}{y}\right)},\qquad(1)$$

* Nielsen, "Zylinder Funktionen, p. 183.　　† Nielsen, *loc. cit.*, p. 197.

if $0 \leq \frac{\varepsilon}{y} \leq 1.$* In this inequality lies the definition of $\varepsilon$. Since $y$ varies from 0 to 1, it is evident that the domain of variation for $\varepsilon$ is

$$0 \leq \varepsilon = \frac{b}{2c} \leq 1. \dagger$$

So from (II″) we will have, after substituting (1),

$$\Phi(\varepsilon) = \frac{2c}{\sqrt{\pi}\,\Pi(\frac{1}{2})} \int_0^\varepsilon d\varepsilon \int_0^1 y\sqrt{1-y^2}\,dy \cdot e^{i \operatorname{arc\,sin}(\frac{\varepsilon}{y})},$$

or

$$= \frac{2c}{\sqrt{\pi}\,\Pi(\frac{1}{2})} \int_0^\varepsilon d\varepsilon \int_0^1 y\sqrt{1-y^2}\left[\cos\left(\operatorname{arc\,sin}\frac{\varepsilon}{y}\right) + i\sin\left(\operatorname{arc\,sin}\frac{\varepsilon}{y}\right)\right]dy. \quad \text{(II‴)}$$

Now we have ‡

$$\sin(\operatorname{arc\,sin} x) = (-1)^n \sin[(-1)^n |\operatorname{arc\,sin} x|]$$

and

$$\cos(\operatorname{arc\,sin} x) = (-1)^n \cos[(-1)^n |\operatorname{arc\,sin} x|].$$

The first of these two functions does not change its sign when $n$ is odd or even; the second changes. But we have

$$\sin(|\operatorname{arc\,sin} x|) = x$$

and

$$\cos(|\operatorname{arc\,sin} x|) = \sqrt{1-x^2}.$$

Therefore

$$\sin(\operatorname{arc\,sin} x) = x$$

and

$$\cos(\operatorname{arc\,sin} x) = \pm\sqrt{1-x^2}.$$

So in our case

$$\sin\left(\operatorname{arc\,sin}\frac{\varepsilon}{y}\right) = \frac{\varepsilon}{y}$$

and

$$\cos\left(\operatorname{arc\,sin}\frac{\varepsilon}{y}\right) = \pm\sqrt{1-\frac{\varepsilon^2}{y^2}}.$$

And (II‴) becomes

$$\Phi(\varepsilon) = \frac{2c}{\sqrt{\pi}\,\Pi(\frac{1}{2})} \int_0^\varepsilon d\varepsilon \int_0^1 y\sqrt{1-y^2} \cdot \left[\pm\sqrt{1-\frac{\varepsilon^2}{y^2}} + i\frac{\varepsilon}{y}\right] dy. \quad \text{(II}^{\text{IV}}\text{)}$$

---

* See Genocchi, "Diff.-Int. Rechnung," p. 170, for $\lim\limits_{\substack{\varepsilon=0\\y=0}} \left(\frac{\varepsilon}{y}\right) \geq 0.$

† Evidently, $M = 1$.

‡ Láska, "Formelsammlung," p. 2.

We will deal separately with the following two integrals:

$$\left.\begin{aligned} I_2 &= \int_0^1 y\sqrt{1-y^2}\sqrt{1-\frac{\varepsilon^2}{y^2}}\,dy, \\ I_3 &= \int_0^1 \sqrt{1-y^2}\,dy, \end{aligned}\right\} 0 \leq \varepsilon \leq 1.$$

After rearrangement we have

$$I_2 = i\varepsilon\int_0^1 \sqrt{1-y^2}\sqrt{1-\frac{y^2}{\varepsilon^2}}\,dy,$$

which is an elliptical integral. Further, we can write

$$I_2 = i\varepsilon\left[\int_0^1 \frac{dy}{\sqrt{W}} - \left(1+\frac{1}{\varepsilon^2}\right)\int_0^1 \frac{y^2\,dy}{\sqrt{W}} + \frac{1}{\varepsilon^2}\int_0^1 \frac{y^4\,dy}{\sqrt{W}}\right], \tag{2}$$

where $\sqrt{W} = \sqrt{(1-y^2)\left(1-\frac{y^2}{\varepsilon^2}\right)}$. Or briefly

$$I_2 = i\varepsilon\left[A - \left(1+\frac{1}{\varepsilon^2}\right)B + \frac{1}{\varepsilon^2}C\right],$$

the meaning of $A$, $B$ and $C$ being clear.

Now we can reduce the $C$ by the well-known reduction formula *

$$(n-1)\,c\int\frac{y^n\,dy}{\sqrt{a+2by^2+cy^4}} = [y^{n-3}\sqrt{W}] - (n-3)\,a\int\frac{y^{n-4}}{\sqrt{W}}\,dy$$
$$-2(n-2)\,b\int\frac{y^{n-2}\,dy}{\sqrt{W}}.$$

Putting here $n=4$, $a=1$, $b=-\frac{1+\frac{1}{\varepsilon^2}}{2}$, $c=\frac{1}{\varepsilon^2}$ and taking into consideration the limits of our integral, we get

$$3\frac{1}{\varepsilon^2}C = -\int_0^1 \frac{dy}{\sqrt{W}} + 2\left(1+\frac{1}{\varepsilon^2}\right)\int_0^1 \frac{y^2\,dy}{\sqrt{W}} = -A + 2\left(1+\frac{1}{\varepsilon^2}\right)B,$$

and so (2) becomes

$$I_2 = i\varepsilon\left[\frac{2}{3}A - \frac{1}{3}\left(1+\frac{1}{\varepsilon^2}\right)B\right]. \tag{2'}$$

But we can evaluate $A$ and $B$. In

$$A = \int_0^1 \frac{dy}{\sqrt{1-y^2}\sqrt{1-\frac{y^2}{\varepsilon^2}}}$$

* Legendre, "Traité des fonctions elliptiques," T. 1, p. 4.

put $y = \varepsilon x$ and we get

$$A = \varepsilon \int_0^{\frac{1}{\varepsilon}} \frac{dx}{\sqrt{1-\varepsilon^2 x^2}\,\sqrt{1-x^2}},$$

which is equal* to

$$A = \varepsilon\,(K + iK'),$$

where the $K$ and $K'$ are complete elliptic integrals of the first species, with the moduli $\varepsilon$ and $\sqrt{1-\varepsilon^2}$ respectively. Further,

$$B = \int_0^1 \frac{y^2\,dy}{\sqrt{1-y^2}\sqrt{1-\dfrac{y^2}{\varepsilon^2}}}$$

becomes by the transformation $y = \varepsilon x$

$$B = \varepsilon \int_0^{\frac{1}{\varepsilon}} \frac{\varepsilon^2 x^2\,dx}{\sqrt{1-\varepsilon^2 x^2}\,\sqrt{1-x^2}} = \varepsilon\left\{\int_0^1 + \int_1^{\frac{1}{\varepsilon}}\right\}.$$

But we can write†

$$B = \varepsilon\,[iE' + K - E],$$

where $E$ and $E'$ are complete elliptic integrals of the second species, with the moduli $\varepsilon$ and $\sqrt{1-\varepsilon^2}$ respectively.

So (2) can be written

$$I_2 = i\varepsilon\left[\frac{2}{3}\varepsilon\,(K + iK') - \frac{1}{3}\left(1 + \frac{1}{\varepsilon^2}\right)\cdot\varepsilon\,(iE' + K - E)\right].$$

We get immediately

$$I_3 = \int_0^1 \sqrt{1-y^2}\,dy = \frac{\pi}{4}.$$

Substituting these values of $I_2$ and $I_3$ in ($\text{II}^{\text{IV}}$), we obtain

$$\Phi(\varepsilon) = \frac{2c}{\sqrt{\pi}\,\Pi(\frac{1}{2})}\int_0^{\varepsilon}\left[\pm\, i\varepsilon\left\{\frac{2}{3}\varepsilon\,(K + iK') - \frac{1}{3}\left(1 + \frac{1}{\varepsilon^2}\right)\cdot\varepsilon\,(iE' + K - E)\right\} + i\,\frac{\pi\varepsilon}{4}\right]d\varepsilon.$$

Separating the real and imaginary parts,

$$\Phi(\varepsilon) = \frac{2c}{\sqrt{\pi}\,\Pi(\frac{1}{2})}\cdot\int_0^{\varepsilon} \pm\left[\frac{1}{3}(1+\varepsilon^2)\,E' - \frac{2\varepsilon^2}{3}K'\right]\cdot d\varepsilon + i\cdot\frac{2c}{\sqrt{\pi}\,\Pi(\frac{1}{2})}\int_0^{\varepsilon}\left[\frac{\pi\varepsilon}{4} \mp \left\{\left(\frac{1-\varepsilon^2}{3}\right)K - \left(\frac{1+\varepsilon^2}{3}\right)E\right\}\right]d\varepsilon.$$

* Durége, "Theorie d. ellipt. Funktionen," p. 91. † Glaisher, *Quart. Journ.*, Vol. XX, p. 315.

Now remembering (II) and (III) and taking into consideration $\Pi(\frac{1}{2}) = \frac{\sqrt{\pi}}{2}$, we get

$$\left.\begin{aligned} I &= \int_0^\infty \left[\frac{J_1(cs)}{s}\right]^2 \cos bs\, ds = \frac{\pm 4c}{3\pi}\left[(1+\epsilon^2)E' - 2\epsilon^2 K'\right] \\ \text{and}\quad I' &= \int_0^\infty \left[\frac{J_1(cs)}{s}\right]^2 \sin bs\, ds = c\epsilon \pm \frac{4c}{3\pi}\left[(1+\epsilon^2)E - (1-\epsilon^2)K\right], \end{aligned}\right\} \quad (\alpha)$$

where $0 \leq \left(\epsilon = \frac{b}{2c}\right) \leq 1$.

In ($\alpha$) we cannot retain the double sign, and we are going to decide which sign is to be used. As for the integration of $\phi(\epsilon)$, we observe that we can use the following formulæ:

$$\left.\begin{aligned} &(n+2)\int \epsilon^n E\, d\epsilon = \epsilon^{n+1}E + \int \epsilon^n K\, d\epsilon \\ \text{and}\quad &(n+2)^2 \int \epsilon^{n+2} K\, d\epsilon = \epsilon^{n+1}E - (n+2)\epsilon^{n+1}(1-\epsilon^2)K + (n+1)^2 \int \epsilon^n K\, d\epsilon. \end{aligned}\right\} \quad (3)$$

If in ($\alpha$) we put $\epsilon = 0$, or, what is the same, $b = 0$, we get

$$\lim_{b=0} \int_0^\infty \left[\frac{J_1(cs)}{s}\right]^2 \cos bs\, ds = \int_0^\infty \left[\frac{J_1(cs)}{s}\right]^2 ds = \frac{\pm 4c}{3\pi},$$

since

$$\Pi(\tfrac{1}{2}) = \Gamma(\tfrac{3}{2}) = \frac{\sqrt{\pi}}{2}, \quad \lim_{\epsilon=0} E' = 1, \quad \lim_{\epsilon=0} \epsilon^2 K' = 0.$$

If we put $\nu = \mu = 1$ and $k = cs$, in the known formula*

$$\int_0^\infty \frac{J_\nu(k)\cdot J_\mu(k)}{k^{\nu+\mu}}\, dk = \frac{\Gamma(\frac{1}{2})\cdot\Gamma(\nu+\mu)}{2^{\nu+\mu}\cdot\Gamma(\nu+\frac{1}{2})\,\Gamma(\mu+\frac{1}{2})\cdot\Gamma(\nu+\mu+\frac{1}{2})},$$

we have

$$\int_0^\infty \left[\frac{J_1(cs)}{s}\right]^2 ds = \frac{+4c}{3\pi}.$$

The same result we obtain from† $Z = \int_v^\infty \frac{J_1^2(z)\, dz}{z\sqrt{z^2 - v^2}}$ by taking the limit for $v = 0$.

We see that $\lim_{\epsilon=0} I > 0$; on the other hand, we can show by simple expansion that for all values of $\epsilon$ the factor $[(1+\epsilon^2)E' - 2\epsilon^2 K']$ in the expression of $I$ is always positive. Therefore, $I$ is positive for all values of $\epsilon$, and we must drop the negative sign from its expression. We will decide later on the choice of sign in $I'$.

---

* Nielsen, *loc. cit.*, p. 104. † Gray-Matthews, "Bessel's Functions," p. 201.

It remains to investigate the case for $\varepsilon = \frac{b}{2c} \geq 1$. To distinguish this case from the previous one, we will let $\varepsilon = \varepsilon'$, and we can write as before

$$\Phi(\varepsilon') = c\int_1^{\varepsilon'}\int_0^{\infty}\left[\frac{J_1(z)}{z}\right]^2 e^{i2\varepsilon' z}\,dz\,d\varepsilon', \tag{IV}$$

where it is sufficient to suppose $|\varepsilon'| > 1$. Again,

$$\text{(I)} = \text{real part of}\left[\frac{d\Phi(\varepsilon')}{d\varepsilon'}\right]. \tag{V}$$

Omitting superfluous calculations, we can write immediately

$$\Phi(\varepsilon') = \frac{4c}{\pi}\int_1^{\varepsilon'} d\varepsilon' \int_0^1 y\sqrt{1-y^2}\,dy \int_0^{\infty} \frac{J_1(2cy\cdot x)}{x}\,e^{ibx}\cdot dx. \tag{IV'}$$

Now

$$I_1 = \int_0^{\infty} \frac{J_1(2cy\cdot x)}{x}\,e^{ibx}\,dx = \int_0^{\infty}\frac{J_1(t)}{t}\,e^{\frac{\varepsilon'}{y}t}\,dt,$$

if $2cy\cdot x = t$ and $\frac{b}{2c} = \varepsilon'$. But*

$$\int_0^{\infty}\frac{J_1(t)}{t}\,e^{ixt}dt = i(x-\sqrt{x^2-1})$$

if $x \geq 1$. So we will have

$$I_1 = i\left[\frac{\varepsilon'}{y} - \sqrt{\left(\frac{\varepsilon'}{y}\right)^2 - 1}\right]$$

if $\frac{\varepsilon'}{y} \geq 1$. Since $y$ varies between 0 and 1, in order that $\varepsilon'$ should satisfy this inequality for all values of $y$, its domain of variation generally must be

$$\varepsilon' \geq 1.\dagger$$

So (IV′) can be written

$$\Phi(\varepsilon') = i\frac{4c}{\pi}\int_1^{\varepsilon'} d\varepsilon' \int_0^1 y\sqrt{1-y^2}\left(\frac{\varepsilon'}{y} - \sqrt{\frac{\varepsilon'^2}{y^2} - 1}\right)dy. \tag{IV''}$$

We have to evaluate the integrals

$$I_2 = \int_0^1 \sqrt{1-y^2}\,dy = \frac{\pi}{4}$$

and

$$I_3 = \int_0^1 y\cdot\sqrt{1-y^2}\cdot\sqrt{\frac{\varepsilon'^2}{y^2}-1}\cdot dy = \varepsilon'\int_0^1\sqrt{1-y^2}\cdot\sqrt{1-\frac{y^2}{\varepsilon'^2}}\cdot dy.$$

---

* Nielsen, *loc. cit.*, p. 197.

† $\frac{\varepsilon'}{y} \geq 1$ could be satisfied, for $|y| < 1$, by $|\varepsilon'| < 1$; since our inequality must be satisfied for all values of $y$, even for $y = 1$, therefore we must take $|\varepsilon'| \geq 1$.

Introducing $\frac{1}{\varepsilon'^2} = \varepsilon''^2 \leq 1$ and applying the previous transformations and reductions, we get finally

$$I_3 = \varepsilon' \cdot \left[\frac{\varepsilon''^2 - 1}{3} K - \frac{\varepsilon''^2 + 1}{3} E\right],$$

where the modulus of $K$ and $E$ is $\frac{1}{\varepsilon'} = \varepsilon'' \leq 1$.

Substituting $I_2$ and $I_3$ in (IV″), we get

$$\phi(\varepsilon') = i\frac{4c}{\pi}\int_1^{\varepsilon'} \varepsilon' \left[\frac{\pi}{4} + \frac{\varepsilon''^2 - 1}{3} K - \frac{\varepsilon''^2 + 1}{3} E\right] d\varepsilon'.$$

We are able to evaluate $\phi(\varepsilon')$ with the aid of formulæ (3), but we are not interested in the explicit value of $\phi(\varepsilon')$; we are interested in $\frac{d\phi(\varepsilon')}{d\varepsilon'}$, in the equality

$$\frac{d\phi(\varepsilon')}{d\varepsilon'} = i\frac{4c}{\pi}\left[\varepsilon'\frac{\pi}{4} + \varepsilon'\frac{\varepsilon''^2 - 1}{3} K - \varepsilon'\frac{\varepsilon''^2 + 1}{3} E\right].$$

The real part of this expression is equal to zero; therefore, by (V),

$$\left.\begin{aligned} Y &= \int_0^\infty \left[\frac{J_1(cs)}{s}\right]^2 \cos bsds = 0 \\ \text{and} \quad Y' &= \int_0^\infty \left[\frac{J_1(cs)}{s}\right]^2 \sin bsds = c\varepsilon' + \frac{4c\varepsilon'}{3\pi}\left[(\varepsilon''^2 - 1)K - (\varepsilon''^2 + 1)E\right], \end{aligned}\right\} \qquad (\beta)$$

where $\left(\varepsilon' = \frac{b}{2c}\right) \geq 1$ and the modulus in $K$ and $E$ is $\frac{1}{\varepsilon'}$.

If, in $Y'$, we put $\varepsilon' = 1$, we get

$$\lim_{\varepsilon'=1} Y' = c - \frac{8c}{3\pi}.$$

On the other hand we have for $I'$ in ($\alpha$) for $\varepsilon = 1$

$$\lim_{\varepsilon=1} I' = c \pm \frac{8c}{3\pi}.$$

Since the integrand in $I'$ is a continuous function, we must have

$$\lim_{\varepsilon'=1} Y' = \lim_{\varepsilon=1} I';$$

further, since, in $I'$, $(1 + \varepsilon^2)E - (1 - \varepsilon^2)K > 0$ for all values of $\varepsilon$, we conclude that we must drop the positive sign in the expression of $I'$.

So we can finally write the formulæ ($\alpha$),

$$\left.\begin{aligned} I &= \int_0^\infty \left[\frac{J_1(cs)}{s}\right]^2 \cos bs\, ds = \frac{4c}{3\pi}\,[(1+\varepsilon^2)E' - 2\varepsilon^2 K'], \\ I' &= \int_0^\infty \left[\frac{J_1(cs)}{s}\right]^2 \sin bs\, ds = c\varepsilon - \frac{4c}{3\pi}\,[(1+\varepsilon^2)E + (1-\varepsilon^2)K], \end{aligned}\right\} \quad (\alpha')$$

where $0 \leq \left(\varepsilon = \frac{b}{2c}\right) \leq 1$.

It is interesting to investigate whether some of the functions $I$, $I'$, $Y$, $Y'$ show maxima or minima. We take $I'$ and write, after putting $c = 1$, the condition for maxima and minima

$$\frac{dI'}{d\varepsilon} = 1 - \frac{4}{3\pi}\left[(1+\varepsilon^2)\frac{dE}{d\varepsilon} + 2\varepsilon E - (1-\varepsilon^2)\frac{dK}{d\varepsilon} + 2\varepsilon K\right] = 0.$$

Or, since*

$$\frac{dK}{d\varepsilon} = -\frac{1}{\varepsilon}\left(K - \frac{1}{1-\varepsilon^2}E\right),$$

and

$$\frac{dE}{d\varepsilon} = -\frac{1}{\varepsilon}(K - E),$$

we have

$$\frac{dI'}{d\varepsilon} = 1 - \frac{4}{\pi}\varepsilon E = 0.$$

Hence

$$\varepsilon E = \frac{\pi}{4} = 0.7854 = \frac{\pi}{2}\varepsilon\left\{1 - \left(\frac{1}{2}\right)^2\varepsilon^2 + 3\left(\frac{1}{2.4}\right)^2\varepsilon^4 - 5\left(\frac{1.3}{2.4.6}\right)^2\varepsilon^6 + \ldots\right\}.$$

By trial we find that $\varepsilon = 0.543$ approximately satisfies this equation. That this is a maximum point, we see immediately from

$$\frac{d^2I'}{d\varepsilon^2} = -\frac{4}{\pi}\,[2E(0.543) - K(0.543)].$$

From tables† we pick out the values of $E(0.543)$ and $K(0.543)$ and get

$$\frac{d^2I'}{d\varepsilon^2} = -\frac{4}{\pi}\,[2\times 1.445 - 1.712] < 0.$$

Similarly we find that neither $I$ nor $Y'$ has any maxima or minima.

What is the value of $\lim_{\varepsilon'=\infty} Y'$? We put $c = 1$ and $\frac{1}{\varepsilon} = \eta$ in $Y'$, and get

$$\lim_{\eta=0} Y' = \lim_{\eta=0} \frac{1}{\eta^3}\left[\eta^2 - \frac{4}{3\pi}\{(1+\eta^2)E(\eta) - (1-\eta^2)K(\eta)\}\right] = \lim_{\eta=0}\frac{N}{D}.$$

* Láska, Formelsammlung, p. 329.

† Jahnke, "Funkt. Tabl.," p. 68.

If we substitute $\eta = 0$, we get the indeterminate form $\frac{0}{0}$. By repeated application of the rules for the determination of such forms, we get finally

$$\lim_{\eta=0} Y' = \lim_{\varepsilon'=\infty} Y' = 0,$$

which shows that the curve representing the function $Y'$ approaches asymptotically the $\varepsilon'$-axis.

The following tables contain numerical values for different values of $\varepsilon$ and $\varepsilon'$:

| $\varepsilon = \sin\theta$ | $I(\varepsilon)$ | $I'(\varepsilon)$ | $\frac{1}{\varepsilon'} = \sin\theta$ | $Y'(\varepsilon')$ |
|---|---|---|---|---|
| $\theta = 0°$ | $\frac{4c}{3\pi} \times 1$ | 0 | $\theta = 0°$ | 0 |
| 5° | " $\times 0.960$ | $0.079c$ | 5° | $0.011c$ |
| 10° | " $\times 0.882$ | $0.140c$ | 10° | $0.012c$ |
| 15° | " $\times 0.776$ | $0.192c$ | 15° | $0.023c$ |
| 20° | " $\times 0.667$ | $0.228c$ | 20° | $0.038c$ |
| 25° | " $\times 0.545$ | $0.249c$ | 25° | $0.047c$ |
| 30° | " $\times 0.435$ | $0.258c$ | 30° | $0.060c$ |
| 40° | " $\times 0.247$ | $0.252c$ | 40° | $0.089c$ |
| 50° | " $\times 0.112$ | $0.225c$ | 50° | $0.106c$ |
| 60° | " $\times 0.040$ | $0.194c$ | 60° | $0.115c$ |
| 70° | " $\times 0.007$ | $0.169c$ | 70° | $0.130c$ |
| 80° | " $\times 0.001$ | $0.154c$ | 80° | $0.146c$ |
| 90° | 0 | $0.150c$ | 90° | $0.150c$ |

Graphically:

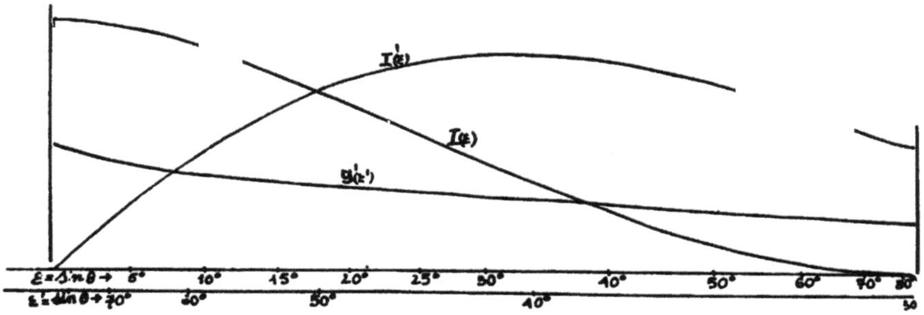


In the figure we take, for unity of the ordinates, $\frac{4c}{3\pi} = 60$ mm. in the case of $I(\varepsilon)$; 200 mm. in the case of $I'(\varepsilon)$ and $Y'(\varepsilon')$. The line of abscissas containing $\varepsilon = \sin\theta$ refers to $I(\varepsilon)$ and $I'(\varepsilon)$; the one containing $\varepsilon' = \sin\theta$ refers to $Y'(\varepsilon')$.

We now come to the end of our investigation. In our problem we are concerned only with the integral

$$\frac{1}{\lambda_b}=2\int_0^\infty\left[\frac{J_1(cs)}{s}\right]^2\cos bsds$$

for

$$0\leq\left(\varepsilon=\frac{b}{2c}\right)\leq 1,$$

or explicitly

$$\frac{1}{\lambda_b}=\frac{8c}{3\pi}\left[(1+\varepsilon^2)E'-2\varepsilon^2K'\right].$$

To every value of $b$ will correspond a value of $\frac{1}{\lambda_b}$. We get the greatest value for $\frac{1}{\lambda_b}$ when $b=0$: $\frac{1}{\lambda_0}=\frac{8c}{3\pi}$. The smallest value of $\frac{1}{\lambda_b}$ is for $b=2c$: $\frac{1}{\lambda_{2c}}=0$. In this case we have no image. For details see Mandelstam's work.

# AMERICAN
# Journal of Mathematics

EDITED BY

**FRANK MORLEY**

WITH THE COÖPERATION OF

A. COHEN, CHARLOTTE A. SCOTT

AND OTHER MATHEMATICIANS

PUBLISHED UNDER THE AUSPICES OF THE JOHNS HOPKINS UNIVERSITY

*Πραγμάτων ἔλεγχος οὐ βλεπομένων*

VOLUME XXXVIII, NUMBER 2

BALTIMORE: THE JOHNS HOPKINS PRESS

LEMCKE & BUECHNER, *New York.* G. E. STECHERT & CO., *New York.* E. STEIGER & CO., *New York.* ARTHUR F. BIRD, *London.* WILLIAM WESLEY & SON, *London.* A. HERMANN, *Paris.* MAYER & MÜLLER, *Berlin.*

APRIL, 1916

Entered as Second-Class Matter at the Baltimore, Maryland, Postoffice.

# *On the Classification and Invariantive Characterization of Nilpotent Algebras.**

BY O. C. HAZLETT.

## CHAPTER I.

### INTRODUCTION.

1. *Relation of Problem to Classification and Invariantive Characterization of General Linear Algebras.* Linear associative algebras in a small number of units, with coordinates ranging over the field $C$ of ordinary complex numbers, have been completely tabulated; that is, their multiplication tables have been reduced to very simple forms.† But if we had before us a linear associative algebra, the chances are that its multiplication table would not be in any of the tabulated forms, nor even in such a form that we could readily ascertain to which standard form it was equivalent. Accordingly the question naturally arises: "Can we find invariantive criteria which will tell us when two algebras are equivalent? or, as we say, which will completely characterize the algebras?"

In a previous paper ‡ we considered the problem of finding invariants which would completely characterize linear associative algebras in two or three units with a modulus, over the field $C$. The terms "invariants" and "characterize," be it understood, are used here in the sense defined by Professor Dickson. § For these special cases, the algebras ‖ can be completely characterized by invariants obtained by the application of the following theorem: "In a general $n$-ary linear algebra over any field $F$, both characteristic determinants are absolute covariants of the algebra; their coefficients are absolute covariants; and the invariants and covariants of the characteristic determinants

* This paper was written when the author was holder of the Fellowship of the Boston Branch of the Association of Collegiate Alumnæ.

† See especially B. Peirce, AMERICAN JOURNAL OF MATHEMATICS, Vol. IV (1881), pp. 97–192; Study, *Göttinger Nachrichten*, 1889, pp. 237–268; G. Scheffers, *Mathematische Annalen*, Vol. XXXIX (1891), pp. 293–390; Hawkes, *Mathematische Annalen*, Vol. LVIII (1904), pp. 361–379; AMERICAN JOURNAL OF MATHEMATICS, Vol. XXVI (1904), pp. 223–242.

‡ Hazlett, *Annals of Mathematics*, Vol. XVI, p. 1.

§ AMERICAN JOURNAL OF MATHEMATICS, Vol. XXXI, p. 337. See § 6 of this paper.

‖ For brevity, we will usually refer to linear algebras simply as algebras; but, unless specified, we do not assume the associative law nor the commutative law.

and of the coefficients of the powers of $\omega$ in these determinants are respectively invariants and covariants of the linear algebra."

If now we proceed to the characterization of the quaternary associative algebras with a modulus, we find that the invariants given us by this theorem only partially characterize these algebras. In other words, the general problem can not be solved by use of this theorem alone.

But fortunately Cartan* and Wedderburn† have some general theorems which bear on the problem.‡ These tell us that we can characterize the general algebras if we can characterize three special kinds of algebras; namely, simple matric algebras, division algebras, and nilpotent algebras. Now in a given number of units there is only one simple matric algebra. Over the field $C$, there is no division algebra other than the algebra of ordinary complex numbers; and in general, the Galois Fields are the only algebras of a finite number of elements such that every number except zero has an inverse,§ whereas, over the field $C$, there is an infinite number of classes of nilpotent algebras.

In this paper, we shall see how nilpotent algebras can be characterized by the aid of certain homogeneous polynomials whose coefficients are constants of multiplication. For algebras in a small number of units, these polynomials are sufficient if we assume the commutative and associative laws; but to characterize the general nilpotent algebra, we need further invariants.

2. *Definitions for Linear Algebras.* If we have given a set of *units* $e_1, \ldots, e_n$ linearly independent with respect to a field $F$, and such that

$$e_i e_j = \sum_{k=1}^{n} \gamma_{ijk} e_k \quad (i, j = 1, \ldots, n), \tag{1}$$

where the *constants of multiplication* $\gamma_{ijk}$ range over $F$; and if the complex numbers of the form $X = \sum_{i=1}^{n} x_i e_i$ combine under addition and subtraction as follows:

$$X \pm Y = \sum_{i}^{1,n} (x_i \pm y_i) e_i ,$$

and if they combine under multiplication according to the distributive law, then the set of complex numbers is said to form a *linear algebra*‖ over the field $F$ with multiplication table (1). The *characteristic right-hand and left-hand determinants* are respectively defined to be

---

* *Annales de Toulouse*, Vol. XII (1898) B.

† *Proceedings of the London Mathematical Society*, Series 2, Vol. VI (1908), pp. 77–118.

‡ See end of § 2.

§ Wedderburn, *Transactions of the American Mathematical Society*, Vol. VI (1905), p. 349.

‖ For other definitions, see Dickson, *Transactions of the American Mathematical Society*, Vol. IV (1903), pp. 21–26.

$$\left.\begin{aligned}\delta(\omega) &\equiv \left|\sum_i \gamma_{ijk} x_i - d_{jk}\omega\right|, \\ \delta'(\omega) &\equiv \left|\sum_i \gamma_{jik} x_i - d_{jk}\omega\right|.\end{aligned}\right\} \tag{2}$$

There may exist in the algebra a number $\varepsilon$ called a *principal unit* or *modulus*, such that, for every number $x$ in the algebra,

$$x\varepsilon = x = \varepsilon x.$$

If such a number exists, it is unique.

*Division*, as a rule, is not unique; that is, if we know the product of two numbers is zero, it does not necessarily follow that one of the numbers is zero. An algebra in which right- and left-hand division, except by zero, is always possible and unique is called a *division algebra* by Dickson,* a *primitive algebra* by Wedderburn.†

Another important family of algebras are the *simple matric algebras*;‡ that is, those which have the multiplication table of the form

$$e_{pq}e_{st} = d_{qs}e_{pt} \quad (p, q, s, t = 1, \ldots, n),$$

where the units are $e_{ij}$ $(i, j = 1, \ldots, n)$.

The *complex*§ $A = (x_1, x_2, \ldots, x_a)$ is defined as the set of all quantities linearly dependent on $x_1, x_2, \ldots, x_a$, and the number of linearly independent elements is called the *order* of the complex. If $A$ and $B$ are two complexes, the complex formed by all quantities linearly dependent on the elements of $A$ and $B$ is called the *sum* of $A$ and $B$, and is denoted by $A+B$. If a complex $C$ is contained in the complex $A$, we write $C \leq A$ or $A \geq C$; similarly, if $x$ is an element of the complex $A$, we write $x < A$. The elements common to two complexes also form a complex. The greatest complex common to two complexes $A$ and $B$ is denoted by $A \cap B$.

If $A$ and $B$ are any two complexes, and if $x$ and $y$ are any elements of $A$ and $B$ respectively, the complex of all elements linearly dependent on those of the form $xy$ is called the *product* of $A$ and $B$, and is written $AB$. If multiplication of all the elements involved is associative, then multiplication of complexes is likewise associative. If the associative law of multiplication is assumed, then the integral powers of a complex $A$ are defined by means of the recursion formula $A \cdot A^n = A^{n+1} = A^n \cdot A$. In general, however, the integral powers of a complex $A$ are defined by the formula

$$A^{m+1} = \sum_{i=1}^{m} A^i A^{m+1-i}.$$

* "Linear Algebras," *Cambridge Tracts in Mathematics and Mathematical Physics*, 1914, p. 66.

† *Loc. cit.*, p. 91.

‡ Wedderburn, *loc. cit.*

§ The definitions in the next two paragraphs are those of Wedderburn, *loc. cit.*, pp. 79–80. For this notion of "complex" compare Kronecker, *Berliner Sitzungsberichte*, 1888, p. 597.

The necessary and sufficient condition that the complex $A$ be an algebra is that $A^2 < A$.

To illustrate these definitions, consider the algebra of four units

$$(e_0, e_1, e_2, e_3),$$

quaternions, over the field $F$, whose multiplication table is

$$e_1 e_2 = -e_2 e_1 = e_3, \quad e_2 e_3 = -e_3 e_2 = e_1, \quad e_3 e_1 = -e_3 e_1 = e_2,$$
$$e_i^2 = -e_0 \ (i=1,2,3), \quad e_j e_0 = e_j = e_0 e_j \ (j=0,1,2,3).$$

If $x_0, x_1, \ldots, y_0, y_1, \ldots$ be marks of $F$, then the totality of all numbers of the form $x_1 e_1 + x_2 e_2 + x_3 e_3$ form a complex $A$ of order 3, but they do not form an algebra. The totality of all numbers of the form $y_0 e_0 + y_1 e_1$ form a complex $B$ of order 2, and they also form an algebra.

$$A+B = (e_0, e_1, e_2, e_3) = AB = A^2 = A^3; \ B^2 = B; \ A \cap B = (e_1).$$

A subcomplex $B$ of a complex $A$ which is such that $AB \leq B$ and $BA \leq B$ is called an *invariant** subcomplex of $A$. $B$ is necessarily an algebra. An algebra which has no invariant subcomplex is said to be *simple.*†

If an algebra $A$ is expressible as the sum of two algebras $A_1 \neq 0$ and $A_2 \neq 0$ which are such that $A_1 A_2 = 0 = A_2 A_1$, then $A$ is said to be *reducible* and to be the *direct sum*‡ of $A_1$ and $A_2$. $A_1$ and $A_2$ can be so chosen that $A_1 \cap A_2 = 0$. Similarly for a complex $A$.

If $C$ and $D$ are any two algebras, such that every element of one is commutative with every element of the other, and if further the order of $A = CD$ is the product of the orders of $C$ and $D$, then the algebra $A$ is said to be the *direct product*§ of $C$ and $D$.

If an algebra $A$ have a modulus, then $A^2 = A$; and in general, since we are dealing with algebras having a finite basis, we must have $A^{a+1} = A^a$ for some integer.

In particular, $A$ may be such that $A^a = 0$; if so, $A$ is said to be ***nilpotent.*** The smallest such integer is called the *index* of the algebra.‖ For a nilpotent algebra, $\delta(\omega) \equiv \omega^n \equiv \delta'(\omega)$. For associative algebras this definition of nilpotent algebras is equivalent to that given by Cartan and others;¶ but for non-associative algebras the two definitions are not equivalent, as can be seen by the

---

* Cartan, *loc. cit.*, p. 57; Molien, *Mathematische Annalen*, Vol. XLI (1893); Frobenius, *Berliner Sitzungsberichte*, 1903, p. 523.

† Cartan, *loc. cit.*, p. 57.

‡ "Sum" was first used in this sense by Scheffers, *loc. cit.*, p. 323.

§ Wedderburn, *loc. cit.*, p. 99.

‖ Wedderburn, *loc. cit.*, p. 87.

¶ Wedderburn, *loc. cit.*, pp. 88–91.

following example, which is nilpotent in Cartan's sense, but not in Wedderburn's, $e_1 e_2 = -e_2 e_1 = e_3$, $e_2 e_3 = -e_3 e_2 = e_1$, $e_3 e_1 = -e_1 e_3 = e_2$.

An algebra which has no nilpotent invariant subalgebra is called *semi-simple.** Such an algebra always has a modulus. Cartan† has shown that an associative algebra $A$, with a modulus, over the field $C$ of ordinary complex numbers is the sum of a semi-simple algebra $S$ and a nilpotent algebra $N$, such that every unit of $N$ is commutative with every unit of $S$.

Wedderburn‡ has proved a similar theorem for associative algebras for a general field $F$; namely: "Any associative algebra can be expressed as the sum of a nilpotent algebra and a semi-simple algebra. The latter algebra is not unique, but any two determinations of it are simply isomorphic. Furthermore, a semi-simple associative algebra can be expressed uniquely as the direct sum of a number of simple algebras; and a simple associative algebra can be expressed as the direct product of a primitive algebra and a simple matric algebra."

## CHAPTER II.

### General Theory.

#### *Preliminary Theorems,* §§ 3–8.

3. *Canonical Form for Nilpotent Algebras.* Wedderburn has the following theorem which gives a very simple form to which any nilpotent algebra can be transformed.

Theorem. *If $E$ be a nilpotent algebra (not necessarily associative) with index $a$, then if $B \equiv E \pmod{E^2}$,§ $E = \sum_i^{1, a-1} B^i$.*

For since $E = B + E^2$, we have $E^2 \leq B^2 + E^3$ and thus $E = B + B^2 + E^3$. By induction, we have in general $E^h \leq B^h + E^{h+1}$ and thus $E = \sum_{i=1}^{a-1} B^i + E^a$. Moreover, for any positive integer $p$, $E^p = \sum_{k=p}^{a-1} B^k$.

If the multiplication table of a nilpotent algebra be in the form given by this theorem, the algebra will be said to be in its *canonical form.* Furthermore, if an algebra be expressible in this form, then it is nilpotent in the sense defined by Wedderburn; and hence, if multiplication be associative, it is also nilpotent in the sense defined by Cartan and others.

---

* This is the definition given by Wedderburn, *loc. cit.*

† *Loc. cit.*, p. 50.

‡ Wedderburn, *loc. cit.*, pp. 86, 94.

§ That is, $B$ is derived from $E$ by considering as equal those elements of $E$ which differ only by an element of $E^2$.

4. *The Group Which Leaves the Canonical Form Unaltered.* Consider a nilpotent algebra $E$ over the field $F$ in canonical form, $E=\sum_{i=1}^{\alpha-1} B^i$, where $B\equiv E \pmod{E^2}$ and where $\alpha$ is the index of $E$. If we have a second algebra $E_1$ over the field $F$ and in canonical form, $E_1=\sum_{i=1}^{\alpha-1} B_1^i$, where $B_1\equiv E_1 \pmod{E_1^2}$, then if $E$ is equivalent to $E_1$, $B_1\leq B+E^2$. Therefore $B_1^2\leq B^2+E^3$; and in general, by induction, we have $B_1^m\leq B^m+E^{m+1}$. Hence, if we denote by $e_i^{(p)}$ a unit of $E$ which is in $B^p$, and similarly for $E_1$, then the transformation which carries $E$ into $E_1$ is of the form

$$e_i'^{(p)}=\sum_{q=p}^{\alpha-1}\sum_j a_{ij}^{(p,q)}e_j^{(q)}, \tag{3}$$

where the $a$'s are in $F$. We shall use $G$ to denote the group of all such transformations.

5. *Invariance of Order of $B^a$ (mod $E^{a+1}$). Modification of Canonical Form.* The equation $E=\sum_{i=1}^{\alpha-1} B^i$ means only that every number of $E$ can be expressed as a linear homogeneous function of elements in $B,\ldots,B^{\alpha-1}$; two of these complexes may overlap, as in the quaternary algebra $e_1e_1=e_2$, $e_1e_2=e_3$, $e_2e_1=e_1e_3=e_4$, where the products not written are zero. Moreover, for $1<i<\alpha$, the number of linearly independent elements in $B^i$ depends on the particular choice of $B$. But the number of linearly independent elements in $B$ is an invariant; and in general, for each $a$, the number of linearly independent elements in $B^a$ reduced modulo $E^{a+1}$ is an invariant.

To show this, consider a fixed $a$, and subject the units of the algebra to a transformation $T$ of the group $G$. Now $T=T_1T_2$, where $T_1$ transforms the units of each power of $B$ among themselves alone, and where $T_2$ at most adds to each unit of $B^i$ some element of $E^{i+1}$, for every $i$. Clearly $T_1$ does not change the number of linearly independent elements in $B^a \pmod{E^{a+1}}$. By $T_2$, let $E$ be transformed into $\bar{E}$ and let the complex $B$ be transformed into $\bar{B}$. Then $B^2$ does not necessarily transform into $\bar{B}^2$, and similarly for the higher powers of $B$. In fact if we consider the multiplication table of $E$ in the square array indicated schematically thus:

| | $B$ | $B^2$ | $B^3$ | $\ldots$ |
|---|---|---|---|---|
| $B$ | $B^2$ | $B^3$ | $B^4$ | $\ldots$ |
| $B^2$ | $B^3$ | $B^4$ | $\ldots$ | $\ldots$ |
| $\ldots$ | $\ldots$ | $\ldots$ | $\ldots$ | $\ldots$ |

then $T_2$ adds to each element in the $j$-th horizontal strip and the $k$-th vertical strip, linear combinations of elements in the $j$-th horizontal strip and below the $j$-th, and in the vertical strips in the $k$-th and to the right of the $k$-th. Hence the statement at the beginning of this section.

Let $n'$ be the order of $B$, and in general let $n^{(i)}$ be the order of $B_i \equiv B^i (\mathrm{mod}\ E^{i+1})$. Then we can assume that $E = (e_1, \ldots, e_n)$ is in such a form that $B_1 = B = (e_1, \ldots, e_{\nu'})$, $B_2 = (e_{\nu'+1}, \ldots, e_{\nu''})$, and in general $B_i = (e_{\nu^{(i-1)}+1}, \ldots, e_{\nu^{(i)}})$ for $i \leq \alpha - 1$, where $\nu^{(j)} = \sum_{k=1}^{j} n^{(k)}$. Now, $n^{(\alpha-1)} \geq 1$, $n = \sum_{i=1}^{\alpha-1} n^{(i)}$ and $E = \sum_{i=1}^{\alpha-1} B_i$. We shall henceforth make this assumption for algebras in the canonical form. The group $G$ of all linear transformations which leave unaltered the canonical form (under this slightly modified definition) is by § 4 seen to be the group of all non-singular transformations which, for every $i$, replace an element of $B_i$ by the sum of an element of $B_i$ and an element of $E^{i+1}$.

6. *Definition of Invariants of Nilpotent Algebras.* Given fixed positive integers $n$ and $\mu^0 = 0, \mu', \ldots, \mu^{(\alpha-1)}$ such that $\sum_i \mu^{(i)} = n$, let $A = (e_1, \ldots, e_n)$ be an algebra over the field $F$ such that, if we denote the complex $(e_{\mu^{(i-1)}+1}, \ldots, e_{\mu^{(i)}})$ by $C_i$ $(i = 1, \ldots, \alpha - 1)$, then the product of two complexes $C_j$ and $C_k$ will be expressible linearly and homogeneously in terms of the elements of $C_l$, where $l \geq j + k$, and where in particular $C_j C_k = 0$ when $j + k \geq \alpha$; but otherwise let the constants of multiplication of the algebra $A$ be undetermined elements of $F$. $A$ is necessarily nilpotent; and furthermore, the particular algebras $A_1, A_2, \ldots$ obtained from $A$ by assigning to the constants of multiplication particular sets of values in the field $F$ include all those $n$-ary nilpotent algebras of index $\alpha$ where, for every $i$, $n^{(i)} = \mu^{(i)} - \mu^{(i-1)}$ is the order of the complex $B_i$ defined in § 5. For convenience, we will say such algebras are of *genus* $(\alpha; n', \ldots, n^{(\alpha-1)})$.

Moreover, if $G$ is the group which leaves unaltered the canonical form of all algebras of genus $(\alpha; n', \ldots, n^{(\alpha-1)})$, then any transformation of the group $G$ will carry any particular one of the algebras $A_1, A_2, \ldots$ into another one of the set, and in particular it will carry the set of all algebras of genus $(\alpha; n', \ldots, n^{(\alpha-1)})$ into itself. Accordingly the particular algebras $A_1, A_2, \ldots$ may be separated into *classes* $\mathfrak{C}_1, \ldots, \mathfrak{C}_\sigma$ such that two of the algebras belong to the same class if and only if they be equivalent with respect to the field $F$ under the group $G$; that is, if and only if there is a transformation of the group $G$ with coefficients in $F$ which will carry one algebra into the other.

A single-valued function $\mathfrak{J}$ of the constants of multiplication of the algebras $A$ is called an *invariant* of $A$ under the group $G$ if, for $j = 1, \ldots, \sigma$, this function has the same value $v_j$ for all algebras in the class $C_j$. A set of invariants $\mathfrak{I}_1, \ldots, \mathfrak{I}_p$ is said to *completely characterize* the algebras of genus $(\alpha; n', \ldots, n^{(\alpha-1)})$ over the field $F$ under the group $G$ when each $\mathfrak{I}_k$ has the same value for two algebras of the genus if and only if they belong to the same class. The *characteristic invariant* $I_k$ for the class $\mathfrak{C}_k$ is defined to be that invariant which has the value unity for algebras of the class $\mathfrak{C}_k$ and the value zero for all other algebras of the same genus.

In order to find invariants in the sense defined above, we will also consider rational (absolute and relative) invariants and covariants for the algebras of a given genus.

7. *Importance of the Complex B for Associative Algebras.* For an associative algebra $E$, the complex $B \equiv E \pmod{E^2}$ in some manner determines the behavior of the whole algebra. If also $B$ is commutative, so is $E$.

On the contrary, if $B$ is such that, when $e_1$ and $e_2$ are any two elements of $B$, $e_1 e_2 = -e_2 e_1$, then by induction it follows that $B^i$ is commutative with $B^j$ unless $i$ and $j$ are both odd; and if $i$ and $j$ are both odd, we have $e^{(i)} e^{(j)} = -e^{(j)} e^{(i)}$, where $e^{(k)}$ is any element in $B^k$.

Moreover, an associative nilpotent algebra $E$ is reducible if and only if $B$ is reducible.* First, if $B$ is reducible, $B = B_1 + B_2$, where $B_1 B_2 = 0 = B_2 B_1$ and $B_1 \cap B_2 = 0$. Therefore, in view of the associative law, $B^i = B_1^i + B_2^i$; and accordingly, if we take $E_1 = \sum_{i=1}^{\alpha-1} B_1^i$ and $E_2 = \sum_{i=1}^{\alpha-1} B_2^i$ (where $\alpha$ is the index of $E$), then $E = E_1 + E_2$, with $E_1 E_2 = 0 = E_2 E_1$, where $E_1$ and $E_2$ are algebras.

Conversely, if $E$ is reducible, with $E = E_1 + E_2$, where $E_1 \cap E_2 = 0$; then $E_1 = \sum_{i=1}^{\alpha-1} B_1^i$ and $E_2 = \sum_{i=1}^{\alpha-1} B_2^i$, where $B_1 \equiv E_1 \pmod{E_1^2}$ and $B_2 \equiv E_2 \pmod{E_2^2}$. Therefore, $B_1 \cap B_2 = 0$, and $B_1 B_2 = 0 = B_2 B_1$, $B_1 + B_2 = B$.

8. "*Special*" *Canonical Form.* As we stated at the beginning of § 5, several of the complexes $B, \ldots, B^{\alpha-1}$ may overlap, but these "overlappings" are more or less restricted in view of one or two simple properties which follow from the formulæ $E^p = \sum_{k=p}^{\alpha-1} B^k$ in § 3. In an associative algebra, for $a < \alpha$, $b < \alpha$, $B^a = B^b$ if and only if $a = b$; more generally, $B^{a+b}$ does not contain $B^a$, where $a < \alpha$ and $b > 0$. Further, we can not have $B^c = B^a + B^b$ for $a, b < \alpha$ unless $a = c$ or $b = c$. Finally, for every exponent $a < \alpha$, there are numbers in $B^a$ which can not be expressed as linear homogeneous functions of elements in $B^{a+1}, \ldots, B^{\alpha-1}$.

Now, if an algebra in canonical form be such that $B^i \cap B^j = 0$ $(i \neq j;\ i, j < \alpha)$, it will be said to be in "*special*" *canonical form.* In particular, in view of the

* Wedderburn, *loc. cit.*, p. 88.

properties mentioned above, all associative algebras in one, two, three or four units are necessarily in "special" canonical form, except for two types of quaternary algebras, namely $n'=2$, $n''=n^{(3)}=1$ and $n'=n''=n^{(3)}=n^{(4)}=1$.

THEOREM. *If $G'$ be the group of all linear transformations with coefficients in $F$ which leaves unaltered the "special" canonical form of the nilpotent algebra $E=B+E^2$ over the field $F$, and if $G''$ be that subgroup of $G'$ which transforms the units of $B$ among themselves alone, the units of $B^2$ among themselves alone, etc., then the invariants of $E$ under the group $G''$ are all the invariants under $G'$. Furthermore, if two nilpotent algebras in "special" canonical form over a field $F$ are equivalent with respect to $F$, they are equivalent under a transformation of the group $G''$.*

To show this, let $T'$ be any transformation of the group $G'$. Then $T'=T_1'T''$, where $T''$ is in $G''$ and where, for every $i$, $T_1'$ at most adds to each unit of $B^i$ an element in $E^{i+1}$. $T_1'$ is in $G'$, and accordingly, from its nature and in view of the definition of "special" canonical form, can not affect the constants of multiplication. Hence, the theorem.

*Invariants for General Nilpotent Algebras in Canonical Form,* §§ 9–12.

9. *Two Classical Methods Which Furnish no Invariants.* Since the characteristic determinants for an $n$-ary nilpotent algebra are $\delta(\omega)\equiv\omega^n\equiv\delta'(\omega)$, all invariants of nilpotent algebras which can be obtained by means of the theorem of § 1 are zero.

Furthermore, no information in regard to nilpotent algebras is furnished by Scheffers' theorem that, for an associative algebra with a modulus, if the right-hand characteristic equation $\delta(\omega)=0$ defines $\omega$ as an $h$-valued function of the coordinates of the general number of the algebra, then we may choose as normalized units $\varepsilon_1, \ldots, \varepsilon_h, \eta_1, \ldots, \eta_k$ satisfying certain conditions. For, if we adjoin a modulus $e_0$ to the nilpotent algebra $(e_1, \ldots, e_n)$, the right-hand characteristic equation $(x_0-\omega)^{n+1}=0$ of the resulting algebra defines $\omega$ as a single-valued function of the $x$'s.

10. *Invariants Obtained from the Parastrophic Matrix.* The argument of § 5 shows that certain matrices obtained from the parastrophic matrix are unaltered under transformations of the group $G$.

If we have an algebra $E=(e_1, \ldots, e_n)$, not necessarily nilpotent, over the field $F$ with multiplication table (1), the *parastrophic matrix** is defined as

$$R\equiv\left(\sum_{k=1}^{n}\gamma_{ijk}\xi_k\right), \tag{4}$$

* This term and the notation are those used by Frobenius, *Berliner Sitzungsberichte,* 1903, p. 522. The covariance of $R$ is also due to him (*loc. cit.*).

where the $\xi_k$ are variables ranging independently over $F$ which are cogredient with the $e_k$. If we subject the units of $E$ to the non-singular transformation

$$\bar{e}_i = \sum_{j=1}^{n} a_{ij}\, e_j \quad (i = 1, \ldots, n),$$

where the $a$'s are in $F$, then the parastrophic matrix $\bar{R}$ for the new algebra $\bar{E}$ is

$$\bar{R} = A\,R\,A', \tag{5}$$

where the prime indicates the conjugate matrix. Thus $R$ is a covariant of the algebra.

In particular, let $E = \sum_{i=1}^{\alpha-1} B_i$ be a nilpotent algebra of genus

$$(\alpha;\; n', \ldots, n^{(\alpha-1)})$$

in canonical form. Then the parastrophic matrix is

$$R = \left|\begin{array}{c|c|c} \displaystyle\sum_{r=\nu'+1}^{n} \gamma_{pqr}\,\xi_r & \displaystyle\sum_{\nu''+1}^{n} \gamma_{pqr}\,\xi_r & \\ (p, q = 1, \ldots, \nu') & \begin{pmatrix} p = 1, \ldots, \nu' \\ q = \nu'+1, \ldots, \nu'' \end{pmatrix} & \cdots\cdots \\ \hline \displaystyle\sum_{\nu''+1}^{n} \gamma_{pqr}\,\xi_r & \displaystyle\sum_{\nu^{(3)}+1}^{n} \gamma_{pqr}\,\xi_r & \\ \begin{pmatrix} p = \nu'+1, \ldots, \nu'' \\ q = 1, \ldots, \nu' \end{pmatrix} & (p, q = \nu'+1, \ldots, \nu'') & \cdots\cdots \\ \hline & & \\ \cdots\cdots & \cdots\cdots & \cdots\cdots \\ & & \end{array}\right|, \tag{6}$$

where $\nu^{(i)} = \sum_{j=1}^{i} n^{(j)} \; (i = 1, \ldots, \alpha - 1)$.

If we subject the algebra to a transformation $T$ of the group $G$ which leaves unaltered the canonical form, then $T = T_2\,T_1$, where $T_1$ transforms the units of $B_1$ among themselves alone, the units of $B_2$ among themselves alone, etc.; and where, for every $i$, $T_2$ at most adds to each unit of $B_i$ an element in $E^{i+1}$. Under the transformation $T_1$, $R$ is subjected to a finite number of elementary transformations in such a way that any "box" of the schematic matrix (6) is subjected to a finite number of elementary transformations; and at the same time the variables $\xi_k$ which occur in any partial sum

$$\sum_{r=\nu^{(i-1)}+1}^{\nu^{(i)}} \gamma_{pqr}\,\xi_r \tag{7}$$

are subjected to a non-singular transformation among themselves alone. By the transformation $T_2$, $R$ is subjected to a finite number of elementary transformations in such a way that to each row of the "box" in the $j$-th horizontal strip and $k$-th vertical strip of the schematic matrix are added linear combinations of the rows from "boxes" which are in the $k$-th vertical strip below the $j$-th horizontal strip, and a similar transformation is then made on the columns. At the same time, by $T_2$ the variables $\xi$ which occur in any partial sum (7) have added to them linear combinations of the variables $\xi$ with a larger subscript. In short, under the transformation $T$, each partial sum (7) in the transformed matrix is affected only by those partial sums (7) in the original matrix having $r$ range from $\nu'+1$ to $\nu^{(i)}$ which are in the same horizontal strip or in the horizontal strips below, and at the same time in the same vertical strip or in the vertical strips to the right.

That is, for every value of $0 \leq l \leq \alpha - 1$, the square matrix obtained from $R$ by erasing the first $l$ horizontal strips and the first $l$ vertical strips of the schematic matrix (6) is subjected to a finite number of elementary transformations under the group $G$. Also, if for any $m$ we delete the last $m$ horizontal strips and the last $m$ vertical strips of the schematic matrix, and if at the same time we erase all partial sums (7) where $i \geq m$, then the resulting matrix is subjected to a finite number of elementary transformations under the group $G$.

Now just as any homogeneous algebraic covariant (or invariant) of a covariant of a quantic is itself a covariant (or invariant) of the quantic, so any homogeneous algebraic covariant (or invariant) of a function of the constants of multiplication and coordinates which is invariantive under the group $G$ for algebras of genus $(\alpha; n', \ldots, n^{(\alpha-1)})$ is itself a covariant (or invariant) under the group $G$ for these algebras. Hence, every homogeneous algebraic covariant (or invariant) and every arithmetic invariant of a matrix obtained from (6) by one of the two methods of deletion described above is in turn a covariant (or invariant) under the group $G$ for the $n$-ary algebras of genus $(\alpha; n', \ldots, n^{(\alpha-1)})$.

*Finally, for every $l$ and $m$, the matrix obtained from* (6) *by combining the two methods of deletion described above is a covariant under the group $G$ for the algebras of genus $(\alpha; n', \ldots, n^{(\alpha-1)})$, and this matrix furnishes further covariants and invariants for such algebras under the group $G$.*

Hence, if two nilpotent algebras $E'$ and $E''$ of index $\alpha$ are equivalent, then if we delete $E'$ by $E'^{\alpha-m}$ and $E''$ by $E''^{\alpha-m}$, the resulting algebras are equivalent, and this is true for every positive integer $m < \alpha$. Similarly for every positive integer $l < \alpha$, if we erase in $E'$ all units in $B', \ldots, B'_l$ and at the same time erase in $E''$ all units in $B'', \ldots, B_i''$, where $B' \equiv E'$ (mod $E'^2$) and $B'' \equiv E''$ (mod $E''^2$), then the resulting algebras are equivalent.

Moreover, there are other invariants of these matrices which are a generalization of the rational invariant factors of a matrix whose elements are polynomials in one variable. Consider a particular matrix $M$ obtained from $R$ by one of the above methods of deletion, and let it be of rank $r$ if and only if the $\xi_k$ take on a set of values linearly dependent on the $t_r$ linearly independent sets $\xi_k = \xi_k^{(r,s)}$ $(s=1, \ldots, t_r)$, where $k$ has the same range as in $M$. The numbers $t_r$ $(r=1, \ldots)$ are unaltered under a transformation of the group $G$.

11. *Invariants of the Complexes $B_i$. Fundamental Quadratics, Cubics, etc.* Consider the complex $B_i = (e_{\nu^{(i-1)}+1}, \ldots, e_{\nu^{(i)}})$ of the algebra $E$ of index $\alpha$, over the field $F$ (see § 5). Modulo $E^{2i+1}$, the square of the general number

$$X^{(i)} = \sum_{p=\nu^{(i-1)}+1}^{\nu^{(i)}} x_p e_p \tag{8}$$

of $B_i$ is

$$X^{(i)2} \equiv \sum_{p,\, q=\nu^{(i-1)}+1}^{\nu^{(i)}} \; \sum_{r=\nu^{(2i-1)}+1}^{\nu^{(2i)}} \gamma_{pqr} x_p x_q e_r \quad (\text{mod } E^{2i+1}). \tag{9}$$

Define

$$\mathcal{Q}^{(i)} = \sum_{r=\nu^{(2i-1)}+1}^{\nu^{(2i)}} \mathcal{Q}_r^{(i)} \eta_r, \tag{10}$$

where

$$\mathcal{Q}_r^{(i)} \equiv \sum_{p,\, q=\nu^{(i-1)}+1}^{\nu^{(i)}} \gamma_{pqr} x_p x_q \quad (r=\nu^{(2i-1)}+1, \ldots, \nu^{(2i)}), \tag{11}$$

and where the $\eta$'s are variables ranging independently over $F$ and such that, when the units of $B_{2i}$ are subjected to the transformation

$$\left.\begin{aligned} e_p' &= \sum_{q=\nu^{(2i-1)}+1}^{\nu^{(2i)}} a_{pq} e_q + \sum_{q>\nu^{(2i)}} a_{pq} e_q \quad (\nu^{(2i-1)}+1 \leq p \leq \nu^{(2i)}), \\ &| a_{pq} | \neq 0 \quad (p, q = \nu^{(2i-1)}+1, \ldots, \nu^{(2i)}), \end{aligned}\right\} \tag{12}$$

then the $\eta$'s are subjected to the transformation

$$\eta_p' = \sum_{q=\nu^{(2i-1)}+1}^{\nu^{(2i)}} a_{pq} \eta_q \quad (\nu^{(2i-1)}+1 \leq p \leq \nu^{(2i)}). \tag{13}$$

The necessary and sufficient condition that there exist in $B_i$ $\tau$ linearly independent numbers such that their squares are all zero (modulo $E^{2i+1}$) is that the $n^{(2i)}$ homogeneous quadratics (11) should have in common $\tau$ linearly independent solutions. Hence the number of linearly independent solutions common to the $\mathcal{Q}_r^{(i)}$ has the invariantive property, and the resultant of the $\mathcal{Q}_r^{(i)}$ is an invariant.

If we subject the units of $E = \sum_{i=1}^{\alpha-1} B_i$ to a non-singular transformation $T$ of the group $G$, the units in $B_{2i}$ are subjected to a transformation of the form (12) and the $\eta$'s to the corresponding transformation (13). Now $T = T_1 \ldots . T_{\alpha-1}$, where $T_j$ leaves unaltered all units except those in $B_j$, and replaces each unit in $B_j$ by the sum of a number in $B_j$ and a number in $E^{j+1}$. The quadratics (11) are affected at most by $T_i$ and $T_{2i}$.

If $e_p$ and $e_q$ be any two units in $B_i$, and if $\gamma'_{pqr}$ are the constants of multiplication in the algebra obtained by applying to $E$ the transformation $T_{2i}$, then by the argument just after (7) we must have

$$\sum_{r=\nu^{(2i-1)}+1}^{\nu^{(2i)}} \gamma_{pqr}\eta_r = \sum_{r=\nu^{(2i-1)}+1}^{\nu^{(2i)}} \gamma'_{pqr}\eta'_r,$$

holding identically in view of the relations between the $\eta$'s and the $\eta'$'s. Hence

$$\mathcal{Q}^{(i)}(\gamma_{pqr};\ x_p;\ \eta_r) \equiv \mathcal{Q}^{(i)}(\gamma'_{pqr};\ x'_p;\ \eta'_r)$$

under the transformation $T_{2i}$.

Finally, if $\gamma''_{pqr}$ be the constants of multiplication in the algebra obtained by applying to $E$ the transformation $T_i$, then since $e_{\nu^{(2i-1)}+1}, \ldots, e_{\nu^{(2i)}}$ are unaltered by $T_i$, the square of $X^{(i)} = \Sigma x''_p e''_p \pmod{E^{2i+1}}$ can be derived from the square of $X^{(i)} = \Sigma x_p e_p \pmod{E^{2i+1}}$ by replacing $\gamma_{pqr}$ by $\gamma''_{pqr}$ and $x_p$ by $x''_p$; and accordingly

$$\mathcal{Q}_r^{(i)}(\gamma_{pqr};\ x_p) \equiv \mathcal{Q}_r^{(i)}(\gamma''_{pqr};\ x''_p) \qquad (\nu^{(2i-1)}+1 \leq r \leq \nu^{(2i)})$$

in view of the relations between the $x_p$ and the $x''_p$. Thus $\mathcal{Q}^{(i)}$ is unaltered under $T_i$.

Combining these results, we see that the quadratic form (10) is unaltered under transformations of the group $G$, and we shall refer to $\mathcal{Q}^{(i)}$ as a covariant of the complex $B_i$; in particular we shall call it the *fundamental quadratic for* $B_i$. Moreover, the discriminant of (10) when we regard the $x$'s as variables and the $\eta$'s as parameters is a function which has the invariantive property for the general algebra of genus $(\alpha;\ n', \ldots, n^{(\alpha-1)})$. Also the invariants and covariants of this discriminant regarded as a function of the $\eta$'s are invariants and covariants for the algebras of genus $(\alpha;\ n', \ldots, n^{(\alpha-1)})$.

Similarly the cubic form

$$\mathcal{C}^{(i)} \equiv \sum_{r=\nu^{(3i-1)}+1}^{\nu^{(3i)}} \mathcal{C}_r^{(i)}\eta_r, \tag{14}$$

where

$$\mathcal{C}_r^{(i)} \equiv \sum_{j,k,l} \left(\sum_m \gamma_{klm}\gamma_{jmr}\right) x_j x_k x_l \quad \begin{pmatrix} \nu^{(i-1)}+1 \leq j,\ k,\ l \leq \nu^{(i)} \\ \nu^{(2i-1)}+1 \leq m \leq \nu^{(2i)} \end{pmatrix},$$

is a covariant for nilpotent algebras of genus $(\alpha;\ n', \ldots, n^{(\alpha-1)})$. Here the $\eta$'s are related to the units in $B_{3i}$ as the $\eta$'s of (10) were related to the units in $B_{2i}$. In an analogous manner we can form a covariantive quartic, quintic, etc., for the complex $B_i$, and continue until we get a $p$-ic form which is identically zero in the parameters $\eta$ and the variables $x$ for all $\gamma$'s. For brevity we shall call these the *fundamental forms for the complex* $B_i$, although they depend also on the complexes $B_{2i}$, $B_{3i}$, etc. In particular, for the complex $B \equiv E (\mathrm{mod}\ E^2)$ there are $\alpha-2$ fundamental forms not identically zero for the algebra of index $\alpha$.

## CHAPTER III.

### CLASSIFICATION OF GENERAL NILPOTENT ALGEBRAS WITH $n \leq 4$.

By using the theory of the previous chapter, we can readily classify the nilpotent algebras (not necessarily associative) in a small number of units. We give below the canonical forms of the multiplication tables of all nilpotent algebras over the field $C$ of ordinary complex numbers, having four units or less, to which any such algebra is equivalent under a non-singular transformation with coefficients in $C$.* Furthermore, no two of the algebras tabulated are equivalent. Where they are not too awkward, we give also the invariantive conditions that a given algebra in canonical form be equivalent to one of the algebras tabulated. Throughout, we shall use $\alpha$ to denote the index of the algebra $E$, and $n^{(i)}$ for the number of linearly independent elements in $B_i \equiv B^i (\mathrm{mod}\ E^{i+1})$, where $B \equiv E\ (\mathrm{mod}\ E^2)$. For convenience, we shall understand that products not written are zero.

12. $n = 1, 2$.

$$n=1: \quad e_1 e_1 = 0.$$
$$n=2: \quad \alpha = 2, \quad e_i e_j = 0 \quad (i, j = 1, 2).$$
$$\alpha = 3, \quad e_1^2 = e_2.$$

13. $n = 3$.

Type A: $n' = 3,\ n'' = n^{(3)} = 0;\ \alpha = 2, \quad e_i e_j = 0 \quad (i, j = 1, 2, 3)$.

Type B: $n' = 2,\ n'' = 1,\ n^{(3)} = 0;\ \alpha = 3$.

$$Q_3 \equiv x_1^2 \gamma_{113} + x_1 x_2 (\gamma_{123} + \gamma_{213}) + x_2^2 \gamma_{223}; \quad D_3 \equiv (\gamma_{123} + \gamma_{213})^2 - 4\gamma_{113}\gamma_{223}.$$

I. $Q_3 \equiv 0$: $e_1 e_2 = -e_2 e_1 = e_3$.

II. $Q_3 \not\equiv 0$; $D_3 \neq 0$: $e_2 e_1 = e_3$, $e_1 e_2 = \lambda e_3$, $\lambda \neq -1$, $|\lambda| \leq 1$.

III. $Q_3 \not\equiv 0$; $D_3 = 0$: $e_1 e_1 = e_3$; $e_1 e_1 = e_3 = e_1 e_2 = -e_2 e_1$.

The classes of Type B (in canonical form) are completely characterized by the ranks of

$$\begin{vmatrix} \gamma_{113} & \dfrac{\gamma_{123}+\gamma_{213}}{2} \\ \dfrac{\gamma_{123}+\gamma_{123}}{2} & \gamma_{223} \end{vmatrix}, \quad \begin{vmatrix} \gamma_{113} & \gamma_{123} \\ \gamma_{213} & \gamma_{223} \end{vmatrix}$$

and the absolute invariant $(\gamma_{113}\gamma_{223} - \gamma_{123}\gamma_{213})/D_3$.

---

*Most of the work, however, holds for any field $F$. These results have been compared and (as far as possible) have been checked with Allen's results for associative nilpotent algebras in 2, 3, 4 units over any field $F$ (*Transactions*, Vol. IX, pp. 203–218).

Type C: $n' = n'' = n^{(3)} = 1$; $\alpha = 4$,

$$e_1 e_1 = e_2,\ e_1 e_2 = e_3,\ e_2 e_1 = \lambda e_3;\ e_1 e_1 = e_2,\ e_2 e_1 = e_3.$$

Here there is a single infinitude of classes characterized among themselves by the values of $r_{1,\lambda}$ and $r_{0,1}$, the ranks of

$$\begin{vmatrix} \gamma_{123} & 1 \\ \gamma_{213} & \lambda \end{vmatrix},\qquad \begin{vmatrix} \gamma_{123} & 0 \\ \gamma_{213} & 1 \end{vmatrix}$$

respectively.

Type D: $n' = n^{(2)} = 1$, $n^{(3)} = 0$, $n^{(4)} = 1$,

$$e_1 e_1 = e_2,\ e_2 e_1 = x e_3,\ e_2 e_2 = e_3.$$

14. $n = 4$.

Type A: $n' = 4$, $n'' = n^{(3)} = n^{(4)} = 0$; $\alpha = 2$, $e_i e_j = 0$ $(i, j = 1, \ldots, 4)$.

Type B: $n' = 3$, $n'' = 1$, $n^{(3)} = n^{(4)} = 0$; $\alpha = 3$.

Let $r$ be the rank of $D_4$, the discriminant of the covariantive quadratic $Q_4 \equiv \sum_{i,j}^{1,2,3} \gamma_{ij4} x_i x_j$; and let $\rho$ be the rank of $R = |\gamma_{ij4}|$ $(i, j = 1, 2, 3)$.

I. $r = 0$, $e_1 e_2 = -e_2 e_1 = e_4$.

II. $r = 1$, (a) $\rho = 3$, $e_1^2 = e_2 e_3 = -e_3 e_2 = e_4$.

(b) $\rho = 2$, $e_1^2 = e_1 e_2 = -e_2 e_1 = e_4$.

(c) $\rho = 1$, $e_1^2 = e_4$.

III. $r = 2$, (a) $\rho = 3$, $e_2 e_1 = e_1 e_3 = -e_3 e_1 = e_2 e_3 = -e_3 e_2 = e_4$.

(b) $\rho = 2$, $e_2 e_1 = e_1 e_3 = -e_3 e_1 = e_4$.

($c_\gamma$) $\rho = 1$, $e_1 e_2 = e_4$, $e_2 e_1 = \gamma e_4$, $|\gamma| \leq 1$.

The algebras of III (c) are characterized completely among themselves as follows: Let $(x_{i1}, x_{i2}, x_{i3})$ $(i = 1, 2, 3)$ be three linearly independent solutions of $Q_4 = 0$, and let $(x_{31}, x_{32}, x_{33})$ be the double solution. Then the class ($c_\gamma$) is characterized by the fact that

$$\begin{vmatrix} 1 & \sum_{k,l}^{1,2,3} x_{1k} x_{2l} \gamma_{kl4} \\ \gamma & \sum_{k,l}^{1,2,3} x_{2k} x_{1l} \gamma_{kl4} \end{vmatrix} \times \begin{vmatrix} 1 & \sum_{k,l}^{1,2,3} x_{2l} x_{1k} \gamma_{kl4} \\ \gamma & \sum_{k,l}^{1,2,3} x_{1k} x_{2l} \gamma_{kl4} \end{vmatrix} = 0.$$

IV. $r = 3$, (a) $\rho = 2$, $-e_2 e_1 = e_3 e_1 = e_2 e_3 = e_3 e_2 = e_4$.

($b_\gamma$) $\rho = 3$, $e_2 e_3 \mp e_3 e_2 = e_1^2 = e_2 e_1 = e_4$, $e_3 e_1 = \gamma e_4$.

The classes of B, IV are completely characterized among themselves by $r$, $\rho$ and the absolute invariant $R/D_4$.

Type C: $n' = 2$, $n'' = 2$, $n^{(3)} = n^{(4)} = 0$; $\alpha = 3$. Here the fundamental quadratic is $Q \equiv Q_3 \eta_3 + Q_4 \eta_4$, where $Q_k \equiv \sum_{i,j}^{1,2} \gamma_{ijk} x_i x_j$ $(k = 3, 4)$. Let $\rho$ be the rank of the discriminant of $|\sum_k^{3,4} \gamma_{ijk}\eta_k|$ $(i, j = 1, 2)$ for general $\eta_3, \eta_4$; and let $\Theta = \frac{\vartheta(\vartheta-1)}{2}$, where $\vartheta$ is the rank of

$$\begin{pmatrix} \gamma_{113} & \gamma_{123}+\gamma_{213} & \gamma_{223} \\ \gamma_{114} & \gamma_{124}+\gamma_{214} & \gamma_{224} \end{pmatrix};$$

and let $R$ be the resultant of $Q_3$ and $Q_4$.

I. $\Theta = 0$, $R = 0$, $e_1 e_2 = e_3$, $e_2 e_1 = e_4$.

II. $\Theta = 1$, $R = 0$, (a) $\rho = 0$, $e_1^2 = e_4$, $e_1 e_2 = e_3$; $e_1^2 = e_4$, $e_2 e_1 = e_3$.

(b$_k$) $\rho = 1$, $e_1^2 = e_4$, $e_2 e_1 = e_3$, $e_1 e_2 = k e_3$ $(k \neq -1, 0)$.

(c) $\rho = 2$, $e_1^2 = e_1 e_2 = e_4$, $e_2 e_1 = e_3$.

The classes in Case II are characterized completely among themselves by $r_{k,1}$ and $r_{1,0}$, the ranks of

$$\begin{pmatrix} A_1 & A_2 & k \\ B_1 & B_2 & 1 \end{pmatrix} \text{ and } \begin{pmatrix} A_1 & A_2 & 1 \\ B_1 & B_2 & 0 \end{pmatrix}$$

respectively, where

$$A_1 = x_1[\gamma_{113} + (x_2 - 1)\gamma_{213}] + x_2[\gamma_{123} + (x_2 - 1)\gamma_{223}],$$
$$A_2 = x_1[\gamma_{114} + (x_2 - 1)\gamma_{214}] + x_2[\gamma_{124} + (x_2 - 1)\gamma_{224}],$$
$$B_1 = x_1[\gamma_{113} + (x_2 - 1)\gamma_{123}] + x_2[\gamma_{213} + (x_2 - 1)\gamma_{223}],$$
$$B_2 = x_1[\gamma_{114} + (x_2 - 1)\gamma_{124}] + x_2[\gamma_{214} + (x_2 - 1)\gamma_{224}],$$

where $(x_1, x_2)$ is the solution $\neq (0, 0)$ common to $Q_3 = 0$ and $Q_4 = 0$.

III. $\Theta = 1$, $R \neq 0$.

(a$_\lambda$) $e_1^2 = e_2^2 = e_4$, $e_2 e_1 = e_3$, $e_1 e_2 = e_3 + \lambda e_4$.

There is a class for every value of $\lambda^2$.

Type D: $n' = 2$, $n'' = 1$, $n^{(3)} = 1$; $\alpha = 4$. Since $\left|\frac{\gamma_{ij3}+\gamma_{ji3}}{2}\right|$ $(i, j = 1, 2)$ is invariantive for classes of this type, we can make use of the results for Type B of the ternary algebras. Let $r$ be the rank of $Q_3 \equiv \sum_{i,j}^{1,2} \gamma_{ij3} x_i x_j$, $\rho$ the rank of

$$\begin{vmatrix} \gamma_{134} & \gamma_{314} \\ \gamma_{234} & \gamma_{324} \end{vmatrix}.$$

I. $r = 0$.

(a) $\rho = 1$.

(1$_k$) $e_1 e_2 = -e_2 e_1 = e_3$, $e_1 e_3 = e_2 e_2 = e_4$, $e_3 e_1 = k e_4$ $(k \neq -1)$.

(2$_k$) $e_1 e_2 = -e_2 e_1 = e_3$, $e_1 e_3 = e_4$, $e_3 e_1 = k e_4$.

(3) $e_1 e_1 = e_1 e_3 = -e_3 e_1 = e_2 e_2 = e_4$, $e_1 e_2 = -e_2 e_1 = e_3$.

(4) $e_1 e_2 = -e_2 e_1 = e_3, \; e_1 e_3 = -e_3 e_1 = e_2 e_2 = e_4.$

(5) $e_1 e_3 = -e_3 e_1 = e_4, \; e_1 e_2 = -e_2 e_1 = e_3, \; e_1 e_1 = e_4.$

(6) $e_1 e_3 = -e_3 e_1 = e_4, \; e_1 e_2 = -e_2 e_1 = e_3.$

(7) $e_1 e_3 = -e_3 e_1 = e_4, \; e_1 e_2 = e_3, \; e_2 e_1 = -e_3 + e_4.$

The classes of Type D, I (a) are characterized completely among themselves by the vanishing or non-vanishing of

$$\gamma_{134}^2 \gamma_{224} - \gamma_{134} \gamma_{234} (\gamma_{124} + \gamma_{214}) + \gamma_{234}^2 \gamma_{114},$$

and by the value of $r_{1,k}$, $r_{0,1}$ and $(2 - r_{1,-1})\,\rho_1$, where $r_{1,k}$, $r_{0,1}$ and $\rho_1$ are respectively the ranks of

$$\begin{pmatrix} \gamma_{134} & \gamma_{234} & 1 \\ \gamma_{314} & \gamma_{324} & k \end{pmatrix}, \quad \begin{pmatrix} \gamma_{134} & \gamma_{234} & 0 \\ \gamma_{314} & \gamma_{324} & 1 \end{pmatrix}, \quad \begin{pmatrix} 2\gamma_{114} & \gamma_{124} + \gamma_{214} \\ \gamma_{124} + \gamma_{214} & 2\gamma_{224} \end{pmatrix}.$$

(b). $\rho = 2.$

(1) $e_1 e_2 = -e_2 e_1 = e_3, \; e_1 e_3 = e_3 e_2 = e_4.$

(2) $e_1 e_2 = -e_2 e_1 = e_3, \; e_1 e_3 = e_3 e_2 = e_1 e_1 = e_4.$

These two classes are distinguished by the vanishing and non-vanishing respectively of

$$\gamma_{114} (\gamma_{234} + \gamma_{324})^2 - (\gamma_{124} + \gamma_{214})(\gamma_{234} + \gamma_{324})(\gamma_{134} + \gamma_{314}) + \gamma_{224} (\gamma_{134} + \gamma_{314})^2.$$

II. $r = 1.$

(a) $\gamma_{123} + \gamma_{213} \neq 0.$

(1) $\mathfrak{M} \equiv \left| \sum_k^{3,4,5} \gamma_{ijk} \xi_k \right| \equiv 0 \; (i, j = 1, 2, 3).$ $\infty^2$ classes.

(a′) $e_1 e_1 = e_1 e_2 = e_3, \; e_2 e_1 = -e_3 + a e_4, \; e_1 e_3 = e_4, \; e_3 e_1 = k e_4 \; (k \neq 0; \; a = 0, 1).$

(b′) $e_1 e_1 = e_3 = e_1 e_2 = -e_2 e_1, \; e_2 e_2 = a e_4, \; e_1 e_3 = e_4 \; (a = 0, 1).$

(c′) $e_1 e_1 = -e_2 e_1 = e_3, \; e_1 e_2 = e_3 + a e_4, \; e_2 e_3 = e_4.$

(d′) $e_1 e_1 = e_1 e_2 = -e_2 e_1 = e_3, \; e_3 e_1 = e_4.$

(e′) $e_1 e_1 = e_1 e_2 = -e_2 e_1 = e_3, \; e_2 e_2 = e_3 e_1 = e_4.$

(f′) $e_1 e_1 = e_1 e_2 = e_3, \; e_2 e_1 = -e_3 + a e_4, \; e_3 e_2 = e_4 \; (a = 0, 1).$

(g′) $e_1 e_1 = e_1 e_2 = e_3, \; e_2 e_1 = -e_3 + a e_4, \; e_2 e_3 = -e_3 e_2 = e_4, \; e_3 e_1 = \lambda e_4 \; (a = 0, 1).$

(h′) $e_1 e_1 = e_1 e_2 = e_3, \; e_2 e_1 = -e_3 + k e_4, \; e_2 e_2 = e_2 e_3 = -e_3 e_2 = e_4, \; e_3 e_1 = \lambda e_4.$

(i′) $e_1 e_1 = e_1 e_2 = -e_2 e_1 = e_3, \; e_3 e_1 = e_3 e_2 = \lambda e_4, \; e_2 e_3 = e_4 \; (\lambda \neq -1).$

(j′) $e_1 e_1 = e_1 e_2 = e_3, \; e_2 e_1 = -e_3 + k e_4, \; e_3 e_1 = e_3 e_2 = \lambda e_4, \; e_2 e_3 = e_4$
$(k = 0, 1; \; \lambda \neq -1).$

(2) $\mathfrak{M}$ a perfect cube not identically zero. $\infty^2$ classes.

(a′) $e_1 e_1 = e_1 e_2 = -e_2 e_1 = e_3, \; e_2 e_2 = e_1 e_3 = e_4, \; e_3 e_1 = \lambda e_4.$

(b′) $e_1 e_1 = e_3, \; e_2 e_3 = -e_3 e_2 = \lambda e_4, \; e_2 e_2 = e_4, \; e_1 e_2 = -e_2 e_1 = e_3 + k e_4, \; e_3 e_1 = e_4.$

(c′) $e_1 e_1 = e_3, \; e_1 e_2 = -e_2 e_1 = e_3 + e_4, \; e_2 e_3 = -e_3 e_2 = \lambda e_4, \; e_3 e_1 = e_4.$

(d′) $e_1 e_1 = e_1 e_2 = -e_2 e_1 = e_3, \; e_3 e_1 = e_2 e_3 = e_4, \; e_3 e_2 = k\, e_4.$

(e′) $e_1 e_1 = e_3, \; e_1 e_2 = -e_2 e_1 = e_3 + e_4, \; e_3 e_1 = e_2 e_3 = e_4, \; e_3 e_2 = k\, e_4.$

(3) $\mathfrak{M}$ not a perfect cube. $\infty^2$ classes.

(a′) $e_1 e_1 = -e_2 e_1 = e_3, \; e_1 e_2 = e_3 + a\, e_4, \; e_2 e_3 = -e_3 e_2 = e_4, \; e_3 e_1 = \lambda\, e_4 \; (a = 0, 1).$

(b′) $e_1 e_1 = -e_2 e_1 = e_3, \; e_1 e_2 = e_3 + a\, e_4, \; e_2 e_2 = e_2 e_3 = -e_3 e_2 = e_4, \; e_3 e_1 = \lambda\, e_4$
$(a = 0, 1; \; \lambda \neq 0).$

(c′) $e_1 e_1 = -e_2 e_1 = e_3, \; e_1 e_2 = e_3 + k\, e_4, \; e_2 e_2 = e_2 e_3 = -e_3 e_2 = e_4.$

(d′) $e_1 e_1 = e_1 e_2 = e_3, \; e_2 e_1 = -e_3 + a\, e_4, \; e_2 e_3 = e_4, \; e_3 e_1 = \lambda\, e_4, \; e_3 e_2 = \mu\, e_4$
$(a = 0, 1; \; \mu \neq 0).$

(e′) $e_1 e_1 = -e_2 e_1 = e_3, \; e_1 e_2 = e_3 + e_4, \; e_2 e_3 = e_4, \; e_3 e_1 = \lambda\, e_4, \; e_3 e_2 = \mu\, e_4.$

(f′) $e_1 e_1 = -e_2 e_1 = e_3, \; e_1 e_2 = e_3 + a\, e_4, \; e_1 e_3 = k\, e_4, \; e_3 e_2 = e_4 \; (a = 0, 1; \; k \neq 0).$

(g′) $e_1 e_1 = e_1 e_2 = e_3, \; e_2 e_1 = -e_3 + a\, e_4, \; e_3 e_1 = k\, e_4, \; e_2 e_3 = e_4 \; (a = 0, 1; \; k \neq 0).$

(b) $\gamma_{123} - \gamma_{213} = 0.$ $\infty^1$ classes.

(1) $e_1 e_1 = e_3, \; e_2 e_1 = a\, e_4, \; e_1 e_3 = e_4, \; e_3 e_1 = \lambda\, e_4 \; (a = 0, 1).$

(2) $e_1 e_1 = e_3, \; e_1 e_2 = a\, e_4, \; e_3 e_1 = e_4 \; (a = 0, 1).$

(3) $e_1 e_1 = e_3, \; e_3 e_1 = e_2 e_3 = -e_3 e_2 = e_4.$

(4) $e_1 e_1 = e_3, \; e_2 e_1 = a\, e_4, \; e_2 e_3 = -e_3 e_2 = e_4 \; (a = 0, 1).$

(5) $e_1 e_1 = e_3, \; e_2 e_2 = e_2 e_3 = -e_3 e_2 = e_4, \; e_2 e_1 = a\, e_4 \; (a = 0, 1).$

(6) $e_1 e_1 = e_3, \; e_2 e_2 = e_2 e_3 = -e_3 e_2 = e_3 e_1 = e_4.$

(7) $e_1 e_1 = e_3, \; e_2 e_1 = a\, e_4, \; e_2 e_2 = e_1 e_3 = e_4, \; e_3 e_1 = \lambda\, e_4 \; (a = 0, 1).$

(8) $e_1 e_1 = e_3, \; e_2 e_2 = e_2 e_3 = -e_3 e_2 = e_4, \; e_2 e_1 = k\, e_4.$

(9) $e_1 e_1 = e_3, \; e_2 e_2 = e_2 e_3 = -e_3 e_2 = e_3 e_1 = e_4.$

(10) $e_1 e_1 = e_3, \; e_2 e_1 = a\, e_4, \; e_1 e_3 = b\, e_4, \; e_3 e_2 = e_4 \; (a, b = 0, 1).$

(11) $e_1 e_1 = e_3, \; e_1 e_2 = a\, e_4, \; e_3 e_1 = b\, e_4, \; e_2 e_3 = e_4, \; e_3 e_2 = k\, e_4 \; (a, b = 0, 1).$

III. $r = 2.$

(a) $\rho = 1.$

(1) $\prod_{i=1,2} (\gamma_{i34} + \gamma_{3i4}) \neq 0.$ $\infty^2$ classes.

(a′) $e_2 e_1 = e_3, \; e_1 e_2 = k\, e_3 + a\, e_4, \; e_3 e_1 = e_4, \; e_3 e_2 = b\, e_4, \; e_i e_3 = \lambda\, e_3 e_i$
$(i = 1, 2), \; (a, b = 0, 1; \; \lambda \neq -1; \; |k| \leq 1 \text{ if } b = 1).$

(b′) $e_2 e_1 = e_3, \; e_1 e_2 = k\, e_3 + a\, e_4, \; e_1 e_3 = e_4, \; e_2 e_3 = b\, e_4$
$(a, b = 0, 1; \; |k| \leq 1 \text{ if } b = 1).$

(2) $\prod_{i=1,2} (\gamma_{i34} + \gamma_{3i4}) = 0.$ $\infty^2$ classes.

(a′) $e_1 e_1 = a\, e_4, \; e_1 e_2 = k\, e_4, \; e_2 e_1 = e_3, \; e_2 e_2 = b\, e_4, \; e_1 e_3 = -e_3 e_1 = e_4$
$(a, b = 0, 1).$

(b′) $e_1 e_1 = a\, e_4, \; e_1 e_2 = k\, e_3, \; e_2 e_1 = e_3, \; e_2 e_2 = \lambda\, e_4, \; e_1 e_3 = -e_3 e_1 = e_4,$
$e_2 e_3 = -e_3 e_2 = c\, e_4$
$(a, c = 0, 1; \; |k| \leq 1 \text{ if } c = 1, \; a = \lambda = 0 \text{ or if } a = c = 1, \; \lambda \neq 0).$

(b) $\rho = 2$.

(1) $\prod_{i=1,2} (\gamma_{i34} + \gamma_{3i4}) \neq 0$. $\infty^3$ classes.

(a′) $e_2 e_1 = e_3,\ e_1 e_3 = e_3 e_2 = e_4,\ e_1 e_2 = k e_3 + a e_4,\ e_3 e_1 = \lambda e_4$ $(a = 0, 1)$.

(b′) $e_2 e_1 = e_3,\ e_1 e_3 = e_2 e_3 = e_4,\ e_1 e_2 = k e_3 + a e_4,\ e_3 e_1 = \lambda e_4,\ e_3 e_2 = \mu e_4$ $(|k| \leq 1;\ a = 0, 1)$.

(2) $\prod_{i=1,2} (\gamma_{i34} + \gamma_{3i4}) = 0$. $\infty^3$ classes.

(a′) $e_2 e_1 = e_3,\ e_1 e_2 = k e_3 + \lambda e_4,\ e_2 e_3 = -e_3 e_2 = e_1 e_3 = e_4,\ e_3 e_1 = \mu e_4$ $(|k| \leq 1)$.

(b′) $e_2 e_1 = e_3,\ e_1 e_2 = k e_3 + a e_4,\ e_2 e_3 = -e_3 e_2 = e_4,\ e_3 e_1 = e_4$ $(|k| \leq 1;\ a = 0, 1)$.

(c′) $e_2 e_1 = e_3,\ e_1 e_2 = k e_3 + \lambda e_4,\ e_2 e_2 = e_1 e_3 = e_2 e_3 = -e_3 e_2 = e_4,$ $e_3 e_1 = \mu e_4$.

(d′) $e_2 e_1 = e_3,\ e_1 e_2 = k e_3 + a e_4,\ e_2 e_2 = e_3 e_1 = e_2 e_3 = -e_3 e_2 = e_4$ $(a = 0, 1)$.

Type E: $n' = n'' = 1,\ n^{(3)} = 2;\ \alpha = 4$.

$$e_1 e_1 = e_2,\ e_1 e_2 = e_3,\ e_3 e_1 = e_4.$$

Type F: $n' = n'' = n^{(3)} = n^{(4)} = 1;\ \alpha = 5$.

I. $\gamma_{224} = 0$. $\infty^2$ classes.

(a) $e_1 e_1 = e_2,\ e_1 e_2 = e_3,\ e_2 e_1 = k e_3 + a e_4,\ e_1 e_3 = e_4,\ e_3 e_1 = \mu e_4$ $(a = 0, 1)$.

(b) $e_1 e_1 = e_2,\ e_2 e_1 = e_3,\ e_1 e_3 = e_4,\ e_3 e_1 = \mu e_4$.

(c) $e_1 e_1 = e_2,\ e_1 e_2 = e_3,\ e_2 e_1 = k e_3 + a e_4,\ e_3 e_1 = e_4$ $(a = 0, 1)$.

(d) $e_1 e_1 = e_2,\ e_1 e_2 = a e_4,\ e_2 e_1 = e_3,\ e_3 e_1 = e_4$ $(a = 0, 1)$.

II. $\gamma_{224} \neq 0$. $\infty^3$ classes.

(a) $\gamma_{123} \gamma_{213} = 0$.

(1) $e_1 e_1 = e_2,\ e_2 e_1 = e_3,\ e_1 e_2 = a e_4,\ e_2 e_2 = e_4,\ e_1 e_3 = -e_4,\ e_3 e_1 = \mu e_4$.

(2) $e_1 e_1 = e_2,\ e_2 e_1 = e_3,\ e_2 e_2 = e_4,\ e_1 e_3 = \mu e_4,\ e_3 e_1 = \nu e_4$ $(\mu \neq -1)$.

(3) $e_1 e_1 = e_2,\ e_1 e_2 = e_3,\ e_2 e_1 = a e_4,\ e_2 e_2 = -e_3 e_1 = e_4,\ e_1 e_3 = \mu e_4$.

(4) $e_1 e_1 = e_2,\ e_1 e_2 = e_3,\ e_2 e_2 = e_4,\ e_3 e_1 = \mu e_4,\ e_1 e_3 = \nu e_4$ $(\mu \neq -1)$.

(b) $\gamma_{123} \gamma_{213} \neq 0$.

$e_1 e_1 = e_2,\ e_1 e_2 = e_3,\ e_2 e_1 = k e_3 + a e_4,\ e_2 e_2 = e_4,\ e_1 e_3 = \mu e_4,$ $e_3 e_1 = \nu e_4$ $(k \neq 0;\ a = 0, 1)$.

The algebras of Type F are completely characterized by the value of $r_{0,0}(r_{1,0} - 1)$, where $r_{0,0}$ and $r_{1,0}$ are respectively the ranks of

$$\begin{pmatrix} c_1 & 0 \\ c_2 & 0 \end{pmatrix},\quad \begin{pmatrix} c_1 & 1 \\ c_2 & 0 \end{pmatrix}$$

$$\begin{cases} c_1 = \gamma_{123}\gamma_{214} - \gamma_{124}\gamma_{213}, \\ c_2 = (\gamma_{123}\gamma_{314} - \gamma_{213}\gamma_{134})(\gamma_{123} + \gamma_{213}) + (\gamma_{123} - \gamma_{213})\gamma_{112}\gamma_{224}, \end{cases}$$

together with the ranks of the following matrices:

$$\begin{pmatrix}\gamma_{123} & 1\\ \gamma_{213} & \kappa\end{pmatrix},\quad \begin{pmatrix}\gamma_{123} & 0\\ \gamma_{213} & 1\end{pmatrix},\quad \begin{pmatrix}\gamma_{134} & 1\\ \gamma_{314} & \mu\end{pmatrix},\quad \begin{pmatrix}\gamma_{134} & 0\\ \gamma_{314} & 1\end{pmatrix},$$

$$\begin{pmatrix}\gamma_{123}\,\gamma_{134} & \kappa'\\ \gamma_{112}\,\gamma_{224} & 1\end{pmatrix},\quad \begin{pmatrix}\gamma_{123}\,\gamma_{134} & 1\\ \gamma_{112}\,\gamma_{224} & 0\end{pmatrix},\quad \begin{pmatrix}\gamma_{123}\,\gamma_{314} & \kappa''\\ \gamma_{112}\,\gamma_{224} & 1\end{pmatrix},\quad \begin{pmatrix}\gamma_{123}\,\gamma_{314} & 1\\ \gamma_{112}\,\gamma_{224} & 0\end{pmatrix},$$

$$\begin{pmatrix}\gamma_{213}\,\gamma_{134} & \lambda'\\ \gamma_{112}\,\gamma_{224} & 1\end{pmatrix},\quad \begin{pmatrix}\gamma_{213}\,\gamma_{134} & 1\\ \gamma_{112}\,\gamma_{224} & 0\end{pmatrix},\quad \begin{pmatrix}\gamma_{213}\,\gamma_{314} & \lambda''\\ \gamma_{112}\,\gamma_{224} & 1\end{pmatrix},\quad \begin{pmatrix}\gamma_{213}\,\gamma_{314} & 1\\ \gamma_{112}\,\gamma_{224} & 0\end{pmatrix}.$$

Type G: $n'=2$, $n^{(2)}=1$, $n^{(3)}=0$, $n^{(4)}=1$.

I. $Q_3 \equiv \sum_{i,j}^{1,2} \gamma_{ij3} x_i x_j$ of rank 0.

(1) $e_1e_2=-e_2e_1=e_3$, $e_3e_3=e_4$.

(2) $e_1e_2=-e_2e_1=e_3$, $e_3e_1=e_3e_3=e_4$.

(3) $e_1e_2=-e_2e_1=e_3$, $e_1e_1=e_3e_3=e_4$, $e_3e_1=\kappa e_4$.

(4) $e_1e_2=-e_2e_1=e_3$, $e_1e_1=e_3e_2=e_3e_3=e_4$.

(5) $e_1e_2=e_3$, $e_2e_1=-e_3+e_4$, $e_3e_1=e_3e_3=e_4$, $e_3e_2=\kappa e_4$.

(6) $e_1e_2=e_3$, $e_2e_1=-e_3+e_4$, $e_3e_3=e_4$, $e_3e_2=ae_4\,(a=0,\,1)$.

II. $Q_3$ of rank 1.

(1) $e_1e_1=e_3$, $e_2e_1=e_2e_2=e_3e_3=e_4$, $e_3e_1=\kappa e_4$, $e_3e_2=\lambda e_4$.

(2) $e_1e_1=e_3$, $e_2e_2=e_3e_3=e_4$, $e_3e_2=ae_4\,(a=0,\,1)$.

(3) $e_1e_1=e_3$, $e_2e_2=e_3e_3=e_3e_1=e_4$, $e_3e_2=\kappa e_4$.

(4) $e_1e_1=e_3$, $e_1e_2=e_3e_2=e_3e_3=e_4$, $e_2e_1=\kappa e_4$.

(5) $e_1e_1=e_3$, $e_1e_2=e_3e_1=e_3e_3=e_4$, $e_2e_1=\kappa e_4$.

(6) $e_1e_1=e_3$, $e_1e_2=e_3e_3=e_4$, $e_2e_1=\kappa e_4$.

(7) $e_1e_1=e_3$, $e_2e_1=e_3e_2=e_3e_3=e_4$.

(8) $e_1e_1=e_3$, $e_2e_1=e_3e_1=e_3e_3=e_4$.

(9) $e_1e_1=e_3$, $e_2e_1=e_3e_3=e_4$.

(10) $e_1e_1=e_3$, $e_3e_2=e_3e_3=e_4$.

(11) $e_1e_1=e_3$, $e_3e_1=e_3e_3=e_4$.

(12) $e_1e_1=e_3$, $e_3e_3=e_4$.

(13) $e_1e_1=e_3$, $e_1e_2=e_3+\kappa e_4$, $e_2e_1=-e_3+\lambda e_4$, $e_3e_2=e_3e_3=e_4$.

(14) $e_1e_1=e_1e_2=e_3$, $e_2e_1=-e_3+\kappa e_4$, $e_2e_2=e_3e_3=e_4$, $e_3e_2=\lambda e_4$.

(15) $e_1e_1=e_1e_2=e_3$, $e_2e_1=-e_3+\kappa e_4$, $e_3e_1=e_3e_3=e_4$, $e_2e_2=\lambda e_4$.

(16) $e_1e_1=e_3$, $e_1e_2=e_3+\kappa e_4$, $e_2e_1=-e_3+\lambda e_4$, $e_3e_1=e_3e_3=e_4$, $e_2e_2=\mu e_4$.

(17) $e_1e_1=e_1e_2=e_3$, $e_2e_1=-e_3+\kappa e_4$, $e_3e_3=e_2e_2=e_4$.

(18) $e_1e_1=e_1e_2=e_3$, $e_2e_1=-e_3+\kappa e_4$, $e_3e_3=e_4$.

III. $Q_3$ of rank 2.

(1) $e_1e_2=e_3$, $e_2e_1=\kappa e_3+\lambda e_4$, $e_1e_1=e_2e_2=e_3e_3=e_4$, $e_3e_1=\mu e_4$,
$e_3e_2=\nu e_4\,(\kappa\neq-1)$.

(2) $e_1e_2=e_3,\ e_2e_1=xe_3+\lambda e_4,\ e_1e_1=e_3e_2=e_3e_3=e_4,\ e_3e_1=\mu e_4\ (x\neq -1)$.

(3) $e_1e_2=e_3,\ e_2e_1=xe_3+\lambda e_4,\ e_1e_1=e_3e_1=e_3e_3=e_4 (x\neq -1)$.

(4) $e_1e_2=e_3,\ e_2e_1=xe_3+\lambda e_4,\ e_1e_1=e_3e_3=e_4 (x\neq -1)$.

(5) $e_1e_2=e_3,\ e_2e_1=xe_3+\lambda e_4,\ e_3e_1=e_3e_2=e_3e_3=e_4 (x\neq -1)$.

(6) $e_1e_2=e_3,\ e_2e_1=xe_3+ae_4,\ e_3e_1=e_3e_3=e_4 (x\neq -1;\ a=0, 1)$.

(7) $e_1e_2=e_3,\ e_2e_1=xe_3+ae_4,\ e_3e_3=e_4 (x\neq -1;\ a=0, 1)$.

Type H: $n'=n^{(2)}=n^{(3)}=1,\ n^{(4)}=0,\ n^{(5)}=1$.

(1) $e_1e_1=e_2,\ e_1e_2=e_3=-e_2e_1,\ e_2e_3=-e_3e_2=e_4,\ e_3e_1=ae_4 (a=0, 1)$.

(2) $e_1e_1=e_2,\ e_1e_2=e_3,\ e_2e_1=-e_3+e_4,\ e_2e_3=-e_3e_2=e_4,\ e_3e_1=xe_4$.

Here there is a class for every value of $x^2$.

(3) $e_1e_1=e_2,\ e_1e_2=e_3,\ e_2e_1=xe_3,\ e_2e_3=-e_3e_2=e_4,\ e_3e_1=ae_4 (a=0, 1)$.

(4) $e_1e_1=e_2,\ e_1e_2=e_3,\ e_2e_1=-e_3+xe_4,\ e_2e_2=e_2e_3=-e_3e_2=e_4,\ e_3e_1=\lambda e_4$.

(5) $e_1e_1=e_2,\ e_1e_2=e_3,\ e_2e_1=xe_3,\ e_2e_2=e_2e_3=-e_3e_2=e_4,\ e_3e_1=\lambda e_4$.

(6) $e_1e_1=e_2,\ e_1e_2=e_3,\ e_2e_1=xe_3,\ e_2e_2=e_3e_1=e_4,\ e_1e_3=ae_4 (a=0, 1)$.

(7) $e_1e_1=e_2,\ e_1e_2=e_3,\ e_2e_1=xe_3,\ e_3e_2=e_4,\ e_1e_3=ae_4 (a=0, 1)$.

(8) $e_1e_1=e_2,\ e_1e_2=e_3,\ e_2e_1=xe_3+e_4,\ e_3e_2=e_4,\ e_3e_1=\lambda e_4,\ e_1e_3=\mu e_4$.

(9) $e_1e_1=e_2,\ e_2e_1=e_3,\ e_3e_2=e_4,\ e_1e_3=ae_4 (a=0, 1)$.

(10) $e_1e_1=e_2,\ e_2e_1=e_3,\ e_3e_1=e_3e_2=e_4,\ e_1e_3=xe_4$.

(11) $e_1e_1=e_2,\ e_1e_2=-e_2e_1=e_3,\ e_2e_3=e_4,\ e_3e_2=xe_4,\ e_3e_1=ae_4$
$(x\neq -1;\ a=0, 1)$.

(12) $e_1e_1=e_2,\ e_2e_1=e_3,\ e_1e_2=e_2e_3=e_4,\ e_1e_3=xe_4,\ e_3e_1=\lambda e_4,$
$e_3e_2=\mu e_4 (\mu\neq -1)$.

(13) $e_1e_1=e_2,\ e_2e_1=e_3,\ e_2e_3=e_4,\ e_3e_2=xe_4,\ e_3e_1=ae_4 (a=0, 1;\ x\neq -1)$.

(14) $e_1e_1=e_2,\ e_2e_1=e_3,\ e_1e_3=e_2e_3=e_4,\ e_3e_1=xe_4,\ e_3e_2=\lambda e_4 (\lambda\neq -1)$.

(15) $e_1e_1=e_2,\ e_1e_2=e_3,\ e_2e_1=xe_3,\ e_2e_3=e_4,\ e_3e_2=\lambda e_4,\ e_3e_1=ae_4$
$(\lambda\neq -1;\ a=0, 1)$.

(16) $e_1e_1=e_2,\ e_1e_2=e_3,\ e_2e_1=xe_3,\ e_1e_3=e_2e_3=e_4,\ e_3e_1=\lambda e_4,\ e_3e_2=\mu e_4$
$(\mu\neq -1, 0;\ x\neq 0)$.

(17) $e_1e_1=e_2,\ e_1e_2=e_3,\ e_2e_1=xe_3+\lambda e_4,\ e_2e_3=e_4,\ e_3e_2=\mu e_4,\ e_3e_1=ae_4$
$(a=0, 1;\ \lambda=1, \text{ if } a=0;\ \mu\neq -1, 0;\ x\neq 0)$.

(18) $e_1e_1=e_2,\ e_1e_2=e_3,\ e_2e_1=xe_3+\lambda e_4,\ e_1e_3=e_2e_3=e_4,\ e_3e_1=\mu e_4,$
$e_3e_2=\nu e_4 (x\neq 0;\ \nu\neq -1, 0)$.

Type I: $n'=n^{(2)}=n^{(3)}=1,\ n^{(4)}=n^{(5)}=0,\ n^{(6)}=1$.

(1) $e_1e_1=e_2,\ e_1e_2=e_3,\ e_2e_1=xe_3+\lambda e_4,\ e_2e_2=e_3e_3=e_4,\ e_3e_1=\mu e_4,\ e_3e_2=\nu e_4$ $(x\neq -1)$. Here there is a class for every set of values of $\lambda^2$, $\mu^2$, $\nu^2$ and $x$.

(2) $e_1e_1=e_2,\ e_1e_2=e_3,\ e_2e_1=xe_3+e_4,\ e_3e_3=e_4,\ e_3e_1=\mu e_4,\ e_3e_2=\nu e_4 (x\neq -1)$. There is a class for every set of values of $\mu^3$, $\nu^3$ and $x$.

(3) $e_1e_1=e_2,\ e_1e_2=e_3,\ e_2e_1=\varkappa e_3,\ e_3e_1=e_3e_3=e_4,\ e_3e_2=\nu e_4\,(\varkappa\neq-1)$. There is a class for every value of $\nu^2$ and $\varkappa$.

(4) $e_1e_1=e_2,\ e_1e_2=e_3,\ e_3e_3=e_4,\ e_2e_1=\varkappa e_3,\ e_3e_2=ae_4\,(a=0,\,1;\ \varkappa\neq-1)$.

(5) $e_1e_1=e_2,\ e_1e_2=e_3,\ e_2e_1=-e_3+\varkappa e_4,\ e_2e_3=e_3e_3=e_4,\ e_2e_2=\lambda e_4,$
$e_3e_1=\mu e_4,\ e_3e_2=\nu e_4$.

(6) $e_1e_1=e_2,\ e_1e_2=e_3,\ e_2e_1=-e_3+\varkappa e_4,\ e_3e_2=e_3e_3=e_4,\ e_2e_2=\lambda e_4,\ e_3e_1=\mu e_4$.

(7) $e_1e_1=e_2,\ e_1e_2=e_3,\ e_2e_1=-e_3+\varkappa e_4,\ e_2e_2=e_3e_3=e_4,\ e_3e_1=\lambda e_4$. There is a class here for every set of values of $\varkappa^2$, $\lambda^2$.

(8) $e_1e_1=e_2,\ e_1e_2=e_3,\ e_2e_1=-e_3+e_4,\ e_3e_3=e_4,\ e_3e_1=\varkappa e_4$. Here there is a class for every value of $\varkappa^3$.

(9) $e_1e_1=e_2,\ e_1e_2=-e_2e_1=e_3,\ e_3e_3=e_4,\ e_3e_1=ae_4\,(a=0,\,1)$.

(10) $e_1e_1=e_2,\ e_2e_1=e_3,\ e_2e_2=e_3e_3=e_4,\ e_1e_2=\lambda e_4,\ e_1e_3=\mu e_4,\ e_2e_3=\nu e_4$. There is a class for every set of values of $\lambda^2$, $\mu^2$, $\nu^2$.

(11) $e_1e_1=e_2,\ e_2e_1=e_3,\ e_1e_2=e_3e_3=e_4,\ e_1e_3=\mu e_4,\ e_2e_3=\nu e_4$. Here there is a class for every set of values of $\mu^3$, $\nu^3$.

(12) $e_1e_1=e_2,\ e_2e_1=e_3,\ e_1e_3=e_3e_3=e_4,\ e_2e_3=\nu e_4$. There is a class for every value of $\nu^2$.

(13) $e_1e_1=e_2,\ e_2e_1=e_3,\ e_3e_3=e_4,\ e_2e_3=ae_4\,(a=0,\,1)$.

Type J: $n'=n^{(2)}=1,\ n^{(3)}=0,\ n^{(4)}=n^{(5)}=1$.

(1) $e_1e_1=e_2,\ e_2e_2=e_3,\ e_1e_3=e_4,\ e_2e_1=ae_4,\ e_3e_1=\varkappa e_4\,(a=0,\,1)$.

(2) $e_1e_1=e_2,\ e_2e_2=e_3,\ e_2e_1=e_3+\varkappa e_4,\ e_3e_1=\lambda e_4$.

(3) $e_1e_1=e_2,\ e_2e_2=e_3,\ e_2e_1=\varkappa e_3+\lambda e_4,\ e_1e_2=e_1e_3=e_4,\ e_3e_1=\mu e_4$. There is a class for every set of values of $\varkappa^2$, $\lambda$ and $\mu$.

(4) $e_1e_1=e_2,\ e_2e_2=e_3,\ e_3e_1=e_4,\ e_1e_3=ae_4\,(a=0,\,1)$.

(5) $e_1e_1=e_2,\ e_2e_2=e_3,\ e_1e_2=e_3+\varkappa e_4,\ e_3e_1=e_4$.

(6) $e_1e_1=e_2,\ e_2e_2=e_3,\ e_2e_1=e_3e_1=e_4,\ e_1e_2=\varkappa e_3+\lambda e_4$. There is a class here for every set of values of $\varkappa^2$, $\lambda$.

Type K: $n'=n^{(2)}=1,\ n^{(3)}=0,\ n^{(4)}=1,\ n^{(5)}=0,\ n^{(6)}=1$.

(1) $e_1e_1=e_2,\ e_2e_2=e_3,\ e_1e_2=ae_3,\ e_2e_1=\varkappa e_3+\lambda e_4,\ e_2e_3=e_4,\ e_3e_1=\mu e_4,$
$e_3e_2=\nu e_4\,(a=0,\,1)$.

(2) $e_1e_1=e_2,\ e_2e_2=e_3,\ e_1e_2=ae_3+e_4,\ e_2e_1=\varkappa e_3+\lambda e_4,\ e_2e_3=e_4,\ e_3e_1=\mu e_4$. There is a class here for every set of values of $a^3$, $\varkappa^3$, $\mu^3$, $\lambda$ and $\nu$.

(3) $e_1e_1=e_2,\ e_2e_2=e_3,\ e_1e_3=\varkappa e_3,\ e_2e_1=\lambda e_3+\mu e_4,\ e_3e_2=e_1e_3=e_4$.

(4) $e_1e_1=e_2,\ e_1e_2=e_2e_2=e_3,\ e_2e_1=\varkappa e_3+\lambda e_4,\ e_3e_2=e_4$.

(5) $e_1e_1=e_2,\ e_2e_2=e_3,\ e_2e_1=e_3+\varkappa e_4,\ e_3e_2=e_4$.

(6) $e_1e_1=e_2,\ e_2e_2=e_3,\ e_2e_1=ae_4,\ e_3e_2=e_4\,(a=0,\,1)$.

Type L: $n'=n^{(2)}=1,\ n^{(3)}=0,\ n^{(4)}=1,\ n^{(5)}=n^{(6)}=n^{(7)}=0,\ n^{(8)}=1$.

(1) $e_1e_1=e_2,\ e_2e_2=e_3,\ e_2e_1=\varkappa e_3+\lambda e_4,\ e_1e_2=e_4=e_3e_3,\ e_3e_1=\mu e_4,\ e_3e_2=\nu e_4,$ $e_2e_3=\sigma e_4$. Here there is a class for every set of values of $\varkappa$, $\lambda^5$, $\mu^5$, $\nu^5$, $\sigma^5$.

(2) $e_1e_1=e_2$, $e_2e_2=e_3$, $e_2e_1=e_3+xe_4$, $e_3e_3=e_4$, $e_3e_1=\lambda e_4$, $e_3e_2=\mu e_4$, $e_2e_3=\nu e_4$.

(3) $e_1e_1=e_2$, $e_2e_2=e_3$, $e_2e_1=e_3e_3=e_4$, $e_3e_1=xe_4$, $e_3e_2=\lambda e_4$, $e_2e_3=\mu e_4$. There is a class here for every set of values of $x$, $\lambda^3$, $\mu^3$.

(4) $e_1e_1=e_2$, $e_2e_2=e_3$, $e_3e_1=e_3e_3=e_4$, $e_2e_3=xe_4$, $e_3e_2=\lambda e_4$. There is a class for every pair of values of $x^3$, $\lambda^3$.

(5) $e_1e_1=e_2$, $e_2e_2=e_3$, $e_2e_3=e_3e_3=e_4$, $e_3e_2=xe_4$. There is a class for every value of $x^2$.

(6) $e_1e_1=e_2$, $e_2e_2=e_3$, $e_3e_3=e_4$, $e_3e_2=ae_4(a=0, 1)$.

## CHAPTER IV.

### Application to Associative Nilpotent Algebras.

15. *Invariantive Characterization of Two Types of General Nilpotent Algebras of Index Three.* There are some types of nilpotent algebras in $n$ units over the field $C$ of ordinary complex numbers which are completely characterized by invariants in a rather interesting manner; and we will now turn our attention to several such types.

Consider a nilpotent algebra $E=(e_1, \ldots, e_n)$ of index three, where $E^2=(e_n)$. Then there are two interesting types of such algebras according as the fundamental quadratic

$$Q=\sum_{i,j}^{1,n'}\gamma_{ijn}x_ix_j\eta_n \quad (n'=n-1)$$

is of rank zero for general $\eta_n$, or of rank one. When $Q$ is of rank zero, we know from §§ 10, 13, 14 that the rank of

$$|\gamma_{ijn}| \quad (i, j=1, \ldots, n'), \qquad (15)$$

is an invariant for algebras of this type which completely characterizes them when $n'=2, 3$. By induction it can readily be shown that

Theorem. *Over the field of ordinary complex numbers the rank of the invariant* (15) *completely characterizes nilpotent algebras $E$ of genus* $(3; n-1, 1)$ *in canonical form, when the fundamental quadratic is of rank zero.*

Similarly by induction from $n=3, 4$ we have the

Theorem. *Over the field of ordinary complex numbers the rank of the invariant* (15) *completely characterizes nilpotent algebras $E$ of genus* $(3; n-1, 1)$ *in canonical form, where the fundamental quadratic is of rank one.*

16. *General Commutative Associative Nilpotent Algebras.* In view of the theorem of § 8, the problem of the equivalence of two commutative $n$-ary algebras $E$ and $E'$ over a field $F$ of index three (therefore associative)—where $n'$ is the order of $B\equiv E$ (mod $E^2$) and of $B'\equiv E'$ (mod $E'^2$)—is the same as the

problem of equivalence of two families of quadratic forms (10), in the field $F$, in $n'$ independent variables $x_1, \ldots, x_{n'}$ and with $n''=n-n'$ independent parameters $\eta_{n'+1}, \ldots, \eta_n$ entering homogeneously.

More generally, the theorem of § 8, together with the invariantive property of the invariants of the fundamental forms of the complex $B\equiv E \pmod{E^2}$, gives us the following

THEOREM. *If we consider over a field F the n-ary commutative, associative nilpotent algebras of genus* $(\alpha; n', \ldots, n^{(\alpha-1)})$ *in "special" canonical form*

$$E=\sum_{i=1}^{\alpha-1} B_i, \text{ where } B\equiv E \pmod{E^2},\ B_i\equiv B^i \pmod{E^{i+1}}, \tag{16}$$

*with multiplication table*

$$e_p e_q = \sum_{r=\nu^{(k+l-1)}+1}^{\nu^{(k+l)}} \gamma_{pqr} e_r \quad \text{for } e_p \text{ in } B_k,\ e_q \text{ in } B_l, \tag{17}$$

*and where*

$$\nu^{(i)}=\sum_{j=1}^{i} n^{(j)}, \quad \sum_{j=1}^{\alpha-1} n^{(j)}=n,$$

*then these algebras are characterized completely by those invariants which completely characterize the following set of $\alpha-2$ families of forms over the field F in $n'$ independent variables:*

$$\left.\begin{aligned} F_i &\equiv \sum_{r=\nu^{(i-1)}+1}^{\nu^{(i)}} F_{ir}\eta_r \quad (i=2, \ldots, \alpha-1), \\ \text{where} \quad F_{2r} &\equiv \sum_{j,k=1}^{n'} \gamma_{jkr} x_j x_k \quad (\nu^{(i-1)}+1 \leq r \leq \nu^{(i)}), \\ F_{3r} &\equiv \sum_{j,k,l}^{1,n'} \Big(\sum_{m=\nu'+1}^{\nu''} \gamma_{klm}\gamma_{jmr}\Big) x_j x_k x_l, \end{aligned}\right\} \tag{18}$$

*and where, in general, $F_p$ is the fundamental p-ic of the complex B. Here the $\eta$'s are parameters independent of the x's.*

17. *Commutative Associative Nilpotent Algebras in n Units with $\alpha=3$, $n''=1$.* In particular, by the theorem of § 16, the problem of the invariantive characterization of commutative nilpotent $n$-ary algebras $E$ over the field $C$ where $E^2=(e_n)$,

$$E:\ e_j e_i = e_i e_j = \gamma_{ijn} e_n \quad (i, j=1, \ldots, n-1),\ e_n e_k = e_k e_n = 0 \quad (k=1, \ldots, n), \tag{19}$$

is essentially the same as the problem of the invariantive characterization of the quadratic forms

$$Q\equiv \sum_{i,j}^{1,n-1} \gamma_{ijn} x_i x_j \tag{20}$$

in the field $C$, where the $x$'s are independent variables—the coordinates of the general number in $E$ (mod $E^2$). Now if $r$ is the rank of $Q$, there is a non-singular transformation on the $x$'s in $C$ which will carry $Q$ into $\sum_{k=1}^{r} x_k^2$; and hence, since $Q$ is a covariant for the algebras considered, and since the $x$'s are contragredient to the units $e_1, \ldots, e_{n-1}$, and in view of the commutativity of $E$, there is a non-singular transformation on the units of $E$ which will carry it into the algebra where $e_k e_k = e_n$ $(k \leq r)$ and where the products not written are zero. Thus we have the

THEOREM. *The n-ary nilpotent commutative algebras of index three where $E^2 = (e_n)$ are completely characterized by the rank of the fundamental quadratics* (20), *where the $\gamma_{ijk}$ are defined by* (19).

18. *Commutative, Associative Nilpotent Algebras in n Units of Genus* (3; $n$—2, 2). Furthermore, § 16 shows that in the field $C$ the problem of the equivalence of two commutative nilpotent $n$-ary algebras $E'$ and $E''$ of index three, where $E'^2 = (e'_{n-1}, e'_n)$ and $E''^2 = (e''_{n-1}, e''_n)$,

$$\left.\begin{aligned} E' : e'_i e'_j &= \sum_k^{1,2} \gamma'_{ijn'+k} e'_{n'+k} \quad (i, j = 1, \ldots, n' = n-2), \\ E'' : e''_i e''_j &= \sum_k^{1,2} \gamma''_{ijn'+k} e''_{n'+k} \quad (i, j = 1, \ldots, n' = n-2), \end{aligned}\right\} \tag{21}$$

is essentially the same as the problem of the equivalence of the two families of quadratic forms

$$\left.\begin{aligned} Q' &= \sum_k^{1,2} \sum_{i,j}^{1,n'} \gamma'_{ijn'+k} x'_i x'_j \eta'_{n'+k}, \\ Q'' &= \sum_k^{1,2} \sum_{i,j}^{1,n'} \gamma''_{ijn'+k} x''_i x''_j \eta''_{n'+k} \end{aligned}\right\} \tag{22}$$

in the field $C$. Here the $x'$'s are independent variables—actually they are the coordinates of the general number in $E'$ (mod $E'^2$)—and the $\eta'$'s are parameters independent of the $x'$'s. Similarly for the $x''$'s and $\eta''$'s.

By § 10 or § 11, a necessary condition that $E'$ be equivalent to $E''$ is that the polynomials

$$\left.\begin{aligned} \left| \sum_k^{1,2} \gamma'_{ijn'+k} \eta'_{n'+k} \right| &= \prod_f (a'_f \eta'_{n-1} + b'_f \eta'_n)^{l'_f} \quad (i, j = 1, \ldots, n'), \\ \left| \sum_k^{1,2} \gamma''_{ijn'+k} \eta''_{n'+k} \right| &= \prod_f (a''_f \eta''_{n-1} + b''_f \eta''_n)^{l''_f} \quad (i, j = 1, \ldots, n') \end{aligned}\right\} \tag{23}$$

be equivalent under a non-singular transformation on the parameters. Let

$$a'_f \eta'_{n-1} + b'_f \eta'_n \quad (f = 1, \ldots, t') \tag{24'}$$

and

$$a''_f \eta''_{n-1} + b''_f \eta''_n \quad (f = 1, \ldots, t'') \tag{24''}$$

be the distinct linear factors of the two polynomials (23) respectively. With each linear factor there belong certain exponents $l'_{j,r}$ (or $l''_{j,r}$) which tell the power to which that expression is a factor common to all $r$-rowed minors $(r=1, \ldots, n')$.

If the two polynomials (23) are equivalent, then $t'=t''$ and a one-to-one correspondence can be set up between the distinct linear factors (24′) and (24″) in such a way that the multiplicities of the corresponding factors are equal. If, further, the algebra $E'$ be equivalent to $E''$, then this correspondence can be set up in such a way that, for every $r$, if a particular factor (24′) occur exactly $l'_{j,r}$ times as a factor common to all $r$-rowed minors of the determinant (23′), then its corresponding factor (24″) will be a factor of all $r$-rowed minors of the determinant (23″) exactly $l'_{j,r}$ times.

Conversely, by the theory of elementary divisors, if such a correspondence can be set up between the distinct linear factors (24′) and the distinct linear factors (24″), the fundamental quadratic (22) for $B'\equiv E' \pmod{E'^2}$ is equivalent to the fundamental quadratic (22) for $B''=E'' \pmod{E''^2}$ under a non-singular transformation on the $x''$s, combined with a non-singular transformation on the $\eta''$s. Thus, since the quadratics (22) are covariants for the algebras of genus $(3;\, n-2,\, 2)$, and since the $x''$s are contragredient to the units $e'_1, \ldots, e'_{n-2}$, and in view of the commutativity, there is a non-singular transformation on the units which will carry $E'$ into $E''$. Hence the following

THEOREM. *Over the field C, two n-ary commutative nilpotent algebras* (21) *of genus* $(3;\, n-2,\, 2)$ *in canonical form are equivalent if and only if the determinants* (23) *are equivalent and such that the elementary divisors of one can be made to correspond to the elementary divisors of the other.*

10. *Commutative, Associative Nilpotent Algebras with* $n<6$. As in the foregoing sections of this chapter, we will consider algebras defined over the field of ordinary complex numbers. Products not written are zero.

$n=1$. There is only one class. $e_1^2=0$.

$n=2$. There are two classes characterized by their index.

$$e_1^2=e_2;\quad e_ie_j=0\quad (i, j=1, 2).$$

$n=3$. (1) $e_ie_j=0\quad (i, j=1, 2, 3)$.

(2) $e_1e_1=e_2e_2=e_3$.

(3) $e_1e_1=e_3$.

(4) $e_1e_1=e_2,\ e_1e_2=e_2e_1=e_3$.

The characteristic invariants here are respectively

$$I_1=\frac{(4-\alpha)(3-\alpha)}{2},$$
$$I_2=(4-\alpha)(\alpha-2)(r-1),$$
$$I_3=(4-\alpha)(\alpha-2)(2-r),$$
$$I_4=\frac{(\alpha-3)(\alpha-2)}{2},$$

where $\alpha$ is the index and $r$ is the rank of $Q_3\equiv\sum_{i,j}^{1,2}\gamma_{ij3}x_ix_j$.

$n=4$. (1) $e_ie_j=0$ $(i, j=1, 2, 3, 4)$.

(2) $e_1e_1=e_2e_2=e_3e_3=e_4$.

(3) $e_1e_1=e_2e_2=e_4$.

(4) $e_1e_1=e_4$.

(5) $e_1e_1=e_3$, $e_1e_2=e_2e_1=e_4$.

(6) $e_1e_1=e_2e_2=e_3$, $e_1e_2=e_2e_1=e_4$.

(7) $e_1e_1=e_3$, $e_1e_3=e_3e_1=e_4$.

(8) $e_1e_1=e_3$, $e_2e_2=e_1e_3=e_3e_1=e_4$.

(9) $e_1e_1=e_3$, $e_2e_3=e_3e_2=e_4$.

(10) $e_1e_2=e_2e_1=e_3$, $e_1e_3=e_3e_1=e_4$.

(11) $e_1e_2=e_2e_1=e_3$, $e_2e_2=e_1e_3=e_3e_1=e_4$.

(12) $e_1e_2=e_2e_1=e_3$, $e_1e_3=e_3e_1=e_2e_3=e_3e_2=e_4$.

(13) $e_1e_1=e_2$, $e_1e_2=e_2e_1=e_3$, $e_2e_2=e_1e_3=e_3e_1=e_4$.

The characteristic invariants for $n=4$ are respectively:

$$I_1=\frac{(5-\alpha)(4-\alpha)(3-\alpha)}{3!},$$

$$I_2=(n'-2)\frac{(5-\alpha)(4-\alpha)(\alpha-2)}{2}(\rho_3-2)(\rho_3-1),$$

$$I_3=(n'-2)\frac{(5-\alpha)(4-\alpha)(\alpha-2)}{2}(3-\rho_3)(\rho_3-1),$$

$$I_4=(n'-2)\frac{(5-\alpha)(4-\alpha)(\alpha-2)}{2}\frac{(3-\rho_3)(2-\rho_3)}{2},$$

$$I_5=(3-n')\frac{(5-\alpha)(4-\alpha)(\alpha-2)}{2}(2-r_2),$$

$$I_6=(3-n')\frac{(5-\alpha)(4-\alpha)(\alpha-2)}{2}(r_2-1),$$

$$I_7=\frac{(5-\alpha)(\alpha-3)(\alpha-2)}{2}(3-r_3)(2-\rho_2),$$

$$I_8=\frac{(5-\alpha)(\alpha-3)(\alpha-2)}{2}(r_3-2)(m-2)(2-\rho_2),$$

$$I_9 = \frac{(5-\alpha)(\alpha-3)(\alpha-2)}{2}(r_3-2)(3-m)(2-\rho_2),$$

$$I_{10} = \frac{(5-\alpha)(\alpha-3)(\alpha-2)}{2}(3-r_3)(\rho_2-1),$$

$$I_{11} = \frac{(5-\alpha)(\alpha-3)(\alpha-2)}{2}(r_3-2)(m-2)(\rho_2-1),$$

$$I_{12} = \frac{(5-\alpha)(\alpha-3)(\alpha-2)}{2}(r_3-2)(3-m)(\rho_2-1),$$

$$I_{13} = \frac{(\alpha-4)(\alpha-3)(\alpha-2)}{3},$$

where $\alpha$ is the index, $\rho_2$ is the rank of $|\gamma_{ij3}|$ $(i, j=1, 2)$, $\rho_3$ is the rank of $|\gamma_{ij4}|$ $(i, j=1, 2, 3)$, $r_2$ is the rank of the discriminant of $|\sum_k^{3,4} \gamma_{ijk}\eta_k|$ $(i, j=1, 2)$, $r_3$ is the rank of the discriminant of $|\sum_k^{3,4} \gamma_{ijk}\eta_k|$ $(i, j=1, 2, 3)$, and $m$ is the maximum multiplicity of a linear factor of $|\sum_k^{3,4} \gamma_{ijk}\eta_k|$ $(i, j=1, 2, 3)$.

$n=5$. (1) $e_i e_j = 0$ $(i, j=1, \ldots, 5)$.

(2) $e_1 e_1 = e_3$, $e_1 e_2 = e_2 e_1 = e_4$, $e_2 e_2 = e_5$.

(3) $e_1 e_1 = e_4$, $e_1 e_2 = e_2 e_1 = e_5$.

(4) $e_1 e_1 = e_2 e_2 = e_4$, $e_1 e_2 = e_2 e_1 = e_5$.

(5) $e_1 e_2 = e_2 e_1 = e_4$, $e_2 e_3 = e_3 e_2 = e_5$.

(6) $e_1 e_1 = e_4$, $e_2 e_2 = e_5$, $e_3 e_3 = e_4 + e_5$.

(7) $e_1 e_1 = e_2 e_2 = e_4$, $e_3 e_3 = e_5$.

(8) $e_1 e_2 = e_2 e_1 = e_4$, $e_2 e_2 = e_3 e_3 = e_5$.

(9) $e_1 e_2 = e_2 e_1 = e_3 e_3 = e_4$, $e_2 e_2 = e_5$.

(10) $e_1 e_2 = e_2 e_1 = e_3 e_3 = e_4$, $e_2 e_3 = e_3 e_2 = e_5$.

(11) $e_1 e_1 = e_2 e_2 = e_3 e_3 = e_4 e_4 = e_5$.

(12) $e_1 e_1 = e_2 e_2 = e_3 e_3 = e_5$.

(13) $e_1 e_1 = e_2 e_2 = e_5$.

(14) $e_1 e_1 = e_5$.

(15) $e_1 e_1 = e_3$, $e_1 e_2 = e_2 e_1 = e_4$, $e_1 e_3 = e_3 e_1 = e_5$.

(16) $e_1 e_1 = e_3$, $e_1 e_2 = e_2 e_1 = e_4$, $e_2 e_2 = e_1 e_3 = e_3 e_1 = e_5$.

(17) $e_1 e_1 = e_3$, $e_1 e_2 = e_2 e_1 = e_4$, $e_2 e_3 = e_3 e_2 = e_1 e_4 = e_4 e_1 = e_5$.

(18) $e_1 e_1 = e_2 e_2 = e_3$, $e_1 e_2 = e_2 e_1 = e_4$, $e_1 e_3 = e_3 e_1 = e_2 e_3 = e_3 e_2 = e_5$,
$e_1 e_4 = e_4 e_1 = e_2 e_4 = e_4 e_2 = e_5$.

(19) $e_1 e_1 = e_2 e_2 = e_3$, $e_1 e_2 = e_2 e_1 = e_4$, $e_1 e_3 = e_3 e_1 = e_2 e_4 = e_4 e_2 = e_5$.

(20) $e_1 e_1 = e_4$, $e_2 e_2 = e_1 e_4 = e_4 e_1 = e_5$.

(21) $e_1 e_1 = e_4$, $e_2 e_2 = e_3 e_3 = e_1 e_4 = e_4 e_1 = e_5$.

(22) $e_1 e_1 = e_3$, $e_1 e_3 = e_3 e_1 = e_4$, $e_1 e_4 = e_4 e_1 = e_3 e_3 = e_5$.

(23) $e_1 e_1 = e_3$, $e_1 e_3 = e_3 e_1 = e_4$, $e_2 e_2 = e_1 e_4 = e_4 e_1 = e_3 e_3 = e_5$.

(24) $e_1 e_1 = e_3$, $e_2 e_2 = e_1 e_3 = e_3 e_1 = e_4$, $e_1 e_4 = e_4 e_1 = e_3 e_3 = e_5$.

(25) $e_1 e_1 = e_2$, $e_1 e_2 = e_2 e_1 = e_3$, $e_2 e_2 = e_1 e_3 = e_3 e_1 = e_4$,
$e_1 e_4 = e_4 e_1 = e_2 e_3 = e_3 e_2 = e_5$.

The characteristic invariants for these classes are respectively:

$$I_1 = \frac{(6-\alpha)(5-\alpha)(4-\alpha)(3-\alpha)}{4!},$$

$$I_2 = \frac{(6-\alpha)(5-\alpha)(4-\alpha)(\alpha-2)}{3!} \frac{(3-n')(4-n')}{2!},$$

$$I_3 = \frac{(6-\alpha)(5-\alpha)(4-\alpha)(\alpha-2)}{3!} (n'-2)(4-n')\tau_1(2-\tau_1)(3-r_3),$$

$$I_4 = \frac{(6-\alpha)(5-\alpha)(4-\alpha)(\alpha-2)}{3!} (n'-2)(4-n') \frac{(\tau_1-1)\tau_1}{2} (3-r_3),$$

$$I_5 = \frac{(6-\alpha)(5-\alpha)(4-\alpha)(\alpha-2)}{3!} (n'-2)(4-n') \frac{(2-\tau_1)(1-\tau_1)}{2} (3-r_3),$$

$$I_6 = \frac{(6-\alpha)(5-\alpha)(4-\alpha)(\alpha-2)}{3!} (n'-2)(4-n')(r_3-2) \frac{(3-m)(2-m)}{2},$$

$$I_7 = \frac{(6-\alpha)(5-\alpha)(4-\alpha)(\alpha-2)}{3!} (n'-2)(4-n')(r_3-2)(3-m)(m-1)(2-\tau_2),$$

$$I_8 = \frac{(6-\alpha)(5-\alpha)(4-\alpha)(\alpha-2)}{3!} (n'-2)(4-n')(r_3-2)(3-m)(m-1)(\tau_2-1),$$

$$I_9 = \frac{(6-\alpha)(5-\alpha)(4-\alpha)(\alpha-2)}{3!} (n'-2)(4-n')(r_3-2) \frac{(m-2)(m-1)}{2} (\tau_2-1),$$

$$I_{10} = \frac{(6-\alpha)(5-\alpha)(4-\alpha)(\alpha-2)}{3!} (n'-2)(4-n')(r_3-2) \frac{(m-2)(m-1)}{2} \tau_2,$$

$$I_{11} = \frac{(6-\alpha)(5-\alpha)(4-\alpha)(\alpha-2)}{3!} \frac{(n'-3)(n'-2)}{2} \frac{(\rho-3)(\rho-2)(\rho-1)}{3!},$$

$$I_{12} = \frac{(6-\alpha)(5-\alpha)(4-\alpha)(\alpha-2)}{3!} \frac{(n'-3)(n'-2)}{2} \frac{(4-\rho)(\rho-2)(\rho-1)}{2},$$

$$I_{13} = \frac{(6-\alpha)(5-\alpha)(4-\alpha)(\alpha-2)}{3!} \frac{(n'-3)(n'-2)}{2} \frac{(4-\rho)(3-\rho)(\rho-1)}{2},$$

$$I_{14} = \frac{(6-\alpha)(5-\alpha)(4-\alpha)(\alpha-2)}{3!} \frac{(n'-3)(n'-2)}{2} \frac{(4-\rho)(3-\rho)(2-\rho)}{3!},$$

$$I_{15} = \frac{(6-\alpha)(5-\alpha)(\alpha-3)(\alpha-2)}{2^2} (3-n')(2-\rho') \frac{(3-r_4)(4-r_4)}{2},$$

$$I_{16}=\frac{(6-\alpha)(5-\alpha)(\alpha-3)(\alpha-2)}{2^2}(3-n')(2-\rho')(r_4-2)(4-r_4),$$

$$I_{17}=\frac{(6-\alpha)(5-\alpha)(\alpha-3)(\alpha-2)}{2^2}(3-n')(2-\rho')\frac{(r_4-3)(r_4-2)}{2},$$

$$I_{18}=\frac{(6-\alpha)(5-\alpha)(\alpha-3)(\alpha-2)}{2^2}(3-n')(\rho'-1)(4-r_4),$$

$$I_{19}=\frac{(6-\alpha)(5-\alpha)(\alpha-3)(\alpha-2)}{2^2}(3-n')(\rho'-1)(r_4-3),$$

$$I_{20}=\frac{(6-\alpha)(5-\alpha)(\alpha-3)(\alpha-2)}{2^2}(n'-2)(4-r_4),$$

$$I_{21}=\frac{(6-\alpha)(5-\alpha)(\alpha-3)(\alpha-2)}{2^2}(n'-2)(r_4-3),$$

$$I_{22}=\frac{(6-\alpha)(\alpha-4)(\alpha-3)(\alpha-2)}{3!}(3-r_3)(4-r_4),$$

$$I_{23}=\frac{(6-\alpha)(\alpha-4)(\alpha-3)(\alpha-2)}{3!}(3-r_3)(r_4-3),$$

$$I_{24}=\frac{(6-\alpha)(\alpha-4)(\alpha-3)(\alpha-2)}{3!}(r_3-2),$$

$$I_{25}=\frac{(\alpha-5)(\alpha-4)(\alpha-3)(\alpha-2)}{4!},$$

where $r_3$ is the rank of $|\sum_k^{4,5}\gamma_{ijk}\eta_k|$ $(i, j=1, 2, 3)$ and $\tau_i$ is the number of linearly independent sets of values $(\eta_4, \eta_5)$ such that this determinant is of rank $i$; $\rho_1$ is the rank of $|\gamma_{ij5}|$ $(i, j=1, \ldots, 4)$; $r_4$ is the rank of $|\sum_k^{3,4,5}\gamma_{ijk}\eta_k|$ $(i, j=1, 2, 3, 4)$; $r_3$ is the rank of $|\sum_k^{3,4}\gamma_{ijk}\eta_k|$ $(i, j=1, 2, 3)$; $\rho'$ is the rank of the discriminant of $|\sum_k^{3,4}\gamma_{ijk}\eta_k|$ $(i, j=1, 2)$; and $m$ is the maximum multiplicity of the linear factors of $|\sum_k^{4,5}\gamma_{ijk}\eta_k|$ $(i, j=1, 2, 3)$.

THE UNIVERSITY OF CHICAGO.

# *Determination of the Order of the Groups of Isomorphisms of the Groups of Order $p^4$, where p is a Prime.*

By Ross W. Marriott.

It is well known that the group of isomorphisms of a group $G$, of prime order $p$, is a group of order $p-1$, and that the group of isomorphisms of a group of order $p^2$ is a group of order $p\ (p-1)$ and cyclic, if the group of order $p^2$ be cyclic. It is isomorphic with the linear homogeneous group on $p^2$ variables (mod $p$), if the group of order $p^2$ be not cyclic.

Western in his paper, "Groups of Order $p^3q$," *Proceedings of London Mathematical Society*, Vol. XXX, treated the isomorphisms of groups of order $p^3$.

A knowledge of the groups of isomorphisms of a group is essential in the work of constructing groups having the given group as a subgroup. The isomorphisms of the groups of order $p$, $p^2$, $p^3$ are of fundamental importance for example in the work of tabulating all groups of the orders $p^2$ *, $p^2q$ †, $p^2qr$ ‡, $p^3$ §, $p^3q$ ||.

It is obvious that the determination of the orders of the groups of isomorphisms of all groups of order $p^4$ must precede the tabulation, for example, of all types of groups of order $p^4q$. This latter tabulation is logically the next problem of this character which remains to be undertaken, the orders $p^5$ ¶, $p^3q^2$ **, $pqrst$ †† having been previously treated.

It is the purpose of this paper to determine the order of the groups of isomorphisms of all groups of order $p^4$. For the sake of convenient references,

---

* Burnside, "Theory of Groups"; Young, "On the Determination of Groups whose Order is the Power of a Prime, American Journal of Mathematics, Vol. XV (1893); Cole and Glover, American Journal of Mathematics, Vol. XV (1893).

† Burnside, "Theory of Groups"; Cole and Glover, American Journal of Mathematics, Vol. XV (1893).

‡ O. E. Glenn, *Trans. of Amer. Math. Soc.*, Vol. VII, No. 7.

§ Burnside, "Theory of Groups"; Young, *loc. cit.*; Cole and Glover, *loc. cit.*

|| Western, *Proc. of London Math. Soc.*, Vol. XXX.

¶ Bagnera, *Armali di Matematica* (1898), pp. 137–228.

** M. O. Tripp, "Groups of Order $p^3q^2$."

†† Hölder, *Göttinger Nachrichten* (1895), pp. 211–229.

I shall recall a few of the well-known definitions and theorems, regarding isomorphisms.

Suppose two groups, $G$ and $G'$, are so related that to each element, $g'$ of $G'$, corresponds one or more elements, $g, g_1, g_2, \ldots$ of $G$; while reciprocally to each element, $g$ of $G$, corresponds one or more elements, $g', g'_1, g'_2, \ldots$ of $G'$. Suppose further, that if $g_i$ and $g_j$ are elements of $G$ corresponding to elements $g'_i$ and $g'_j$ of $G'$, $g_i g_j$ is an element of $G$ corresponding to $g'_i\, g'_j$ of $G'$. Then, $G$ and $G'$ are said to be isomorphic.

If $l$ elements of $G$ correspond to each element of $G'$, and $k$ elements of $G'$ correspond to each element of $G$, $G$ and $G'$ are said to have an $(l, k)$ isomorphism with each other. The most important case is that in which $l=k=1$. Then $G$ and $G'$ are called simply isomorphic. Or, what is the same thing, if $G$ and $G_0$ be two groups of equal order, and if a correspondence can be established between the operations of $G$ and $G_0$, so that to every operation of $G$ there corresponds a single operation of $G_0$, and to every operation of $G_0$ there corresponds a single operation of $G$, while to the product $PQ$ of any two operations of $G$, there corresponds the product $P_0 Q_0$ of the corresponding operations of $G_0$, the groups $G$ and $G'$ are said to be simply isomorphic.

Again, if the correspondence between the operations of a group $G$ be such that to every operation $P$ there corresponds a single operation $P_0$, while to the product $PQ$ of two operations there corresponds the product $P_0 Q_0$ of the corresponding operations, it is said to define an isomorphism of the group $G$ with itself.

The elements $P, Q, R, S, T, \ldots$ of a group are called independent if no one of them can be expressed as a product of powers of the others,—thus $P$ is not expressible in the form $Q^a R^b S^c \ldots$

A group is said to be determined abstractly when we know the number of elements it contains, and the way in which any two combine by multiplication. Hence, the group is given abstractly by a set of generating elements, and certain independent and mutually consistent relations which they satisfy. Two groups, $G$ and $G_0$, having the same number of elements, combining according to the same laws abstractly considered, are one and the same group.

THEOREM. *The order of the group of isomorphisms of a group $G$ is equal to the number of ways that the group $G$ can be made simply isomorphic with itself, and this is equal to the number of ways the generators can be chosen from the elements of $G$ so that they will obey those laws, and only those laws which define the group abstractly.*

Let us suppose $P, Q, R, S, \ldots$ are generators of a group $G$, satisfying certain relations, and $P_0, Q_0, R_0, S_0, \ldots$ are elements of $G$ satisfying precisely similar relations but no relations independent of these. Let these define the group $G_0$. Then an isomorphism $(G_0 \sim G)$ of $G$ exists in which $P$ corresponds to $P_0$, $Q$ to $Q_0$, $R$ to $R_0$, etc.

For $G$ and $G_0$ having the proper correspondence between their elements are simply isomorphic, but as the elements of $G_0$ are also elements of $G$, this simply isomorphic property exhibits an isomorphism $(G_0 \sim G)$ of $G$ with itself. *Hence an isomorphism of $G$ exists*, where $P$ corresponds to $P_0$, $Q$ to $Q_0$, etc.

Now to each way of choosing the elements $P_0, Q_0, R_0, \ldots.$ of $G_0$, corresponds a distinct isomorphism of $G$. The symbol of an isomorphism defines an operation simply isomorphic with a substitution performed upon the elements of the group $G$, the totality of these operations forms a group * simply isomorphic with the group of isomorphisms of the group $G$. Hence, the order of the group of isomorphisms is the number of ways of selecting the generators $P_0, Q_0, R_0$, etc.

Our general method of procedure in the present problem is to take for the generating elements the most general operations in the group $G$, of order $p^4$; *i. e.*, to assume the new generators in the form:

$$\begin{aligned} P_0 &= P^x Q^y R^z S^w \ldots., \\ Q_0 &= P_1^x Q_1^y R_1^z S_1^w \ldots., \\ R_0 &= P_2^x Q_2^y R_2^z S_2^w \ldots., \\ S_0 &= P_3^x Q_3^y R_3^z S_3^w \ldots., \\ &\ldots\ldots\ldots\ldots\ldots, \end{aligned}$$

where $P, Q, R, \ldots.$ are the generators of $G$, to impose on the exponents the conditions necessary to make the new generators obey the defining relations of $G$, to subject the exponents to the conditions such that the new generators have no other relations existing among them. These conditions will be sufficient to insure that the new generators are independent, and so may be taken to generate the group.

This method leads always to a set of congruences, the number of solutions of which, is the number of ways the new generators may be chosen and therefore gives the order of the group of isomorphisms of the group under consideration.

The groups are considered in the order in which they are tabulated in Burnside, "Theory of Groups," pp. 87, 88, 89. Nine of the first ten groups require no separate investigation when $p$ is an even prime. Number 7, how-

* Burnside, "Theory of Groups," 156.

ever, requires an independent procedure when $p$ is 2. The types of $G_{p^4}$ which exist only when $p$ is even are treated in the latter part of the paper.

The transformation formulæ have been developed by mathematical induction, and are given for each non-abelian group.

Certain applications to the theory of numbers in connection with the determinants which enter into the congruences for certain of the types is another interesting feature of the details of the paper.

With the exception of three types, every group of order $p^4$ has a group of isomorphisms of an order of the general form $N=p^a(p-1)^b(p+1)^c$. Two of the exceptions are of the form $N=2p^a(p-1)^b(p+1)^c$.

The notation in the paper is practically the same as in Burnside, "Theory of Groups."

*Section* 1.

$$P^{p^4}=1.$$

Let $P_0=P^x$ be a new generator. It will generate the group provided $x$ is prime to $p$. The group contains $p^3(p-1)$ operations of order $p^4$. Hence,

$$N=p^3(p-1).^*$$

*Section* 2.

$$P^{p^3}=1,\ Q^p=1,\ Q^{-1}PQ=P.$$

Let $P_0=P^xQ^y$, $Q_0=P^{x_1}Q^{y_1}$. Then,

$$P_0^{p^3}=1,\ \therefore\ x\not\equiv 0\ (\text{mod}\ p),\dagger \qquad (1)$$

$$Q_0^p=1,\ \therefore\ x_1\equiv 0\ (\text{mod}\ p^2). \qquad (2)$$

The conditions (1) and (2) are sufficient to insure the independence of the generators so far as their powers prime to $p$ are concerned.

Suppose $P_0^hQ_0^k=1$, where $k\not\equiv 0$ (mod $p$),

$$\therefore \begin{cases} hx+kx_1\equiv 0\ (\text{mod}\ p^3), \\ hy+ky_1\equiv 0\ (\text{mod}\ p). \end{cases}$$

Since $x\not\equiv 0$ (mod $p$) and $x_1\equiv 0$ (mod $p^2$), we must have $h\equiv 0$ (mod $p^2$), and hence, also, $y_1\equiv 0$.

Now, if $y_1\not\equiv 0$ (mod $p$), the generators $P_0$ and $Q_0$ are independent,

$$\therefore\ y_1\not\equiv 0\ (\text{mod}\ p). \qquad (3)$$

There being no restrictions on $y$, there are $p^4(p-1)^2$ ways of choosing the generators,

$$\therefore\ N=p^4(p-1)^2.$$

* Throughout the paper $N$ denotes the order of the group of isomorphisms.

† And no less power of $P_0$ is unity. This will be understood hereafter.

*Section 3.*

$$P^{p^2}=1,\ Q^{p^2}=1,\ Q^{-1}PQ=P.$$

Let $P_0=P^xQ^y$, $Q_0=P^{x_1}Q^{y_1}$. Then,

$x$ and $y$ cannot be simultaneously $\equiv 0 \pmod{p}$,
$x_1$ and $y_1$ cannot be simultaneously $\equiv 0 \pmod{p}$.

Suppose $P_0^hQ_0^k=1$, $h$ and $k$ prime to $p$. This gives:

$$\left.\begin{matrix} hx+kx_1\equiv 0 \\ hy+ky_1\equiv 0 \end{matrix}\right\} \pmod{p^2},$$

$$\therefore \begin{vmatrix} x & x_1 \\ y & y_1 \end{vmatrix} \not\equiv 0 \pmod{p^2}. \qquad (1)$$

Condition (1) insures the independence of $P_0$ and $Q_0$ so far as their powers prime to $p$ are concerned. Suppose now

$$P_0^{h'p}Q_0^{k'p}=1,\ h' \text{ and } k' \text{ prime to } p. \text{ This gives:}$$

$$\left.\begin{matrix} h'px+k'px_1\equiv 0 \\ h'py+k'py_1\equiv 0 \end{matrix}\right\} \pmod{p^2},$$

$$\therefore \begin{vmatrix} x & x_1 \\ y & y_1 \end{vmatrix} \not\equiv 0 \pmod{p}. \qquad (2)$$

Conditions (1) and (2) are sufficient to insure that $P_0$ and $Q_0$ generate the group. The number of solutions which satisfy (1) is

$$p^3(p^2-1)(p-1)(p^2+p-1),$$

and this is the number of ways a pair of generators of order $p^2$ may be chosen. Of these solutions, the number which fail to satisfy condition (2) is

$$p^3(p-1)^3(p+1).$$

The number of ways of choosing the generators is

$$N=p^3(p^2-1)(p-1)(p^2+p-1)-p^3(p-1)^3(p+1)=p^5(p-1)^2(p+1).$$

*Section 4.*

$$P^{p^2}=1,\ Q^p=1,\ R^p=1,\ Q^{-1}PQ=P,\ R^{-1}PR=P,\ R^{-1}QR=Q.$$

Let $P_0$, $Q_0$, $R_0$ be defined as above,

$$P_0^{p^2}=1,\ \therefore\ x\not\equiv 0 \pmod{p}, \qquad (1)$$
$$Q_0^p=1,\ \therefore\ x_1\equiv 0 \pmod{p}, \qquad (2)$$
$$R_0^p=1,\ \therefore\ x_2\equiv 0 \pmod{p}. \qquad (3)$$

Let $x_1 = x'p$ and $x_2 = x''p$. Then,

$$\left.\begin{array}{l} x',\ y_1 \text{ and } z_1 \text{ are not simultaneously} \equiv 0 \\ x'',\ y_2 \text{ and } z_2 \text{ are not simultaneously} \equiv 0 \end{array}\right\} \pmod{p}.$$

For, if they were, $Q_0$ and $R_0$ would be unity.

Suppose $P_0^h Q_0^k R_0^m = 1$, $k$ and $m$ prime to $p$,

$$\therefore \left\{\begin{array}{l} hx + kpx' + mpx'' \equiv 0 \pmod{p^2}, \\ \left.\begin{array}{l} hy + ky_1 + my_2 \equiv 0 \\ hz + kz_1 + mz_2 \equiv 0 \end{array}\right\} \pmod{p}. \end{array}\right.$$

That this may be satisfied requires $h \equiv 0$, and also $\begin{vmatrix} y_1 & y_2 \\ z_1 & z_2 \end{vmatrix} \equiv 0 \pmod{p}$.
Hence, we must have

$$\begin{vmatrix} y_1 & y_2 \\ z_1 & z_2 \end{vmatrix} \not\equiv 0 \pmod{p}. \tag{4}$$

The conditions (1), (2), (3), (4) are sufficient to insure the generators $P_0$, $Q_0$ and $R_0$ independent.

There are $p(p-1)^2(p+1)$ solutions of (4); $x$ has $p(p-1)$ values, $x'$, $x''$, $y$ and $z$, each has $p$ values,

$$\therefore N = p^6(p-1)^3(p+1).$$

*Section* 5.

$P^p = Q^p = R^p = S^p = 1$, $PQ = QP$, $RP = PR$, $PS = SP$, $QR = RQ$, $QS = SQ$, $RS = SR$.

As before, $P_0$, $Q_0$, $R_0$, $S_0$ are new generators. Since every operation of the group is of order $p$, these new generators satisfy the defining relations without condition. Suppose there is a relation $P_0^h Q_0^k R_0^l S_0^m = 1$. We get

$$\left.\begin{array}{l} hx + kx_1 + lx_2 + mx_3 \equiv 0 \\ hy + ky_1 + ly_2 + my_3 \equiv 0 \\ hz + kz_1 + lz_2 + mz_3 \equiv 0 \\ hw + kw_1 + lw_2 + mw_3 \equiv 0 \end{array}\right\} \pmod{p}.$$

If these congruences do not co-exist, the new operations are independent, and will generate the group. The condition is fulfilled if

$$\begin{vmatrix} x & x_1 & x_2 & x_3 \\ y & y_1 & y_2 & y_3 \\ z & z_1 & z_2 & z_3 \\ w & w_1 & w_2 & w_3 \end{vmatrix} \not\equiv 0. \tag{1}$$

The number of sets of values which satisfy (1) is

$$N = (p^4-1)(p^4-p)(p^4-p^2)(p^4-p^3).$$

### *Section* 6.

$$P^{p^3}=1,\ Q^p=1,\ Q^{-1}PQ=P^{1+p^2}.$$

The transformation formulæ for this group are:

$$P^x Q^y = Q^y P^{x(1+p^2)^y}, \tag{1}$$
$$Q^y P^x = P^{x(1-p^2y)} Q^y, \tag{2}$$
$$[P^x Q^y]^k = P^{kx - \frac{1}{2}k(k-1)xyp^2} Q^{ky}, \tag{3}$$
$$P_0^{p^3}=1,\ \therefore\ x \not\equiv 0 \pmod{p}, \tag{4}$$
$$Q_0^p=1,\ \therefore\ x_1 \equiv 0 \pmod{p^2}. \tag{5}$$

Let $x_1 = x' p^2$.

$$Q_0^{-1} P_0 Q_0 = P_0^{1+p^2},\ \therefore\ y_1 \equiv 1 \pmod{p},\ \textit{i. e.},\ y_1 = 1. \tag{6}$$

Suppose now $P_0^h Q_0^k = 1$.

The application of transformation equation (3) leads to a set of congruences inconsistent with those obtained from the defining relations. Hence,

$$N = p^4(p-1),\ \text{whether } p \text{ is odd or even.}$$

### *Section* 7.

$$P^{p^2}=1,\ Q^p=1,\ R^p=1,\ R^{-1}QR=QP^p,\ QP=PQ,\ RP=PR.$$

The transformation formulæ for this group are:

$$Q^y R^z = R^z Q^y P^{yzp}, \tag{1}$$
$$P^x Q^y R^z = P^{x+yzp} R^z Q^y, \tag{2}$$
$$[P^x Q^y R^z]^k = P^{kx - \frac{1}{2}k(k-1)yzp} Q^{ky} R^{kz}. \tag{3}$$

Let $P_0$, $Q_0$ and $R_0$ be new generators.

$$P_0^{p^2}=1,\ \therefore\ x \not\equiv 0 \pmod{p}, \tag{4}$$
$$Q_0^p=1,\ \therefore\ x_1 \equiv 0 \pmod{p}, \tag{5}$$
$$R_0^p=1,\ \therefore\ x_2 \equiv 0 \pmod{p}, \tag{6}$$
$$R_0^{-1} P_0 R_0 = P_0,\quad \therefore\ \begin{vmatrix} y & y_2 \\ z & z_2 \end{vmatrix} \equiv 0 \pmod{p}, \tag{7}$$
$$Q_0^{-1} P_0 Q_0 = P_0,\quad \therefore\ \begin{vmatrix} y & y_1 \\ z & z_1 \end{vmatrix} \equiv 0 \pmod{p}, \tag{8}$$
$$R_0^{-1} Q_0 R_0 = Q_0 P_0^p,\ \therefore\ \begin{vmatrix} y_1 & z_1 \\ y_2 & z_2 \end{vmatrix} \equiv x \pmod{p}. \tag{9}$$

Let now $P_0^h Q_0^k R_0^l = 1$, and apply the transformation formulæ (1), (2), (3). There results a congruence contrary to (9). Thus the defining relations of the group determine the conditions which insure that $P_0$, $Q_0$ and $R_0$ are independent. We have now to find the number of solutions of the congruences (4)–(9) inclusive. (4) and (9) are the sufficient conditions that (7) and (8) have only zero solutions. Hence, the number of solutions is, when $p$ is odd,

$$N=p^4(p-1)^2(p+1).$$

Suppose now that $p=2$. In this case we have the congruences

$$x\not\equiv 0,\ x_1-y_1z_1\equiv 0,\ x_2-y_2z_2\equiv 0,\ x\equiv z_2y_1-y_2z_1,\ y\equiv z\equiv 0,\ \text{all (mod 2)},$$

$$\therefore\ N=3\cdot 2^3.$$

*Section* 8.

$$P^{p^2}=1,\ Q^{p^2}=1,\ Q^{-1}PQ=P^{1+p}.$$

The transformation formulæ for this group are:

$$P^xQ^y=Q^yP^{x(1+p)^y}, \tag{1}$$

$$Q^yP^x=P^{x(1+p)-y}Q^y, \tag{2}$$

$$[P^xQ^y]^k=P^{kx-\frac{1}{2}k(k-1)xyp}Q^{ky}. \tag{3}$$

Neither $x$ and $y$ nor $x_1$ and $y_1$ can be simultaneously $\equiv 0 \pmod p$. Else $P_0$ and $Q_0$ are not of order $p^2$.

$P_0^{p^2}=1$ and $Q_0^{p^2}=1$, without any condition on the exponents.

$$Q_0^{-1}P_0Q_0=P_0^{1+p},\ \therefore\ y\equiv 0,\ x\equiv xy_1-yx_1 \pmod p). \tag{4}$$

Let $P_0^hQ_0^k=1$. This leads to a set of conditions inconsistent with the conditions obtained from the defining relations,

$$\therefore\ N=p^5(p-1).$$

*Section* 9.

$$P^{p^2}=Q^p=R^p=1,\ PR=RP^{1+p},\ PQ=QP,\ RQ=QR.$$

The transformation formulæ are, for this group:

$$P^xR^z=R^zP^{x(1+p)^z}, \tag{1}$$

$$R^zP^x=P^{x-xzp}R^z, \tag{2}$$

$$P^xQ^yR^z=R^zP^{x(1+p)^z}Q^y, \tag{3}$$

$$[P^xQ^yR^z]^k=R^{kz}P^{kx\left[1+\frac{pz(k+1)}{2}\right]}Q^{ky}. \tag{4}$$

$$P_0^{p^2}=1.\quad \text{Hence, } x\not\equiv 0 \pmod p), \tag{5}$$

$$Q_0^p=1.\quad x_1\equiv 0 \pmod p), \tag{6}$$

$$R_0^p=1.\quad x_2\equiv 0 \pmod p), \tag{7}$$

$$R_0^{-1}P_0R_0=P_0^{1+p}.\quad z_2\equiv 1 \pmod p), \tag{8}$$

$$P_0^{-1}Q_0P_0=Q_0.\quad z_1\equiv 0 \pmod p), \tag{9}$$

$R_0^{-1}Q_0R_0=Q_0$. This requires no further conditions.

Let $P_0^hQ_0^kR_0^m=1$.

The transformation formulæ show that we must impose the further condition $y_1 \not\equiv 0$, if the generators $P_0$, $Q_0$, $R_0$ are to be independent.

Hence, we have

$$y_1 \not\equiv 0 \pmod{p}. \tag{10}$$

The number of solutions of the above conditions is

$$N = p^6(p-1)^2.$$

*Section* 10.

$$P^{p^2} = Q^p = R^p = 1, \ R^{-1}PR = PQ, \ PQ = QP, \ RQ = QR,$$

The transformation formulæ for this group are:

$$P^x R^z = R^z P^x Q^{xz}, \tag{1}$$
$$R^z P^x = P^x Q^{-xz} R^z, \tag{2}$$
$$P^x Q^y R^z = R^z P^x Q^{y+xz}, \tag{3}$$
$$[P^x Q^y R^z]^k = R^{kz} P^{kx} Q^{ky+\frac{k(k+1)xz}{2}}, \tag{4}$$
$$P_0^{p^2} = 1, \ \therefore \ x \not\equiv 0 \pmod{p}, \tag{5}$$
$$Q_0^p = 1, \quad x_1 \equiv 0 \pmod{p}, \tag{6}$$
$$R_0^p = 1, \quad x_2 \equiv 0 \pmod{p}. \tag{7}$$

Let $x_1 = x' p$ and $x_2 = x'' p$,

$$Q_0^{-1} P_0 Q_0 = P_0, \ \therefore \ z_1 \equiv 0 \pmod{p}, \tag{8}$$

$$R_0^{-1} P_0 R_0 = P_0 Q_0, \ \therefore \begin{cases} y_1 \equiv x z_2 \pmod{p}, & (9) \\ x' p \equiv 0 \pmod{p^2}. & (10) \end{cases}$$

Since $z_1 \equiv 0 \pmod{p}$ and $x_1 = 0$, we must have

$$y_1 \not\equiv 0 \pmod{p}, \tag{11}$$

and hence

$$z_2 \not\equiv 0 \pmod{p}. \tag{12}$$

Let $P_0^h Q_0^k R_0^l = 1$. The transformation formulæ applied to this give a set of impossible congruences. The above generators then are independent and whether $p$ is odd or even

$$N = p^5(p-1)^2.$$

*Sections* 11, 12, 13, $p > 3$.

$$P^{p^2} = 1, \ Q^p = 1, \ R^p = 1, \ Q^{-1}PQ = P^{1+p}, \ RPQ = PR, \ R^{-1}QR = P^{rp} Q.$$

The transformation formulæ are:

$$\left.\begin{aligned} Q^y R^z &= R^z Q^y P^{ryzp} \\ R^z Q^y &= P^{-ryzp} Q^y R^z \end{aligned}\right\} \tag{1}$$

$$\left.\begin{aligned} P^x Q^y &= Q^y P^{x(1+p)^y} \\ Q^y P^x &= P^{x(1+p)^{-y}} Q^y \end{aligned}\right\} \tag{2}$$

$$\left.\begin{aligned} P^x R^z &= R^z P^{x-\frac{zx(z-1)}{2}p+\frac{z(z-1)}{2}rxp} Q^{xz} \\ R^z P^x &= P^{x+z\frac{x(x-1)}{2}p+\frac{z(z+1)}{2}rxp} Q^{-xz} R^z \end{aligned}\right\} \quad (3)$$

$$R^z P^x Q^y = P^{x+z\frac{x(x-1)}{2}p+\frac{z(z+1)}{2}rxp-ryzp} Q^{y-xz} R^z \quad (4)$$

$$[P^x Q^y R^z]^k = P^{kx-\frac{k(k-1)}{2}xyp+\frac{k(k-1)(k-2)}{\lfloor 3}x^2zp+\frac{k(k-1)}{2}\cdot z\frac{x(x-1)}{2}p+\frac{rxp}{2}\left[\frac{k(k-1)(2k-1)}{\lfloor 3}z^2+\frac{k(k-1)}{2}z\right]}$$

$$Q^{ky-\frac{k(k-1)}{2}xz} R^{kz}. \quad (5)$$

The above formulæ hold for groups 11, 12 and 13, according as $r$ is a non-residue (mod $p$) unity or zero.

(1) $r$ is a non-residue (mod $p$). $P_0$, $Q_0$ and $R_0$ being the new generators:

$$P_0^{p^2}=1, \ \therefore \ x \not\equiv 0 \pmod{p}, \quad (1)$$

$$Q_0^p=1, \ \therefore \ x_1 \equiv 0 \pmod{p}, \quad (2)$$

$$R_0^p=1, \ \therefore \ x_2 \equiv 0 \pmod{p}. \quad (3)$$

Let $x_1 = x'p$ and $x_2 = x''p$.

$$R_0^{-1} P_0 R_0 = P_0 Q_0, \ \therefore \ y_1 \equiv x z_2 \pmod{p}. \quad (4)$$

$$xy_2 - rzy_2 - z_2\frac{x(x-1)}{2} + \frac{z_2(z_2-1)}{2}rx + z_2ry \equiv x' - rzy_1 \pmod{p}. \quad (5)$$

$$Q_0^{-1} P_0 Q_0 = P_0^{1+p}, \ \therefore \ \begin{cases} z_1 \equiv 0 \pmod{p}, & (6) \\ x \equiv xy_1 - rzy_1 \pmod{p}, & (7) \end{cases}$$

$$R_0^{-1} Q_0 R_0 = P_0^{rp} Q_0, \ \therefore \ x \equiv y_1 z_2 \pmod{p}. \quad (8)$$

The above conditions are sufficient to insure that the new generators be independent. For a consideration of $P_0^h Q_0^k R_0^m = 1$, leads to a set of congruences inconsistent with the above. The above congruences yield $2p^4(p-1)$ sets of values. Hence,

$$N = 2p^4(p-1).$$

(2) $r=1$. Set $r=1$, in the above congruences, and again,

$$N = 2p^4(p-1).$$

(3) $r=0$. Set $r=0$, in the above congruences, and

$$N = 2p^5.$$

*Sections* 11, 12, 13, $p=3$.

$$P^{p^2}=1, \ Q^p=1, \ R^p=P^{pa}, \ Q^{-1}PQ=P^{1+p}, \ R^{-1}PR=PQ^{\beta}, \ R^{-1}QR=Q.$$

The transformation formulæ for these groups are:

$$\left.\begin{aligned} P^x Q^y &= Q^y P^{x(1+p)^y} \\ Q^y P^x &= P^{x(1+p)^{-y}} Q^y \end{aligned}\right\} \quad (1)$$

$$\left.\begin{aligned} P^x R^z &= R^z P^{x-z\beta p\frac{x(x-1)}{2}} Q^{z\beta x} \\ R^z P^x &= P^{x+z\beta p\frac{x(x-1)}{2}} Q^{y-z\beta x} R^z \end{aligned}\right\} \quad (2)$$

$$P^x Q^y R^z = R^z P^{x - z\beta p \frac{z(z-1)}{2}} Q^{y+\beta zx}, \tag{3}$$

$$[P^x Q^y R^z]^k = P^{kx - \frac{k(k-1)}{2} zyp + \frac{k(k-1)(k-2)}{6} z^2 x\beta p + \frac{k(k-1)}{2} z\beta p \frac{x(x-1)}{2}} Q^{ky - \frac{k(k-1)}{2}\beta zx} R^{kz}. \tag{4}$$

The above formulæ hold for groups 11, 12 and 13, when $p=3$.

(1) $p=3$, $\alpha=0$, $\beta=1$. $P_0$, $Q_0$ and $R_0$ being the new generators.

$$P_0^9=1, \ \therefore \ x+x^2 z \not\equiv 0 \pmod 3, \tag{1}$$

$$Q_0^3=1, \ \therefore \ x_1+x_1^2 z_1 \equiv 0 \pmod 3), \tag{2}$$

$$R_0^3=1, \ \therefore \ x_2+x_2^2 z_2 \equiv 0 \pmod 3). \tag{3}$$

$$Q_0^{-1} P_0 Q_0 = P_0^{1+3}, \ \therefore \ x z_1 - x_1 z \equiv 0 \pmod 3, \tag{4}$$

$$x+x^2 z \equiv x y_1 - x_1 y + z\frac{x_1(x_1-1)}{2} - z_1\frac{x(x-1)}{2} \pmod 3, \tag{5}$$

$$R_0^{-1} P_0 R_0 = P_0 Q_0, \ \therefore \ z_1 \equiv 0 \pmod 3, \tag{6}$$

$$y_1 - x_1 z \equiv z_2 x - x_2 z \pmod 3, \tag{7}$$

$$x_1 - 3x_1 y + 3z\frac{x_1(x_1-1)}{2} \equiv 3xy_2 - 3x_2 y + 3z\frac{x_2(x_2-1)}{2} - 3z_2\frac{x(x-1)}{2} \pmod 9, \tag{8}$$

$$R_0^{-1} Q_0 R_0 = Q_0, \ \therefore \ x_1 z_2 \equiv 0 \pmod 3, \tag{9}$$

$$x_1 y_2 - x_2 y_1 - z_2 x_1 \frac{(x_1-1)}{2} \equiv 0 \pmod 3. \tag{10}$$

Before solving the above congruences can be reduced to the equilent teavs

$$x\not\equiv 0, \ z_1\equiv 0, \ x_1\equiv 0, \ x_2\equiv 0, \ z_2\not\equiv 0, \ x+x^2 z \equiv x y_1, \ y_1 \equiv x z_2, \ y_1\not\equiv 0,$$

$$x' \equiv x y_2 + z_2 x\frac{(x-1)}{2}, \text{ all taken } \pmod 3.$$

The congruences obtained from $P_0^h Q_0^k R_0^l=1$ are not consistent with those obtained from the defining relations. Hence, $P_0$, $Q_0$ and $R_0$ are independent, without further condition

$$N = 2^2 3^4.$$

(2) $p=3$, $\alpha=1$, $\beta=-1$.

$$P^9=Q^3=1, \ R^3=P^3, \ Q^{-1}PQ=P^{1+3}, \ R^{-1}PR=PQ^{-1}, \ QR=RQ.$$

$$P_0^9=1, \quad \therefore \ x-x^2 z+z \not\equiv 0 \pmod 3, \tag{1}$$

$$Q_0^3=1, \quad \therefore \ x_1 - x_1^2 z + z_1 \equiv 0 \pmod 3, \tag{2}$$

$$R_0^3=P_0^3, \ \therefore \ x_2 - x_2^2 z_2 + z_2 \equiv x - x^2 z + z \pmod 3, \tag{3}$$

$$Q_0^{-1} P_0 Q_0 = P_0^{1+3}, \ \therefore \ z_1 x - x_1 z \equiv 0 \pmod 3, \tag{4}$$

$$x - x^2 z + z \equiv x y_1 - x_1 y + z x_1 \frac{(x_1-1)}{2} - z_1 x\frac{(x-1)}{2} \pmod 3, \tag{5}$$

$$R_0^{-1} P_0 R_0 = P_0 Q_0^{-1}, \ \therefore \ z_1 \equiv 0 \pmod 3, \tag{6}$$

$$y_1 \equiv z_2 x - x_2 z \pmod 3, \tag{7}$$

$$-(x_1+z_1) \equiv 3xy_2 - 3x_2 y + 3z_2 x\frac{(x-1)}{2} - 3zx_2\frac{(x_2-1)}{2} \pmod 9, \tag{8}$$

$$R_0^{-1} Q_0 R_0 = Q_0, \ \therefore \ x_2 y_1 \equiv 0 \pmod 3, \tag{9}$$

$$\therefore \ x_2 \equiv 0 \text{ and } y_1 \not\equiv 0 \pmod 3. \tag{10}$$

Let $z_1 = 3z'$, $x_1 = 3x'$, $x_2 = 3x''$.

A consideration of $P_0^k Q_0^h R_0^m = 1$ gives no further conditions to be imposed on the exponents to insure the independence of $P_0, Q_0, R_0$.

The above congruences can be reduced to the equivalent set $y_1 \not\equiv 0$, $x_2 \equiv 0$, $x_1 \equiv 0$, $z_1 \equiv 0$, $z_2 \not\equiv 0$, $x \not\equiv 0$, $x - x^2 z + z \equiv z_2$, $x y_1 \equiv x - x^2 z + z \equiv z_2$, $y_1 \equiv z_2 x$, $x' + z' \equiv -x y_2 - z_2 x \frac{(x-1)}{2}$, all (mod 3),

$$\therefore N = 2 \cdot 3^5.$$

(3) $p=3$, $\alpha=-1$, $\beta=-1$.

$$P^9=1,\ Q^3=1,\ R^3=P^{-3},\ Q^{-1}PQ=P^{1+3},\ R^{-1}PR=PQ^{-1},\ RQ=QR.$$

$$P_0^9=1, \qquad x - x^2 z - z \not\equiv 0 \pmod 3, \tag{1}$$

$$Q_0^3=1, \qquad x_1 - x_1^2 - z \equiv 0 \pmod 3, \tag{2}$$

$$R_0^3=P_0^3, \qquad x_2 - x_2^2 - z_2 \equiv -x + x^2 z + z \pmod 3, \tag{3}$$

$$Q_0^{-1} P_0 Q = P_0^{1+3}, \quad x y_1 \equiv x - x^2 z - z \pmod 3, \tag{4}$$

$$R_0^{-1} P_0 R_0 = P_0 Q_0^{-1}, \ \therefore\ z_1 \equiv 0 \pmod 3, \tag{5}$$

$$y_1 \equiv z_2 x - x_2 z \pmod 3, \tag{6}$$

$$z' - x' \equiv x y_2 - x_2 y + z_2 x \frac{(x-1)}{2} - z x_2 \frac{(x_2-1)}{2} \pmod 3, \tag{7}$$

$$R_0^{-1} Q_0 R_0 = Q_0, \ \therefore\ x_2 y_1 \equiv 0 \pmod 3. \tag{8}$$

Now, $y_1 \not\equiv 0 \pmod 3$, else $P_0$ and $Q_0$ are permutable operations,

$$\therefore\ x_2 \equiv 0,\ z_2 \not\equiv 0.$$

The relation $P_0^h Q_0^k R_0^l = 1$ gives no further conditions. Hence, the above relations are sufficient for the independence of $P_0, Q_0, R_0$. The above congruences reduce to the simpler set $x_1 \equiv 0$, $z_1 \equiv 0$, $x_2 \equiv 0$, $z_2 \not\equiv 0$, $y_1 \not\equiv 0$, $x \not\equiv 0$, $y_1 \equiv z_2 x$, $z_2 \equiv x - x^2 z - z$, $x y_1 \equiv z_2 \equiv x - x^2 z - z$, $z' - x' \equiv x y_2 + z_2 x \frac{(x-1)}{2}$, all (mod 3), and where $x_1 = 3x'$, $x_2 = 3x''$, $z_1 = 3z'$. We easily deduce from the above $x^2 \equiv 1$, $z_2 \equiv x + z$, $y_1 \equiv 1 + xz$, and find the number of solutions to be

$$N = 2^2 \cdot 3^7.$$

## *Section* 14.

$$P^p=1,\ Q^p=1,\ R^p=1,\ S^p=1,\ R^{-1}PR=P,\ R^{-1}QR=Q,\ Q^{-1}PQ=P,\ S^{-1}PS=P,$$
$$S^{-1}QS=Q,\ S^{-1}RS=RP.$$

The transformation formulæ are:

$$R^z S^w = S^w R^z P^{wz}, \tag{1}$$

$$S^w R^z = P^{-wz} R^z S^w, \tag{2}$$

$$P^x Q^y R^z S^w = P^{x+wz} Q^y S^w R^z, \tag{3}$$

$$[P^x Q^y R^z S^w]^k = P^{kx - k\frac{(k-1)}{2}wz} Q^{ky} R^{kz} S^{kw}. \tag{4}$$

Every operation of the group is of order $p$, $p$ being an odd prime.

$P_0^p = Q_0^p = R_0^p = S_0^p = 1$ without restriction on the exponents.

$$S_0^{-1} R_0 S_0 = R_0 P_0 \quad \therefore \quad \begin{cases} w \equiv z \equiv y \equiv 0, & (5) \\ x \equiv w_3 z_2 - z_3 w_2 \not\equiv 0, & (6) \end{cases}$$

$$S_0^{-1} Q_0 S_0 = Q_0, \quad \therefore \quad w_3 z_1 - z_3 w_1 \equiv 0, \qquad (7)$$

$$S_0^{-1} P_0 S_0 = P_0, \quad \therefore \quad w_3 z - z_3 w \equiv 0, \qquad (8)$$

$$R_0^{-1} Q_0 R_0 = Q_0, \quad \therefore \quad w_2 z_1 - z_2 w_1 \equiv 0, \qquad (9)$$

$$R_0^{-1} P_0 R_0 = P_0, \quad \therefore \quad w_2 z - z_2 w \equiv 0, \qquad (10)$$

$$Q_0^{-1} P_0 Q_0 = P_0, \quad \therefore \quad w_1 z - z_1 w \equiv 0. \qquad (11)$$

The relation $P_0^h Q_0^k R_0^l S_0^m = 1$ gives a set of congruences which cannot be satisfied if we impose the further condition

$$y_1 \not\equiv 0, \qquad (12)$$

This last condition, $y_1 \not\equiv 0$, together with those obtained from the defining relations, insures $P_0$, $Q_0$, $R_0$, $S_0$ independent. The number of their solutions is

$$N = p^6 (p-1)^3 (p+1).$$

*Section* 15, $p > 3$.

$$P^p = Q^p = R^p = S^p = 1, \; S^{-1}RS = RQ, \; S^{-1}QS = QP, \; S^{-1}PS = P, \; R^{-1}QR = Q,$$
$$R^{-1}PR = P, \; Q^{-1}PQ = P.$$

The transformation formulæ are:

$$Q^y S^w = S^w Q^y P^{wy}, \qquad (1)$$

$$R^z S^w = S^w R^z Q^{wz} P^{w\frac{(w-1)}{2}z}, \qquad (2)$$

$$[P^x Q^y R^z S^w]^k = P^{kx - k\frac{(k-1)}{2}wy + \frac{z}{2}\left[k\frac{(k-1)(2k-1)}{6}w^2 + k\frac{(k-1)}{2}w\right]} Q^{ky - k\frac{(k-1)}{2}wz} R^{kz} S^{kw}, \qquad (3)$$

$$P_0^p = Q_0^p = R_0^p = S_0^p = 1, \text{ without restrictions.}$$

$$S_0^{-1} P_0 S_0 = P_0, \quad \therefore \quad w_1 \equiv z_1 \equiv y_1 \equiv w_4 z_4 \equiv 0, \; x \not\equiv 0, \qquad (4)$$

$$R_0^{-1} Q_0 R_0 = Q_0, \quad \therefore \quad w_2 \equiv z_2 \equiv y_2 w_3 \equiv 0, \qquad (5)$$

$$S_0^{-1} Q_0 S_0 = Q_0 P_0, \quad \therefore \quad x_1 \equiv y_2 w_4, \qquad (6)$$

$$S_0^{-1} R_0 S_0 = R_0 Q_0, \quad \therefore \quad \begin{cases} y_2 \equiv w_4 z_3, & (7) \\ x_2 \equiv y_3 w_4 + w_4 \dfrac{(w_4+1)}{2} z_3. & (8) \end{cases}$$

$Q_0^{-1} P_0 Q_0 = P_0$. This is satisfied without restriction.

The relation $P_0^h Q_0^k R_0^l S_0^m = 1$ is impossible from the above conditions. The generators $P_0$, $Q_0$, $R_0$, $S_0$ are independent, and the number of solutions of the above conditions is

$$N = p^4 (p-1)^2.$$

*Section* 15, $p = 3$.

$$P^{p^2}=Q^p=R^p=1,\ Q^{-1}PQ=P,\ R^{-1}PR=PQ,\ R^{-1}QR=P^{p(p-1)}Q.$$

The transformation formulæ are:

$$Q^y R^s = R^s P^{syp(p-1)} Q^y, \tag{1}$$

$$P^x R^s = R^s P^{x+xs\frac{(s-1)}{2}p(p-1)} Q^{xs}, \tag{2}$$

$$P^x Q^y R^s = R^s P^{x+xs\frac{(s-1)}{2}p(p-1)+syp(p-1)} Q^{y+xs}, \tag{3}$$

$$[P^x Q^y R^s]^k = P^{kx+p\frac{(p-1)}{2}xs^2k\frac{(k-1)(2k-1)}{6}+p\frac{(p-1)}{2}xsk\frac{(k-1)}{2}-p(p-1)ysk\frac{(k-1)}{2}} Q^{ky-xsk\frac{(k-1)}{2}} R^{ks}. \tag{4}$$

$$P_0^p=1,\ \therefore\ x\not\equiv 0, \tag{5}$$

$$Q_0^3=1,\ \therefore\ x_1\equiv 0,\ x_1=3x', \tag{6}$$

$$R_0^3=1,\ \therefore\ x_2\equiv 0,\ x_2=3x'', \tag{7}$$

$$R_0^{-1}Q_0R_0=P_0^{-3},\ \therefore\ 2(y_1z_2-y_2z_1)\equiv x(2+z^2), \tag{8}$$

$$R_0^{-1}P_0R_0=P_0Q_0,\ \therefore\ \begin{cases} z_1\equiv 0, & (9)\\ y_1\equiv xz_2, & (10)\\ xz_2(z_2-1)+2[yz_2-zy_2+y_1z]\equiv x', & (11)\end{cases}$$

$$Q_0^{-1}P_0Q_0\equiv P_0,\ \therefore\ y_1z\equiv 0. \tag{12}$$

The relation $P_0^h Q_0^k R_0^l=1$ leads to a set of congruences inconsistent with those obtained from the defining relations of the group.

The above congruences are all (mod 3). The number of solutions is

$$N = 540.$$

*Section* 16.

The order of the groups of isomorphisms of the abelian groups is the same whether $p$ is an odd or even prime, and so requires no further investigation. The order of the groups of isomorphisms of the non-abelian groups in Sections 6, 8, 9, 10 holds when $p$ is odd or even. The group Section 7 required independent investigation when $p$ is even. The following sections are given to the investigation of groups which exist only when $p=2$.

*Section* 17.

$$P^4=1,\ Q^4=1,\ R^2=1,\ Q^{-1}PQ=P^{-1},\ Q^2=P^2,\ QR=RQ,\ PR=RP.$$

The following transformation formulæ hold:

$$P^x Q^y = Q^y P^{x(1+2y)}, \tag{1}$$

$$[P^x Q^y]^k = P^{kx-2xyk\frac{(k-1)}{2}} Q^{ky}, \tag{2}$$

$$[P^x Q^y R^z]^k = P^{kx-2xyk\frac{(k-1)}{2}} Q^{ky} R^{kz}. \tag{3}$$

$$P_0^4=Q_0^4=1, \text{ without condition.}$$

$$P_0^2=Q_0^2,\ \therefore\ x-xy+y\equiv x_1-x_1y_1+y_1\not\equiv 0 \pmod 2, \tag{4}$$

$$R_0^2=1, \quad \therefore\ x_2-x_2 y_2+y_2 \equiv 0 \pmod 2, \tag{5}$$
$$Q_0^{-1} P_0 Q_0=P_0^{-1}, \quad \therefore\ x y_1-x_1 y \equiv x-x y+y \pmod 2, \tag{6}$$
$$R_0^{-1} Q_0 R_0=Q_0, \quad \therefore\ x_1 y_2-x_2 y_1 \equiv 0 \pmod 2, \tag{7}$$
$$R_0^{-1} P_0 Q_0=P_0, \quad \therefore\ x y_2-x_2 y \equiv 0 \pmod 2. \tag{8}$$

Now, (6) is the condition that (7) and (8) have only zero solutions for $x_2$ and $y_2$,

$$\therefore\ x_2 \equiv 0,\ y_2 \equiv 0 \pmod 2. \tag{9}$$

Suppose the relation to exist

$$P_0^h Q_0^k R_0^l=1.$$

This requires $z_2 \equiv 0 \pmod 2$. Hence, if $z_2 \not\equiv 0$, the above generators are independent,

$$\therefore\ z_2 \not\equiv 0 \pmod 2. \tag{10}$$

The number of solutions of these congruences is:

$$N=3 \cdot 2^7.$$

*Section* 18.

$$P^8=1,\ Q^2=1,\ Q^{-1}PQ=P^{-1}=P^7.$$
$$P^x Q^y=Q^y P^{x(1+4y+2y^2)}, \tag{1}$$
$$Q^y P^x=P^{x(1-4y+2y^2)} Q^y, \tag{2}$$
$$[P^x Q^y]^k=P^{kx+2xy\left[\frac{k(k-1)(2k-1)}{6}y-k(k-1)\right]} Q^{ky}. \tag{3}$$

There are four operations of order 8, and they are the odd powers of $P$. Let

$$P_0=P^x,$$
$$Q_0=P^{x_1} Q,$$

and it is easy to see that

$$N=2^5.$$

*Section* 19.

$$P^8=1,\ Q^2=1,\ Q^{-1}PQ=P^8.$$
$$P^x Q^y=Q^y P^{x(1+2y^2)}, \tag{1}$$
$$Q^y P^x=P^{x(1+2y^2)} Q^y, \tag{2}$$
$$[P^x Q^y]^k=P^{kx+2k\frac{(k-1)(2k-1)}{6}xy^2} Q^{ky}. \tag{3}$$

There are only four operations of order 8, and they are the powers of $P$, not congruent to zero. Let

$$P_0=P^x,$$
$$Q_0=P^{x_1} Q^{y_1}.$$
$$P_0^8=[P^x]^8=P^{8x}=1, \quad \therefore\ x \not\equiv 0 \pmod 2. \tag{4}$$
$$Q_0^2=1, \quad \therefore\ x_1(1+y_1^2) \equiv 0 \pmod 4.$$

Since $y_1 \not\equiv 0$, for if it were $P_0$ and $Q_0$ would be dependent, we must have $x_1 \equiv 0 \pmod 2$ and $y_1 \equiv 1 \pmod 2$. Hence,

$$N = 2^4.$$

*Section* 20.

$$P^8=1,\ Q^4=1,\ Q^{-1}PQ=P^{-1}=P^7,\ Q^2=P^4.$$

$$P^x Q^y = Q^y P^{x(1+4y+2y^2)}, \tag{1}$$

$$Q^y P^x = P^{x(1-4y+2y^2)} Q^y, \tag{2}$$

$$[P^x Q^y]^k = P^{kx+2xy\left[\frac{k(k-1)(2k-1)}{6}y-k(k-1)\right]} Q^{ky}. \tag{3}$$

There are four operations of order 8, and they are the odd powers of $P$. Let

$$P_0 = P^x,$$

$$Q_0 = P^{x_1} Q,$$

$$P_0^8 = 1,\ \therefore\ x \not\equiv 0 \pmod 2, \tag{4}$$

$Q_0^4 = 1$. This is satisfied without condition.

$$Q_0^2 = P_0^4,\ \therefore\ x \equiv 1 \pmod 2. \tag{5}$$

$$N = 2^5.$$

# *Correspondences Determined by the Bitangents of a Quartic.*

By J. R. Conner.

## Part I.

### *Introduction.*

The connection of the plane quartic curve with the figure of the general net of quadrics in space was first pointed out by Hesse *; he considered the net of quadrics in the form

$$(a\lambda)(\alpha x)^2=0\,(3\ \lambda\text{'s},\ 4\ x\text{'s})\,; \tag{1}$$

writing the discriminant of (1) considered as a form in $x$,

$$(a\lambda)(b\lambda)(c\lambda)(d\lambda)\,|\alpha\beta\gamma\delta|^2=0, \tag{2}$$

and regarding the $\lambda$'s as coordinates of a point in a plane $\pi$, (2) is the equation of a quartic curve in $\pi$, and the points of this curve are in (1, 1) correspondence with nodal quadrics of the system (1). In other words, (1) defines a univocal correspondence between points of $\pi$ and quadrics of the net; corresponding to the quadrics of a pencil chosen from the net is a line of $\pi$, and to the four nodal quadrics of the pencil the points of intersection of the line with the quartic (2). Hesse showed that if (1) is a general net, (2) is a general quartic. To a double tangent of (2) corresponds a pencil of quadrics (1) which has the property that the nodal quadrics coincide in pairs; such a pencil is that containing one line of the twenty-eight joining in pairs the base-points of the net (1). If $a_i$ are these base-points, the base-curve of the pencil containing $a_ia_j$ is $a_ia_j$ taken with the cubic curve on the other six $a$'s; the two quadric cones corresponding to the points of contact of the associated double tangent are the cones projecting this cubic from its two points of intersection with $a_ia_j$. The double tangents of (2) are thus associated in a one-one way with the twenty-eight lines joining the eight base-points of (1), and this is the basis for the duad notation for the double tangents of a quartic, each being represented by two symbols chosen from eight.

* Hesse, *Journal für Mathematik*, Vol. XLIX (1855).

In what follows we prefer to regard the equation (1) as determined in the following manner*: isolating one of the base-points of the net, say $a_0$, there is determined a one-to-one relation between the quadrics of the net and the bundle of planes on $a_0$; thus, a given quadric has a definite tangent plane at $a_0$, and a given plane on $a_0$ has a definite tangent quadric at $a_0$; this being true without exception, there is a collinear relation between the system of quadrics and the system of planes. The planes of the bundle on $a_0$, or, better, their lines of intersection with an arbitrary plane $\omega$, take the place of the plane $\pi$ in Hesse's discussion. It is our object, by means of the space figure, to prove the existence of certain Cremona transformations and other noteworthy correspondences associated with sets of the nodes of the quartic of lines thus obtained. The intimate connection between the plane and space figures makes many geometrical relations in both intuitive which might otherwise be difficult to prove. From our point of view there is another special convenience in determining the quartic in this way as a quartic of lines, for the correspondences arising appear as point to point correspondences. The method is almost exclusively geometrical, though the geometrical notions lead to analytical results whose possibilities we use to some extent, but make no attempt to exhaust.

In the first part of the paper attention is given to the correspondences themselves as they arise from the figure in space, and independently of their connection with the nodes of the quartic; then by means of these correspondences certain of their singular points are identified with the nodes of the quartic determined in the manner above specified. The correspondences are closely associated with configurations arising from the grouping of the nodes, and this feature is considered. Especial attention may be called to the Cremona involution of order 15 in space discussed in § 1 and to its use in the sequel. It is the analogue of the Bertini 17-ic transformation† in the plane, and is interesting for its own sake. We take occasion also to show how there arises from the figure in space a Kummer surface which has the property that its enveloping cone from $a_0$ is exactly our quartic of lines.

## § 1. *The* 15-*ic Cremona Transformation in Space.*

Let seven points, $a_0, a_1, \ldots, a_6$, be given in general position in space. They determine a 15-ic Cremona transformation which, as has been remarked in the Introduction, is the extension to space of the involutory 17-ic transfor-

* Compare Frobennis, *Journal für Mathematik*, Vol. XCIX, p. 275.

† Sturm, "Die Lehre von den geometrischen Verwandtschaften," Vol. IV, p. 105.

mation in the plane. In order that the analogy may be seen more clearly we summarize the latter briefly. The 17-ic transformation is determined when its eight singular points $\alpha_i$ are given. There is a pencil of cubic curves "$c$" on the points $\alpha_i$ and $\infty^3$ sextics "$s$" with nodes at these points. There is a unique sextic $s$ with triple point at one of the $\alpha$'s. Sextics $s$ meet a definite cubic $c$ in points of a $g_2^1$ on this cubic; thus, if $x$ is a general point of the plane, sextics $s$ on $x$ pass through a definite point $y$, the partner of $x$ in the $g_2^1$ on the unique cubic $c$ on $x$; $x$ and $y$ may easily be seen to be partners in an involutory 17-ic Cremona transformation which has $\alpha_i$ as six-fold singular point, the corresponding fundamental curve being the unique sextic $s$ with triple point at $\alpha_i$. The Jacobian, $j$, of one sextic $s$ and two cubics $c$ is the locus of further possible nodes of sextics $s$; any node of a proper sextic $s$ not at $\alpha_i$ must lie on $j$. The curve $j$ is of order 9 and has triple points at $\alpha_i$ with the same tangents at $\alpha_i$ as the sextic $s$ with triple point there. It is a locus of fixed points of the transformation. The transformation has besides an isolated fixed point, the ninth base-point of cubics $c$.

In space, we have a net of quadrics on the seven points $a$; we call these "quadrics $Q$" and indicate the eighth base-point by $a$. The general quartic surface in space has thirty-four constants, and since it is four linear conditions on a surface in space to have a node at a given point, there is a linear system of $\infty^{34-7\cdot 4}$ or $\infty^6$ quartic surfaces with nodes at $a_i$; we call these "quartics $C$." It is ten linear conditions on a surface to have a triple point at a given point, and hence there is a unique quartic $C$ with triple point at $a_i$; we call this surface $C_i$. There are seven such surfaces. The $\infty^2$ elliptic quartic curves on $a_i$ (and $a$) we call "curves $E$;" these are base curves of pencils of quadrics $Q$, and every quadric $Q$ may be regarded as a locus of curves $E$,—it contains $\infty^1$ such curves. The degenerate curves $E$ are of special importance for us. A line joining two points $a$ taken with the norm curve on the other six $a$'s (this curve is bisecant to the line) is a degenerate curve $E$. There are twenty-eight of these degenerate curves. The norm curves on $a_0$ taken with five of the other $a$'s are conveniently separated into a set of six and a set of fifteen. We indicate the curve on $a_0, a_2, \ldots, a_6$ by $r_1$ (there are six such curves) and the curve on $a_0, a_3, \ldots, a_6$ by $r_{12}$, there are fifteen of these. We add $r_0$, the norm curve on $a_1, a_2, \ldots, a_6$.

A curve $E$ is met by quartics $C$ in sixteen points of which fourteen fall at $a_i$; the two variable points are pairs of a $g_2^1$ on $E$, either point of such a pair determining the other uniquely. Since there is a unique quartic $E$ on the general point, $x$, of space, it is seen that *all quartics $C$ on $x$ pass through a second* *point $y$, $x$ and $y$ lying on the same curve $E$ and forming a pair of a definite $g_2^1$ on this curve.* The points $x$ and $y$ are partners in the Cremona transformation in space which we wish to consider. This transformation we call $T$.

From the way in which $T$ is defined follows the invariance of curves $E$ under $T$; since quadrics $Q$ are loci of curves $E$ they are invariant, and the invariance of quartics $C$ follows also from the definition. If

$$C=0,\ Q_i=0$$

are the equations of a general quartic $C$ and three linearly independent quadrics $Q$, then any surface of the form

$$(\alpha Q)^{n-2}C+(\beta Q)^n=0 \tag{1}$$

is invariant under $T$. The surface (1) is of order $2n$ with $n$-fold points at $a_i$, though not the most general such surface; it is impossible to express the surface $J$ to be mentioned presently in this form; $J$ is, however, invariant under $T$, and every sextic surface with triple point at $a_i$ may be built from $J$ and the form (1) for $n=3$. Thus every sextic surface with triple point at $a_i$ is invariant under $T$.

It is unnecessary for us to pursue further the general question of surfaces invariant under $T$; certain identities exist among powers of forms like $J$, $Q$ and $C$ which make the general treatment of the subject rather long. We shall have special use later for those invariant surfaces of the form $(\beta Q)^n=0$.

Recalling the definition of $T$ by means of quartics $C$, observe now that quartics $C$ on a general point of a line $a_i\,a_j$ (we call this $p_{ij}$) contain $p_{ij}$ entirely; that quartics $C$ on a general point of a curve $r_i$ contain $r_i$ entirely. Also, bearing in mind the relation of curves $E$ to $T$, note that curves $E$ meet a surface $C_i$ in only one variable point, the other intersection falling always at $a_i$.

Indicating for the moment a point of a surface of curve $S$ by $[x, S]$, and keeping in mind the remarks of the above paragraph, we have the following relations which characterize $T$:

$$T\,a_i=C_i, \tag{1}$$

$$T\,[x, p_{ij}]=p_{ij}, \tag{2}$$

$$T\,[x, r_i]\ =r_i, \tag{3}$$

$$T\,E=E, \tag{4}$$

$$T\,Q=Q, \tag{5}$$

$$T\,C=C. \tag{6}$$

Equations (1), (2), (3) define all possible singular points of $T$. The points $a_i$ are quartic singular points, the corresponding fundamental surfaces being $C_i$;

points of $p_{ij}$ and $r_{ij}$ are singular points such that to a point of such a curve corresponds the whole curve.*

Equations (4), (5), (6) tell the rest of the story. From (1) we have that the transform of a general line passes four times through $a_i$, and that from the transform of a surface passing through $a_i$, $C_i$ factors. If $n$ is the order of the transformation, we have from (5)

$$2n - 7 \cdot 4 = 2,$$

whence $n = 15$. If $m$ is the order of multiplicity at $a_i$ of the transform of a plane, we have from (4)

$$15 \cdot 4 - 7m = 4,$$

whence $m = 8$. Hence,

a) *The transformation $T$ is an involutory Cremona transformation of order* 15. *The points $a_i$ are singular points with $C_i$ as corresponding fundamental surfaces; points of $p_{ij}$ are singular points with $p_{ij}$ as corresponding lines; points of $r_i$ are singular points with $r_i$ as corresponding curve. The transform of a general plane passes* 8-*fold through $a_i$, contains $r_i$ triply and $p_{ij}$ singly. If a plane passes through $a_i$, $C_i$ factors from its transform. Analogous theorems are true for the transforms of surfaces of any order. The transform of a curve of order $n$ is in general of order* $15n$ *with* $4n$-*fold points at $a_i$. The order of the transform is reduced by one, three or eight, respectively, if the curve meets $p_{ij}$ or $r_{ij}$, or passes through $a_i$.*†

Consider two corresponding points of $T$, $x$, $y$ say. The curve $E$ through $x$ and $y$ carries a pencil of quadrics $Q$, one of which contains the line $xy$ entirely. It follows that

b) *Lines joining pairs of corresponding points of $T$ are in the cubic complex of generators of quadrics $Q$.*

The conditions that two points $\zeta$ and $\eta$ be apolar to all quadrics $Q$ determine the well-known cubic Cremona involution in space, $Z$ say, and the lines joining corresponding points of $Z$ are generators of quadrics $Q$. If $l$ is a line of the complex of generators, curves $E$ bisecant to $l$ (of course these curves lie on the quadric $Q$ containing $l$) mark on $l$ an involution of which the fixed points, points of contact of curves $E$ with $l$, are the points $\zeta$, $\eta$ on $l$. Hence,

c) *The points $xy$, $\xi\eta$ on a line of the complex of generators of quadrics $Q$ are harmonic pairs, where $Tx = y$, $Z\xi = \eta$.*

---

* Sturm, "Die Lehre von den geometrischen Verwandtschaften," Vol. IV, p. 415.

† It seems unnecessary to insert here the usual arguments by which this conclusion is reached from the preceding; compare the volume of Sturm already referred to.

If a quartic $C$ have a node at point $x$, the correspondent, $y$, of $x$ by $T$ falls at $x$. The point $x$ is then either at $a$ or on the Cayley* dianodal surface,

$$J \equiv \begin{vmatrix} \frac{\partial C}{\partial x_1}, & \frac{\partial Q_1}{\partial x_1}, & \frac{\partial Q_2}{\partial x_1}, & \frac{\partial Q_3}{\partial x_1} \\ \frac{\partial C}{\partial x_2}, & \frac{\partial Q_1}{\partial x_2}, & \cdots & \cdots \\ \frac{\partial C}{\partial x_3}, & \frac{\partial Q_1}{\partial x_3}, & \cdots & \cdots \\ \frac{\partial C}{\partial x_4}, & \frac{\partial Q_1}{\partial x_4}, & \cdots & \cdots \end{vmatrix} = 0,$$

where $C=0$ is a general quartic $C$ and $Q_i=0$ are three linearly independent quadrics $Q$. It follows from the theory of Jacobians that $J$ is a sextic surface with triple points at $a_i$ and having at $a_i$ the same tangent cone as $C_i$. These facts which are important in what follows, we state in the theorem

d) *The sextic dianodal surface $J$ is a locus of fixed points of $T$ and $a$ is an isolated fixed point. The surface $J$ has triple points at $a_i$ with the same tangent cubic cones as $C_i$ at these points.*

The plane $\pi_{123}$ on $a_1$, $a_2$, $a_3$ is transformed into a surface of order $15-3\cdot 4=3$; this is the cubic surface with nodes at $a_0$, $a_4$, $a_5$, $a_6$ and passing through $a_1$, $a_2$ and $a_3$. The surfaces $C_1$, $C_2$ and $C_3$ which factor out of the transform of $\pi_{123}$ make up the necessary multiplicity at $a_i$ eight of the transform of a plane by $T$. The cubic surface meets $\pi_{123}$ in a cubic curve which is invariant under $T$. The plane $\pi_{123}$ and its transform together make up a quartic $C$ which may be taken as the $C$ in the expression for $J$.†

A set of lines invariant under $T$ are the lines $a_i a$. For such a line meets $C_i$ at one and only one point not at $a_i$, and this means that its transform passed once through $a_i$. The line is bisecant to $r_i$. On it $a$ is fixed under $T$. Its transform is of order $15-8-2\cdot 3=1$ (Th. a)); hence, it is invariant under $T$. Corresponding points of $T$ on $a_i a$ are pairs of an involution, the points $l_i$ and $m_i$ in which $r_i$ meets $a_i a$ forming such a pair. The double points of the involution are $a$ and its harmonic conjugate as to $l_i$ and $m_i$, a point in general different from $a_i$ since $a_i a$, $l_i m_i$ are not in general harmonic pairs. We state this in another way more suitable for our purpose. If the second double point on $a_i a$ is at $a_i$, $a_i a$ must touch $J$ at $a_i$, and conversely. Therefore,

e) *The lines $a_i a$ are invariant under $T$. They do not touch $J$ at $a_i$.*

* Cayley, *Proceedings of the London Mathematical Society*, Vol. III, p. 19.

† Pascal, "Repertorium der höheren Mathematik," Vol. II (1902), p. 297.

The following may be of interest in passing: The Geiser transformation in the plane, namely, the octavic correspondence of eighth and ninth base-points of pencils of cubics on seven given points, may be obtained from the Geiser transformation in space, the 7-ic correspondence of seventh and eighth base-points of quadrics on six given points; it arises from one of the invariant quadrics on the six points by simple projection of this quadric on a plane from one of the six points. Similarly, the 17-ic Bertini transformation may be obtained from $T$ by projection of the correspondence set up by $T$ on an invariant quadric $Q$ from, say $a_0$.

## § 2. *The Quartic Surfaces $C_i$.*

A quartic surface with a triple point, $a_0$, in a given space is represented in a one-to-one way on a plane $\pi$ in the same space by simple projection of its points from $a_0$. There are twelve lines through $a_0$ which lie on the surface; these are cut out by the cubic cone $\Gamma_0$ tangent to the surface at $a_0$. On $\pi$ the plane sections of $C_0$ appear as quartic curves on the points of intersection of a quartic and a cubic curve $\Gamma$. The cubic $\Gamma$ is the trace on $\pi$ of the tangent cubic cone $\Gamma_0$, and the quartic may be taken to be the intersection of $C_0$ and $\pi$. The quartic surface $C_0$ with triple point at $a_0$ is thus mapped from a plane $\pi$ in its space by quartic curves on the intersections of a quartic and a cubic in $\pi$, and it is easily proved conversely that any such mapping scheme gives a quartic surface with a triple point. If two base-points of the system of quartic curves in $\pi$ coincide at a point $p$, $C_0$ acquires a node, plane sections of $C_0$ through the node corresponding to the net of curves of the mapping system which have a node at $p$. In general this is the only way in which a node of $C_0$ can be represented. It follows from the mutual relations of $\pi$ and the space figure that

f) *If a quartic surface have a triple point and a node, it has the same tangent plane at all points along the line joining the triple point and node.*

This result is so important for later developments that it seems advisable to add a direct geometrical proof. Let the surface have a triple point at $a_0$ and a double point at $a_1$ and consider the tangent plane at any point, $x$, of the line joining $a_0$ and $a_1$. The tangent plane meets the surface in the line $a_0a_1$ and a residual cubic curve which has a node at $a_0$ and passes through $x$ and $a_1$; thus this residual cubic meets $a_0a_1$ in four points, and the line $a_0a_1$ must factor from it. The tangent plane at $x$ therefore meets $C_0$ in the line $a_0a_1$ counted doubly, and hence must touch $C_0$ at all points along $a_0a_1$. Similarly, it may be proved that *if a surface of order $n$ have two multiple points of multiplicities whose sum is $n+1$ it has the same tangent plane at all points along the line joining the multiple points.* The cubic surface with more than one node is a case in point. Likewise, *if a surface of order $n$ have two points whose multiplicities total $n+2$ the line joining the points is a double line of the surface.* And so on.

In our mapping scheme the pencil of lines on $\pi$ through $p$ corresponds to plane sections of $C_0$ through triple point and node, and the tangent line to $\Gamma$ at $p$ corresponds to the conic cut from $C_0$ by the tangent plane along the line joining triple point and node.

For the surface $C_0$ of § 1 the section by a plane $\pi$ is a quartic curve $C_{0\pi}$ touching at six points the section $\Gamma_{0\pi}$ of the cubic cone $\Gamma_0$ tangent to $C_0$ at $a_0$, the six points being the intersections of $\pi$ with $p_{0i}$ $(i=1, \ldots, 6)$. Hence,

g) *A plane $\pi$ meets $C_0$ in a quartic curve which touches the intersection $\Gamma_{0\pi}$ of $\pi$ with $\Gamma_0$ at the points $p_{0i}\pi$ $(i=1, \ldots, 6)$. The surface $C_0$ is then mapped from $\pi$ by means of quartic curves touching $\Gamma_{0\pi}$ at the points $p_{0i}\pi$, and a quartic curve of this system in $\pi$ is the projection of its corresponding plane section from $a_0$.*

A curve $r_i (i \neq 0)$ meets $C_0$ in all multiple points but the node $a_i$; these count for $5 \cdot 2 + 3 = 13$ intersections. Hence,

g′) *The surface $C_0$ contains the six cubic curves $r_i (i \neq 0)$. These are maps from $\pi$ of conics on five out of the six contacts of the quartic curves with $\Gamma_{0\pi}$ on $\pi$.*

The latter statement is obvious from the space figure. Similarly,

g″) *The surface $C_i$ contains the cubic curves $r_j (j \neq i)$.*

We indicate by $q_{0,1}$ the tangent line to $r_1$ at $a_0$ and by $q_{0,12}$ the tangent line to $r_{12}$ at $a_0$.

Since $C_1$ contains the line $p_{01}$ and the curves $r_2, \ldots, r_6$ it follows that the lines $p_{01}, q_{0,2}, \ldots, q_{0,6}$ are lines of a quadric cone with vertex at $a_0$, this cone being the cone tangent to $C_1$ at $a_0$. Again, projecting $r_1$ from $a_0$ we obtain a quadric cone containing the lines $q_{0,1}, p_{02}, \ldots, p_{06}$. For later reference we state the above in the theorem

h) *The lines $p_{0i}, q_{0,i}$ passing through $a_0$ are twelve generators of the cubic cone $\Gamma_0$ tangent to $C_0$ at $a_0$. The lines $q_{0,1}, p_{02}, \ldots, p_{06}$ are generators of a quadric cone, and the lines $p_{01}, q_{0,2}, \ldots, q_{0,6}$ are generators of a quadric cone. Analogous results follow by symmetry.*

## § 3. *The (1, 4) Quadratic Correspondence.**

In what follows we shall be concerned mainly with the geometry of the bundle of lines through $a_0$; in order to attain clearer statement we shall regard

* The reader will, of course, recognize in this the (1, 2) connex; cf. Clebsch-Lindemann, "Vorlesungen über Geometrie," pp. 1007–1014.

this bundle as cut by a plane $\pi$. In this bundle it is important to distinguish between *directions* about $a_0$ and *lines* through $a_0$; these we shall denote by the generic symbols $d$ and $e$, respectively. The symbols $d$ and $e$ will also be used to indicate the corresponding points in $\pi$, and the transformations $P$, $R$, $R_{ijk}$, $R_{ij}$ to be defined are always to be taken as meaning the replacing of a $d$ by one or more $e$'s; their inverses replace $e$'s by $d$'s.

The most general (1, 4) quadratic correspondence in the plane is given by a set of equations of the form

$$\rho y_i = f_i(x), \tag{1}$$

where $f_i$ are quadratic functions of $x_1$, $x_2$, $x_3$. From (1) we have for a given $x$ a unique $y$, and this without exception if the conics $f_i=0$ have no common point; if $f_i=0$ pass through a point $a$, $y$ is indeterminate for this point, and $a$ is a singular point of the correspondence. Fixing $y$ we have, eliminating $\rho$ from (1),

$$\left\| \begin{matrix} y_1 & y_2 & y_3 \\ f_1 & f_2 & f_3 \end{matrix} \right\| = 0,$$

or if $K_1 = y_2 f_3 - y_3 f_2$, etc.,

$$K_1 = K_2 = K_3 = 0,$$

three conics of a pencil since

$$y_1 K_1 + y_2 K_2 + y_3 K_3 = 0;$$

the base-points of this pencil are the correspondents of $y$, four in number. The fixed points of (1) are given by the matrix

$$\left\| \begin{matrix} x_1 & x_2 & x_3 \\ f_1 & f_2 & f_3 \end{matrix} \right\| = 0,$$

of order 7. Hence,

i) *The general* (1, 4) *quadratic correspondence in the plane has seven fixed points.*

These fixed points can not all lie on a line; for this to occur the line would necessarily contain a singular point, and singular points do not exist in general. It follows by symmetry that no three fixed points can lie on a line. If three fixed points are chosen as the vertices of a triangle of reference, and the conics corresponding to the three sides of this triangle are $\phi_i = 0$, the transformation appears in the form

$$\rho y_i = \phi_i(x) \tag{2}$$

where $\phi_i = 0$ are now conics passing through the vertices of the triangle of reference by twos. The $\phi$'s are three linearly independent conics generating the net of conics corresponding to lines $(\eta y) = 0$ of the plane.

Returning now to the figures in space, attention being fixed on the geometry about $a_0$, and on the section of the figures arising by a cutting plane, it is clear that there is a unique curve $E$, without exception, having a given direction $d$ at $a_0$, this direction fixing on a plane $\pi$ a point. The direction $d$, or the point in $\pi$, determines four lines $e_i$, the lines to the points where the corresponding curve $E$ meets $\pi$. There is determined a $(1, 4)$ correspondence in $\pi$; to a given $d$ correspond four $e$'s. Call this $P$. In order to determine the order of $P$ we consider a plane $\alpha$ through $a_0$. There is a unique quadric $Q_\alpha$ touching $\alpha$ at $a_0$, and all curves $E$ with directions $a_0\alpha$ lie on $Q_\alpha$; the contrary would mean too many intersections with $Q_\alpha$. Hence, the transform by $P$ of the line $l$ which $\alpha$ marks on $\pi$ is the intersection with $\pi$ of the quadric $Q_\alpha$, a conic.

In this and the sections immediately following we call the intersections of $p_{0i}$, $a_0a$, $q_{0,i}$, $q_{0,ij}$ with a cutting plane $\pi$, simply $p_i$, $p$, $q_i$, $q_{ij}$, respectively, where no ambiguity can arise. If the direction $d$ at $a_0$ is $p_{0i}$ the curve $E$ determined is $p_{0i}$ taken with the cubic curve on all $a$'s but $a_0$ and $a_i$. The four points corresponding to $p_i$ are the intersections of $\pi$ with this curve and $p_i$ itself. The points $p_i$ and similarly $p$ are therefore fixed points of $P$. We have

j) *The correspondence $P$ determined on a plane $\pi$ by making to correspond to a point $x$ in $\pi$ the four points of intersection of $\pi$ with the curve $E$ tangent to $a_0x$ at $a_0$, is a quadratic* $(1, 4)$ *correspondence having the seven fixed points $p_1, \ldots, p_6, p$. The net of conics corresponding to lines $(\eta y)=0$ of $\pi$ is determined as the conics of intersection of $\pi$ with quadrics $Q$.*

In the general $(1, 4)$ quadratic correspondence (1) let us suppose given seven pairs of corresponding points, say $x^{(i)}$, $y^{(i)}$, and let it be further required that the conics corresponding to three lines $\alpha^{(i)}$ pass through three definite points $a^{(i)}$, given in general position. This gives the set of twenty-four linear equations

$$\rho_j y_i = f_i(x^{(j)}), \quad j=1, \ldots, 7, \quad i=1, \ldots, 3$$
$$\sum_{i=1}^{3} \alpha_i^{(j)} f_i(x^{(j)}) = 0, \qquad j=1, \ldots, 3.$$

This set of linear equations determines uniquely the ratios of the eighteen coefficients in $f_i$ and the seven proportionality factors $\rho_j$. Hence,

k) *The most general* $(1,4)$ *quadratic correspondence in the plane is uniquely determined when* (1) *seven pairs of corresponding points are given, and* (2) *the correspondents of three given lines are required to pass respectively through three points given in general position.* In particular,

k′) *The most general* $(1, 4)$ *quadratic correspondence in the plane is uniquely determined when its seven fixed points are given, three of these points,*

*$p_1p_2p_3$ are isolated, and it is required that the correspondent of $p_ip_j$ pass through three points $a_k$ in general position.*

The figure of eight base-points of a net of quadrics in space taken with a general plane $\pi$ has $21+3-15=9$ absolute invariants. The complete figure determined in the plane $\pi$ has therefore nine absolute invariants and $8+9=17$ constants. On the other hand the most general (1,4) quadratic correspondence in the plane has seventeen constants (the ratios of the eighteen coefficients in equations (1), and we would expect the correspondence $P$ which we have determined to be general. It will now be proved that it is general, and a still more important result follows from the proof. First, the set of seven points $p$ in $\pi$ is general. For, choosing a plane $\pi$ in space and seven general points $p_ip$ on it, choose the point $a_0$; the lines $p_{01}, \ldots, p_{06}, p_0=a_0a$ are then determined. Choose $a_4, a_5, a_6, a$ arbitrarily on $p_{04}, p_{05}, p_{06}, p_0$. There are then three unique quadrics $Q_1, Q_2, Q_3$ containing $p_{02}p_{03}a_4, \ldots, a$, etc., respectively. The quadrics $Q_1, Q_2$ and $Q_3$ meet $p_{01}, p_{02}, p_{03}$ in three points $a_1, a_2, a_3$, respectively, and the points $a$ thereby determined are the base-points of a net of quadrics of which $Q_i$ are three linearly independent surfaces.* Hence,

l) *The set of points $p$ in a plane $\pi$ not passing through $a_0$ is general. The geometry in the bundle of lines at $a_0$ is identical*† *with the geometry of seven points in the plane.*

Let us now assume a general (1,4) quadratic correspondence $P'$ in a plane $\pi$ in the form (2); The curves $\phi_i=0$ are conics passing through two out of three vertices of the reference triangle. Let these vertices be $p_1, p_2, p_3$. When the conics $\phi_i$ are given the equations (2) are not definite, but they are absolutely fixed when a fourth fixed point, say $p_4$, is given. The transformation $P'$ is thereby uniquely determined. Choose now a point $a_0$ not on $\pi$ and a point $a_4$ on $a_0p_4$. There is a unique quadric $Q_1$ containing the lines $a_0p_1, a_0p_2, a_0p_3$, the conic $\phi_1$, and passing through $a_4$. Two other quadrics, $Q_2, Q_3$, are similarly determined. These quadrics are linearly independent. They pass through $a_0$, three points $a_1, a_2, a_3$ on $a_0p_1, a_0p_2, a_0p_3$ respectively, through $a_4$, and through three further points $a_5, a_6, a$. The (1,4) correspondence determined on $\pi$ by the points $a$ with $a_0$ isolated must now have exactly the equations (2), and must coincide with $P'$. Hence,

m) *The geometry of the* (1,4) *quadratic correspondence in the plane is completely identical with that of the plane section of the figure of eight base-*

---

* Compare A. C. Dixon, *Quarterly Journal*, Vol. XLI (1910), p. 210 ff.

† The word "identical" is used here in a slightly strained sense, but its meaning will not be misunderstood.

*points of a net of quadrics when one of these base-points is isolated. The domains of rationality are in both cases the same.* Or,

m′) *The geometry of the* (1, 4) *quadratic correspondence is identical with that of the configuration which is the plane section of a complete seven-point in space. The fixed points of the correspondence pair with the seven points.*

It follows easily from the argument above that *the* (1, 4) *quadratic correspondence is uniquely determined when the net of conics corresponding to lines of the plane and four fixed points are given.* For, if $p_1, \ldots, p_4$ are the four fixed points, $p_1, \ldots, p_3$ may be chosen as vertices of the reference triangle; then there is a unique conic of the net passing through $p_2\, p_3$; this is the $\phi_1$ of (2). The conics $\phi_2$ and $\phi_3$ are similarly determined. The fourth fixed point $p_4$ fixes the arbitrary multipliers of the $\phi$'s in (2), and the correspondence is uniquely determined. It is obviously immaterial which three $p$'s are chosen as vertices of the triangle of reference. This may be stated in another way.

m″) *The* (1, 4) *quadratic correspondence is uniquely determined by a cubic curve and four points in general position given as fixed points of the correspondence;* the polars of the cubic, of course, furnishing the net of conics.

It is of interest to see how the configurations of theorems m) and m′) determine the correspondence without reference to space. For brevity, we omit proofs of the statements made; little further use is to be made of them. The configuration of theorem m′) consists of thirty-five lines and twenty-one points, in which every line carries three points and every point five lines. The points and lines may be named by symbols $ij$ and $ijk$, respectively, chosen from seven indices, the point $ij$ being incident with the line $ijk$. Now, if five indices are chosen from the seven and we consider only those points and lines whose names involve indices among these, we obtain a Desargues configuration (configuration $B$). There are twenty-one such configurations contained in the larger configuration. A configuration $B$ is self-polar with regard to a definite conic $N$; we obtain in this way twenty-one conics $N_{ij}$, and these conics are in a linear system, they lie in a web of conics. To this web is apolar a net of conics, and this net is the net of our correspondence. The seven fixed points may be linearly constructed from our configuration, and the correspondence $P$ is thereby determined.

It is easy to see from the space figure how, conversely, the configuration is determined when $P$ is given. We show in the next section how the points $q_i$ and $q_{ij}$ are determined by the fixed points of $P$ alone. Let us consider a point 12, say, of the configuration, plane and space figures being supposed given. The line $a_1a_2$ marks on $\pi$ the point 12. This line taken with $r_{12}$ is a degenerate curve $E$, and the tangent to $r_{12}$ at $a_0$ marks on $\pi$ the point $q_{12}$. Thus *the transform by $P$ of $q_{12}$ is the point* 12 *of the configuration taken with the three points in which $r_{12}$ meets $\pi$.* The points of the configuration thus appear as transforms by $P$ of the points $q_i$, $q_{ij}$, whose importance will be abundantly evident in what follows. This may be stated purely from the point of view of the plane correspondence. Given $P$ the seven fixed points are given and the points $q_i$, $q_{ij}$ are determined by them. To $q_{12}$, $q_{23}$, $q_{31}$ correspond by $P$ three sets of four points. Three points may be selected, one from each set of four, and in only one way, such that the three points lie on a line. These points are the points 12, 23, 31 of the configuration.

In the figure in space, as we have remarked, there is a one-one correspondence between quadrics $Q$ and planes on $a_0$, every quadric having a unique tangent plane at $a_0$ and every plane on $a_0$ having a unique tangent quadric. On the quadric is a pair of lines which lie in the tangent plane. Projecting from $a_0$, the curves $E$ on $Q$ give a pencil of cubics in $\pi$ on the points $p$ and $x, y$, —the traces on $\pi$ of the pair of lines in the tangent plane. In the plane $\pi$ the points $x, y$ are pairs of the Geiser involution, $G$, of order 8, having the $p$'s as triple singular points, the correspondent of $p_1$ say, being the cubic with node at $p_1$ and on $p_2, \ldots, p$. There is a unique pair of points $x, y$—pairs of $G$—on any line; this, of course, may be seen from the space figure.* But since a quadric $Q$ marks on $\pi$ the transform by $P$ of the trace of its tangent plane at $a_0$ (Th. j)) and the points $x, y$ are on this line, we have

n) *The pair of points in which a line meets its correspondent conic in a* (1, 4) *quadratic correspondence, $P$, are partners in the Geiser involution determined by the seven fixed points of $P$.*

This is true without exception; on a line through $p_1$ say, $p_1$ and the point where the line meets the corresponding cubic are partners. This is still true, if the line passes through another singular point, $p_2$ say. On $p_1p_2$, $p_1$ and $p_2$ are to be regarded as partners.

## § 4. *The Transformations $R_{ijk}$ and $R_{ij}$.*

We saw in the previous section that a (1, 4) quadratic correspondence is determined in the bundle at $a_0$ or on a plane $\pi$, and if $\pi$ is arbitrary (or in general position), the (1, 4) correspondence is general. This may be stated in another way. A cutting plane $\pi$ may be fixed once for all, and all possible (1, 4) correspondences with the seven general fixed points may be studied by

* Compare Dixon, *loc. cit.*

regarding them as determined by a second plane $\pi'$ in any position, and by choosing $\pi'$ in a special position we obtain a specialization of $P$. After all we are studying merely the geometry at $a_0$, the planes of space figuring as auxiliaries. In this section we examine the most interesting of these specializations.*

The general quadratic Cremona transformation in the plane is uniquely determined when three singular points (the three of either set) and its four fixed points are given, for, if the singular points are taken as reference points, the equations of the transformations are of the form

$$\rho y_i = \sum_j \alpha_j^{(i)}/x_j,$$

and when four fixed points are given we have from this twelve equations to determine the ratios of the nine $\alpha$'s and the four proportionality factors. Thus, it is clear, *a priori*, that given seven general points in the plane, a set of thirty-five quadratic Cremona transformations is determined by requiring three of the seven points to be singular and the other four fixed. We consider the set of thirty-five transformations determined in this way by the seven points $p$ in the plane $\pi$. For the immediate use which we make of these transformations, it is convenient to separate them into a set of twenty, and a set of fifteen. We call the transformation with singular points $p_1, p_2, p_3$ and fixed points $p_4, p_5, p_6, p$, $R_{123}^{-1}$, and the transformation with singular points $p_1, p_2, p$ and fixed points $p_3, p_4, p_5, p_6$, $R_{12}^{-1}$. There are twenty transformations of the first type and fifteen transformations of the second. The inverse notation is here chosen in order to conform to the notation of the previous section. These transformations are specializations of the $(1, 4)$ correspondence $P$.

Consider the section of the space figure by the plane $\pi_{123}$ on $a_1$, $a_2$ and $a_3$. Lines in the bundle on $a_0$ give us points on $\pi_{123}$, thus we have from $p_{0i}a_0a$ a set of seven points $p_1, p_2, p_3$ $(= a_1, a_2, a_3)$, $p_4, p_5, p_6, p$; from $q_{0,i}$ a set of six points $q_i$, and from $q_{0,ij}$ a set of fifteen points $q_{ij}$. Given a direction $d$ at $a_0$ a curve $E$ is determined which meets $\pi_{123}$ in $p_1, p_2, p_3$ $(a_1, a_2, a_3)$ and a fourth point $e$. The correspondence $(d, e)$ is here one-to-one; call it $R_{123}$. It appears here geometrically as a special case of the correspondence $P$ of the previous section, and is therefore a quadratic Cremona transformation. The transform $R_{123}\,l$, where $l$ is a line in $\pi_{123}$, is a conic on $p_1, p_2, p_3$, and hence, $p_1, p_2, p_3$ are singular points of $R_{123}^{-1}$; this transformation is identical with the $R_{123}$ of the previous paragraph. The singular points of $R_{123}$ are the correspondents by

---

* Compare Clebsch-Lindemann, *loc. cit.* See also Timerding, *Mathematische Annalen*, Vol. LIII, p. 193 ff.

$R_{123}^{-1}$ of the lines $p_2p_3$, $p_3p_1$, $p_1p_2$. Now, $p_2p_3$ taken with $r_{23}$ is a degenerate curve $E$; hence, the correspondent of $p_2p_3$ by $R_{123}^{-1}$ is marked on $\pi_{123}$ by the tangent $q_{0,23}$ to $r_{23}$ at $a_0$; that is, it is the point $q_{23}$. This gives us

o) *The transformation $R_{123}$ is a quadratic Cremona transformation with singular points $q_{23}$, $q_{31}$, $q_{12}$; $R_{123}^{-1}$ has singular points $p_1$, $p_2$, $p_3$. The transform $R_{123}\,l$ where $l$ is any line meets $l$ in a pair of the Geiser involution $G$ determined by the seven points $p_1, \ldots, p_6, p$. It has the four fixed points $p_4, p_5, p_6, p$.*

The two latter statements of this theorem are consequences of the fact that $R_{123}$ is a specialization of $P$. For the first see Theorem j), for the second, Theorem n). Of course, they may be proved directly by the methods employed in the proof of the theorems cited. There are twenty transformations $R_{ijk}$ determined in a manner similar to the above; $R_{ijk}$ has singular points $q_{jk}$, $q_{ki}$, $q_{ij}$, with corresponding fundamental lines $p_jp_k$, $p_kp_i$, $p_ip_j$: $R_{ijk}^{-1}$ has singular points $p_1$, $p_2$, $p_3$ with corresponding fundamental lines $q_{12}$, $q_{13}$, etc. The transformation has all $p$'s, except $p_i$, $p_j$, $p_k$ as fixed points.

We consider now the three points $R_{123}q_i$ $(i=1, 2, 3)$. The point $q_i$ is the trace on $\pi_{123}$ of the tangent to $r_i$ at $a_0$. Hence $R_{123}q_i$ is the third point, *i. e.*, the point not at $p_j$ or $p_k$, in which $r_i$ meets $\pi_{123}$. We call this point $q_i^{123}$; there are three such points. Since $r_i$ lies on the quartic surface $C_0$ (see p. 162), it follows that the points $q_i^{123}$ lie on the section of $C_0$ by $\pi_{123}$; this is a rational quartic with nodes at $p_1$, $p_2$ and $p_3$, $C_{123}$ say. Now $C_0$ contains the lines $a_0a_i$, hence, $C_{123}$ passes through the points $p_4$, $p_5$, $p_6$—fixed points of $R_{123}$. Transforming $C_{123}$ by $R_{123}^{-1}$ and remembering that $q_i^{123}$ passes into $q_i$ we obtain the theorem:

p) *The six points $q_1$, $q_2$, $q_3$, $p_4$, $p_5$, $p_6$ are six points of a conic, the transform $R_{123}^{-1}C_{123}$. Similarly the six points $q_i$, $q_j$, $q_k$, $p_l$, $p_m$, $p_n$ are six points of a conic, the transform $R_{ijk}^{-1}C_{lmn}$.*

The twelve points $p_1, \ldots, p_6, q_1, \ldots, q_6$ are seen from Theorems h) and p) to lie by sixes on thirty-two conics, Theorem h) furnishing twelve, and Theorem p) furnishing twenty. Observe that there are sixteen pairs of conics, any pair containing the whole set of twelve points. These facts will appear later in another light.

After the preceding discussion of $R_{123}$ the transformations which we call $R_{ij}$ are easily disposed of. We consider the plane $\pi_{12}$ on $a_1$, $a_2$, $a$. As before, we indicate the projections of $a_i$, $a$ on $\pi_{12}$ by $p_i$, $p$. Directions $d$ determine on $\pi_{12}$ sets of four points $e$ of which three are at $p_1$, $p_2$, $p$; the fourth is variable. Again we have a quadratic Cremona transformation, a special case of $P$, which we call $R_{12}$. By an argument similar to that employed in the case of $R_{123}$ we have

q) *The quadratic Cremona transformation* $R_{12}$ *has the singular points* $q_1, q_2, q_{12}$ *and the fixed points* $p_3, p_4, p_5, p_6$. $R_{12}^{-1}$ *has singular points* $p_1, p_2, p$. *The points* $q_1, q_2, q_{12}$ *pair respectively with* $p_1, p_2, p$.

## § 5. *The Quintic Cremona Transformation R.*

It is for its usefulness in connection with the transformation which is the subject of this section that the division of the quadratic transformations of the previous section into two sets is adopted for the present; of course all thirty-five are a symmetrical set determined by seven points.

The 15-ic Cremona transformation $T$ of § 1 has the singular point $a_0$, and to $a_0$ corresponds the quartic surface $C_0$; this surface has a triple point at $a_0$ and double points at $a_1, \ldots, a_6$ and, as we saw, is uniquely determined by these conditions. To directions about $a_0$ correspond univocally points of $C_0$; to a direction $d$ at $a_0$ corresponds a definite line $e$ on $a_0$, the line joining $a_0$ to the correspondent $d$ on $C_0$. Conversely, given the line $e$ on $a_0$, it meets $C_0$ at only one point not at $a_0$ since $a_0$ is a triple point of $C_0$, and this point furnishes a unique correspondent $d$. We obtain thus a Cremona transformation $(d, e)$ in a cutting plane $\pi$ which we denote by $R$.

Quadrics $Q$ are invariant under $T$. The directions along a quadric $Q$ at $a_0$ must therefore pass by $T$ into the curve of intersection of the quadric with $C_0$. Let $l$ be a line in a plane $\pi$ and let $Q_l$ be the unique quadric $Q$ touching the plane $a_0 l=\alpha$ at $a_0$. Since $\alpha$ touches $Q_l$ at $a_0$ and therefore defines there the same set of directions, it follows that to the directions $a_0\alpha$ correspond points on the curve of intersection of $Q_l$ and $C_0$, a curve of order 8 with triple point at $a_0$ and double points at $a_1, \ldots, a_6$. Projecting this octavic curve from $a_0$, its triple point, we obtain in $\pi$ the curve $Rl$, a quintic curve with nodes at $p_1, \ldots, p_6$; $R$ is, therefore, a quintic Cremona transformation and $p_1, \ldots, p_6$ are the singular points of $R^{-1}$. Since $Q_l$ touches $\alpha$ at $a_0$ the curve of intersection of $Q_l$ and $C_0$ touches $\alpha$ along three directions at $a_0$, and these directions lie along the generators of $\Gamma_0$, the cubic cone tangent to $C_0$ at $a_0$. Since $J$, the surface of invariant points of $T$, has at $a_0$ the tangent cone $\Gamma_0$ (Th. d)) $d$ and $e$ must coincide along these directions, and the transformation $R$ in $\pi$ has $\Gamma$, the trace of $\Gamma_0$ on $\pi$, as a curve of fixed points.

The curves $r_i$ lie entirely on $C_0$ and to a point of such a curve corresponds by $T$ the whole curve. To the direction $d$ at $a_0$ along $q_{0,i}$, the tangent to $r_i$ at $a_0$, corresponds the whole curve $r_i$, and hence to the point $q_i$ in $\pi$ corresponds by $R$ the projection of $r_i$ from $a_0$, the conic on the five points $p_j(j \neq i)$. The points $q_1, p_2, \ldots, p_6$ are on a conic (Th. h)), $u_1$ say. Five other conics are similarly defined. The points $q_i$ thus fixed on $\Gamma$ when the $p$'s are given are singular points of $R$. The corresponding fundamental curves are of course the conics $u_i$, but it is also especially to be noted that the conic $q_1, \ldots, q_5 = v_6$, say, passes through $p_6$ (Th. h)). It thus appears that the points $p_i$ and $q_i$ are two symmetrical sets, although our determination of them is asymmetrical. That the two sets are symmetrical will be proved presently. Summarizing the above we have the theorem:

r) *The transformation $R$ determined on a plane $\pi$ by the space figure is quintic. $R$ has the singular points $q_1, \ldots, q_6$, and $R^{-1}$ the singular points $p_1, \ldots, p_6$. The twelve points $p_1, \ldots, p_6, q_1, \ldots, q_6$ lie on a cubic curve $\Gamma$ which is a curve of fixed points of $R$, and $R$ has besides the fixed point $p$ not on $\Gamma$ (Th. e)). It has no other fixed points. The transformation $R$ is the most general quintic Cremona transformation in the plane to within a collineation on the $q$'s. The conic $u_1$ on $p_2, \ldots, p_6$ passes through $q_1$, and the conic $v_1$ on $q_2, \ldots, q_6$ passes through $p_1$. There are two sets of six such conics which are fundamental curves corresponding to the singular points of $R$ and $R^{-1}$.*

Two statements in this theorem require proof. That $R$ is general to within a collineation on the $q$'s follows from the fact that the set of points $p_i$ is general. The statement that $R$ has no fixed points other than those cited may be proved as follows: The general (1, 25) quintic correspondence in the plane may be given in the form

$$\rho y_i = f_i(x), \tag{1}$$

where $f_i$ are general homogeneous functions of the coordinates $x_i$ of order 5. The fixed points of this correspondence are the roots of the matrix

$$\left\| \begin{matrix} x_1, & x_2, & x_3 \\ f_1, & f_2, & f_3 \end{matrix} \right\| = 0, \tag{2}$$

thirty-one in number. If (1) is a Cremona transformation the three quintics $f_i$ have six common nodes, and the three sextics of the matrix (2) have the same nodes. The number of fixed points in the case of the general quintic Cremona transformation is, therefore, $31 - 6 \cdot 4 = 7$. If (1) in addition has a cubic curve of fixed points $\Gamma$—passing necessarily through both sets of singular points—the curve $\Gamma$ is a factor of the three determinants (2). The two cubics

$$\frac{x_1 f_2 - x_2 f_1}{\Gamma} = 0, \quad \frac{x_1 f_3 - x_3 f_1}{\Gamma} = 0$$

meet in the six singular points, the two points $x_1 = f_1 = 0$ (the three points of

$x_1=f_1=0$ on $\Gamma$ are to be excluded) and in one further point. There is, therefore, only one fixed point of the transformation which does not lie on $\Gamma$. It is here especially to be emphasized that

s) *The transformation R is the most general quintic Cremona transformation in the plane that has a cubic curve of fixed points.*

If $\mu_i$ are the elliptic parameters of $p_i$ on the curve $\Gamma$, these parameters must be taken to be connected by the relation (Th. g))

$$\Sigma\mu\equiv\omega,$$

where $\omega$ is a half period of the elliptic functions on $\Gamma$. If $\nu_i$ are the parameters of $q_i$, we have, since $q_i$ and the five points $p_j$ $(j\neq i)$ are on a conic,

$$\omega-\mu_i+\nu_i\equiv 0,$$

whence

$$\nu_i\equiv\mu_i-\omega,$$

an involutory relation connecting $\mu_i$ and $\nu_i$. Also

$$\Sigma\nu_i\equiv\Sigma\mu_i\equiv\omega,$$

and the symmetry of the relations between the $p$'s and $q$'s is put definitely in evidence, either set given on $\Gamma$ determining the other uniquely and in the same way.

The line $p_1p_2$ meets $\Gamma$ again in the point with parameter $-(\mu_1+\mu_2)$ and the line $q_1q_2$ passes through the same point. Therefore,

t) *The lines $p_ip_j$ and $q_iq_j$ meet on the cubic curve $\Gamma$.*

It may easily be verified that the lines $p_1q_2$ and $p_2q_1$ meet on the cubic $\Gamma$, and that there are thus obtained fifteen points like 12, lying by threes on fifteen lines like 12|34|56, this being the line through the three points 12, 34, and 56. The configuration thus obtained is the section by $\pi$ of the fifteen lines and fifteen tritangent planes of a cubic surface left when a double six is omitted. The transforms of these lines by $R$ bear especially interesting relations to the lines themselves.*

## § 6. *Analytical Form of the Transform of a Curve by R.*

It is possible, and in a unique way, to build up from the quadrics $Q$ a surface of order $2n$ having the points $a$ as $n$-fold points when the tangent cone at one of these points is given arbitrarily. Let the $a$ chosen be $a_0$ and choose $a_0$ as the vertex $x_1=x_2=x_3=0$ of the tetrahedron of reference, the opposite plane being chosen as the plane $\pi$ of former sections. Let $(ax)^n=0$ be a given cone at $a_0$, — this cone is of course uniquely determined by the curve $(ax)^n=0$ in $\pi$, $(ax)^n=0$ not involving $x_4$. Let $Q_1$, $Q_2$, $Q_3$ be the three quadrics $Q$

* See Morley, *Johns Hopkins Circulars*, February, 1912, p. 69.

touching the three reference planes at $a_0$. These quadrics are linearly independent. Then

$$(\alpha Q)^n = 0 \tag{1}$$

is the surface above cited having the tangent cone $(\alpha x)^n = 0$ at $a_0$. For, the $Q$'s are of the form

$$\left.\begin{aligned} Q_1 &\equiv x_1 x_4 - k_1, \\ Q_2 &\equiv x_2 x_4 - k_2, \\ Q_3 &\equiv x_3 x_4 - k_3, \end{aligned}\right\} \tag{2}$$

where $k_i$ are quadratic functions of $x_1$, $x_2$, $x_3$. Substituting (2) in (1), we find that the coefficient of $x_4^n$, the highest power of $x_4$ occurring, is $(\alpha x)^n$, and hence, (1) is the surface built on the $Q$'s of order $2n$ and with $n$-fold point at $a_0$ and the tangent cone $(\alpha x)^n = 0$.

The $k$'s in (2) are significant as being the intersections of the quadrics $Q_i$ with $\pi (x_4 = 0)$. Since $Q_i$ touches the plane $x_i$ at $a_0$ $(i = 1, 2, 3)$ we have

$$P x_i = k_i, \tag{3}$$

where $P$ is the quadratic (1, 4) correspondence of § 3.

The surface $(\alpha Q)^n = 0$ may be defined geometrically in such a way as to bring out more clearly the matters with which we are dealing. Let, in the first place, a curve $(\alpha x)^n = 0$ be given in $\pi$, and let it be regarded as defining a set of directions at $a_0$. The curves $E$ with these directions lie on a surface whose intersection with $\pi$ is $P(\alpha x)^n$, a curve of order $2n$. This surface must coincide with $(\alpha Q)^n = 0$. Incidentally,

u) *Every locus of curves $E$ is of the form $(\alpha Q)^n = 0$, and, conversely, every surface of this form is a locus of curves $E$.*

This might also have been inferred from the facts that, 1) a curve $E$ meets $(\alpha Q)^n = 0$ in $8n$ points; 2) that these $8n$ intersections are accounted for at the points $a$; 3) that the sum of the elliptic parameters on any curve $E$ at the points $a$ is a period; 4) that, therefore, if a curve $E$ meets $(\alpha Q)^n = 0$ at a point not an $a$, it lies entirely on the surface.

The invariance under $T$ of the surface $(\alpha Q)^n = 0$ follows from the fact that it is a locus of curves $E$, and of course this is also obvious from its form. In order to determine the transform by $R$ of $(\alpha x)^n = 0$ we have only to project from $a_0$ the curve of intersection of $(\alpha Q)^n = 0$ with $C_0$, where the $Q$'s are of the form (2). We have from (2)

$$(\alpha Q)^n \equiv [(\alpha x) x_4 - (\alpha k)]^n = 0, \tag{4}$$

or, taking account of (3), and writing

$$\alpha_1 P x_1 + \alpha_2 P x_2 + \alpha_3 P x_3 = (\alpha P x) = P(\alpha x), \quad (\alpha Q)^n \equiv [(\alpha x) x_4 - P(\alpha x)]^n. \tag{5}$$

Now, $C_0$ is of the form

$$C_0 \equiv \Gamma x_4 + C = 0, \tag{6}$$

where $C=0$ is the quartic curve of intersection of $C_0$ with $\pi$ $(x_4=0)$; then the projection of the curve of intersection of $C_0$ and $(\alpha Q)^n=0$, or the transform $R(\alpha x)^n$ is obtained by eliminating $x_4$ from (5) and (6). Substituting in (5) the value of $x_4$ from (6), we obtain

$$R(\alpha x)^n \equiv (\alpha x)^n C^n + \binom{n}{1}(\alpha x)^{n-1} P(\alpha x) C^{n-1}\Gamma + \binom{n}{2}(\alpha x)^{n-2} P(\alpha x)^2 C^{n-2}\Gamma^2 + \ldots + P(\alpha x)^n \Gamma^n = 0.^* \tag{7}$$

We state this in the theorem

v) *Given the transformations $R$, $R^{-1}$ having the singular points $q_i$, $p_i$, the curve of fixed points $\Gamma$ and the fixed point $p$, and choosing any one of the $\infty^8$ (1, 4) quadratic correspondences, say $P$, with the fixed points $p_i p$, the transform $R(\alpha x)^n$, where $(\alpha x)^n=0$ is any curve in the plane, is expressible in terms of the cubic $\Gamma$, a quartic $C$ determined uniquely by $P$, and the set of covariants of $(\alpha x)^n$ and $P$ which are of the form*

$$(\alpha x)^{n-R} P(\alpha x)^R = 0, \quad (R = 0, 1, \ldots, n). \tag{8}$$

*These curves are loci of points $x$ which lie on the correspondent by $P$ of their $(n-R)$-th polars as to $(\alpha x)^n=0$.*

The last statement is easily proved. Fixing $x$, the $(n-R)$-th polar of $x$ to $(\alpha x)^n=0$ is $(\alpha x)^{n-R}(\alpha y)^R=0$, where $y$ is variable; its transform by $P$ is $(\alpha x)^{n-R} P(\alpha y)^R = 0$, and the condition that $x$ should lie on this curve is $(\alpha x)^{n-R} P(\alpha x)^R = 0$.

Applying (7) to the cubic $\Gamma \equiv (\gamma x)^3$ which we know to be invariant under $R$, and remembering that if a curve passes through a singular point $q_i$ of $R$, the conic $u_i$ factors from its transform, we obtain

$$R\Gamma \equiv u_1 u_2 \ldots u_6 \Gamma \equiv \Gamma C^3 + 3(\gamma x)^2 P(\gamma x)\Gamma C^2 + 3(\gamma x) P(\gamma x)^2 \Gamma C^2 + P\Gamma \cdot \Gamma^3,$$

whence the interesting identity

$$u_1 \ldots u_6 \equiv C^3 + 3(\gamma x)^2 P(\gamma x) C^2 + 3(\gamma x) P(\gamma x)^2 C\Gamma + P\Gamma \cdot \Gamma^2. \tag{8'}$$

The curve $P\Gamma$ is a sextic passing through the points $p_i$, since $p_i$ are fixed points of $P$. The curve $u_i$ meets $C=0$, a quartic touching $\Gamma$ at the points $p_i$, at three points not on $\Gamma$, $u_i=0$ passing through the five points $p_j$ $(j \neq i)$ of $\Gamma$. The identity (8') tells us that the eighteen points thus obtained from $u_i$ are the $24-6=18$ intersections of $P\Gamma$ and $C$ which are not on $\Gamma$. This may be seen in another way, which incidentally serves to identify these six triads of points. The transform of $q_i$ by $P$ is obtained from the direction $d$ which $q_{0,i}$ determines at $a_0$, the corresponding curve $E$ being $r_i$ taken with the line $a_i a$. This meets

* No attempt is here made to fix the possible constant multipliers of the forms occurring in this formula. This would become necessary if several choices of $P$ were made and exact results were required from combination of the resulting expressions. It is sufficient for our present purpose that the formula (7) gives a system of curves from which $R(\alpha x)^n$ may be linearly built.

$\pi$ in a point of the configuration of theorem m′) and in the three points in which $r_i$ meets $\pi$. Since $r_i$ lies on $C_0$ and is projected from $a_0$ into $u_i$, the inferences from the identity are obvious. The curve $C$ is of course determined by these conditions.

As a special case of theorem v) we have

v′) *The transform of a line $l$ of $\pi$ by $R$ is of the form*

$$Rl \equiv Cl + \Gamma Pl, \tag{9}$$

*where $P$ is any one of the $\infty^8$ (1,4) correspondences having $p_i$, $p$ as fixed points.*

We infer from (9) that, 1) $Rl$ meets $\Gamma$ at points of $C$, — $C$ touches $\Gamma$ at $p_i$ and $Rl$ has nodes there; 2) $Rl$ meets $l$ at its three points of intersection with $\Gamma$ and its two points of intersection with $Pl$, —the former is obvious from the fact that points of $\Gamma$ are fixed under $R$, the latter furnishes additional information; 3) $Rl$ meets $C$ at its eight intersections with $Pl$. We have here the new facts:

w) *The transform $Rl$ meets $l$ in five points, three of which are on $\Gamma$ and are fixed points of $R$; the other two are the intersections of $l$ with its transform by $P$. These two points are the pair of the Geiser involution determined by the points $p_i$, $p$ which lie on $l$* (Th. n)). *It meets $Pl$ in these two points and in its intersections with $C$, the quartic curve associated with $P$.*

Since the quadratic transformation $R_{ijk}$ and $R_{ij}$ are special cases of the correspondence $P$, we have

x) *The transform $Rl$ may be written in twenty ways in the form*

$$C_{ijk}\, l + \Gamma R_{ijk}\, l = 0,$$

*and in fifteen ways in the form*

$$C_{ij}\, l + \Gamma R_{ij}\, l = 0.$$

Theorems v′) and x) lead to an endless variety of algebraic identities, one of which we use in the next section.

### § 7. *The Points $q_{ij}$ and the Quadratic Transformations S.*

The transformation $R$ sends points on the line $q_iq_j$ into points on the line $p_ip_j$, and a correspondence is thus determined between the points of the two lines which is necessarily projective since it is univocal. The lines $p_ip_j$ and $q_iq_j$ meet in a point of $\Gamma$ (Th. t)), a fixed point of $R$, and hence their intersection is self-corresponding. The ranges determined on $p_ip_j$ and $q_iq_j$ by $R$ are, therefore, in perspective position. It will presently be seen that the point $q_{ij}$ is the center of perspective, this theorem showing exactly how the points $q_{0j}$ may be determined directly by $R$ without using the quadratic transformations $R_{ijk}$.

The transform $Rl$ may be written in the form (Th. x))

$$Rl \equiv C_{123}l + \Gamma R_{123}l,$$

where $C_{123}$ is the rational quartic cut from $C_0$ by the plane on $a_1$, $a_2$, $a_3$ (or its projection); $C_{123}$ touches $\Gamma$ at $p_4$, $p_5$, $p_6$ and has nodes at $p_1$, $p_2$, $p_3$, and is thereby uniquely determined: $l$ is any line of the plane and $R_{123}l$ its transform by the quadratic transformation of § 4. $R_{123}$ has singular points $q_{23}$, $q_{31}$, $q_{12}$ and $R_{123}^{-1}$ singular points $p_1$, $p_2$, $p_3$. Take $l$ now as the line $q_{12}q_{13}$. We have

$$R\overline{q_{12}q_{13}} \equiv C_{123} \cdot \overline{q_{12}q_{13}} + \Gamma \cdot \overline{p_1 p_2} \cdot \overline{p_1 p_3}, \tag{1}$$

since $\overline{q_{12}q_{13}}$ contains the two singular points $q_{12}$, $q_{13}$, and hence

$$R_{123}\overline{q_{12}q_{13}} \equiv \overline{p_1 p_2} \cdot \overline{p_1 p_3}. \tag{2}$$

We see from the identity (1) that $R\overline{q_{12}q_{13}}$ meets $\overline{p_1 p_2}$ at its intersection with $\overline{p_{12} p_{13}}$, and this point must be the transform by $R$ of the point in which $\overline{q_{12}q_{13}}$ meets $\overline{q_1 q_2}$. We have then that the line $\overline{q_{12}q_{13}}$ through $q_{12}$ marks on $\overline{p_1 p_2}$ and $\overline{q_2 q_1}$ correspondents by $R$. Similarly, it may be proved that $\overline{q_{12}q_{14}}$, $\overline{q_{12}q_{15}}$, $\overline{q_{12}q_{16}}$ mark on these lines correspondents by $R$, and hence $q_{12}$ is the center of perspective of these ranges. We state this in the theorem

y) *The fifteen points $q_{ij}$ are symmetrically related to the two sets of seven points $p_1, \ldots, p_6, p$, $q_1, \ldots, q_6, p$, either set determining them in the same way. The point $q_{ij}$ is the center of perspective of the correspondence determined between the lines $\overline{p_i p_j}$ and $\overline{q_i q_j}$ by the quintic Cremona transformation $R$ which is itself determined by either set of seven points.*

From the symmetry of the relations of the points $q_{ij}$ to the two sets of seven points $p_i$, $p$, $q_i$, $p$, we have

z) *There exist twenty quadratic Cremona transformations $S_{123}$, etc.; $S_{123}$ has singular points $q_{23}$, $q_{31}$, $q_{12}$; $S_{123}^{-1}$ has singular points $q_1$, $q_2$, $q_3$, and $S_{123}$ has fixed points $q_4$, $q_5$, $q_6$, $p$. There exist fifteen quadratic Cremona transformations $S_{12}$, etc.; $S_{12}$ has the singular points $q_{12}$, $p_1$, $p_2$, and $S_{12}^{-1}$, the singular points $p_1$, $q_1$, $q_2$. The transformation $S_{12}$ has the fixed points $q_3$, $q_4$, $q_5$, $q_6$.*

In the part of this paper which is to follow the correspondences here treated will be discussed in connection with their relation to the nodes of a quartic of lines. The lines of the quartic referred to are marked on $\pi$ by planes tangent at $a_0$ to nodal quadrics $Q$. This curve has the twenty-eight points $p_i$, $p$, $q_i$, $q_{ij}$ as nodes. This known fact we shall prove by means of the transformations whose theory we have developed. The existence of a vast number of other transformations follows by symmetry, and these transformations have what seems to the author an interesting bearing on certain configurations of points and lines associated with the nodes, apart from the fact that they seem to add geometrical interest to their abstract grouping.

BALTIMORE, *January*, 1916.

## *Infinite Groups Generated by Conformal Transformations of Period Two (Involutions and Symmetries).**

BY EDWARD KASNER.

Our main problem is to find all conformal transformations of period 2 and the infinite group generated by such transformations. The result is quite simple and is given in Theorem IV below.

We shall consider only those conformal transformations of the plane which convert the origin into itself and are regular at the origin. Such regular transformations are expressed by power series of the two forms

$$Z=\alpha z+\beta z^2+\gamma z^3+\ldots, \qquad \alpha\neq 0, \tag{1}$$
$$Z=\alpha' z_0+\beta' z_0^2+\gamma' z_0^3+\ldots, \qquad \alpha'\neq 0, \tag{2}$$

where $z=x+iy$, $z_0=x-iy$ and the coefficients are arbitrary complex members.† The *direct* (or proper) conformal transformations (1) form a continuous infinite group $G$; if we add the *reverse* (or improper) conformal transformations (2), we have a mixed group $G'$.

It is easy to determine all regular transformations of period 2. In the direct type $Z=f(z)$ the functional equation is $f(f(z))\equiv z$, that is, $f^2=1$; in the reverse type $Z=f(z_0)$ the functional equation is $f(f_0(z))\equiv z$, that is, $ff_0=1$, where $f_0$ denotes the series whose coefficients are the conjugates of the coefficients of series $f$. We shall call a transformation the former type (*excluding* the identical transformation) a *conformal involution,* and one of the latter type a *conformal symmetry.*

It can be shown without difficulty (by undetermined coefficients and otherwise) that every involution can be reduced conformally to the simple form $Z=-z$ (symmetry with respect to the origin); and that every symmetry can be reduced to $Z=z_0$ (symmetry with respect to the axis of reals). Conformal symmetry is thus the same as Schwarzian reflexion in an analytic curve.

---

* Presented to the American Mathematical Society, September, 1914 and December, 1915.

† All the work of the present paper is formal; we shall not take into account the convergence of the series. Hence, we are discussing groups of formal power series.

THEOREM I. *The only transformations of period 2 in the general conformal group $G'$ are conformal involutions and conformal symmetries. These are reducible to the normal forms $Z=-z$, $Z=z_0$, respectively.*

The set of all involutions, likewise the set of all symmetries, is in number $\infty^\infty$. Roughly speaking, the functional equations allow half of the coefficients in the series (1) and (2) to be arbitrary. Neither set by itself, nor the two together, constitute a group.

The question then arises, what types of transformation are obtained by combining any number of involutions or symmetries. The results are given in the following theorems:

THEOREM II. *The only transformations which can be obtained as products of conformal involutions are of the two types*

$$Z=z+\beta z^2+\gamma z^3+\ldots, \qquad \gamma-\beta^2=0, \tag{3}$$
$$Z=-z+\beta' z^2+\gamma' z^3+\ldots, \qquad \gamma'+\beta'^2=0. \tag{4}$$

*These form a mixed group which we denote by $G'_{inv.}$ (the group generated by involutions).*

Any one of these transformations can be factored into involutions in an infinitude of ways, of which at least one will contain four or fewer factors. When the number of factors is even we have the type (3), forming a continuous subgroup $G_{inv}$.

THEOREM III. *The only transformations which can be obtained as products of conformal symmetries are of the two types*

$$Z=\alpha z+\beta z^2+\gamma z^3+\ldots, \qquad |\alpha|=1, \tag{5}$$
$$Z=\alpha' z_0+\beta' z_0^2+\gamma' z_0^3+\ldots, \qquad |\alpha'|=1. \tag{6}$$

*These form a mixed group which we denote by $G'_{sym.}$ (the group generated by symmetries).*

Any one of these transformations (5), (6) can be factored into symmetries in an infinitude of ways, of which at least one will contain four or fewer factors. In the direct type (5) the number of factors is, of course, always even. This type forms a continuous subgroup $G_{sym.}$.

It will be observed that (3) and (4) are both included as special cases in (5). It follows that every product of any number of involutions can be factored into a product of symmetries. In particular, *every involution can be factored into symmetries.* The converse is obviously not true; for in a symmetry (or product of symmetries) the leading coefficient may be any complex number whose absolute value is unity, while in the involution case we are restricted to the two values $\pm 1$.

Every product of symmetries and involutions in any order can be written as a product of symmetries. This gives our

FUNDAMENTAL THEOREM IV. *The group generated by all conformal transformations of period 2 is identical with $G'_{sym.}$. It is represented by* (5) *and* (6), *that is, it consists of all regular transformations for which the modulus of the leading coefficient is unity.*

*Discussion of Products of Involutions.*

The chief question here is to find the form of those transformations which can be obtained as the product of two involutions. Given a function $f(z)$ we inquire whether it is possible to find two functions $\phi(z)$, $\psi(z)$ such that, symbolically,

$$f=\psi\phi, \quad \phi^2=1, \quad \psi^2=1. \tag{7}$$

The direct treatment by the method of undetermined coefficients is somewhat long and involved. We may simplify the discussion by noticing that every involution $Z=\phi(z)$, $\phi^2=1$ can be written in the implicit form

$$Z+z=a(z-Z)^2+b(z-Z)^4+c(z-Z)^6+\ldots., \tag{8}$$

where all the coefficients are entirely arbitrary. (In the explicit form (1) half of the coefficients are arbitrary and the others are determined.) The product of two involutions will necessarily have its leading coefficient unity. Our question is to find when the equation

$$Z=z+c_2z^2+c_3z^3+\ldots. \tag{9}$$

can result from the elimination of the variable $w$ from the two equations

$$\left.\begin{aligned} z+w&=a(z-w)^2+b(z-w)^4+\ldots., \\ w+Z&=A(w-Z)^2+B(w-Z)^4+\ldots., \end{aligned}\right\} \tag{9'}$$

defining two arbitrary involutions.

It is convenient to write $c_\kappa=2^\kappa e_\kappa$ and to introduce

$$t=z-w. \tag{9''}$$

Eliminating $z, w, Z$ from the four equations (9), (9′), (9″), we find the following equation in the single variable $t$,

$$\begin{aligned} &-t+P+e_2P^2+e_3P^3+\ldots. \\ &\qquad =A\{t+e_2P^2+e_3P^3+\ldots.\}^2+B\{t+\ldots.\}^4+\ldots., \end{aligned} \tag{10}$$

where

$$P\equiv t+at^2+bt^4+ct^6+\ldots..$$

Arranging (10) in powers of $t$ and equating corresponding coefficients, we obtain the infinite sequence of equations which must be discussed.

The first two of these equations are

$$a+e_2=A, \quad 2ae_2+e_3=2Ae_2.$$

This gives a necessary relation $e_3-2e_2^2=0$, that is, $c_3-c_2^2=0$. We shall not write down the higher equations of the sequence (corresponding to $t^3$, $t^4$, etc.), but merely state that, if we assume $c_2\neq 0$, there are no further relations between the coefficients of (9). Therefore,

THEOREM V. *Every transformation of the form*

$$Z=z+c_2z^2+c_3z^3+\ldots, \quad c_2\neq 0, \quad c_3-c_2^2=0 \tag{11}$$

*can be factored into two involutions. This can always be done in* $\infty^1$ *ways.*

Assume next $c_2=0$. The first two equations of the sequence give $a=A$, $c_3=0$. The equations corresponding to $t^4$, $t^5$, and $t^7$ now take such a form that we can eliminate $a$, $b$ and $A$, $B$, obtaining a necessary relation

$$4c_4^4-c_5^3+2c_4c_5c_6-c_4^2c_7=0. \tag{12'}$$

No additional relations exist if we assume $c_4\neq 0$. Hence, transformations of the form

$$Z=z+c_4z^4+c_5z^5+\ldots, \qquad c_4\neq 0, \tag{12}$$

where (12′) holds, can be factored into two involutions.

A new type arises when $c_4=0$, and so on. The final result is

THEOREM VI. *All regular conformal transformations which can be obtained as the product of two involutions are of the form*

$$Z=z+c_{2\kappa}z^{2\kappa}+c_{2\kappa+1}z^{2\kappa+1}+\ldots, \qquad c_{2\kappa}\neq 0, \tag{13}$$

*where* $\kappa=1, 2, 3, \ldots,$ *and a single rational relation*

$$R_\kappa(c_{2\kappa}, c_{2\kappa+1}, \ldots, c_{4\kappa-1})=0 \tag{13'}$$

*holds between the coefficients.*

The form of this relation, as well as the number of coefficients involved, changes with the integer $\kappa$, but in all cases $c_{4\kappa-1}$ can be expressed as a rational function of the previous coefficients.

For each value of the integer $\kappa$ we have a certain set of transformations which can be factored into two involutions. Thus, for $\kappa=1$, we have the set (11); and for $\kappa=2$, we have the set defined by (12) and (12′). *No one of these sets has the group property.* The same is true of the totality of all the sets.

All the transformations in question are, however, included in the larger class,

$$Z=z+c_2z^2+c_3z^3+\ldots, \qquad c_3-c_2^2=0, \tag{14}$$

which *does* constitute a group, as may be verified immediately.

A simple example of a transformation of the group (14), which is not of the form specified in Theorem VI, and hence cannot be factored into *two* involutions, is $Z=z+z^4$. We shall now show, however, that *every transformation of the group* (14) *can surely be factored into four* (*or fewer*) *involutions.*

If, in (14), $c_2$ does not vanish, we already know (Theorem V) that two involutions are sufficient. Consider, therefore, any transformation, $T$, of our group (14) for which $c_2$ does vanish. Of course, then, $c_3=0$ on account of the relation $c_3-c_2^2=0$. We can factor $T$ into two transformations, $T'$ and $T''$, both of the form (14), and such that the coefficients $c_2'$ and $c_2''$ do not vanish. This is seen from the fact that the product $T'T''$ is of the form

$$z+(c_2'+c_2'')z^2+\ldots,$$

and, hence, the coefficient of $z^2$ can be made to vanish without taking either $c_2'$ or $c_2''$ equal to zero. We already know that $T'$ and $T''$ can each be factored into two involutions. Hence, $T$ can be factored into four involutions.

The type (14) is what we have written as (3) in Theorem II. To complete the proof of this theorem, we merely multiply (11) by a general involution and find the type (4). The latter type can, therefore, surely be factored into three (or fewer) involutions.

### *Discussion of Products of Symmetries.*

A conformal symmetry we have defined as a reverse conformal transformation, $Z=f(z_0)$, whose square is the identical transformation; that is, such that $f(f_0(z))\equiv z$. The conditions that

$$Z=\alpha z_0+\beta z_0^2+\gamma z_0^3+\ldots \tag{15}$$

shall be a symmetry are therefore that the coefficients of the conjugate series

$$Z_0=\alpha_0 z+\beta_0 z^2+\gamma_0 z^3+\ldots \tag{15'}$$

shall be the same as the coefficients of the series obtained by reverting the original series (15). Therefore,

$$\alpha_0=\frac{1}{\alpha}, \quad \beta_0=-\frac{\beta}{\alpha^3}, \quad \gamma_0=\frac{2\beta^2-\alpha\gamma}{\alpha^5}, \quad \ldots\ldots \tag{15''}$$

The first of the conditions gives $|\alpha|=1$.

We now examine the conditions that a direct transformation (1) shall be expressible as the product of two symmetries. An obviously necessary condition is $|\alpha|=1$. The question is whether transformations of the form

$$Z=\alpha z+\beta z^2+\gamma z^3+\ldots, \qquad |\alpha|=1, \tag{16}$$

can be factored into symmetries.

Since we have two arbitrary symmetries at our disposal, each involving an infinitude of arbitrary constants, we might, at first sight, expect that the coefficients $\beta, \gamma, \ldots$ in (16) are entirely arbitrary. But the detailed examination, by the method of undetermined coefficients, shows that this is so only "in general," namely, if we assume that the amplitude $\theta$ of the leading coefficient $\alpha$ is irrational; that is, $\theta/\pi$ is an irrational number. This distinction between the *irrational* and the *rational* cases is analogous to that arising in the writer's treatment of another problem of conformal geometry, namely, the invariants of curvilinear (analytic) angles.*

For the irrational case we have this simple result:

THEOREM VII. *Transformations of the form*

$$Z=\alpha z+\beta z^2+\gamma z^3+\ldots, \tag{17}$$

*where the modulus of $\alpha$ is unity, and the amplitude of $\alpha$ is irrational (so that $\alpha$ is not a root of unity), are always factorable (formally) into two symmetries.*

In the rational case such a factoring is *not* always possible. This may be verified most easily by taking the case $\alpha=1$; that is,

$$Z=z+\beta z^2+\gamma z^3+\ldots. \tag{18}$$

We find, by the method of undetermined coefficients, that there is now a necessary condition on the higher coefficients, namely,

$$|\gamma|-|\beta|^2=0. \tag{18'}$$

If this relation does not hold, factoring into two symmetries is impossible. If it does hold, and if $\beta\neq 0$, the factoring is possible. Higher relations arise when $\beta=0$, etc. Analogous results are obtained when $\alpha$ is a root of unity.

*Transformations* (17), *for which $\alpha$ is a root of unity (rational case), are not factorable into two symmetries unless the higher coefficients are subjected to certain restrictions. The form of these higher relations and the number of coefficients involved is different for different rational angles $\theta$.*

---

* See "Conformal Geometry," *Proceedings of the Fifth International Congress of Mathematicians*, Cambridge (1912), Vol. II, pp. 81–87. Also G. A. Pfeiffer, AMERICAN JOURNAL OF MATHEMATICS, Vol. XXXVII.

It will not be necessary, however, to employ these higher relations in order to solve our group problem. Transformations factorable into *two* symmetries do not, by themselves, constitute a group. The larger class defined by $|\alpha|=1$ do constitute a group, and we shall show that every one of its transformations can be factored into *four* (or fewer) symmetries.

In the irrational case we already know that two are sufficient. If, now, $\theta$ is rational, we may break it up into two parts, $\theta'$ and $\theta''$, both of which are irrational. (This, of course, can be done in an infinitude of ways.) Hence, the given transformation $T$,

$$Z=\alpha z+\beta z^2+\ldots, \qquad \alpha=e^{i\theta},$$

can be decomposed into two transformations, $T'$, $T''$,

$$Z=\alpha' z+\beta' z^2+\ldots, \qquad \alpha'=e^{i\theta'},$$
$$Z=\alpha'' z+\beta'' z^2+\ldots, \qquad \alpha''=e^{i\theta''},$$

such that $T'$ and $T''$ both come under the irrational type. (This can be done in $\infty^\infty$ ways, since $\beta'$, $\gamma'$, .... can be taken arbitrarily, $\beta''$, $\gamma''$, .... being then determined.) Since, by Theorem VII, we can factor $T'$ and $T''$ each into two symmetries, it follows that $T$ can be factored into four symmetries.

It follows also that all products with an even number of symmetries as factors make up the continuous group of direct transformations (5). To complete the proof of Theorem III, stated earlier, we need merely observe the reverse type (6) is obtained from (5) by multiplying with the simple symmetry $Z=z_0$. This shows that (6) can be factored into five (or fewer) symmetries. By a simple device we can, however, always reduce the number of factors to three.

If, in the general regular transformation (1) or (2), the leading coefficient $\alpha$ or $\alpha'$ has an absolute value different from unity, then factoring into symmetries or any regular transformations of period 2 will, of course, be impossible. We can, however, reduce the modulus (which expresses the stretching ratio at the origin) to unity by means of a homothetic transformation

$$Z=mz,$$

where $m$ is real. This gives the following interesting result:

*Any regular conformal transformation of the plane (converting the origin into itself) can be decomposed into a homothetic transformation together with a finite number (not exceeding four) of conformal symmetries.*

That is, it is possible to find four or fewer regular analytic arcs through the origin (analytic elements), such that successive Schwarzian reflexion in these arcs, followed by a stretch, is identical with the arbitrary transformation (1) or (2).

Homothetic transformations combined with involutions generate a continuous subgroup defined by (1) with the single relation $\beta^2-\alpha\gamma=0$.

We point out, in conclusion, that our results are valid in the complex (four-dimensional) plane, as well as in the usual real or Gaussian (two-dimensional) plane.*

COLUMBIA UNIVERSITY, NEW YORK.

* See *Trans. Amer. Math. Soc.*, Vol. XVI (1915), pp. 333–349, for the distinction between real and complex conformal geometry.

# *On the Solutions of Linear Homogeneous Difference Equations.**

BY R. D. CARMICHAEL.

*Introduction.*

Let us consider the linear homogeneous difference equation of order $n$,

$$F(x+n)+\bar{a}_1(x)F(x+n-1)+\bar{a}_2(x)F(x+n-2)+\ldots+\bar{a}_n(x)F(x)=0, \quad (1)$$

in which the coefficients are analytic at infinity or have poles there. By means of a transformation of the form

$$F(x)=x^{\mu x}f(x),$$

we arrive at the new equation,

$$f(x+n)+a_1(x)f(x+n-1)+a_2(x)f(x+n-2)+\ldots+a_n(x)f(x)=0, \quad (2)$$

where

$$a_k(x)=(x+n-k)^{\mu(x+n-k)}(x+n)^{-\mu(x+n)}\bar{a}_k(x), \qquad k=1, 2, \ldots, n.$$

It is obvious that integers $\mu$ exist such that each of the functions $a_k(x)$ is analytic at infinity. For the value of $\mu$ we choose the least integer such that the functions $a_k(x)$ are all analytic at infinity.† Let us write

$$a_k(x)=a_{k0}+\frac{a_{k1}}{x}+\frac{a_{k2}}{x^2}+\ldots, \qquad k=1, 2, \ldots, n, \ |x|\geqq R. \quad (3)$$

We shall confine our attention to the case in which the constants $a_{10}$, $a_{20}$, $\ldots$, $a_{n0}$ are such that the *characteristic algebraic equation* ‡

$$\alpha^n+a_{10}\alpha^{n-1}+a_{20}\alpha^{n-2}+\ldots+a_{n0}=0, \quad (4)$$

has its roots $\alpha_1, \alpha_2, \ldots, \alpha_n$, different from each other and from zero. In this

* Read before the American Mathematical Society at Chicago, December 29, 1914.

† It is easy to see that $x^{-\mu k}\bar{a}_k(x)$ is analytic at infinity for every value of $k$ from 1 to $n$ and that $\mu$ is the least integer for which this is true. We shall say that the value of $x^{-\mu k}\bar{a}_k(x)$ at infinity is $\bar{a}_{k0}$. Comparing with equation (3) we see that $a_{k0}=e^{-\mu k}\bar{a}_{k0}$.

‡ It will be said that this is the characteristic algebraic equation *associated* either with (2) or with (1).

case, as is well-known, equation (2) has $n$ formal power series solutions of the form *

$$f_i(x)=a_i^x x^{\mu_i}\left(1+\frac{c_{i1}}{x}+\frac{c_{i2}}{x^2}+\ldots\right),\ i=1,\ 2,\ \ldots,\ n. \tag{5}$$

These are in general divergent.

In this paper I shall show how the formal solutions (5) afford a means by which equation (2) may be separated into two members so that the method of successive approximations, when applied in its usual form, will lead to an actual solution of (2) and by aid of this single solution shall develop means by which the most important fundamental systems of solutions of (2), and hence of (1), are found.

This paper contains no new results, so that its novelty lies entirely in the new methods employed. The theorems demonstrated are essentially equivalent to those obtained by Birkhoff in the *Transactions of the American Mathematical Society*, Vol. XII (1911), pp. 242-284, and hence are more complete than the similar ones previously demonstrated by me and published in the volume just referred to, pp. 99-134. The method of the present paper is more closely related to that of my previous paper than to that of Birkhoff's later memoir. It shares with both of them the desirable property of being a direct means for attaining the end in view; but it is essentially different from either of them and in several respects is to be preferred to them. I have found it convenient to use the Birkhoff modification of the integral earlier employed by me, since this modified integral serves the purpose of securing a rather simpler fundamental system of solutions, especially as regards the asymptotic character of the functions in the solution.

For other methods of dealing with the problem of difference equations see Nörlund, *Acta Mathematica*, Vol. XL, pp. 191–249, and the papers to which he refers.

## §1. *Formal Solutions by Successive Approximation.*

Let us denote by $l_i(x)$ the functions

$$l_i(x)=a_i^x x^{\mu_i}\left(1+\frac{c_{i1}}{x}+\frac{c_{i2}}{x^2}+\ldots+\frac{c_{im}}{x^m}\right),\ i=1,\ 2,\ \ldots,\ n.$$

Form the equation

* The value of the constants $\mu_i$, $c_{i1}$, $c_{i2}$, .... may be determined by substituting the value of $f_i(x)$ from (5) into (2) and directly reckoning out the constants so that the resulting equation shall be a formal identity in $x$.

$$\begin{vmatrix} g_1(x+n) & g_1(x+n-1)\ldots\ldots g_1(x) \\ l_1(x+n) & l_1(x+n-1)\ldots\ldots l_1(x) \\ l_2(x+n) & l_2(x+n-1)\ldots\ldots l_2(x) \\ \ldots\ldots & \ldots\ldots\ldots\ldots \\ l_n(x+n) & l_n(x+n-1)\ldots\ldots l_n(x) \end{vmatrix}=0. \tag{6}$$

This is obviously a linear homogeneous difference equation of order $n$ at most; and it is of order $n$ if the minors of $g_1(x+n)$ and $g_1(x)$ are neither identically zero. These minors are $D(x)$ and $D(x+1)$ respectively, where

$$D(x)=\begin{vmatrix} l_1(x+n-1) & l_1(x+n-2)\ldots\ldots l_1(x) \\ l_2(x+n-1) & l_2(x+n-2)\ldots\ldots l_2(x) \\ \ldots\ldots & \ldots\ldots\ldots\ldots \\ l_n(x+n-1) & l_n(x+n-2)\ldots\ldots l_n(x) \end{vmatrix}. \tag{7}$$

Divide the first row of this determinant by $\alpha_1^x x^{\mu_1}$, the second row by $\alpha_2^x x^{\mu_2},\ldots.,$ the last by $\alpha_n^x x^{\mu_n}$. Each element of the resulting determinant is analytic at infinity; and hence the function $\bar{D}(x)$ represented by this resulting determinant is analytic at infinity. It is easy to see that the value of $\bar{D}(x)$ at infinity is

$$\bar{D}(\infty)=\begin{vmatrix} \alpha_1^{n-1} & \alpha_1^{n-2}\ldots\ldots\alpha_1 & 1 \\ \alpha_2^{n-1} & \alpha_2^{n-2}\ldots\ldots\alpha_2 & 1 \\ \ldots & \ldots\ldots & . \\ \alpha_n^{n-1} & \alpha_n^{n-2}\ldots\ldots\alpha_n & 1 \end{vmatrix};$$

or

$$\bar{D}(\infty)=\prod_{r,s}(\alpha_r-\alpha_s), \quad r<s.$$

Since the numbers $\alpha_1, \alpha_2, \ldots., \alpha_n$ are all different by hypothesis, we see that $\bar{D}(x)$ is not identically zero; in fact, it is different from zero at infinity. Hence, $D(x)$ is not identically zero: and therefore equation (6) is of order $n$.

In the determinant in equation (6) divide the second row by $\alpha_1^x x^{\mu_1}$, the third row by $\alpha_2^x x^{\mu_2},\ldots.,$ the last row by $\alpha_n^x x^{\mu_n}$, and expand the resulting determinant in terms of the elements of the first row. The coefficients of the quantities $g_1(x+n), g_1(x+n-1), \ldots., g_1(x)$ are obviously all analytic at infinity, that of $g_1(x+n)$, namely, $D(x)$, being different from zero at infinity. Hence we have from (6) an equation of the form

$$g_1(x+n)+\lambda_1(x)g_1(x+n-1)+\lambda_2(x)g_1(x+n-2)+\ldots.+\lambda_n(x)g_1(x)=0, \tag{8}$$

in which $\lambda_1(x), \lambda_2(x), \ldots., \lambda_n(x)$ are analytic at infinity. This equation has the solutions $l_1(x), l_2(x), \ldots., l_n(x)$, as we see readily from equation (6). Moreover, these form a fundamental system of solutions, since $D(x)$ is not identically zero.

By means of the fact that equation (8) has the solution $l_1(x)$, $l_2(x)$, ...., $l_n(x)$ and the relation which exists between $l_i(x)$, $i=1, 2, \ldots, n$, and the formal solution (5) of equation (2), it may be shown by direct computation that

$$\lambda_i(x)=a_{i0}+\frac{a_{i1}}{x}+\ldots+\frac{a_{i,m+1}}{x^{m+1}}+\frac{a_{i,m+2}}{x^{m+2}}+\ldots;$$

that is, that the expansion of $\lambda_i(x)$ in a descending power series in $x$ coincides with that of $a_i(x)$ as far as the term in $1/x^{m+1}$.

Now, write equation (2) in the form

$$\sum_{k=0}^{n}\lambda_k(x)f(x+n-k)=\sum_{k=1}^{n}\Psi_k(x)f(x+n-k), \qquad \lambda_0(x)=1. \tag{9}$$

Then the function $\Psi_k(x)$ may be expanded as follows:

$$\Psi_k(x)=\frac{\beta_{k,m+2}}{x^{m+2}}+\frac{\beta_{k,m+3}}{x^{m+3}}+\ldots, \qquad k=1, 2, \ldots, n. \tag{10}$$

Form the system of equations

$$\left.\begin{array}{l}\sum_{k=0}^{n}\lambda_k(x)g_1(x+n-k)=0,\\ \sum_{k=0}^{n}\lambda_k(x)g_2(x+n-k)=\sum_{k=1}^{n}\Psi_k(x)g_1(x+n-k),\\ \ldots\ldots\ldots\ldots\ldots\ldots\ldots\ldots\ldots\ldots\ldots,\\ \sum_{k=0}^{n}\lambda_k(x)g_i(x+n-k)=\sum_{k=1}^{n}\Psi_k(x)g_{i-1}(x+n-k),\\ \ldots\ldots\ldots\ldots\ldots\ldots\ldots\ldots\ldots\ldots\ldots\end{array}\right\} \tag{11}$$

By means of these, used as equations for successive approximation, we shall be able to obtain formal solutions of equation (9) or (2).

The first equation in (11) is homogeneous and has the fundamental system of solutions $l_1(x)$, $l_2(x)$, ...., $l_n(x)$, as we saw above. The remaining equations in (11) are of the form

$$\sum_{k=0}^{n}\lambda_k(x)g(x+n-k)=\eta(x). \tag{12}$$

We shall next find two formal solutions of this typical equation. Our method is the classic one of variation of parameters. We assume a solution of the form

$$g(x)=\sum_{k=1}^{n}t_k(x)l_k(x), \tag{13}$$

and determine the functions $t_1(x)$, $t_2(x)$, ...., $t_n(x)$ in a convenient way so that they shall formally satisfy equation (12). Thus we may write

$$g(x+1)=\sum_{k=1}^{n} t_k(x)l_k(x+1), \quad \text{if} \quad \sum_{k=1}^{n} \Delta t_k(x)l_k(x+1)=0;$$

$$g(x+2)=\sum_{k=1}^{n} t_k(x)l_k(x+2), \quad \text{if} \quad \sum_{k=1}^{n} \Delta t_k(x)l_k(x+2)=0;$$

$$\cdots\cdots\cdots\cdots\cdots\cdots\cdots\cdots\cdots\cdots;$$

$$g(x+n-1)=\sum_{k=1}^{n} t_k(x)l_k(x+n-1), \quad \text{if} \quad \sum_{k=1}^{n} \Delta t_k(x)l_k(x+n-1)=0;$$

$$g(x+n)=\sum_{k=1}^{n} t_k(x)l_k(x+n)+\sum_{k=1}^{n} \Delta t_k(x)l_k(x+n).$$

So far we have imposed $n-1$ conditions on the $n$ functions $t_1(x)$, $t_2(x)$, ...., $t_n(x)$. We obtain an $n$-th condition, namely,

$$\sum_{k=1}^{n} \Delta t_k(x)l_k(x+n)=\eta(x),$$

by substituting the preceding values of $g(x)$, $g(x+1)$, ...., $g(x+n)$ in equation (12), remembering that $l_1(x)$, $l_2(x)$, ...., $l_n(x)$ satisfy the equation obtained from (12) by replacing $\eta(x)$ by zero.

The $n$ equations of condition now placed upon the functions $t(x)$ may readily be solved by determinants for the quantities $\Delta t_k(x)$, $k=1, 2, \ldots, n$. Thus we have

$$\Delta t_k(x)=(-1)^{n+k}\eta(x)M_k(x+1)/M(x+1), \qquad k=1, 2, \ldots, n, \tag{14}$$

where

$$M(x)=\begin{vmatrix} l_1(x) & l_2(x) & \ldots l_n(x) \\ l_1(x+1) & l_2(x+1) & \ldots l_n(x+1) \\ \cdots & \cdots & \cdots \\ l_1(x+n-1) & l_2(x+n-1) & \ldots l_n(x+n-1) \end{vmatrix}$$

and $M_k(x)$ is the minor of the element $l_k(x+n-1)$ in $M(x)$.

Now, the function

$$M(x)/l_1(x)l_2(x)\ldots l_n(x)$$

is analytic at infinity and is different from zero at that point, as one may see readily by considering its value in the form of a determinant obtained from that for $M(x)$ by dividing the columns in order by $l_1(x)$, $l_2(x)$, ...., $l_n(x)$; for each element of the resulting determinant is analytic at infinity and the determinant has at infinity the same value as the determinant $\bar{D}(x)$ considered above. Likewise it may be shown that the function

$$M_k(x)/l_1(x)\ldots l_{k-1}(x)l_{k+1}(x)\ldots l_n(x)$$

is analytic at infinity and is different from zero at that point. Hence, we may write

$$(-1)^{n+k}M_k(x)/M(x)=A_k(x)/l_k(x),$$

where $A_k(x)$ is analytic at infinity and is different from zero at that point. Hence, we have from (14)

$$\Delta t_k(x) = \eta(x) A_k(x+1)/l_k(x+1), \qquad k=1, 2, \ldots, n. \tag{15}$$

Equations (15) have the following two formal solutions:

$$t_k(x) = -\sum_{i=0}^{\infty} \eta(x+i) A_k(x+i+1)/l_k(x+i+1), \qquad k=1, 2, \ldots, n;$$

$$t_k(x) = \sum_{i=1}^{\infty} \eta(x-i) A_k(x-i+1)/l_k(x-i+1), \qquad k=1, 2, \ldots, n.$$

Substituting these values in (13), we have the following two particular formal solutions of equation (12):

$$g^+(x) = -\sum_{i=0}^{\infty} \sum_{k=1}^{n} \eta(x+i) A_k(x+i+1) l_k(x)/l_k(x+i+1); \tag{16}$$

$$g^-(x) = \sum_{i=1}^{\infty} \sum_{k=1}^{n} \eta(x-i) A_k(x-i+1) l_k(x)/l_k(x-i+1). \tag{17}$$

These formal solutions will be denoted by

$$S_x(\eta) \quad \text{and} \quad T_x(\eta),$$

respectively.

Let $p_1(x), p_2(x), \ldots, p_n(x)$ be $n$ arbitrary periodic functions of $x$ of period 1. Then for the formal solutions of the successive equations in (11) we may conveniently select either of the following sets:

$$g_1^+(x) = \sum_{k=1}^{n} p_k(x) l_k(x),$$

$$g_2^+(x) = \sum_{k=1}^{n} p_k(x) l_k(x) + S_x \Big\{ \sum_{i_1=1}^{n} \Psi_{i_1}(x) \sum_{k=1}^{n} p_k(x) l_k(x+n-i_1) \Big\},$$

$$g_3^+(x) = \sum_{k=1}^{n} p_k(x) l_k(x) + S_x \Big\{ \sum_{i_1=1}^{n} \Psi_{i_1}(x) \sum_{k=1}^{n} p_k(x) l_k(x+n-i_1) \Big\}$$
$$+ S_x \Big[ \sum_{i_2=1}^{n} \Psi_{i_2}(x) S_{x+n-i_2} \Big\{ \sum_{i_1=1}^{n} \Psi_{i_1}(x) \sum_{k=1}^{n} p_k(x) l_k(x+n-i_1) \Big\} \Big],$$

. . . . . . . . . . . . . . . . . . . . . . . . . . . . . . . . . . . . . . . . . . . . . . . . . . . . . . . . ;

$$g_1^-(x) = \sum_{k=1}^{n} p_k(x) l_k(x),$$

$$g_2^-(x) = \sum_{k=1}^{n} p_k(x) l_k(x) + T_x \Big\{ \sum_{i_1=1}^{n} \Psi_{i_1}(x) \sum_{k=1}^{n} p_k(x) l_k(x+n-i_1) \Big\},$$

$$g_3^-(x) = \sum_{k=1}^{n} p_k(x) l_k(x) + T_x \Big\{ \sum_{i_1=1}^{n} \Psi_{i_1}(x) \sum_{k=1}^{n} p_k(x) l_k(x+n-i_1) \Big\}$$
$$+ T_x \Big[ \sum_{i_2=1}^{n} \Psi_{i_2}(x) T_{x+n-i_2} \Big\{ \sum_{i_1=1}^{n} \Psi_{i_1}(x) \sum_{k=1}^{n} p_k(x) l_k(x+n-i_1) \Big\} \Big],$$

. . . . . . . . . . . . . . . . . . . . . . . . . . . . . . . . . . . . . . . . . . . . . . . . . . . . . . . .

We are thus led to the following two formal expansions:

$$f^+(x)=\sum_{k=1}^{n} p_k(x)l_k(x)+S_x\{\sum_{i_1=1}^{n}\Psi_{i_1}(x)\sum_{k=1}^{n}p_k(x)l_k(x+n-i_1)\}$$
$$+S_x[\sum_{i_2=1}^{n}\Psi_{i_2}(x)S_{x+n-i_2}\{\sum_{i_1=1}^{n}\Psi_{i_1}(x)\sum_{k=1}^{n}p_k(x)l_k(x+n-i_1)\}]+\ldots, \quad (18)$$

$$f^-(x)=\sum_{k=1}^{n} p_k(x)l_k(x)+T_x\{\sum_{i_1=1}^{n}\Psi_{i_1}(x)\sum_{k=1}^{n}p_k(x)l_k(x+n-i_1)\}$$
$$+T_x[\sum_{i_2=1}^{n}\Psi_{i_2}(x)T_{x+n-i_2}\{\sum_{i_1=1}^{n}\Psi_{i_1}(x)\sum_{k=1}^{n}p_k(x)l_k(x+n-i_1)\}]+\ldots. \quad (19)$$

It is easy to verify that these functions afford formal solutions of equation (9). It is sufficient to this purpose to make a direct substitution of the expansions into the equation, reducing the result by means of the relations

$$\sum_{k=0}^{n}\lambda_k(x)S_{x+n-k}(\eta)=\eta,\quad \sum_{k=0}^{n}\lambda_k(x)T_{x+n-k}(\eta)=\eta.$$

Each of these solutions is in general illusory on account of the divergence of certain series contained in the terms of the expansions. Let $\alpha_r$ and $\alpha_s$ be any two $\alpha$'s such that $|\alpha_r|>|\alpha_s|$. Put $p_i(x)=0$, $i\neq r$; $p_r(x)=1$. Then the second term in the series in (18) takes the form

$$S_x\{\sum_{i_1=1}^{n}\Psi_{i_1}(x)l_r(x+n-i_1)\}=-\sum_{i=0}^{\infty}\sum_{k=1}^{n}\Psi_{i_1}(x+i)A_k(x+i+1)$$
$$l_r(x+n+i-i_1)l_k(x)/l_k(x+i+1).$$

It is easy to see that the series in the second member of this equation is not in general convergent. A similar discussion may be made of the second term in the series in (21); it turns out that the series denoted by this term is not in general convergent.

## § 2. *Two Actual Solutions.*

From the last result of the preceding section we see that the series in (18) and (19) do not in general represent actual solutions of equation (9). We shall now show that an appropriate choice of the periodic functions $p_1(x)$, $p_2(x),\ldots, p_n(x)$ will lead to actual solutions of this equation.

First, we shall consider the case of the series in (18). We assume that the notation for the roots $\alpha_1, \alpha_2, \ldots, \alpha_n$ of the characteristic algebraic equation is chosen so that

$$|\alpha_1|\leq|\alpha_2|\leq\ldots\leq|\alpha_n|. \quad (20)$$

Then put

$$p_1(x)=1,\quad p_2(x)=0=p_3(x)=\ldots=p_n(x).$$

Denote the resulting formal solution (18) by $f_1^+(x)$. Then we may write

$$\begin{aligned} f_1^+(x) = l_1(x) &+ S_x\{\sum_{i_1=1}^{n}\Psi_{i_1}(x)l_1(x+n-i_1)\} \\ &+ S_x[\sum_{i_2=1}^{n}\Psi_{i_2}(x)S_{x+n-i_2}\{\sum_{i_1=1}^{n}\Psi_{i_1}(x)l_1(x+n-i_1)\}] \\ &+ S_x\Big(\sum_{i_3=1}^{n}\Psi_{i_3}(x)S_{x+n-i_3}[\sum_{i_2=1}^{n}\Psi_{i_2}(x)S_{x+n-i_2}\{\sum_{i_1=1}^{n}\Psi_{i_1}(x)l_1(k+n-i_1)\}]\Big) \\ &+ \cdots\cdots\cdots\cdots \qquad (21)\end{aligned}$$

It is convenient to introduce the term right $D$-region to denote a region of the complex plane defined as follows: Consider the curve made up of a semicircle $C$ with center at the point zero and terminated at each end on the axis of imaginaries, its position being to the right of that axis and its extremities being $A$ and $B$, together with the straight lines $A\infty$ and $B\infty$ drawn from the points $A$ and $B$, respectively, to infinity in a direction parallel to the axis of reals and lying in the left half-plane. This curve $\infty ACB\infty$ divides the plane into two parts, one containing the point 0 and the other not containing this point. The part not containing the point 0 is called a right $D$-region. It is understood that this $D$-region does not contain the point infinity.

Now it is clear that there exists a right $D$-region, call it $D_1$, and constants $M_1, M_2, M_3$ such that each of the following inequalities is true so long as $x$ is in $D_1$:

$$\left.\begin{aligned} &|\Psi_k(x)| && < M_1|x|^{-m-2}, \\ &\left|\frac{A_k(x+i+1)l_k(x)}{l_k(x+i+1)}\right| && < M_2|a_k^{-i}|, \\ &|l_1(x+n-k)| && < M_3|l_1(x)|, \\ &|l_1(x+i)| && < 2|l_1(x)|\cdot|a_1^i|. \end{aligned}\right\} k=1, 2, \ldots, n. \qquad (22)$$

It will also be assumed that $D_1$ is chosen so that the functions $\Psi_k(x)$, $A_k(x)$, $\{l_k(x)\}^{-1}$ are analytic in $D_1$.

The first term in the second member of equation (21) is a function which is analytic throughout the finite plane except at $x=0$. The second term denotes a series. It is obvious that

$$|S_x\{\sum_{i_1=1}^{n}\Psi_{i_1}(x)l_1(x+n-i_1)\}| \leq \bar{S}_x\{\sum_{i_1=1}^{n}\Psi_{i_1}(x)l_1(x+n-i_1)\},$$

where $\bar{S}_x$ denotes the series obtained from $S_x$ by replacing each term in the latter by its absolute value. If we confine $x$ to the region $D_1$, and employ inequalities (22) and (20), we come readily to the following relations among infinite series, these relations being valid term by term:

$$\bar{S}_x\Big\{\sum_{i_1=1}^{n}\Psi_{i_1}(x)l_1(x+n-i_1)\Big\}\Big| < \sum_{i=0}^{\infty} M_3|l_1(x+i)|\cdot nM_1|x+i|^{-m-2}\cdot M_2\sum_{k=1}^{n}|\alpha_k^{-i}|$$

$$< \sum_{i=0}^{\infty}|l_1(x)|\cdot 2n^2M_1M_2M_3|x+i|^{-m-2}. \qquad (23)$$

If we further confine $x$ to lie in some closed region $S$ in $D_1$, then it is clear that a constant $\bar{M}$ exists such that the last series above is term by term less than the series of constant terms

$$\sum_{i=0}^{\infty}\bar{M}(i+1)^{-m-2}.$$

This series being convergent, it now follows by the theorem of Weierstrass that the series denoted by

$$S_x\Big\{\sum_{i_1=1}^{n}\Psi_{i_1}(x)l_1(x+n-i_1)\Big\} \qquad (24)$$

is uniformly convergent in $S$. Furthermore, each term of this series is analytic in $S$. Hence, the sum of the series is a function of $x$ which is analytic throughout $S$. It is therefore analytic at every point in $D_1$, since $S$ is any closed region in $D_1$.

By means of (23) we shall now obtain more convenient inequalities governing the character of the function in (24). We have

$$\Big|S_x\Big\{\sum_{i_1=1}^{n}\Psi_{i_1}(x)l_1(x+n-i_1)\Big\}\Big| < 2n^2M_1M_2M_3|l_1(x)|\sum_{i=0}^{\infty}|x+i|^{-m-2}.$$

If $x=u+v\sqrt{-1}$ and $u$ and $v$ are real it is easy (cf. Birkhoff, *loc. cit.*, p. 248) to show that we have

$$\sum_{i=0}^{\infty}\frac{1}{|x+i|^{m+2}} < \frac{\pi}{|x|^{m+1}}, \quad |x|>1, \quad u \geqq 0;$$

$$\sum_{i=0}^{\infty}\frac{1}{|x+i|^{m+2}} < \frac{2\pi}{|v|^{m+1}}, \quad |v|>1.$$

Hence, if we assume that the semicircle used in defining $D_1$ has a radius greater than unity, we have the following inequalities valid in $D_1$:

$$\Big|S_x\Big\{\sum_{i_1=1}^{n}\Psi_{i_1}(x)l_1(x+n-i_1)\Big\}\Big| < \begin{cases} 2\pi n^2M_1M_2M_3|l_1(x)|\cdot|x|^{-m-1}, & u \geqq 0, \\ 4\pi n^2M_1M_2M_3|l_1(x)|\cdot|v|^{-m-1}, & |v|>1. \end{cases} \qquad (25)$$

It is obvious that the regions in which the two inequalities in (25) are separately valid overlap and that the two regions make up the whole of $D_1$.

We may proceed in like manner to the study of the function defined by the third term in equation (21). It is obvious that the series denoted by $S_x$ in the expression

$$S_z[\sum_{i_2=1}^{n}\Psi_{i_2}(x)S_{z+n-i_2}\{\sum_{i_1=1}^{n}\Psi_{i_1}(x)l_1(x+n-i_1)\}] \tag{26}$$

is term by term equal to or less in absolute value than the series denoted by $\bar{S}_z$ in

$$\bar{S}_z[\sum_{i_2=1}^{n}\Psi_{i_2}(x)\bar{S}_{z+n-i_2}\{\sum_{i_1=1}^{n}\Psi_{i_1}(x)l_1(x+n-i_1)\}]. \tag{27}$$

It is now convenient to make a separation into two cases.

Let us consider first the case in which $u \geqq 0$. By employing (25) we may readily see that the series denoted by $\bar{S}_z$ in (27) is term by term less than the series

$$\bar{S}_z[\sum_{i_2=1}^{n}\Psi_{i_2}(x)2\pi n^2M_1M_2M_3|l_1(x+n-i_2)|\cdot|x+n-i_2|^{-m-1},$$

and that this in turn is term by term less than the series

$$\bar{S}_z[2\pi n^2M_1M_2M_3nM_1|x|^{-m-2}M_3|l_1(x)|\cdot|x|^{-m-1}]. \tag{28}$$

From this point we may proceed as in the discussion of the second term of the series in (21). Thus we may show that the series denoted by $S_z$ in (26) is uniformly convergent in any closed region $S$ lying in $D_1$ and having no point to the left of the axis of imaginaries, and therefore that (26) denotes a function which is analytic at every point in $D_1$ and not to the left of the axis of imaginaries. Moreover, through an examination of (28) it is easy to see that

$$|S_z[\sum_{i_2=1}^{n}\Psi_{i_2}(x)S_{z+n-i_2}\{\sum_{i_1=1}^{n}\Psi_{i_1}(x)l_1(x+n-i_1)\}]|$$
$$< (2\pi n^2M_1M_2M_3)^2|l_1(x)|\cdot|x|^{-2m-2}, \quad u \geqq 0. \tag{29}$$

For the next discussion we confine $x$ to $D_1$ and to that part of $D_1$ in which $|v|>1$. Then the second inequality in (27) becomes available. Making use of this and proceeding by the method just employed in the preceding paragraph, we may show that the function (26) is analytic at every point in the region in consideration and that moreover the following inequality is satisfied:

$$|S_z[\sum_{i_2=1}^{n}\Psi_{i_2}(x)S_{z+n-i_2}\{\sum_{i_1=1}^{n}\Psi_{i_1}(x)l_1(x+n-i_1)\}]|$$
$$< (4\pi n^2M_1M_2M_3)^2|l_1(x)|\cdot|v|^{-2m-2}, \qquad |v|>1. \tag{30}$$

It is now easy to see that the method used above may be further employed in an investigation of the successive terms of the series in (21) and that the following results will thus emerge: Every term of the second member in (21) represents a function which is analytic at every point in $D_1$; this series is term by term less in absolute value than each of the following series:

$$\left.\begin{array}{l} |l_1(x)|+|l_1(x)|\dfrac{M}{|x|^{m+1}}+|l_1(x)|\left(\dfrac{M}{|x|^{m+1}}\right)^2+\dots, \\ \qquad\qquad\qquad\qquad M=2\pi n^2M_1M_2M_3, \quad u \gtrless 0; \\ |l_1(x)|+|l_1(x)|\dfrac{2M}{|v|^{m+1}}+|l_1(x)|\left(\dfrac{2M}{|v|^{m+1}}\right)^2+\dots, \quad |v|>1, \end{array}\right\} \tag{31}$$

each inequality being valid for that part of $D_1$ for which the corresponding adjoined condition $u \gtrless 0$ or $|v|>1$ is satisfied.

Let us put on the region $D_1$ a further restriction consistent with those which we have previously placed upon it, namely, that it shall have the property that for any $x$ in $D_1$, where $x=u+v\sqrt{-1}$, the following two inequalities shall be true:

$$M|x|^{-m-1}<1 \text{ when } u \gtrless 0; \quad 2M|v|^{-m-1}<1 \text{ when } |v|>\beta,$$

$\beta$ being an appropriately chosen constant. Now let $S$ be any closed region lying in the modified $D_1$. Then, from the result associated with (31) above, we see readily that the series in (21) is uniformly convergent in $S$; whence it follows that the sum $f_1^+(x)$ is analytic at every point in the region $D_1$, since the terms of that series are analytic at all such points. Moreover, it is clear that $f_1^+(x)$ satisfies the following inequalities, valid in $D_1$:

$$\left.\begin{array}{ll} |f_1^+(x)-l_1(x)|<|l_1(x)|\dfrac{M}{|x|^{m+1}-M}, & u \gtrless 0; \\ |f_1^+(x)-l_1(x)|<|l_1(x)|\dfrac{2M}{|v|^{m+1}-2M}, & |v|>\beta, \end{array}\right\} \tag{31'}$$

as one sees readily by transposing the first term in the series in (21) to the first member of the equation and comparing the resulting second member term by term with the series in (31), exclusive of the first term in each, and finally summing the last series (with first term removed).

From the last inequality we have readily the following limits:

$$\left.\begin{array}{l} \lim\limits_{x=\infty} x^m \left\{\dfrac{f_1^+(x)}{l_1(x)}-1\right\}=0, \\ \lim\limits_{v=\pm\infty} v^m\left\{\dfrac{f_1^+(x)}{l_1(x)}-1\right\}=0, \end{array}\right\} \tag{32}$$

the first being valid for any approach of $x$ to infinity in the right half-plane, and the second being valid for the approach of the real variable $v$ to either $+\infty$ or $-\infty$. Since $v=|x|\sin(\arg x)$, it is easy to see from these two limits that

$$\lim_{x=\infty} x^m\left\{\frac{f_1^+(x)}{l_1(x)}-1\right\}=0, \tag{33}$$

provided that $x$ approaches infinity in any way so as to remain always without any sector determined by two rays from zero to infinity and including between them the negative part of the axis of reals.

By actual substitution it may now readily be shown that the function $f_1^+(x)$ satisfies equation (2) for all $x$ in $D_1$. By means of the equation

$$f(x) = -\frac{1}{a_n(x)} f(x+n) - \frac{a_1(x)}{a_n(x)} f(x+n-1) - \ldots - \frac{a_{n-1}(x)}{a_n(x)} f(x+1),$$

which is obtained from (2) in an obvious manner, one may further define the solution $f_1^+(x)$ in the part of the finite plane exterior to $D_1$. It may thus be seen that the extended solution $f_1^+(x)$ is analytic throughout the finite plane except at the singularities of the functions

$$\frac{1}{a_n(x)}, \quad \frac{a_1(x)}{a_n(x)}, \ldots, \quad \frac{a_{n-1}(x)}{a_n(x)} \tag{34}$$

and points congruent to them on the left. (A point $\alpha - i$ is said to be congruent to $\alpha$ on the left, if $i$ is a positive integer; similarly, we say that $\alpha + i$ is congruent to $\alpha$ on the right, if $i$ is a positive integer.) It is further obvious that the singularities of $f_1^+(x)$ are poles in case the functions (34) are rational and that the complete set of numbers each of which is the order of a pole of $f_1^+(x)$ is a bounded set.

In obtaining the solution whose properties we have just developed, we began with the series (21) which at first was only known to have formal validity and by means of it arrived at the actual solution $f_1^+(x)$. Similarly, one might start from the formal solution

$$f_n^-(x) = l_n(x) + T_s\{\sum_{i_1=1}^{n} \Psi_{i_1}(x) l_n(x+n-i_1)\}$$
$$+ T_s[\sum_{i_2=1}^{n} \Psi_{i_2}(x) T_{s+n-i_2}\{\sum_{i_1=1}^{n} \Psi_{i_1}(x) l_n(x+n-i_1)\}] + \ldots, \tag{35}$$

obtained from (19) in a way similar to that by which (21) was obtained from (18), and show that we are thus led to an actual solution $f_n^-(x)$ of (9). There would be no essential modification in the style of the argument. Instead of a right $D$-region we would employ a left $D$-region, such a region being defined by saying that its points are obtained from those of a right $D$-region by reflection through the axis of imaginaries. In fact, practically the only modification of the argument necessary is that which comes of an interchange of the rôles of the right and left sides of the plane.

Combining the principal conclusions which we reached in the treatment of equation (21) and those which will emerge from a discussion of (35) in the manner indicated, we have the following result:

PRELIMINARY THEOREM. *If the characteristic algebraic equation associated with the difference equation* (2) *has its roots* $\alpha_1, \alpha_2, \ldots, \alpha_n$ *different from each other and from zero and these are ordered so that* $|\alpha_1| \leqq |\alpha_2| \leqq \ldots \leqq |\alpha_n|$, *then this difference equation has two particular solutions* $f_1^+(x)$ *and* $f_n^-(x)$, *possessing the following properties:*

1) *The solution* $f_1^+(x)$ *is analytic throughout the finite plane except at the singularities of the functions* (34) *and points congruent to them on the left. The solution* $f_n^-(x)$ *is analytic throughout the finite plane except at the points congruent on the right (at a distance of n units or more) to the singularities of the functions* $a_1(x), a_2(x), \ldots, a_n(x)$. *Moreover, the singularities of each solution in the finite plane are poles provided that* $a_1(x), a_2(x), \ldots, a_n(x)$ *are rational functions, and the complete set of numbers each of which is the order of one of these poles is a bounded set.*

2) *There exists a right D-region RD* [*left D-region LD*] *in which the solution* $f_1^+(x)$ [$f_n^-(x)$] *is representable by the series* (21) [(35)], *this series being absolutely and uniformly convergent throughout any closed region S lying entirely in RD* [*LD*]. *Equation* (2) *itself then suffices to determine the solution in that part of the plane not in the region RD* [*LD*].

3) *If x approaches infinity in any way so as to remain always without any sector (however small) determined by two rays from zero to infinity and including between them the negative* [*positive*] *part of the axis of reals, then*

$$\lim x^m \left\{ \frac{f_1^+(x)}{l_1(x)} - 1 \right\} = 0 \qquad \left[ \lim x^m \left\{ \frac{f_n^-(x)}{l_n(x)} - 1 \right\} = 0 \right],$$

$l_1(x)$ [$l_n(x)$] *being the function denoted by that symbol in the first equation in* § 1. *Also,*

$$\lim_{v=\pm\infty} v^m \left\{ \frac{f_1^+(x)}{l_1(x)} - 1 \right\} = 0 \qquad \left[ \lim_{v=\pm\infty} v^m \left\{ \frac{f_n^-(x)}{l_n(x)} - 1 \right\} = 0 \right], \quad x = u + v\sqrt{-1}.$$

**REMARK.** It is important for later use to point out the fact that, in so far as any statement in the conclusion of the preliminary theorem refers to a region *RD* [*LD*] or any part of it, it is independent of the nature of the coefficients $\Psi_1(x), \Psi_2(x), \ldots, \Psi_n(x)$ of (9), and hence of the coefficients $a_1(x), a_2(x), \ldots, a_n(x)$ of (2), in the part of the plane exterior to *RD* [*LD*]. More precisely, *if the functions* $\Psi_1(x), \Psi_2(x), \ldots, \Psi_n(x)$ *are analytic in some right* [*left*] *D-region and throughout such a region are represented asymptotically by the expansions in* (10) *or throughout such a region merely satisfy an inequality of the form*

$$|\Psi_k(x)| < \frac{\bar{\mu}}{|x|^{m+2}},$$

*where $\bar{\mu}$ is a constant, then there exists a solution $f_1^+(x)$ $[f_n^-(x)]$ and a right [left] D-region RD [LD] such that $f_1^+(x)$ $[f_n^-(x)]$ has in RD [LD] the same properties as the solution $f_1^+(x)$ $[f_n^-(x)]$ given in the theorem. Likewise the inequalities derived above for such solutions in such D-regions remain valid under the less restrictive hypotheses.*

It should be observed, at this point, that the two solutions obtained above were gotten by the aid of certain functions $l_1(x), l_2(x), \ldots, l_n(x)$, these being obtained from the formal expansions (5) by breaking them off at the $(m+1)$-th term. It follows therefore that the functions $f_1^+(x)$ and $f_n^-(x)$ are now to be treated as functions of $m$ as well as of $x$; or, more exactly, we are to look upon them as possibly functions of $m$. Ultimately we shall show that they are indeed independent of $m$ (at least when $m$ is sufficiently large), this proof being readily made after we have obtained a fundamental set of solutions, each solution being a possible function of $m$. As soon as this independence of $m$ is proved, we shall be able to obtain readily the asymptotic character of $f_1^+(x)$ and $f_n^-(x)$. It is clear that the limits in the above theorem are sufficient for this.

## § 3. *The First System of Intermediate Solutions.*

For the difference equation (2) we shall now obtain four sets of $n$ particular solutions each, the solutions in each set being independent with respect to periodic multipliers of period 1, and therefore constituting a fundamental system of solutions.

We shall first start from the particular solution $f_1^+(x)$ obtained in the preceding section. By means of the substitution

$$f(x) = f_1^+(x)\,g(x), \tag{36}$$

equation (2) goes over into the form

$$g(x+n) + ka_1(x) f_1^+(x+n-1) g(x+n-1) + \ldots + ka_n(x) f_1^+(x) g(x) = 0, \tag{37}$$

where $k = 1/f_1^+(x+n)$. If we put

$$G(x) = \Delta g(x),$$

the preceding equation becomes

$$G(x+n-1) + B_1(x)\,G(x+n-2) + \ldots + B_{n-1}(x)\,G(x) = 0, \tag{38}$$

where

$$\begin{aligned}
B_1(x) &= 1 + ka_1(x)\,f_1^+(x+n-1),\\
B_2(x) &= 1 + ka_1(x)\,f_1^+(x+n-1) + ka_2(x)\,f_1^+(x+n-2),\\
&\ldots\ldots\ldots\ldots\ldots\ldots\ldots\ldots\ldots\ldots,\\
B_{n-1}(x) &= 1 + ka_1(x)\,f_1^+(x+n-1) + \ldots + ka_{n-1}(x)\,f_1^+(x+1).
\end{aligned}$$

From the way in which equation (37) was found, it is clear that its formal power series solutions $g_i(x)$ may be determined from the relations

$$g_i(x)=\frac{\alpha_i^x x^{\mu_i}(1+c_{i1}x^{-1}+c_{i2}x^{-2}+\ldots.)}{\alpha_1^x x^{\mu_1}(1+c_{11}x^{-1}+c_{12}x^{-2}+\ldots.)},\quad i=1,2,\ldots.,n, \tag{39}$$

by formal division in the second member for each value of $i$. Furthermore, the formal power series solutions $G_i(x)$ of (38) may be obtained from these by reckoning out formally the differences $\Delta g_i(x)$, $i=1,2,\ldots.,n$. Thus we have

$$G_i(x)=\left(\frac{\alpha_{i+1}}{\alpha_1}\right)^x x^{\mu_{i+1}-\mu_1}\left(\frac{\alpha_{i+1}}{\alpha_1}-1\right)\left(1+\frac{d_{i1}}{x}+\frac{d_{i2}}{x^2}+\ldots.\right),\quad i=1,2,\ldots.,n-1. \tag{40}$$

Let us denote by $L_i(x)$, $i=1,2,\ldots.,n-1$, the functions

$$L_i(x)=\left(\frac{\alpha_{i+1}}{\alpha_1}\right)^x x^{\mu_{i+1}-\mu_1}\left(\frac{\alpha_{i+1}}{\alpha_1}-1\right)\left(1+\frac{d_{i1}}{x}+\ldots.+\frac{d_{im}}{x^m}\right),\quad i=1,2,\ldots.,n-1, \tag{41}$$

where $m$ has the same value as in the preceding section.

An examination of equation (38) with reference to the first remark following the theorem of the preceding section is sufficient to bring out the fact that there exists a right $D$-region in which the coefficients $B_1(x),\ldots.,$ $B_{n-1}(x)$ satisfy the conditions requisite for the application of that part of the theorem which relates to the right $D$-regions. In order to set up in detail the inequalities by which this may be proved, it is necessary to separate equation (38) into two members after the manner in which equation (2) was separated into the two members in (9), and then to employ the inequalities by which we have found $f_1^+(x)$ to be restricted. There is nothing involved in this more than the straightforward algebra of inequalities, and one can even see beforehand what must be the result of the computation, so that it is unnecessary to perform it.

By means of the theorem of the preceding section we now have the following result: There exists a right $D$-region $RD$ and a solution $G_1^+(x)$ of equation (38) such that $G_1^+(x)$ is analytic throughout $RD$, and is representable in such a region by a series of type (21) which is absolutely and uniformly convergent throughout any closed region $S$ lying entirely in $RD$. If $x$ approaches infinity in any way so as to remain always without any sector (however small) determined by two rays from 0 to $\infty$ and including between them the negative part of the axis of reals, then

$$\lim x^m\left\{\frac{G_1^+(x)}{L_1(x)}-1\right\}=0.$$

Also,

$$\lim_{v=\pm\infty} v^m\left\{\frac{G_1^+(x)}{L_1(x)}-1\right\}=0.$$

These limits may be replaced by the somewhat stronger inequalities by which they were derived, namely, those corresponding to (31′). Then for appropriately chosen constants $K$ and $\beta$ we have

$$\left.\begin{array}{ll} |G_1^+(x)-L_1(x)| < |L_1(x)|\cdot K|x|^{-m-1}, & u \gtreqless 0, \\ |G_1^+(x)-L_1(x)| < |L_1(x)|\cdot K|v|^{-m-1}, & |v|>\beta, \end{array}\right\} \tag{42}$$

where $x=u+v\sqrt{-1}$ and lies in the region $RD$.

Now, equation (37) has particular solutions $g_1^+(x)$ such that

$$\Delta g_1^+(x)=G_1^+(x). \tag{43}$$

There are two solutions of (43) from each of which we are led to interesting particular solutions of (37) and hence of (2). Of these we shall treat one in detail in this section and the other in a later section.

Equation (43) has the obvious formal solution

$$\bar{g}_1^+(x)=G_1^+(x-1)+G_1^+(x-2)+G_1^+(x-3)+\ldots.. \tag{44}$$

In order to treat conveniently the matter of convergence of this series, it will be necessary to put on the further restriction that $|\alpha_1|<|\alpha_2|$. (Compare equation (20).) Then, from the second inequality in (42), it follows readily that the series in equation (44) converges for every value of $x$ such that $x-1$' $x-2$, $x-3$, .... all lie in $RD$; that is, it converges for every $x$ such that $|v|\geq R_1$, where $R_1$ is the radius of the semicircle by means of which the region $RD$ is defined. Moreover, if $S$ is any closed region throughout which $|v|\geq R_1$, then the series in (44) is clearly uniformly convergent in $S$. But each term of this series is analytic in $S$. Hence, $\bar{g}_1^+(x)$ is analytic in $S$; it is therefore analytic at every point of the finite plane for which $|v|\geq R_1$.

We may also obtain an upper bound to the absolute value of $\bar{g}_1^+(x)$ when $|v|\geq R_1$. We observe that $R_1$ should be chosen greater than $2K$ and greater than $\beta$ so as to make valid all our inequalities. Define a function $S(x)$ by means of the series

$$S(x)=L_1(x-1)+L_1(x-2)+L_1(x-3)+\ldots.. \tag{45}$$

It is easy to see that this series is absolutely and uniformly convergent when $x$ is confined to a closed region $S$ throughout which $|v|\geq R_1$. Then, from (42), (44) and (45) we have

$$|\bar{g}_1^+(x)-S(x)|<\sum_{i=1}^{\infty}|L_1(x-i)|\cdot K|v|^{-m-1}, \quad |v|\geq R_1.$$

Now, positive constants $N_1$ and $N_2$ exist such that

$$N_1\left|\frac{\alpha_2}{\alpha_1}\right|^{-i}<\frac{|L_1(x-i)|}{|L_1(x)|}<N_2\left|\frac{\alpha_2}{\alpha_1}\right|^{-i},$$

provided that $|v|$ is sufficiently large. Hence,

$$|\bar{g}_1^+(x)-S(x)|<N_2K|L_1(x)|\frac{1}{|v|^{m+1}}\sum_{i=1}^{\infty}\left|\frac{\alpha_2}{\alpha_1}\right|^{-i}.$$

From (45) we see that $|S(x)|/|L_1(x)|$ is bounded away from zero.* From this fact and the last relation it follows readily that a positive constant $N_3$ exists such that

$$\left|\frac{\bar{g}_1^+(x)}{S(x)}-1\right|<\frac{N_3}{|v|^{m+1}}. \tag{46}$$

Referring to the definitions of $L_1(x)$, $l_1(x)$ and $l_2(x)$, we see that

$$L_1(x)=\frac{l_2(x+1)}{l_1(x+1)}-\frac{l_2(x)}{l_1(x)}+\left(\frac{\alpha_2}{\alpha_1}\right)^x x^{\mu_2-\mu_1}\left(\frac{\gamma_{m+1}}{x^{m+1}}+\frac{\gamma_{m+2}}{x^{m+2}}+\ldots\right), \tag{46a}$$

where $\gamma_{m+1},\ldots$ are constants, the expansion being valid for sufficiently large values of $|x|$. From this relation and (45) it follows that

$$S(x)=\frac{l_2(x)}{l_1(x)}+\sum_{i=1}^{\infty}\left(\frac{\alpha_2}{\alpha_1}\right)^{x-i}(x-i)^{\mu_2-\mu_1}\left(\frac{\gamma_{m+1}}{(x-i)^{m+1}}+\ldots\right). \tag{45'}$$

On transposing the term $l_2(x)/l_1(x)$ and dividing the resulting equation through by this term, we have a relation from which it is easy to show that a positive constant $N_4$ exists such that

$$\left|\frac{S(x)l_1(x)}{l_2(x)}-1\right|<\sum_{i=1}^{\infty}\left|\frac{\alpha_2}{\alpha_1}\right|^{-i}\frac{N_4}{|x-i|^{m+1}}<\frac{N_4}{|v|^{m+1}}\sum_{i=1}^{\infty}\left|\frac{\alpha_2}{\alpha_1}\right|^{-i}.$$

In showing this, it is necessary to observe that

$$\frac{l_2(x)}{l_1(x)}=\left(\frac{\alpha_2}{\alpha_1}\right)^x x^{\mu_2-\mu_1}(1+\ldots),$$

the quantity in parenthesis being a descending power series in $x$. Hence, a positive constant $N_5$ exists such that

$$\left|\frac{S(x)l_1(x)}{l_2(x)}-1\right|<\frac{N_5}{|v|^{m+1}}. \tag{47}$$

Multiply relations (46) and (47) together member by member and combine the resulting inequality with (46) and (47); thus we have

$$\left|\frac{\bar{g}_1^+(x)l_1(x)}{l_2(x)}-1\right|<\frac{N_6}{|v|^{m+1}}, \tag{48}$$

where $N_6$ is a positive constant.

* Or this may more readily be proved by aid of relation (47) and the fact that $L_1(x)l_1(x)/l_2(x)$ approaches $(\alpha_2/\alpha_1)-1$ as $x$ approaches infinity, the latter fact being readily demonstrated by aid of equation (46a).

Now, a solution $\bar{f}_2^+(x)$ of equation (2) is given by the relation

$$\bar{f}_2^+(x) = f_1^+(x)\bar{g}_1^+(x),$$

as one sees from (36). This solution is analytic for $|v|$ sufficiently large. From (31′) we have a relation of the form

$$\left|\frac{f_1^+(x)}{l_1(x)} - 1\right| < \frac{N_7}{|v|^{m+1}},$$

where $N_7$ is a positive constant. From this relation and (48) it follows readily that a constant $N$ exists such that

$$\left|\frac{\bar{f}_2^+(x)}{l_2(x)} - 1\right| < \frac{N}{|v|^{m+1}}, \quad |v| \geqq \bar{R}, \tag{49}$$

where $\bar{R}$ is a positive constant. This is one of the fundamental inequalities for $\bar{f}_2^+(x)$.

A similar one may be found corresponding to the first inequality in (42). It may be written

$$\left|\frac{\bar{f}_2^+(x)}{l_2(x)} - 1\right| < \frac{\mu}{|x|^{m+1}}, \quad u \gtrless 0, \quad |v| \geqq \bar{R}, \tag{50}$$

where $\mu$ is a positive constant. In view of (49) it is obviously sufficient to prove this for the case when $x$ is further limited to lie in a sector made by two rays from zero to infinity and lying entirely in the right half-plane. Furthermore, it is clear that $|x|$, and hence $u$, may be taken as large as one pleases.

We start from equation (44). Choose $x$ so that $u$ is large and denote by $k$ the greatest integer such that $2k \leqq u$. Write (44) in the form

$$\bar{g}_1^+(x) = \sum_{i=1}^{k} G_1^+(x-i) + \sum_{i=k+1}^{\infty} G_1^+(x-i).$$

Then

$$\bar{g}_1^+(x) - S(x) = \sum_{i=1}^{k} \{G_1^+(x-i) - L_1(x-i)\} + \sum_{i=k+1}^{\infty} \{G_1^+(x-i) - L_1(x-i)\},$$

so that, in view of (42), we have

$$|\bar{g}_1^+(x) - S(x)| < \sum_{i=1}^{k} |L_1(x-i)| \frac{K}{|x-i|^{m+1}} + \sum_{i=k+1}^{\infty} |L_1(x-i)| \frac{K}{|v|^{m+1}}, \tag{51}$$

since it is clear that the first inequality in (42) is applicable to the first summation and the second inequality to the second.

Now, it is clear that a positive constant $\mu_1$ exists such that

$$\left|\frac{L_1(x-i)}{L_1(x)}\right| < \mu_1 \left|\frac{\alpha_2}{\alpha_1}\right|^{-i}, \quad i = 1, 2, \ldots, k.$$

Then from (51) we have

$$|\bar{g}_1^+(x)-S(x)|<\mu_1|L_1(x)|\frac{2K}{|x|^{m+1}}\sum_{i=1}^{k}\left|\frac{\alpha_2}{\alpha_1}\right|^{-i}+\mu_1|L_1(x)|\frac{K}{|v|^{m+1}}\sum_{i=k+1}^{\infty}\left|\frac{\alpha_2}{\alpha_1}\right|^{-i}.$$

If $|x|$ increases indefinitely, it is clear that the second term in the second member becomes indefinitely small with respect to the first term; to see this, it is sufficient to observe the rôle of the exponential quantity $|\alpha_2/\alpha_1|^{-k}$. Therefore, a positive constant $\mu_2$ exists such that

$$|\bar{g}_1^+(x)-S(x)|<|L_1(x)|\cdot\mu_2|x|^{-m-1}.$$

From this point forward the march of the argument is like that in the preceding treatment. As in the previous case it may readily be shown that $|S(x)|/|L_1(x)|$ is bounded away from zero; hence we see from the last inequality that a positive constant $\mu_3$ exists such that

$$\left|\frac{\bar{g}_1^+(x)}{S(x)}-1\right|<\mu_3|x|^{-m-1}.$$

By a method similar to that by which (47) was proved, it may be shown that a positive constant $\mu_4$ exists such that

$$\left|\frac{S(x)l_1(x)}{l_2(x)}-1\right|<\mu_4|x|^{-m-1}.$$

In the demonstration of this result it will be necessary to separate the series for $S(x)$ in (45′) into two series, one of $k$ terms and the other of the remaining terms. The argument is similar to that just employed; it may readily be supplied by the reader.

Making use of the last two relations and continuing precisely as in the argument following (47) one may readily complete the proof of (50). Thus, we have the requisite properties of the second solution $\bar{f}_2^+(x)$ of equation (2) in case $|\alpha_1|<|\alpha_2|$.

In case $|\alpha_1|=|\alpha_2|$ the method by which $f_1^+(x)$ was found is obviously valid for a second solution $f_2^+(x)$. We now call it $\bar{f}_2^+(x)$. It has the properties expressed in (49) and (50).

Now, apply to equation (38) the result which we have just derived for equation (2). We see that (38) has a second solution $G_2^+(x)$ such that

$$|G_2^+(x)-L_2(x)|<|L_2(x)|\cdot K_1|x|^{-m-1},\quad u\gtreqless 0;$$
$$|G_2^+(x)-L_2(x)|<|L_2(x)|\cdot K_1|v|^{-m-1},$$

where $K_1$ is a positive constant, these relations being valid when $|v|$ is sufficiently large. If we start from this solution and proceed by a method similar to that employed in the discussion associated with $G_1^+(x)$, it is clear that we shall be led to a third solution $\bar{f}_3^+(x)$ of equation (2), that this solution will be analytic for $|v|$ sufficiently large and that it will verify the following relations:

$$\left.\begin{aligned} &\left|\frac{\bar{f}_3^+(x)}{l_3(x)}-1\right| < \frac{\mu'}{|x|^{m+1}}, \quad u \gtrless 0; \\ &\left|\frac{\bar{f}_3^+(x)}{l_3(x)}-1\right| < \frac{\mu'}{|v|^{m+1}}, \end{aligned}\right\} \tag{52}$$

when $|v|$ is sufficiently large, $\mu'$ being a constant.

By continuing this process of interaction between equations (2) and (38), we shall be led finally to a set of $n$ solutions,

$$\bar{f}_1^+(x) = f_1^+(x),\ \bar{f}_2^+(x),\ \ldots,\ \bar{f}_n^+(x), \tag{53}$$

of equation (2), each solution being analytic when $|v|$ is sufficiently large. From the inequalities such as (52), by which these functions are bounded, we have the following relations:

$$\left.\begin{aligned} &\lim x^m \left\{\frac{\bar{f}_i^+(x)}{l_i(x)}-1\right\} = 0, \quad i=1, 2, \ldots, n, \\ &\lim_{v=\pm\infty} v^m \left\{\frac{\bar{f}_i^+(x)}{l_i(x)}-1\right\} = 0, \quad i=1, 2, \ldots, n, \end{aligned}\right\} \tag{54}$$

the first relation being valid for $x$ approaching infinity in any way so as to satisfy the two conditions that $|v|$ is sufficiently large and that $x$ remains without a sector formed by two rays from zero to infinity and including between them the negative part of the axis of reals.

That these solutions are linearly independent with respect to periodic multipliers of period 1, and hence constitute a fundamental set of solutions of (2), is easily shown. It is sufficient to prove that the determinant

$$D(x) = \begin{vmatrix} \bar{f}_1^+(x) & \bar{f}_2^+(x) & \ldots\ \bar{f}_n^+(x) \\ \bar{f}_1^+(x+1) & \bar{f}_2^+(x+1) & \ldots\ \bar{f}_n^+(x+1) \\ \ldots & \ldots & \ldots \\ \bar{f}_1^+(x+n-1) & \bar{f}_2^+(x+n-1) & \ldots\ \bar{f}_n^+(x+n-1) \end{vmatrix}$$

is not identically equal to zero. Divide the $k$-th column of this determinant by $\alpha_k^x x^{\mu_k}$, $k=1, 2, \ldots, n$. It is sufficient to prove that the resulting determinant is not identically zero. But in view of (54) it may be seen that this determinant approaches the value $\bar{D}(\infty)$ as $|v|$ approaches infinity, $\bar{D}(\infty)$ having the same meaning as in the first part of § 1. This is different from zero, since $\alpha_1, \ldots, \alpha_n$ are all different. Therefore, the set (53) is a fundamental system of solutions of equation (2).

We have seen that the function $\bar{f}_1^+(x)$ was possibly dependent on $m$. The other functions in (53) were obtained by aid of $\bar{f}_1^+(x)$, and therefore they are possibly functions of $m$. We shall now show that they are indeed independent of $m$, at least if $m$ is sufficiently large.

More generally, let $f_k(x)$ be any solution of equation (2) such that

$$\lim x^m \left\{ \frac{f_k(x)}{l_k(x)} - 1 \right\} = 0 \tag{55}$$

when $x, = u + v\sqrt{-1}$, approaches infinity in such a way that $u$ approaches $+\infty$, and $|v|$ becomes sufficiently large, say $|v| \gtreqless V$.

Since $f_k(x)$ is a solution of equation (2) it may be written in the form

$$f_k(x) = p_1(x)\bar{f}_1^+(x) + p_2(x)\bar{f}_2^+(x) + \ldots + p_n(x)\bar{f}_n^+(x), \tag{56}$$

where $p_1(x), p_2(x), \ldots, p_n(x)$ are suitably determined periodic functions of $x$ of period 1. Then, from (55) and the first equation in (54), we see that

$$\lim x^m \left\{ p_1(x)\frac{l_1(x)}{l_k(x)} + p_2(x)\frac{l_2(x)}{l_k(x)} + \ldots + p_n(x)\frac{l_n(x)}{l_k(x)} - 1 \right\} = 0.$$

Let $x_0, = u_0 + v_0\sqrt{-1}$, be any value of $x$ such that $|v_0| \geqq V$, let $t$ be a variable running over the set 0, 1, 2, 3, ...., and let $s$ be a non-negative integer. Then on replacing $x$ by $x_0 + t + s$ in the preceding limit and reducing, we have the relations

$$\lim_{t=\infty} t^m \{ r_{k1} u_{k1}^t t^{\mu_1 - \mu_k} \cdot u_{k1}^s (1 + \varepsilon_{1s}) + \ldots + r_{kn} u_{kn}^t t^{\mu_n - \mu_k} \cdot u_{kn}^s (1 + \varepsilon_{ns}) \} = 0,$$

where $u_{kl} = \alpha_l / \alpha_k$, $\varepsilon_{ls}$ approaches zero as $t$ approaches infinity, and

$$r_{kl} = p_l(x_0) u_{kl}^{x_0}, \qquad l \neq k, \quad r_{kk} = p_k(x_0) - 1.$$

It is to be observed that no two of the quantities $u_{kl}$ are equal and that each of them is different from 0.

Now suppose that $m$ is not less than the real part of any of the differences $\mu_1 - \mu_k, \mu_2 - \mu_k, \ldots, \mu_n - \mu_k$. Then from the preceding limit we have the following:

$$\lim_{t=\infty} \{ r_{k1} u_{k1}^t t^{v_1} \cdot u_{k1}^s (1 + \varepsilon_{1s}) + \ldots + r_{kn} u_{kn}^t t^{v_n} \cdot u_{kn}^s (1 + \varepsilon_{ns}) \} = 0, \tag{57}$$

where the real part of $v_l$ is zero or positive for every value of $l$. This relation is valid for any non-negative integral value of $s$.

Consider the set of relations obtained from (57) by giving to $s$ the values $s = 0, 1, 2, \ldots, n-1$. By multiplying these in order by the cofactors (in order) of the elements in the $l$-th column of the determinant

$$\Delta = \begin{vmatrix} 1 + \varepsilon_{10} & 1 + \varepsilon_{20} & \ldots 1 + \varepsilon_{n0} \\ u_{k1}(1 + \varepsilon_{11}) & u_{k2}(1 + \varepsilon_{21}) & \ldots u_{kn}(1 + \varepsilon_{n1}) \\ \ldots & \ldots & \ldots \\ u_{k1}^{n-1}(1 + \varepsilon_{1,n-1}) & \ldots & \ldots \end{vmatrix}$$

and adding the resulting relations, we have

$$\lim_{t=\infty} r_{kl} u_{kl}^t t^{v_l} \Delta = 0.$$

Since

$$\lim_{t=\infty} \Delta = \begin{vmatrix} 1 & 1 & \dots & 1 \\ u_{k1} & u_{k2} & \dots & u_{kn} \\ u_{k1}^2 & u_{k2}^2 & \dots & u_{kn}^2 \\ \dots & \dots & \dots & \dots \\ u_{k1}^{n-1} & u_{k2}^{n-1} & \dots & u_{kn}^{n-1} \end{vmatrix} \neq 0,$$

we have

$$\lim_{t=\infty} r_{kl} u_{kl}^t t^{v_l} = 0.$$

This relation can be true only if $|u_{kl}| < 1$ or $r_{kl} = 0$, since the real part of $v_l$ is not less than 0. But $|u_{kl}| = |\alpha_l/\alpha_k| \geqq 1$ unless $l < k$. Therefore $r_{kl} = 0$ if $l \geqq k$. Hence, $p_k(x_0) = 1$ and $p_l(x_0) = 0$ if $l > k$. Therefore, we see from (56) that

$$f_k(x) = p_1(x) \bar{f}_1^+(x) + \dots + p_{k-1}(x) \bar{f}_{k-1}^+(x) + \bar{f}_k^+(x), \qquad |v| \geqq V, \quad (58)$$

where $p_1(x), \dots, p_{k-1}(x)$ are suitably determined periodic functions.

From the particular case when $k=1$ we see that a solution $f_1(x)$ of equation (2) which satisfies the relation (55) for $k=1$ coincides with $\bar{f}_1^+(x)$, at least if $m$ is greater than the real part of each of the quantities $\mu_2 - \mu_1$, $\mu_3 - \mu_1$, $\dots, \mu_n - \mu_1$. From this it follows readily that each of the solutions of (2) in the set (53) is independent of $m$, provided that $m$ is greater than the real part of every one of the differences $\mu_i - \mu_j$, $j = 1, 2, \dots, n$, $i = j+1, \dots, n$. In order to see this, it is only necessary to observe the manner in which these solutions were obtained.

It should be observed that the only use made of the fact that the real part of $v_l$ is not less than zero was to insure that $u_{kl}^t t^{v_l}$ does not approach zero, for $t$ becoming infinite, when $l \geqq k$. This condition is obviously satisfied without regard to the value of $v_l$, provided that the roots $\alpha_1, \alpha_2, \dots, \alpha_n$ of the characteristic algebraic equation are different in absolute value, for then $|u_{kl}| > 1$ when $l > k$. Hence we may take $m = 0$ for the case when the $\alpha$'s are all different in absolute value.

We shall now determine largely the asymptotic character of the solutions (53) under the hypothesis that $m$ is greater than the real part of each of the differences $\mu_i - \mu_j$, $j = 1, 2, \dots, n$, $i = j+1, \dots, n$, or without restriction on $m$ when the roots of the characteristic algebraic equation are different in absolute value.

In this connection it is convenient to distinguish (cf. Birkhoff, *loc. cit.*, p. 248) between two kinds of asymptotic representation of a function by a series of the form

$$S(x)=x^{sx}\alpha^{x}x^{\mu}\left(c_0+\frac{c_1}{x}+\frac{c_2}{x^2}+\ldots.\right).$$

Let $g(x)$ be the given function; if for each $m$ the difference

$$d(x)=g(x)x^{-sx}\alpha^{-x}x^{-\mu}-\left(c_0+\frac{c_1}{x}+\ldots.+\frac{c_m}{x^m}\right)$$

becomes uniformly small of order $m$; that is, if $\lim x^m d(x)=0$ for $x$ approaching infinity in a certain region, we say that $g(x)$ is asymptotically represented by $S(x)$ in that region, *with respect to x*; if, on the other hand, we have $\lim\limits_{v=\pm\infty} v^m d(x)=0$, we say that $g(x)$ is asymptotically represented by $S(x)$ in that region, *with respect to v.*

Making use of relation (54) for increasingly large values of $m$ and remembering that $\overline{f}_i^+(x)$ is independent of $m$, we see readily that $\overline{f}_i^+(x)$ is asymptotic to

$$\alpha_i^x x^{\mu_i}\left(1+\frac{c_{i1}}{x}+\frac{c_{i2}}{x^2}+\ldots.\right)$$

with respect to $v$ and also with respect to $x$, provided that $x$ approaches infinity so as to satisfy the conditions that $|v|$ is sufficiently large and that $x$ remains without a sector formed by two rays from zero to infinity and including between them the negative part of the axis of reals.

The fundamental properties of the functions $\overline{f}_1^+(x),\ldots.,\overline{f}_n^+(x)$ are now determined throughout a region in which $|v|$ is sufficiently great, say $|v|\geqq V$, and $x$ is to the right of some line parallel to the axis of imaginaries. Their properties in the remainder of that part of the plane for which $|v|\geqq V$ may be determined, as in the case of $f_1^+(x)$ in § 2, by means of equation (2) itself.

The principal results indicated in this section may be put together in the form of the following theorem:

First System of Intermediate Solutions.* *If the characteristic algebraic equation associated with the difference equation* (2) *has its roots different from each other and from zero, then the difference equation has the fundamental system of solutions*

---

* The existence of these solutions was first pointed out in my memoir, *loc. cit.*, p. 119. Their properties were first developed in Birkhoff's memoir, already referred to.

$$\bar{f}_1^+(x), \bar{f}_2^+(x), \ldots, \bar{f}_n^+(x), \tag{53 bis}$$

***determined above, and these solutions have the following properties:***

***1) A constant $V$ exists such that each function in the set is analytic when $|v| \geq V$, where $v$ is defined by writing $x = u + v\sqrt{-1}$, $u$ and $v$ being real;***

***2) The asymptotic relations***

$$\bar{f}_i^+(x) \sim a_i^\pm x^{\mu_i}\left(1 + \frac{c_{i1}}{x} + \frac{c_{i2}}{x^2} + \ldots\right), \qquad i = 1, 2, \ldots, n,$$

***exist and are valid with respect to $x$ in that part* of any right half-plane for which $|v| \geq V$ and with respect to $v$ in the entire plane.***

**(By a right [left] half-plane is meant that part of the plane which is to the right [left] of a line parallel to the axis of imaginaries.)**

**Let us retain for $m$ the value 0 if the roots of the characteristic equation are different in absolute value and in the contrary case a value not less than the real part of any of the differences $\mu_i - \mu_j$, $j = 1, 2, \ldots, n$, $i = j+1, \ldots, n$. Denote by $l_i(x)$, $i = 1, 2, \ldots, n$, the functions so represented in the first part of § 1. Then a solution $f(x)$ of equation (2),**

$$f(x) = p_1(x)\bar{f}_1^+(x) + p_2(x)\bar{f}_2^+(x) + \ldots + p_n(x)\bar{f}_n^+(x), \tag{59}$$

**where $p_1(x), p_2(x), \ldots, p_n(x)$ are periodic functions of $x$ of period 1, has an important set of properties obtainable from the first relation in (54), namely,**

$$\left.\begin{aligned}
&\lim_{t=\infty} t^m\left\{\frac{f(x_0+t)}{l_n(x_0+t)} - p_n(x_0)\right\} = 0, \\
&\lim_{t=\infty} t^m\left\{\frac{f(x_0+t) - p_n(x_0)\bar{f}_n^+(x_0+t)}{l_{n-1}(x_0+t)} - p_{n-1}(x_0)\right\} = 0, \qquad |v| \geq V, \\
&\ldots\ldots\ldots\ldots\ldots\ldots\ldots\ldots\ldots\ldots, \\
&\lim_{t=\infty} t^m\left\{\frac{f(x_0+t) - p_n(x_0)\bar{f}_n^+(x_0+t) - \ldots - p_2(x_0)\bar{f}_2^+(x_0+t)}{l_1(x_0+t)} - p_1(x_0)\right\} = 0.
\end{aligned}\right\} \tag{60}$$

**The proof of this is so far similar in character to the argumentation associated with equations (55) to (58) that it may readily be supplied by the reader. Furthermore, any solution of (2) having the properties (60) is easily seen to be identical with the solution $f(x)$ defined by (59). Consequently, the relations (60) may be looked upon as a sort of set of "initial conditions" at infinity, defining uniquely a solution of the given equation (2).**

**In general these relations are complicated. It may be observed, however, that the single "initial condition"**

---

*** The asymptotic character of the first solution $\bar{f}_1^+(x)$ is maintained with respect to $x$ in any unrestricted right half-plane.**

$$\lim_{t=\infty} t^m\left\{\frac{f(x_0+t)}{l_1(x_0+t)}-p(x_0)\right\}=0$$

is sufficient to define uniquely a solution $f(x)=p(x)\bar{f}_1^+(x)$. In this case alone is the solution defined in a simple way by the initial conditions.

## § 4. *Second System of Intermediate Solutions.*

The fundamental set of solutions of the preceding section was obtained by means of a class of transformations depending on the solution $f_1^+(x)$ of § 2. If one starts from the solution $f_n^-(x)$ of § 2 and proceeds by a method in evéry respect similar to that employed in § 3 he will obviously be led to a new fundamental system of solutions related to those in § 3 as $f_n^-(x)$ is related to $f_1^+(x)$. It is unnecessary to give the argumentation. We merely state the result:

Second System of Intermediate Solutions. *If the characteristic algebraic equation associated with the difference equation* (2) *has its roots different from each other and from zero, then the difference equation has a fundamental system of solutions,*

$$\bar{f}_1^-(x),\ \bar{f}_2^-(x),\ \ldots,\ \bar{f}_n^-(x),$$

*possessing the following properties:*

1) *A constant V exists such that each function in the set is analytic if $|v|\geq V$, where v is defined by writing $x=u+v\sqrt{-1}$, u and v being real;*

2) *The asymptotic relations*

$$\bar{f}_i^-(x)\sim\alpha_i^x x^{\rho_i}\left(1+\frac{c_{i1}}{x}+\frac{c_{i2}}{x^2}+\ldots\right),\qquad i=1,2,\ldots,n,$$

*exist and are valid with respect to x in that part* of any left half-plane for which $|v|\geq V$ and with respect to v in the entire plane.*

The second property implies a set of relations analogous to those in (54). In fact, it is by means of such relations that this part of the theorem is demonstrated. From these follows a set of "initial conditions" analogous to those in (60), the limits in the present case being taken for $t$ approaching $-\infty$. The reader may readily supply them.

## § 5. *The First System of Principal Solutions.*

By means of the particular solution $f_1^+(x)$ of equation (2), obtained in § 2, we proceeded in § 3 to build up a fundamental set of solutions. In doing this

* The asymptotic character of $\bar{f}_n^-(x)$ is maintained with respect to $x$ in any unrestricted left half-plane.

we made a transformation which introduced a new function $g_1^+(x)$ satisfying the relation

$$\Delta g_1^+(x) = G_1^+(x), \tag{43 bis}$$

$G_1^+(x)$ being a particular solution of equation (38), this latter equation being obtained from equation (2) by means of the transformation (36). For $g_1^+(x)$ we took the function $\bar{g}_1^+(x)$, defined in equation (44), this function satisfying equation (43). We stated that another solution of (43) would also lead to interesting solutions of (2). These latter solutions we shall now find.

Throughout this discussion we shall assume that the integer $m$ has been so chosen that $f_1^+(x)$ is independent of $m$, this being possible, as was shown in § 3. Then we have not merely the limits affecting $f_1^+(x)$, which were stated in the preliminary theorem of § 2, but also the further property expressed in the asymptotic relation

$$f_1^+(x) \sim \alpha_1^x x^{\mu_1}\left(1+\frac{c_{11}}{x}+\frac{c_{12}}{x^2}+\ldots\right),$$

valid with respect to $x$ in the entire right half-plane and with respect to $v$ in the entire plane. Similarly, if we make use of this extended property of $f_1^+(x)$, then, instead of the inequality (42), we have the asymptotic relation

$$G_1^+(x) \sim \left(\frac{\alpha_2}{\alpha_1}\right)^x x^{\mu_2-\mu_1}\left(\frac{\alpha_2}{\alpha_1}-1\right)\left(1+\frac{d_{11}}{x}+\frac{d_{12}}{x^2}+\ldots\right),$$

valid with respect to $x$ in the entire right half-plane and with respect to $v$ in the entire plane. Furthermore, the function $\bar{g}_1^+(x)$, defined by equation (44), is asymptotically represented by

$$\bar{g}_1^+(x) \sim \left(\frac{\alpha_2}{\alpha_1}\right)^x x^{\mu_2-\mu_1}(1+\ldots) \equiv \frac{\alpha_2^x x^{\mu_2}(1+c_{21}x^{-1}+\ldots)}{\alpha_1^x x^{\mu_1}(1+c_{11}x^{-1}+\ldots)},$$

as one sees readily from its properties as developed in § 3; the range being the same as before, except that $|v|$ must now be restricted to be not less than a certain fixed quantity.

A solution $g_1^+(x)$ of (43) in the form of a contour integral, the path of integration going to infinity along two parallel lines, will be found to serve our purpose. We form the function

$$g_1^+(x) = \int_L \rho(x-z)\, p(x, z)\, G_1^+(z)\, dz, \tag{61}$$

where

$$\rho(x-z) = \{1-e^{2\pi\sqrt{-1}(z-x)}\}^{-1},$$

and where the path of integration $L$ is yet to be chosen and the function $p(x, z)$ is to be determined subject to the condition that it is analytic with respect to $x$

and also with respect to $z$ throughout the finite planes of these variables and that

$$p(x+1, z) = p(x, z).$$

Let $RD$ be a right $D$-region throughout which $f_1^+(x)$ and $G_1^+(x)$ are both analytic and verify the relations (31′) and (42), respectively, for every value of $m$. Since $G_1^+(x)$ is analytic throughout $RD$, it is clear that all the infinities of the integrand in (61) are at the points $z = x \pm r$, where $r$ is zero or a positive integer. Let $AB$ be the straight line containing all these points. Let the path of integration $L$ or $CKRH$ lie entirely within $RD$ and be formed in the following manner: It consists of three parts: a part $RH$ lying entirely above $AB$ and extending to infinity in a negative direction $RH$ parallel to the axis of reals; a part $KC$ lying below $AB$ and extending to infinity in a negative direction $KC$ parallel to the axis of reals; a finite part $KR$ which crosses $AB$ once between $x-1$ and $x$ and at no other point. Moreover, the path $CKRH$ is such that the part of the finite plane not in $RD$ lies entirely to the left of this path.

In order that the integral along such a path shall exist, it is necessary that the function $p(x, z)$ shall have certain requisite properties. The value $p(x, z) = 1$ would satisfy this requirement, provided that $|\alpha_1| < |\alpha_2|$; and it would indeed lead to solutions of our equation (2) (compare my memoir, *loc. cit.*, p. 119). Simpler solutions may, however, be obtained by a different choice (compare Birkhoff, *loc. cit.*, p. 264).

For the present we assume that $|\alpha_1| < |\alpha_2|$ and choose for $p(x, z)$ the function

$$p(x, z) = e^{2\pi\lambda\sqrt{-1}(x-z)},$$

where $\lambda$ is the least integer not less than the real part of

$$\frac{1}{2\pi\sqrt{-1}}(\log \alpha_2 - \log \alpha_1).$$

Then for $g_1^+(x)$ we have the value

$$g_1^+(x) = \int_L \rho(x-z)\, e^{2\pi\lambda\sqrt{-1}(x-z)} G_1^+(z)\, dz. \tag{62}$$

It is evident that this function is analytic throughout $RD$. Furthermore, the value of $g_1^+(x)$, for a particular $x$, is unaffected by any deformation of the path of integration, provided that during the deformation it has always the properties stated in its definition.

We show that $g_1^+(x)$, as defined in (62), is indeed a solution of equation (43). We have

$$g_1^+(x+1) = \int_{L'} \rho(x-z)\, e^{2\pi\sqrt{-1}\lambda(x-z)} G_1^+(z)\, dz,$$

where $L'$ is a path $CKTSRH$ which differs from $CKRH$ only in crossing $AB$ between $x$ and $x+1$ instead of between $x-1$ and $x$. Then

$$\int_{L'}-\int_{L}=\int_{KTSRK}.$$

By Cauchy's theorem on residues the last integral is equal to $2\pi\sqrt{-1}$ times the residue of the integral at the point $z=x$; that is, it is equal to $G_1^+(x)$. Hence,

$$g_1^+(x+1)-g_1^+(x)=G_1^+(x);$$

or, $g_1^+(x)$ satisfies equation (43). It is this particular solution of (43) which we shall denote by $g_1^+(x)$ in the remaining discussion.

We shall now determine a bound to the magnitude of $|g_1^+(x)|$ for certain ranges of variation for $x$. For this purpose it is convenient to break up the path of integration into two parts $L_1$ and $L_2$, as follows: The contour $L_1$ is the fixed contour $\infty A_1B_1\infty$ lying entirely in the region $RD$ and containing within it that part of the finite plane which is exterior to $RD$. Moreover, it consists of three parts, each of which is a straight line, a finite part $A_1B_1$ perpendicular to the axis of reals, and two infinite parts $A_1\infty$ and $B_1\infty$ parallel to the axis of reals. The contour $L_2$ lies entirely without the region inclosed by $L_1$. If $x$ lies above $B_1\infty$, $L_2$ consists of a loop circuit to infinity lying above $B_1\infty$ and including within it the points $x-1, x-2, x-3, \ldots.$, but not the points $x, x+1, x+2, \ldots.$. Moreover, in the neighborhood of infinity the bounding lines of $L_2$ are parallel to the axis of reals. If $x$ lies below $A_1\infty$, $L_2$ consists of a loop circuit to $\infty$ of the same nature as when $x$ lies above $B_1\infty$. If $x$ lies between $B_1\infty$ and $A_1\infty$, then $L_2$ consists of a finite loop containing within it the points $x-1, x-2, \ldots., x-l$, but no other point congruent to $x$, $x-l$ being the leftmost point congruent to $x$ and not on or within the contour $L_1$. For the integrand in (62) it is clear that the integral along $L$ is equal to the sum of two integrals, one along $L_1$ and the other along $L_2$, provided that $x$ is not congruent to a point on the contour $L_1$. In symbols

$$I_L=I_{L_1}+I_{L_2},$$

where $I_L$ denotes the integral in (62) and $I_{L_1}$ and $I_{L_2}$ similar integrals taken along the paths $L_1$ and $L_2$, the integrand being the same in all three cases.

Let us consider first the integral $I_{L_1}$ It denotes a periodic function of $x$ of period 1 since the contour is fixed and the integrand is a periodic function of $x$ of period 1. It may be written in either of the forms:

$$I_{L_1}=e^{2\pi\lambda\sqrt{-1}x}\int_{L_1}\frac{e^{-2\pi\lambda\sqrt{-1}z}G_1^+(z)\,dz}{1-e^{2\pi\sqrt{-1}(z-x)}},$$

$$I_{L_1}=e^{2\pi(\lambda-1)\sqrt{-1}x}\int_{L_1}\frac{e^{-2\pi(\lambda-1)\sqrt{-1}z}G_1^+(z)\,dz}{e^{-2\pi\sqrt{-1}(z-x)}-1}.$$

If $x, =u+v\sqrt{-1}$, approaches infinity so that $v$ becomes positively infinite, the integral in the first one of these equations is bounded. If $v$ becomes negatively infinite, the integral in the second one of these equations is bounded. If $x$ approaches infinity while $v$ remains bounded, then $I_{L_1}$ is itself bounded. Hence,

$$I_{L_1}=\begin{cases} e^{2\pi\lambda\sqrt{-1}x}q(x), \\ q(x), \\ e^{2\pi(\lambda-1)\sqrt{-1}x}q(x), \end{cases}$$

according as $x$ lies above $B_1\infty$, between $B_1\infty$ and $A_1\infty$, or below $A_1\infty$, the function $q(x)$ in each case being periodic in $x$ of period 1 and bounded. This result is strictly valid only when $x$ is not congruent to a point of the contour $L_1$. For all such $x$ as this, it is obvious that a single modified contour might be used and that a similar form for $I_L$ would be obtained, the periodic functions $q(x)$ being different in this case, but still being bounded.

Let us next consider the integral $I_{L_2}$. We separate the treatment into two cases.

In the first place, let us suppose that $x$ lies above $B_1\infty$ or below $A_1\infty$. In this case the integral $I_{L_2}$ may be evaluated as a sum of residues multiplied by $2\pi\sqrt{-1}$; thus, we have

$$I_{L_2}=G_1^+(x-1)+G_1^+(x-2)+G_1^+(x-3)+\ldots..$$

Hence, in this case,

$$I_{L_2}=\bar{g}_1^+(x),$$

a function whose asymptotic character is known.

In the second place, let us suppose that $x$ lies between $B_1\infty$ and $A_1\infty$ and to the right of $A_1B_1$. Then, evaluating as before by means of residues, we have

$$I_{L_2}=\sum_{k=1}^{l} G_1^+(x-k).$$

The asymptotic character of $I_{L_2}$ with respect to $x$ in the right half-plane can now be determined in precisely the same way as that by which the corresponding bounds to the increase of $|\bar{g}_1^+(x)|$ were determined in § 3. The only modification necessary arises from the fact that we now have a finite series instead of the infinite series in (44), so that when this series is separated into two parts, as in the earlier discussion, the second part as well as the first consists of only a finite number of terms. But it was precisely this second part which did not affect the asymptotic form, so that the modification in the argument will not affect the conclusion.

Taking these two cases together, then, and remembering the asymptotic form of $\bar{g}_1^+(x)$, we find that

$$I_{L_2} \sim \left(\frac{\alpha_2}{\alpha_1}\right)^x x^{\mu_2-\mu_1}(1+\ldots\ldots) \equiv \frac{\alpha_2^x x^{\mu_2}(1+c_{21}x^{-1}+\ldots\ldots)}{\alpha_1^x x^{\mu_1}(1+c_{11}x^{-1}+\ldots\ldots)}$$

throughout the right half-plane.

Now let us define a second solution $f_2^+(x)$ of equation (2) by means of the relation

$$f_2^+(x) = g_1^+(x) f_1^+(x) = (I_{L_1} + I_{L_2}) f_1^+(x).$$

From the known form of $I_{L_1}$ and the known asymptotic character of $I_{L_2}$ and $f_1^+(x)$ we see that

$$\left.\begin{aligned}
f_2^+(x) &\sim e^{2\pi\lambda\sqrt{-1}x} q(x)\left\{\alpha_1^x x^{\mu_1}\left(1+\frac{c_{11}}{x}+\ldots\ldots\right)\right\} + \alpha_2^x x^{\mu_2}\left(1+\frac{c_{21}}{x}+\ldots\ldots\right),\\
f_2^+(x) &\sim q(x)\left\{\alpha_1^x x^{\mu_1}\left(1+\frac{c_{11}}{x}+\ldots\ldots\right)\right\} + \alpha_2^x x^{\mu_2}\left(1+\frac{c_{21}}{x}+\ldots\ldots\right),\\
f_2^+(x) &\sim e^{2\pi(\lambda-1)\sqrt{-1}x} q(x)\left\{\alpha_1^x x^{\mu_1}\left(1+\frac{c_{11}}{x}+\ldots\ldots\right)\right\} + \alpha_2^x x^{\mu_2}\left(1+\frac{c_{21}}{x}+\ldots\ldots\right),
\end{aligned}\right\} \quad (63)$$

according as $x$ lies above $B_1\infty$, between $B_1\infty$ and $A_1\infty$ or below $A_1\infty$, the asymptotic representation being valid with respect to $x$ in any right half-plane and with respect to $v$ in the entire plane.

From this result we may determine, in simpler form, the asymptotic character of $f_2^+(x)$ with respect to $x$ in a right half-plane. Suppose first that $x$ lies above $B_1\infty$. The dominating term in the above representation of $f_2^+(x)$ depends upon the dominating term of the set

$$e^{(2\pi\lambda\sqrt{-1}+\log\alpha_1)x}, \quad e^{(\log\alpha_2)x}.$$

If we divide both of these quantities by the last, the exponents are

$$2\pi\sqrt{-1}\left\{\lambda + \frac{\log\alpha_1 - \log\alpha_2}{2\pi\sqrt{-1}}\right\}x, \quad 0.$$

If the real part of the first exponent is negative the latter of the two quantities predominates. From the definition of $\lambda$ and the fact that we now have $|\alpha_1| < |\alpha_2|$ it follows readily that the real part of this first exponent is indeed negative if $u$ (where $x = u + v\sqrt{-1}$) is positive and also if $u$ is merely bounded below, provided that the real part of $(\log\alpha_2 - \log\alpha_1)/(2\pi\sqrt{-1})$ is not the integer $\lambda$ and $v$ is sufficiently large. Therefore,

$$f_2^+(x) \sim \alpha_2^x x^{\mu_2}\left(1+\frac{c_{21}}{x}+\ldots\ldots\right) \quad (64)$$

for $x$ above $B_1\infty$ and to the right of the axis of imaginaries; and, unless the real part of $(\log \alpha_2 - \log \alpha_1)/(2\pi\sqrt{-1})$ is an integer, the asymptotic representation is valid for $x$ above $B_1\infty$ and to the right of any line parallel to the axis of imaginaries.

Suppose next that $x$ lies below $A_1\infty$. The dominating term in the asymptotic representation (63) of $f_2^+(x)$ now depends upon the dominating term of the set

$$e^{\{2\pi(\lambda-1)\sqrt{-1}+\log \alpha_1\}x}, \quad e^{(\log \alpha_2)x}.$$

The second of these predominates, provided that the real part of

$$2\pi\sqrt{-1}\left\{\lambda - 1 + \frac{\log \alpha_1 - \log \alpha_2}{2\pi\sqrt{-1}}\right\}x$$

is negative; and this real part is in fact negative if $u$ is positive or if $u$ is merely bounded below and $-v$ is sufficiently large. Hence, the asymptotic representation (64) is valid for $x$ below $A_1\infty$ and to the right of any line parallel to the axis of imaginaries.

Finally, suppose that $x$ lies between $A_1\infty$ and $B_1\infty$. Then the dominating term in the asymptotic representation (63) of $f_2^+(x)$ is clearly the last term, since now $|\alpha_1| < |\alpha_2|$, so that the asymptotic representation (64) is valid for $x$ between $A_1\infty$ and $B_1\infty$.

Combining these results, we see that the asymptotic representation (64) is certainly valid with respect to $x$ for $x$ approaching infinity in any way so as to remain to the right of the axis of imaginaries when in the upper half-plane and to the right of any line parallel to the axis of imaginaries when in the lower half-plane. Moreover, if the real part of $(\log \alpha_2 - \log \alpha_1)/(2\pi\sqrt{-1})$ is not an integer, then the asymptotic representation is valid with respect to $x$ in any right half-plane.

The above result was obtained on the supposition that $|\alpha_1| < |\alpha_2|$. If $|\alpha_1| = |\alpha_2|$, we form a second solution $f_2^+(x)$ in the same way as $f_1^+(x)$ was formed in § 2 (see the preliminary theorem), this being obviously possible since $\alpha_1$ and $\alpha_2$ then play exactly parallel rôles. This solution $f_2^+(x)$ has the asymptotic representation (64) with respect to $x$ in any right half-plane.

Thus, in any event, we have two solutions of equation (2). We proceed further as in § 3. Equation (38) is now known to have a second solution and its asymptotic character is determined in the same way as that of $f_2^+(x)$. By means of this second solution of equation (38) we are led to a third solution of equation (2). Then by a further reaction between equations (2) and (38) we find another solution of (2), and so on until $n$ solutions of (2) are found.

These are analytic in a suitable right half-plane and are represented asymptotically by the formal power series solutions of (2). By means of their asymptotic form it may be shown (as in the case of the solutions obtained in § 3) that they are linearly independent with respect to periodic multipliers of period 1, and hence form a fundamental system of solutions of equation (2). Since they are analytic in a suitable right half-plane, the possible position of their singularities may be determined as in the case of $f_1^+(x)$ in § 2. Combining the various results which we thus have, we obtain the following theorem:

FIRST SYSTEM OF PRINCIPAL SOLUTIONS. *If the characteristic algebraic equation associated with the difference equation* (2) *has its roots different from each other and from zero, then this difference equation has the fundamental system of solutions*

$$f_1^+(x),\ f_2^+(x),\ \ldots,\ f_n^+(x),$$

*determined above and possessing the following properties:*

1) *Each function in the system is analytic throughout the finite plane, except at the singularities of the functions* (34) *and points congruent to them on the left. Moreover, the singularities, in the finite plane, of each function in the solution are poles, provided that* $a_1(x), a_2(x), \ldots, a_n(x)$ *are rational functions and the complete set of numbers, each of which is the order of one of these poles, is a bounded set.*

2) *The functions* $f_i^+(x)$, $i=1, 2, \ldots, n$, *have with respect to* $x$ *the asymptotic representation*

$$f_i^+(x) \sim \alpha_i^x x^{\mu_i}\left(1+\frac{c_{i1}}{x}+\frac{c_{i2}}{x^2}+\ldots\right), \qquad i=1, 2, \ldots, n,$$

*this representation being valid for* $x$ *approaching infinity in any way so as to remain to the right of the axis of imaginaries when in the upper half-plane and to the right of any line parallel to the axis of imaginaries when in the lower half-plane.* Moreover, this representation is valid in any right half-plane provided that no one of the quantities*

$$\frac{\log \alpha_i - \log \alpha_j}{2\pi\sqrt{-1}}, \qquad j=1, 2, \ldots, n;\ i=j+1, \ldots, n,$$

*has its real part equal to an integer.*

---

* The upper and lower half-planes enter here unsymmetrically. Their rôles may be interchanged by a different choice of the quantities $\lambda$ entering into the discussion. See the paragraph containing equation (62).

From results previously obtained we know, furthermore, that $f_1^+(x)$ always maintains its asymptotic character with respect to $x$ in any right half-plane and with respect to $v$ in the entire plane.

The solutions mentioned in the foregoing theorem are called principal solutions on account of the important properties which they possess.

If $f_k(x)$ is any solution of (2) which is analytic in some right half-plane and is there represented asymptotically with respect to $x$ by

$$a_k^x x^{\mu_k}\left(1+\frac{c_{k1}}{x}+\frac{c_{k2}}{x^2}+\ldots\right),$$

then it may be shown (as in a similar case in § 3, beginning with equation (55)) that*

$$f_k(x)=p_1(x)f_1^+(x)+\ldots+p_{k-1}(x)f_{k-1}^+(x)+f_k^+(x),$$

where $p_1(x),\ldots,p_{k-1}(x)$ are suitably determined periodic functions of $x$ of period 1. By means of this result one again has a theory of "initial conditions" analogous to that associated with equation (60).

### § 6. *The Second System of Principal Solutions.*

In the preceding section a fundamental system of solutions was obtained by starting with the solution $f_1^+(x)$, obtained in § 2, and carrying out a certain process of transformation of equation (2) and interaction between it and the transformed equation. Similarly, one may start from the solution $f_n^-(x)$, obtained in § 2, and by a similar process arrive at another fundamental system of solutions of equation (2). We state merely the result of this work in the form of the following theorem:

Second System of Principal Solutions. *If the characteristic algebraic equation associated with the difference equation* (2) *has its roots different from each other and from zero, then this difference equation has a fundamental system of solutions*

$$f_1^-(x),\ f_2^-(x),\ \ldots,\ f_n^-(x),$$

*possessing the following properties:*

1) *Each function in the system is analytic throughout the finite plane except at points congruent on the right* (*at a distance of $n$ units or more*) *to the singularities of the functions $a_1(x)$, $a_2(x),\ldots,a_n(x)$. Moreover, the singularities of each solution, in the finite plane, are poles provided that $a_1(x)$,*

* If $f_k(x)$ is replaced by the $k$-th solution of the first fundamental set of associated solutions, then it is easy to see that we continue to have the same relation as that given in the text, this relation certainly being valid if $|v|$ is sufficiently large.

*$a_2(x), \ldots, a_n(x)$ are rational functions and the complete set of numbers, each of which is the order of one of these poles, is a bounded set.*

2) *The functions $f_i^-(x)$, $i=1, 2, \ldots, n$, have with respect to $x$ the asymptotic representation*

$$f_i^-(x) \sim a_i^x x^{\mu_i}\left(1+\frac{c_{i1}}{x}+\frac{c_{i2}}{x^2}+\ldots\right), \qquad i=1, 2, \ldots, n,$$

*this representation being valid for $x$ approaching infinity in any way so as to remain to the left of the axis of imaginaries when in the lower half-plane and to the left of any line parallel to the axis of imaginaries when in the upper half-plane.* Moreover, this representation is valid in any left half-plane, provided that no one of the quantities*

$$\frac{\log \alpha_i - \log \alpha_j}{2\pi\sqrt{-1}}, \; j=1, 2, \ldots, n; \; i=j+1, \ldots, n,$$

*has its real part equal to an integer.*

Further remarks, similar to those at the close of the preceding section, may be made with reference to the solutions $f_1^-(x), \ldots, f_n^-(x)$.

## § 7. *Other Results.*

The methods by which the foregoing results are demonstrated are essentially distinct from any at present existent in the literature of difference equations. There are three other results which are necessary to complete the first fundamentals of a general theory of equation (1). These are due to Birkhoff (*loc. cit.*), who has demonstrated them for a system of $n$ equations of the first order rather than for a single equation of order $n$. Birkhoff's proof, with merely formal modifications, will apply to equation (1) and may be based on the theorems demonstrated above. It is therefore unnecessary to repeat the proof. It seems desirable, however, to have a careful statement of these in a form to be directly applicable to equation (1) and for the sake of completeness of the fundamental results assembled in this paper. Consequently, a statement of these theorems, without proof, is given below:

1. If the characteristic algebraic equation associated with the difference equation (1) has its roots different from each other and from zero, and if the coefficients $\bar{a}_1(x), \bar{a}_2(x), \ldots, \bar{a}_n(x)$ are polynomials in $x$ and if further we write

* Compare the foot-note to the previous theorem.

$$F_i^-(x) = p_{i1}(x)\,F_1^+(x) + p_{i2}(x)\,F_2^+(x) + \ldots\ldots + p_{in}(x)\,F_n^+(x), \quad i=1, 2, \ldots\ldots, n,$$

where

$$F_i^+(x) = x^{\mu_i x} f_i^+(x), \quad F_i^-(x) = x^{\mu_i x} f_i^-(x), \quad i=1, 2, \ldots\ldots, n,$$

then the fundamental periodic functions $p_{ij}(x)$ are of the form

$$p_{ii}(x) = 1 + c_{ii}^{(1)}\,e^{2\pi\sqrt{-1}\,x} + c_{ii}^{(2)}\,e^{4\pi\sqrt{-1}\,x} + \ldots\ldots + e^{2\pi\mu_i\sqrt{-1}}\,e^{2\pi\mu\sqrt{-1}\,x},$$

$$p_{ij}(x) = e^{2\pi\lambda_{ij}\sqrt{-1}\,x}\left[c_{ij}^{(0)} + c_{ij}^{(1)}\,e^{2\pi\sqrt{-1}\,x} + \ldots\ldots + c_{ij}^{(\mu-1)}\,e^{2(\mu-1)\pi\sqrt{-1}\,x}\right], \quad i \neq j,$$

where $\lambda_{ij}$ denotes the least integer not less than the real part of

$$\frac{\log \alpha_i - \log \alpha_j}{2\pi\sqrt{-1}}.$$

If the functions $\bar{a}_1(x)$, $\bar{a}_2(x)$, ...., $\bar{a}_n(x)$ are restricted merely to be rational, then the periodic functions $p_{ij}(x)$ are rational functions of $e^{2\pi\sqrt{-1}\,x}$.

2. Let the characteristic algebraic equation associated with the difference equation (1) have its roots different from each other and from zero, and let the coefficients $\bar{a}_1(x)$, $\bar{a}_2(x)$, ...., $\bar{a}_n(x)$ be polynomials in $x$. Write

$$2\pi\sqrt{-1}\left(\frac{\log \alpha_j - \log \alpha_i}{2\pi\sqrt{-1}} + \lambda_{ij}\right) = \varepsilon_{ij},$$

where $\lambda_{ij}$ is the least integer not less than the real part of the negative of the first term in parenthesis. Mark the two sets of points

$$\varepsilon_{in},\ \varepsilon_{i,n-1},\ \ldots\ldots,\ \varepsilon_{i,i-1},\ 0 \ \text{ and } \ \varepsilon_{in} - 2\pi\sqrt{-1},\ \ldots\ldots,\ \varepsilon_{i,i-1} - 2\pi\sqrt{-1},\ 0,$$

and call them, respectively,

$$P_n,\ P_{n-1},\ \ldots\ldots,\ P_i \ \text{ and } \ Q_n,\ Q_{n-1},\ \ldots\ldots,\ Q_i.$$

Construct the convex broken line $P_n\,P_\sigma \ldots\ldots P_i$ above which all the remaining points $P$ lie, and likewise construct the convex broken line $Q_n\,Q_\theta \ldots\ldots Q_i$ below which all the remaining points $Q$ lie. Let the acute angles which the successive sides of $P_n P_\sigma \ldots\ldots P_i$ and $Q_n Q_\theta \ldots\ldots Q_i$ make with the axis of reals be $\phi_n, \phi_\sigma, \ldots\ldots$ and $\psi_n, \psi_\theta, \ldots\ldots$, respectively. The critical rays between $\arg x = 0$ and $\arg x = 2\pi$ are then

$$\frac{\pi}{2} - \phi_n,\ \frac{\pi}{2} - \phi_\sigma,\ \ldots\ldots,\ -\frac{\pi}{2} + \psi_\theta,\ -\frac{\pi}{2} + \psi_n,$$

in angular order. The asymptotic form of $F_i^-(x)$ changes from

$$e^{2\pi\lambda_{ti}\sqrt{-1}\,x}\,c_{ti}^{(0)}\,x^{\mu x}\,\alpha_t^x\,x^{\mu_t}\left(1 + \frac{c_{t1}}{x} + \frac{c_{t2}}{x^2} + \ldots\ldots\right)$$

to

$$e^{2\pi\lambda_{si}\sqrt{-1}x}c_{si}^{(0)}x^{\mu x}\alpha_s^x x^{\mu_s}\left(1 + \frac{c_{s1}}{x} + \frac{c_{s2}}{x^2} + \ldots\ldots\right),$$

along the critical ray $\arg x = \frac{1}{2}\pi - \phi_t$, and likewise from*

$$e^{2\pi(\lambda_{it}+\delta_{it}-1)\sqrt{-1}x} c_{it}^{(\mu+\delta_{it}-1)} e^{-2\pi\mu_t\sqrt{-1}} x^{\mu x} \alpha_t^x x^{\mu_t}\left(1+\frac{c_{t1}}{x}+\ldots.\right),$$

to

$$e^{2\pi(\lambda_{is}+\delta_{is}-1)\sqrt{-1}x} c_{is}^{(\mu+\delta_{is}-1)} e^{-2\pi\mu_s\sqrt{-1}} x^{\mu x} \alpha_s^x x^{\mu_s}\left(1+\frac{c_{s1}}{x}+\ldots.\right),$$

along the critical ray $\arg x = -\frac{1}{2}\pi + \psi_t$. Between the last critical ray of the first set and the first of the last set the asymptotic form is given by

$$x^{\mu x} \alpha_i^x x^{\mu_i}\left(1+\frac{c_{i1}}{x}+\ldots.\right).$$

A similar theorem exists for the functions $F_i^+(x)$ in case equation (1) is such that it may be written in the form

$$a_0'(x)F(x+n)+\ldots.+a_{n-1}'(x)F(x+1)+F(x)=0,$$

where $a_0'(x)$, $a_1'(x)$, ...., $a_{n-1}'(x)$ are polynomials

3. Let $F_1^+(x)$, ...., $F_n^+(x)$ and $F_1^-(x)$, ...., $F_n^-(x)$ be two sets of single-valued functions which are analytic throughout the finite plane, except for poles, and which have the further property that

$$\lim_{x=\infty} F_i^+(x)x^{-\mu x}\alpha_i^{-x}x^{-\mu_i}=1, \quad \lim_{x=\infty} F_i^-(x)x^{-\mu x}\alpha_i^{-x}x^{-\mu_i}=1,$$

$\mu$ being an integer, the constants $\alpha_1, \alpha_2, \ldots., \alpha_n$ being different from each other and from zero and the range of $x$ for the limits being as follows: In the first [second] case $x$ may approach infinity in any way so as to remain to the right [left] of any line parallel to the axis of imaginaries when in the upper [lower] half-plane and to the right [left] of any line parallel to the axis of imaginaries when in the lower [upper] half-plane. Furthermore, let these two sets of functions be connected by the relations

$$F_i^-(x)=p_{i1}(x)F_1^+(x)+\ldots.+p_{in}(x)F_n^+(x), \qquad i=1, 2, \ldots., n,$$

where the functions $p_{ij}(x)$ are periodic in $x$ of period 1. Then the sets $F_1^+(x)$, ...., $F_n^+(x)$ and $F_1^-(x)$, ...., $F_n^-(x)$ are fundamental systems of solutions of a difference equation (1) in which the coefficients $\bar{a}_1(x)$, ...., $\bar{a}_n(x)$ are rational functions of $x$.

UNIVERSITY OF ILLINOIS.

* Here $\delta_{ij}$ is equal to unity or zero, according as $i$ is or is not equal to $j$.

# AMERICAN
# Journal of Mathematics

EDITED BY

**FRANK MORLEY**

WITH THE COÖPERATION OF

A. COHEN, CHARLOTTE A. SCOTT

AND OTHER MATHEMATICIANS

PUBLISHED UNDER THE AUSPICES OF THE JOHNS HOPKINS UNIVERSITY

*Πραγμάτων ἔλεγχος οὐ βλεπομένων*

VOLUME XXXVIII, NUMBER 3

BALTIMORE: THE JOHNS HOPKINS PRESS

LEMCKE & BUECHNER, *New York.* E. STEIGER & CO., *New York.* A. HERMANN, *Paris.*
G. E. STECHERT & CO., *New York.* ARTHUR F. BIRD, *London.* MAYER & MÜLLER, *Berlin.*
WILLIAM WESLEY & SON, *London.*

JULY, 1916

Entered as Second-Class Matter at the Baltimore, Maryland, Postoffice.

# *A Class of Asymptotic Orbits in the Problem of Three Bodies.*

By L. A. H. Warren.

Contents.



## I. *Introduction.*

If we have a system consisting of two arbitrary finite spherical bodies revolving in circles about their common center of gravity, Lagrange* has shown that there are three points on the straight line passing through the centers of the finite bodies such that if an infinitesimal body be placed at one of them and be projected so as to be instantaneously fixed relatively to the revolving system it will always remain fixed relatively to the system. These three equilibrium points, as they are called, are separated by the finite bodies, whose masses are denoted by $\mu$ and $1-\mu$, $(0<\mu\leq\frac{1}{2})$. The point beyond the mass $\mu$ is called $(a)$, that between the finite bodies is called $(b)$, and that beyond the mass $1-\mu$ is called $(c)$.

In the present paper, it is shown that in the neighborhood of each of the equilibrium points $(a)$, $(b)$, $(c)$ there exists a class of orbits, in which the infinitesimal body approaches the equilibrium points as $t$ becomes infinite. Such orbits are called asymptotic orbits.†

* Lagrange, "Collected Works," Vol. VI, pp. 229–324; Tisserand, "Mécanique Céleste," Vol. I, Chapter VIII; Moulton, "Introduction to Celestial Mechanics" (New Edition), Chapter VIII.

† Poincare, "Les Méthodes Nouvelles de la Mécanique Céleste."

In § II the equations of motion of the infinitesimal body are given with reference to the system of rotating axes, passing through the center of gravity of the finite masses, the $x$-axis coinciding with the line joining their centers.

In § III there is a detailed explanation of what is meant by asymptotic solutions of systems of differential equations, and asymptotic orbits of a moving body; and it is shown that in the three-body problem under consideration the only existing orbits which are asymptotic to any of the equilibrium points $(a)$, $(b)$, or $(c)$ lie wholly in the plane of revolution of the finite masses.

In § IV the equations of the asymptotic orbits are developed as power series in exponential functions of the time, and in the following article it is shown that these power series expansions are convergent for $t$ sufficiently large.

In § VI an alternative method is given for building up the equations of these orbits.

In § VII there is a discussion of some of the principal properties of these asymptotic orbits, their position relative to the rotating axes, and the change in their direction of approach to the equilibrium points as $\mu$ varies from zero to $\frac{1}{2}$.

In § VIII it is shown how the orbits can be continued by the method of mechanical quadratures beyond the range of convergence of the solutions of the differential equations. A special case $\mu=0.02$ is discussed in detail, and one of the corresponding asymptotic orbits for point $(a)$ is continued by this process.

## II. *The Differential Equations of Motion of the Infinitesimal Body.*

In the following discussion we consider a system consisting of two finite bodies revolving in circles about their common center of mass, and of an infinitesimal body subject to their attractions. Let the constant distance between the finite bodies be unity. Denote the masses of the finite bodies by $\mu$ and $1-\mu$, where $0<\mu\leq\frac{1}{2}$, so that the sum of the masses shall be unity. Choose the unit of time so that the gravitational constant $k^2$ shall be unity. With the units so chosen the time of revolution of the finite bodies will be *unity*.

Take the origin of coordinates at the center of mass of the finite bodies, and refer the motion of the bodies to a system of axes rotating in the plane of motion of the finite body, in such a way that the $\xi$-axis always passes through the centers of the finite bodies. If $\xi$, $\eta$, $\zeta$ denote the coordinates of the infinitesimal body, then the differential equations of motion for the infinitesimal body are:*

$$\left.\begin{aligned}
&\frac{d^2\xi}{dt^2}-2\frac{d\eta}{dt}=\xi-(1-\mu)\frac{(\xi-\xi_1)}{r_1^3}-\mu\frac{(\xi-\xi_2)}{r_2^3},\\
&\frac{d^2\eta}{dt^2}+2\frac{d\xi}{dt}=\eta-(1-\mu)\frac{\eta}{r_1^3}-\mu\frac{\eta}{r_2^3},\\
&\frac{d^2\zeta}{dt^2}\qquad\quad=-(1-\mu)\frac{\zeta}{r_1^3}-\mu\frac{\zeta}{r_2^3},
\end{aligned}\right\}\qquad(1)$$

where $(\xi_1, 0, 0)$, $(\xi_2, 0, 0)$ are the coordinates of the bodies $1-\mu$ and $\mu$ respectively; $r_1=\sqrt{(\xi-\xi_1)^2+\eta^2+\zeta^2}$ and $r_2=\sqrt{(\xi-\xi_2)^2+\eta^2+\zeta^2}$.

Let $(\xi_0, 0, 0)$ denote the coordinates of one of the equilibrium points on the $\xi$-axis. If, then, by the transformation $\xi=\xi_0+x$, $\eta=y$, $\zeta=z$, we move the origin to one of these points, the equations of motion take the form†

$$\left.\begin{aligned}
&x''-2y'=(1+2A)x+\frac{3}{2}B[-2x^2+y^2+z^2]+2C[2x^3-3xy^2-3xz^2]\ldots.,\\
&y''+2x'=(1-A)y+3Bxy+\frac{3}{2}Cy[-4x^2+y^2+z^2]+\ldots.,\\
&z''\qquad\;\;=-Az+3Bxz+\frac{3}{2}Cz[-4x^2+y^2+z^2]+\ldots.,\\
&\text{where}\\
&A=\frac{1-\mu}{[(\xi_0+\mu)^2]^{\frac{3}{2}}}+\frac{\mu}{[(\xi_0-1+\mu)^2]^{\frac{3}{2}}}=\frac{1-\mu}{r_1^{(0)^3}}+\frac{\mu}{r_2^{(0)^3}},\\
&B=\frac{+}{\pm}\frac{1-\mu}{r_1^{(0)^4}}\pm\frac{\mu}{r_2^{(0)^4}},\\
&C=\;\frac{1-\mu}{r_1^{(0)^5}}+\frac{\mu}{r_2^{(0)^5}},
\end{aligned}\right\}\qquad(2)$$

where in the expression for $B$ the upper, middle, or lower signs are to be taken according as the orbits in the vicinity of the point $(a)$, $(b)$, or $(c)$ are being treated.

In what follows we shall have to deal chiefly with the first two equations of (2), and it will be more convenient to have them in a normal form.

The linear terms of the first two equations of (2) are

$$\left.\begin{aligned}
&x''-2y'-(1+2A)x=0,\\
&y''+2x'-(1-\;A)y=0.
\end{aligned}\right\}\qquad(3)$$

---

* Moulton, "Introduction to Celestial Mechanics" (New Edition), § 152.

† Moulton, "Periodic Orbits," pp. 156, 168.

The general solution of these equations is

$$\left.\begin{aligned} x&=K_1e^{\sigma\sqrt{-1}t}+K_2e^{-\sigma\sqrt{-1}t}+K_3e^{\rho t}+K_4e^{-\rho t},\\ y&=n\sqrt{-1}[K_1e^{\sigma\sqrt{-1}t}-K_2e^{-\sigma\sqrt{-1}t}]+m[K_3e^{\rho t}-K_4e^{-\rho t}].\end{aligned}\right\} \quad (4)$$

Where $\sigma\sqrt{-1}$, $-\sigma\sqrt{-1}$, $\rho$, $-\rho$ are the roots of the bi-quadratic equation

$$\lambda^4+(2-A)\lambda^2+(1+2A)(1-A)=0. \quad (5)$$

If then we make the transformation

$$\left.\begin{aligned} x&=u_1+u_2+u_3+u_4,\\ x'&=\sigma\sqrt{-1}(u_1-u_2)+\rho(u_3-u_4),\\ y&=n\sqrt{-1}(u_1-u_2)+m(u_3-u_4),\\ y'&=-n\sigma(u_1+u_2)+m\rho(u_3+u_4),\end{aligned}\right\} \quad (6)$$

the first two equations of (3) assume the normal form,*

$$\left.\begin{aligned} u_1'&=\sigma\sqrt{-1}u_1+\frac{m}{2(m\sigma-n\rho)\sqrt{-1}}\Big[\tfrac{3}{2}B(-2x^2+y^2+z^2)\\ &\qquad\qquad +2C(2x^3-3xy^2-3xz^2)+\ldots\Big]\\ &\quad -\frac{1}{2(m\rho+n\sigma)}\Big\{3Bxy+\tfrac{3}{2}Cy(-4x^2+y^2+z^2)+\ldots\Big\},\\ u_2'&=-\sigma\sqrt{-1}u_2-\frac{m}{2(m\sigma-n\rho)\sqrt{-1}}\Big[\text{``}\Big]-\frac{1}{2(m\rho+n\sigma)}\Big\{\text{``}\Big\},\\ u_3'&=\ \rho u_3-\frac{n}{2(m\sigma-n\rho)}\Big[\text{``}\Big]+\frac{1}{2(m\rho+n\sigma)}\Big\{\text{``}\Big\},\\ u_4'&=-\rho u_4+\frac{n}{2(m\sigma-n\rho)}\Big[\text{``}\Big]+\frac{1}{2(m\rho+n\sigma)}\Big\{\text{``}\Big\}.\end{aligned}\right\} \quad (7)$$

Equation (7) may be put in the form

$$\left.\begin{aligned} u_1'&=\ \sigma\sqrt{-1}u_1+H_1^{(2)}+H_1^{(3)}+H_1^{(4)}+\ldots,\\ u_2'&=-\sigma\sqrt{-1}u_2+H_2^{(2)}+H_2^{(3)}+H_2^{(4)}+\ldots,\\ u_3'&=\ \rho u_3+H_3^{(2)}+H_3^{(3)}+H_3^{(4)}+\ldots,\\ u_4'&=-\rho u_4+H_4^{(2)}+H_4^{(3)}+H_4^{(4)}+\ldots,\end{aligned}\right\} \quad (8)$$

where $H_k^{(r)}$ denotes all the terms on the right-hand side of the $k$-th equation which are of degree $r$ in $x$, $y$, $z$, and therefore of degree $r$ in $u_1$, $u_2$, $u_3$, $u_4$.

## III. *Asymptotic Orbits Defined and Proof that They all Lie in the xy-Plane.*

It has been shown by Poincaré† and Picard‡ that certain systems of differential equations of the form $\frac{dx_i}{dt}=X_i(x,t)$ $(i=1,\ldots,n)$ admit of solu-

* Moulton, "Periodic Orbits," pp. 161–162.

† Poincaré, "Les Méthodes Nouvelles de la Mécanique Céleste," Vol. I, Chap. VIII.

‡ Picard, "Traité D'Analyse," Vol. III, Chap. VIII, § V.

tions as power series in $A_i e^{\lambda_i t}$ $(i=1, \ldots, n)$, where the $A_i$ are arbitrary constants, and where the fixed constants $\lambda_i$, which are the roots of the characteristic equation, are called the "characteristic exponents." It has been shown further* that if there are $k$ of the $\lambda_i$ $(i=1, \ldots, n)$ which are represented by $k$ points on the complex plane all of which lie on the same side of a straight line passing through the origin, and which are such that none of the relations $\sum_{j=1}^{k} p_j\lambda_j - \lambda_i = 0$ $(i=1, \ldots, n)$ holds for any positive integral values of the $p_j$ such that $\sum_{j=1}^{k} p_j \geqq 2$, then the solutions as power series in $A_i e^{\lambda_i t}$ $(i=1, \ldots, k)$ will be convergent for $|A_i e^{\lambda_i t}|$ sufficiently small. In particular, if we put equal to zero the $A_i$ corresponding to those $\lambda_i$ whose real parts are zero or positive, the solutions as power series in the remaining $A_i e^{\lambda_i t}$ will be convergent for all values of $t$ which are sufficiently great; and if the $A_i$ involved in these latter expansions are taken sufficiently small the convergence will hold for all values of $t$ from $t=0$ to $t=\infty$.

Again, if we build up solutions as power series in those $A_j e^{\lambda_j t}$ where the real parts of $\lambda_j$ are positive, the exponentials $e^{\lambda_j t}$ approach zero as $t$ approaches $-\infty$. Such expansions will be convergent for $t$ sufficiently large and negative; and if the $A_j$ are sufficiently small they will be convergent for all negative values of $t$. Such solutions are said to be "asymptotic" to the solutions obtained by putting all the $A_i$ $(i=1, \ldots, n)$ equal to zero, and they are called "Asymptotic Solutions of the System of Differential Equations."

In the problem under consideration we shall show that the differential equations of motion of the infinitesimal body have asymptotic solutions such as have just been described. We shall see also that the infinitesimal body, moving in an orbit defined by one of these solutions, will approach asymptotically one of the equilibrium points $(a)$, $(b)$, or $(c)$ as $t$ becomes infinitely great.

We proceed to show, first of all, that all orbits which are asymptotic to one of the equilibrium points $(a)$, $(b)$, or $(c)$ lie entirely in the plane of revolution of the finite bodies, that is, in the $xy$-plane.

If in the equations of motion (2), we make the transformation $x=x\varepsilon$, $y=y\varepsilon$, $z=z\varepsilon$, we obtain, on dividing through by $\varepsilon$,

$$\left.\begin{aligned} &x''-2y'-(1+2A)x=\varepsilon X_2(x^2, y^2, z^2)+\varepsilon^2 X_3(x^3, xy^2, xz^2)+\varepsilon^3 X_4(\ldots\ldots)+\ldots\ldots, \\ &y''+2x'-(1-\ A)x=\varepsilon y Y_2(x)+\varepsilon^2 y Y_3(x^2, y^2, z^2)+\varepsilon^3 y Y_4(\ldots\ldots)+\ldots\ldots, \\ &z''+\qquad\qquad\quad Az=\varepsilon z Z_2(x)+\varepsilon^2 z Z_3(x^2, y^2, z^2)+\varepsilon^3 z Z_4(\ldots\ldots)+\ldots\ldots \end{aligned}\right\} \quad (9)$$

---

* Poincaré, "Les Méthodes Nouvelles de la Mécanique Céleste," Vol. I, Chap. VIII, § 105.

Since the right-hand members of (2) converge, so also will the right members of (9) converge for all $\epsilon$ $(0<\epsilon\leq 1)$. It follows, therefore, that the equations (9) have a solution of the form

$$\left.\begin{aligned} x &= \sum_{j=0}^{\infty} x_j(t)\epsilon^j, \\ y &= \sum_{j=0}^{\infty} y_j(t)\epsilon^j, \\ z &= \sum_{j=0}^{\infty} z_j(t)\epsilon^j. \end{aligned}\right\} \quad (10)$$

In order that $z$ may be zero for all $t$ sufficiently large, and for all values of $\epsilon$, it follows that $z_j(t)$ $(j=0, \ldots . \infty)$ must each be zero for all $t$ sufficiently large. On substituting (10) in (9) and equating coefficients of like powers of $\epsilon$ on both sides of the resulting equations, we obtain sets of differential equations from which the values of $x_j$, $y_j$, $z_j$ $(j=0, \ldots . \infty)$ can be obtained sequentially.

From the terms independent of $\epsilon$, we have

$$\left.\begin{aligned} & x_0''-2y_0'-(1+2A)x_0=0, \\ & y_0''+2x_0-(1-\ A)y_0=0, \\ & z_0'' \qquad\qquad +\ Az_0\ =0. \end{aligned}\right\} \quad (11)$$

The general solution of equations (11) is

$$\left.\begin{aligned} & x_0=K_1^{(0)}e^{\sigma\sqrt{-1}t}+K_2^{(0)}e^{-\sigma\sqrt{-1}t}+K_3^{(0)}e^{\rho t}+K_4^{(0)}e^{-\rho t}, \\ & y_0=n\sqrt{-1}\,[K_1^{(0)}e^{\sigma\sqrt{-1}t}-K_2^{(0)}e^{-\sigma\sqrt{-1}t}]+m\,[K_3^{(0)}e^{\rho t}-K_4^{(0)}e^{-\rho t}], \\ & z_0\ =c_1^{(0)}\cos\sqrt{A}\,t+c_2^{(0)}\sin\sqrt{A}\,t. \end{aligned}\right\} \quad (12)$$

In order that $z_0(t)$ shall approach zero for $t$ infinite, we see that $c_1^{(0)}=c_2^{(0)}=0$, and, therefore

$$z_0(t)=0. \quad (13)$$

When we equate the terms in the first power of $\epsilon$, we get

$$\left.\begin{aligned} & x_1''-2y_1''-(1+2A)x_1=X_2(x_0^2,\ y_0^2,\ z_0^2), \\ & y_1''+2x_1''-(1-\ A)y_1=y_0Y_2(x_0), \\ & z_1'' \qquad\qquad +\ \ Az_1=z_0Z_2(x_0). \end{aligned}\right\} \quad (14)$$

If we substitute the value $z_0=0$ in (14) the third equation becomes $z_1''+Az_1=0$, which is of the same form as the third equation in (11). In order that $z_1(t)$ shall approach zero for $t$ infinite we see, therefore, that $z_1(t)\underset{t}{\equiv}0$. Similarly, we can show that $z_2(t)\underset{t}{\equiv}0$: Suppose we have proved sequentially by this method that $z_j(t)\underset{t}{\equiv}0$, for $j=0, 1, \ldots, n$. Since the right-hand member of the third equation of (9) carries $z$ as a factor, and since the factor $\epsilon$ has been removed from this equation, it follows that the right member of the differential equation which defines $z_{n+1}$ will consist of terms which carry as a factor one or more of the $z_j(t)$ $(j=0, 1, \ldots, n)$ but will not contain any $z_j$, $j>n$. Hence, the right member of this equation is zero, and, therefore, $z_{n+1}(t) \underset{t}{\equiv} 0$. We see, therefore, that $z_j(t) \underset{t}{\equiv} 0$ for $j=0, 1, \ldots \infty$, from which it follows that $z(t) \underset{t}{\equiv} 0$. In order, therefore, that the infinitesimal body shall come to rest at one of the equilibrium points $(a)$, $(b)$, $(c)$ its whole orbit must be in the $xy$-plane.

## IV. *Formal Construction of the Solutions.*

We proceed to show that equations (7) admit of solutions defining asymptotic orbits in the $xy$-plane. These will be found as power series in $e^{-\rho t}$, and $e^{+\rho t}$, the former convergent for $t$ sufficiently large and positive, the latter for $t$ sufficiently large and negative.

### (A) *Solutions as Power Series in* $e^{-\rho t}$.

If, in equations (8), we make a transformation on the independent variable by putting $\omega = e^{-\rho t}$ these equations take the form

$$\left.\begin{aligned}
-\rho\omega\frac{\partial u_1}{\partial \omega} &= \sigma\sqrt{-1}u_1 + H_1^{(2)} + H_1^{(3)} + \ldots,\\
-\rho\omega\frac{\partial u_2}{\partial \omega} &= -\sigma\sqrt{-1}u_2 + H_2^{(2)} + H_2^{(3)} + \ldots,\\
-\rho\omega\frac{\partial u_3}{\partial \omega} &= \rho u_3 + H_3^{(2)} + H_3^{(3)} + \ldots,\\
-\rho\omega\frac{\partial u_4}{\partial \omega} &= -\rho u_4 + H_4^{(2)} + H_4^{(3)} + \ldots.
\end{aligned}\right\} \quad (15)$$

It is required to find solutions of (15) as power series in $\omega$, convergent for $\omega$ sufficiently small.

By Maclaurin's expansion

$$u_i(\omega) = u_i(0) + \omega\left(\frac{\partial u_i}{\partial \omega}\right)_{\omega=0} + \frac{\omega^2}{2!}\left(\frac{\partial^2 u_i}{\partial \omega^2}\right)_{\omega=0} + \ldots, \quad (i=1, \ldots, 4). \quad (16)$$

Since the body is to be at rest at the origin at $\omega=0$, it follows from (6) that $u_i(0)=0$, $(i=1, \ldots, 4)$. By repeated differentiation of equations (15) with regard to $\omega$, and putting $\omega=0$, we can build up the coefficients of the successive powers of $\omega$ in the expansion (16). Since

$$\left.\begin{aligned}
x &= u_1 + u_2 + u_3 + u_4,\\
y &= n\sqrt{-1}(u_1 - u_2) + m(u_3 - u_4),
\end{aligned}\right\} \quad (17)$$

it follows that the coefficients of successive powers of $\omega$ in the expansions for $x$ and $y$ will thus be known also.

The first equation of (15) written at length is

$$-\rho\omega\frac{\partial u_1}{\partial\omega}=\sigma\sqrt{-1}\,u_1+\left[\frac{3mB}{4(m\sigma-n\rho)\sqrt{-1}}(-2x^2+y^2)-\frac{3B}{2(m\rho+n\sigma)}(xy)\right]$$
$$+\left[\frac{mC}{(m\sigma-n\rho)\sqrt{-1}}(2x^3-3xy^2)-\frac{3C}{4(m\rho+n\sigma)}(-4x^3y+y^3)\right]$$
$$+\text{(terms of higher degree in } x \text{ and } y).$$

On differentiating this with regard to $\omega$, we get

$$-\rho\frac{\partial u_1}{\partial\omega}-\rho\omega\frac{\partial^2 u_1}{\partial\omega^2}=\sigma\sqrt{-1}\frac{\partial u_1}{\partial\omega}$$
$$+\left[\frac{3mB}{4(m\sigma-n\rho)\sqrt{-1}}\left(-4x\frac{\partial x}{\partial\omega}+2y\frac{\partial y}{\partial\omega}\right)-\frac{3B}{2(m\rho+n\sigma)}\left(y\frac{\partial x}{\partial\omega}+x\frac{\partial y}{\partial\omega}\right)\right]$$
$$+\left[\frac{mC}{(m\sigma-n\rho)\sqrt{-1}}\left(6x^2\frac{\partial x}{\partial\omega}-6xy\frac{\partial y}{\partial\omega}-3y^2\frac{\partial x}{\partial\omega}\right)-\frac{3C}{4(m\rho+n\sigma)}\right.$$
$$\left.\left(-8xy\frac{\partial x}{\partial\omega}-4x^2\frac{\partial y}{\partial\omega}+3y^2\frac{\partial y}{\partial\omega}\right)\right]+\ldots\ldots \tag{18}$$

On putting $\omega=0$, and therefore $x=y=0$, this becomes

$$(-\rho-\sigma\sqrt{-1})\left(\frac{\partial u_1}{\partial\omega}\right)_{\omega=0}=0. \tag{19}$$

Since $A>1$* for each of the equilibrium points for $0\gtreqless\mu\leqq\frac{1}{2}$, it can be readily seen that two of the roots of equation (5) are real and equal numerically but opposite in sign, and that the other two are conjugate imaginaries. Further, none of the roots of (5) is zero. It follows that none of the relations,

$$p\rho=0;\quad p\rho=-\rho;\quad p\rho=\pm\sigma\sqrt{-1},$$

can hold for $p\gtreqless 1$ ($p$ a positive integer). Hence, from (19) we see that

$$\left(\frac{\partial u_1}{\partial\omega}\right)_{\omega=0}=0. \tag{20}$$

Similarly, by differentiating the second, third, and fourth equations of (15) with regard to $\omega$, we obtain, respectively,

$$(-\rho+\sigma\sqrt{-1})\left(\frac{\partial u_2}{\partial\omega}\right)_{\omega=0}=0,\quad -2\rho\left(\frac{\partial u_3}{\partial\omega}\right)_{\omega=0}=0,\text{ and } (-\rho+\rho)\left(\frac{\partial u_4}{\partial\omega}\right)_{\omega=0}=0. \tag{21}$$

* Moulton, "Periodic Orbits," p. 159.

It follows then, that

$$\left(\frac{\partial u_2}{\partial \omega}\right)_{\omega=0}=0;\quad \left(\frac{\partial u_3}{\partial \omega}\right)_{\omega=0}=0;\quad \left(\frac{\partial u_4}{\partial \omega}\right)_{\omega=0}=\text{arbitrary}=c.$$

Hence,

$$\left(\frac{\partial x}{\partial \omega}\right)_{\omega=0}=c;\quad\text{and}\quad \left(\frac{\partial y}{\partial \omega}\right)_{\omega=0}=-mc. \tag{22}$$

If we differentiate equations (18) and the three corresponding equations in $u_2$, $u_3$, and $u_4$ obtained from (15), and put $\omega=0$, we obtain in succession, by the aid of (20) and (21)

$$\left.\begin{aligned}
\left(\frac{\partial^2 u_1}{\partial \omega^2}\right)_{\omega=0}&=-\frac{1}{2\rho+\sigma\sqrt{-1}}\left[\frac{3m(m^2-2)B}{2(m\sigma-n\rho)\sqrt{-1}}+\frac{3mB}{m\rho+n\sigma}\right]c^2,\\
\left(\frac{\partial^2 u_2}{\partial \omega^2}\right)_{\omega=0}&=\frac{1}{2\rho-\sigma\sqrt{-1}}\left[\frac{3m(m^2-2)B}{2(m\sigma-n\rho)\sqrt{-1}}-\frac{3mB}{m\rho+n\sigma}\right]c^2,\\
\left(\frac{\partial^2 u_3}{\partial \omega^2}\right)_{\omega=0}&=\frac{1}{3\rho}\left[\frac{3n(m^2-2)B}{2(m\sigma-n\rho)}+\frac{3mB}{m\rho+n\sigma}\right]c^2,\\
\left(\frac{\partial^2 u_4}{\partial \omega^2}\right)_{\omega=0}&=-\frac{1}{\rho}\left[\frac{3n(m^2-2)B}{2(m\sigma-n\rho)}-\frac{3mB}{m\rho+n\sigma}\right]c^2.
\end{aligned}\right\} \tag{23}$$

On reduction, then, we obtain

$$\left.\begin{aligned}
\left(\frac{\partial^2 x}{\partial \omega^2}\right)_{\omega=0}&=c^2\left[\frac{3m(m^2-2)B\sigma}{(m\sigma-n\rho)(4\rho^2+\sigma^2)}-\frac{n(m^2-2)B}{(m\sigma-n\rho)\rho}\right.\\
&\qquad\left.-\frac{12mB\rho}{(m\rho+n\sigma)(4\rho^2+\sigma^2)}+\frac{4mB}{(m\rho+n\sigma)\rho}\right],\\
\left(\frac{\partial^2 y}{\partial \omega^2}\right)_{\omega=0}&=-mc^2\left[\frac{6n(m^2-2)B\rho}{(m\sigma-n\rho)(4\rho^2+\sigma^2)}-\frac{2n(m^2-2)B}{(m\sigma-n\rho)\rho}\right.\\
&\qquad\left.+\frac{6nB\sigma}{(m\rho+n\sigma)(4\rho^2+\sigma^2)}+\frac{2mB}{(m\rho+n\sigma)\rho}\right].
\end{aligned}\right\} \tag{24}$$

By repeating the process and applying (20) and (21) at each step, we can find in succession the values of $\left(\frac{\partial^k u_i}{\partial \omega^k}\right)_{\omega=0}$ $(i=1,\ \ldots,\ 4;\ k=3,\ \ldots\ \infty)$, and thence we can readily find $\left(\frac{\partial^k x}{\partial \omega^k}\right)_{\omega=0}$ and $\left(\frac{\partial^k y}{\partial \omega^k}\right)_{\omega=0}$. We notice from (22) and (24) that the first and second partial derivatives carry the arbitrary parameter $c$ to the first and second powers respectively. We see further that, at each step in the differentiating, the right-hand members of the equations are homogeneous in the orders of the partial derivatives in each term. It follows then that $\left(\frac{\partial^k u_i}{\partial\omega^k}\right)_{\omega=0}$ and, therefore, also $\left(\frac{\partial^k x}{\partial\omega^k}\right)_{\omega=0}$ and $\left(\frac{\partial^k y}{\partial\omega^k}\right)_{k=0}$ each carry a factor $c^k$, after the values of the partial derivatives of orders lower than $k$ have been substituted in the right-hand members. Hence, the terms of the expansion in (16) carry $c$ and $\omega$ as factors to the same power. On substituting the values of the partial derivatives in Maclaurin's expansions for $x$ and $y$, and replacing $\omega$ by its value $e^{-\rho t}$, we obtain a set of solutions of equations (8) in the form

$$\left.\begin{aligned} x= &\quad ce^{-\rho t}+\frac{1}{2!}\left[\frac{3m(m^2-2)B\sigma}{(m\sigma-n\rho)(4\rho^2+\sigma^2)}-\frac{n(m^2-2)B}{(m\sigma-n\rho)\rho}-\frac{12mB\rho}{(m\rho+n\sigma)(4\rho^2+\sigma^2)}\right.\\ &\left.+\frac{4mB}{(m\rho+n\sigma)\rho}\right]c^2e^{-2\rho t}+\frac{1}{3!}[\ldots\ldots]c^3e^{-3\rho t}+\ldots\ldots,\\ y= &-mce^{-\rho t}-\frac{m}{2!}\left[\frac{6n(m^2-2)B\rho}{(m\sigma-n\rho)(4\rho^2+\sigma^2)}-\frac{2n(m^2-2)B}{(m\sigma-n\rho)\rho}+\frac{6nB\sigma}{(m\rho+n\sigma)(4\rho^2+\sigma^2)}\right.\\ &\left.+\frac{2mB}{(m\rho+n\sigma)\rho}\right]c^2e^{-2\rho t}-\frac{m}{3!}[\ldots\ldots]c^3e^{-3\rho t}+\ldots\ldots \end{aligned}\right\}\quad(25)$$

## (B) *Solutions in Powers of $e^{+\rho t}$.*

If, in equations (8), we were to transform our independent variable by writing $\omega=e^{+\rho t}$, we could build up a second set of solutions of form similar to (25), by a process exactly parallel to that used in section (A). We shall show, however, that this new set of solutions can be obtained directly from solutions (25) by changing the sign of $t$ throughout, and changing the sign of $y$ in the result.

The first two equations of (1), with $z$ dropped from the right-hand members, can be written in the form

$$\left.\begin{aligned} x''-2y'&=F_1(x, y^2),\\ y''+2x'&=yF_2(x, y^2). \end{aligned}\right\}\quad(26)$$

If we suppose the initial conditions are

$$x(0)=\alpha,\quad x'(0)=\alpha_1,\quad y(0)=\beta,\quad y'(0)=\beta_1, \tag{27}$$

then the solutions of (26) have the form

$$x=f_1(t),\quad x'=\phi_1(t),\quad y=f_2(t),\quad y'=\phi_2(t). \tag{28}$$

If in equations (26) we put $x=x$, $y=-\eta$, and $t=-\tau$, we get equations of identically the same form in $x$, $\eta$, and $\tau$ as (26) are in $x$, $y$, and $t$. These equations are

$$\left.\begin{aligned} \ddot{x}-2\dot{\eta}&=F_1(x, \eta^2),\\ \ddot{\eta}+2\dot{x}&=\eta F_2(x, \eta^2), \end{aligned}\right\}\quad(29)$$

where the dot denotes differentiation with regard to $\tau$. If we impose the same initial conditions in these new variables, viz.:

$$x(0)=\alpha,\quad \dot{x}(0)=\alpha_1,\quad \eta(0)=\beta,\quad \dot{\eta}(0)=\beta_1, \tag{30}$$

then the solutions of (29) are

$$x=f_1(\tau),\quad \dot{x}=\phi_1(\tau),\quad \eta=f_2(\tau),\quad \dot{\eta}=\phi_2(\tau); \tag{31}$$

that is $x$, $\dot{x}$, $\eta$, and $\dot{\eta}$ are the same functions of $\tau$ as $x$, $x'$, $y$ and $y'$ were of $t$ before. But the initial conditions (30) are the same as

$$x(0)=\alpha,\quad x'(0)=-\alpha_1,\quad y_0=-\beta,\quad y'(0)=\beta_1. \tag{32}$$

Again, the solutions (31) are the same as

$$x=f_1(-t),\quad x'=-\phi_1(-t),\quad y=-f_2(-t),\quad y'=\phi_2(-t).$$

It is readily seen, therefore, that for initial conditions (32) equations (8) admit of a solution which can be derived from (25) simply by changing the sign of $t$ throughout and changing the sign of $y$ in the result. These solutions, therefore, have the form

$$\left.\begin{aligned} x=&\ c_1e^{\rho t}+\frac{1}{2!}\left[\frac{3m(m^2-2)B\sigma}{(m\sigma-n\rho)(4\rho^2+\sigma^2)}-\frac{n(m^2-2)B}{(m\sigma-n\rho)\rho}-\frac{12mB\rho}{(m\rho+n\sigma)(4\rho^2+\sigma^2)}\right.\\ &\left.+\frac{4mB}{(m\rho+n\sigma)\rho}\right]c_1^2e^{2\rho t}+\frac{1}{3!}[\ldots.]c_1^3e^{3\rho t}+\ldots.,\\ y=&\ mc_1e^{\rho t}+\frac{m}{2!}\left[\frac{6n(m^2-2)B\rho}{(m\sigma-n\rho)(4\rho^2+\sigma^2)}-\frac{2n(m^2-2)B}{(m\sigma-n\rho)\rho}+\frac{6nB\sigma}{(m\rho+n\sigma)(4\rho^2+\sigma^2)}\right.\\ &\left.+\frac{2mB}{(m\rho+n\sigma)\rho}\right]c_1^2e^{2\rho t}+\frac{m}{3!}[\ldots.]c_1^3e^{3\rho t}+\ldots.. \end{aligned}\right\} \tag{33}$$

### V. *Proof of the Convergence of the Solutions.*

It is necessary to prove the convergence of the series (25), which we have deduced from equations (8), or their transformed equivalents (9). This we shall do by a method analogous to that used by Picard.* Write equations (9) in the following form

$$\left.\begin{aligned} -\rho\omega\frac{\partial u_1}{\partial\omega}-\sigma\sqrt{-1}\,u_1&=H_1^{(2)}+H_1^{(3)}+\ldots.,\\ -\rho\omega\frac{\partial u_2}{\partial\omega}+\sigma\sqrt{-1}\,u_2&=H_2^{(2)}+H_2^{(3)}+\ldots.,\\ -\rho\omega\frac{\partial u_3}{\partial\omega}-\rho u_3\qquad&=H_3^{(2)}+H_3^{(3)}+\ldots.,\\ -\rho\omega\frac{\partial u_4}{\partial\omega}+\rho u_4\qquad&=H_4^{(2)}+H_4^{(3)}+\ldots.. \end{aligned}\right\} \tag{34}$$

---

* Picard, "Traité D'Analyse," Vol. III, Chap. I, § 12.

The coefficients of any of the partial derivatives of $u_i$ $(i=1,\ldots,4)$ for $\omega=0$ are respectively of the form

$$p(-\rho)-\sigma\sqrt{-1},\quad p(-\rho)-(-\sigma\sqrt{-1}),\quad p(-\rho)-\rho,\quad p(-\rho)-(-\rho), \tag{35}$$

all of which are, in absolute value, greater than $\varepsilon(p-1)$ where $\varepsilon$ is a real quantity. In this case $\varepsilon$ is a fixed quantity smaller than the absolute value of the smallest of $\pm\rho$ and $\pm\sigma\sqrt{-1}$.

It has been shown by Moulton* that the expansions in the right-hand members of (14) converge within and on the boundary of a circle of radius $\alpha$ $(\alpha>0)$ about each of the equilibrium points; the value of $\alpha$ depending upon which of the points $(a)$, $(b)$, or $(c)$ is being considered.

Let $M$ be the maximum modulus of the expressions in the right members of (34) for all values of the variables $u_1$, $u_2$, $u_3$, $u_4$, in this circle of radius $\alpha$.

Consider the comparison set of differential equations

$$\varepsilon\left[\omega\frac{\partial v_i}{\partial\omega}-v_i\right]=\frac{M}{1-\dfrac{v_1+v_2+v_3+v_4}{\alpha}}-M-M\frac{v_1+v_2+v_3+v_4}{\alpha},\quad (i=1,\ldots,4). \tag{36}$$

It is evident that equations (36) dominate (34); and it can readily be shown that the solutions of (36) dominate the solutions of (34). The terms on the right of (36) are all positive, and it is obvious from the method by which the solutions are built up that all the terms of the solution of (36) are positive. In building up the solutions of (34) and (36) as power series in $\omega$, whose coefficients are the values of the successive derivatives of the $u_i$ and the $v_i$, respectively, for $\omega=0$, we see by (35) that the absolute value of the coefficient of any partial derivative on the left obtained from (34) is greater than the coefficient of the corresponding partial derivative on the left obtained from (36). But each term on the right-hand side of (36) dominates the corresponding term on the right of (34), and therefore any partial derivative of the right members of (36) is greater than the absolute value of the corresponding partial derivative of the right members of (34). It follows, therefore, that the values of the successive partial derivatives obtained from (36) are greater respectively than the absolute values of the corresponding partial derivatives obtained from (34). Hence, each term in the solutions of (36) is greater than the absolute value of the corresponding term in the solutions of (34); or the solutions of (36) dominate those of (34).

---

* Moulton, "Periodic Orbits," p. 154.

It remains to be shown that the solutions of (36) are convergent for $\omega$ sufficiently small. From the symmetry of (36) in the $v_i$ $(i=1, \ldots, 4)$, all the $v_i$ are equal. We can, therefore, replace equations (36) by a single equation in one variable, viz.:

$$\varepsilon\left[\omega\frac{\partial v}{\partial\omega}-v\right]=\frac{M}{1-\frac{4v}{\alpha}}-M-M\frac{4v}{\alpha},$$

which reduces to the form

$$\omega\frac{\partial v}{\partial\omega}-v=\frac{M'v^2}{\alpha-4v}, \text{ where } M'=\frac{16M}{\alpha\varepsilon}; \tag{37}$$

whence,

$$\frac{\partial\omega}{\omega}=\frac{\partial v}{v}-\frac{M'\partial v}{\alpha+(M'-4)v}.$$

On integrating this equation, we have

$$\log c\omega=\log\frac{v}{[\alpha+(M'-4)v]^{\frac{M'}{M'-4}}},$$

or

$$c\omega=\frac{v}{\left(\alpha+\frac{M'}{\varkappa}v\right)^{\varkappa}}, \text{ where } \varkappa=\frac{M'}{M'-4}.$$

Therefore,

$$v-c\omega\left(\alpha^{\varkappa}+\varkappa\alpha^{\varkappa-1}\frac{M'}{\varkappa}v+\ldots\right)=0.$$

From this it follows, by the theory of implicit functions, that $v$ can be expressed as a power series in $\omega$, vanishing with $\omega$, and convergent for $\omega$ sufficiently small. Thus, we see that the solutions of (36) converge for $\omega$ sufficiently small; and it has been shown that the solutions of (36) dominate the solutions of (34). Hence the solutions of (34) are convergent for $\omega$ sufficiently small, that is for $t$ sufficiently large.

## VI. *Alternative Method of Constructing the Solutions.*

The solutions of equations (8) may be built up by a process quite different from that used in § IV.

If, in equations (2), we make the transformation $x=x\varepsilon$, $y=y\varepsilon$, the first two equations, considered for $z=0$, give, on dividing through by $\varepsilon$,

$$\left.\begin{aligned}x''-2y'&=(1+2A)x+\tfrac{3}{2}B[-2x^2+y^2]\varepsilon+2C[2x^3-3xy^2]\varepsilon^2+\ldots,\\ y''+2x'&=(1-A)y+3Bxy\varepsilon+\tfrac{3}{2}Cy[-4x^2+y^2]\varepsilon^2+\ldots.\end{aligned}\right\} \tag{38}$$

Since, as was pointed out in § V, the right-hand members of the original equations converge within and on the boundary of a circle of radius $\alpha$ ($\alpha > 0$) about each of the equilibrium points, it follows that the right-hand members of (38) converge for $\varepsilon$ sufficiently small ($0 < \varepsilon \overline{\gtrless} 1$). Such a system of differential equations, with constant coefficients on the left, can be solved for $x$ and $y$ as power series in $\varepsilon$, of the form

$$\left. \begin{aligned} x &= \sum_{j=0}^{\infty} x_j(t)\varepsilon^j, \\ y &= \sum_{j=0}^{\infty} y_j(t)\varepsilon^j, \end{aligned} \right\} \quad (39)$$

where the coefficients $x_j$ and $y_j$ are to be determined. Since (39) are the solutions of (38), the latter must be satisfied identically when $x$ and $y$ are replaced by the power series in (38). When we make this substitution, from terms independent of $\varepsilon$, we have

$$\left. \begin{aligned} x_0'' - 2y_0' - (1+2A)x_0 &= 0, \\ y_0'' + 2x_0' - (1-\ A)y_0 &= 0. \end{aligned} \right\} \quad (40)$$

The general solution of these equations is

$$\left. \begin{aligned} x_0 &= K_1 e^{\sigma\sqrt{-1}t} + K_2 e^{-\sigma\sqrt{-1}t} + K_3 e^{\rho t} + K_4 e^{-\rho t}, \\ y_0 &= L_1 e^{\sigma\sqrt{-1}t} + L_2 e^{-\sigma\sqrt{-1}t} + L_3 e^{\rho t} + L_4 e^{-\rho t}, \end{aligned} \right\} \quad (41)$$

where the $K_i$ and $L_i$ are constants of integration, the $K_i$ ($i=1, \ldots, 4$) being arbitrary, that is dependent upon the initial conditions imposed which are arbitrary, and the $L_i$ depending on the $K_i$, the relations between them being *

$$\left. \begin{aligned} L_1 &= \sqrt{-1}\,\frac{\sigma^2+1+2A}{2\sigma}K_1 = \sqrt{-1}\,nK_1 = -\frac{K_1}{K_2}L_2, \\ L_3 &= \frac{\rho^2-1-2A}{2\rho}K_3 = mK_3 = -\frac{K_3}{K_4}L_4. \end{aligned} \right\} \quad (42)$$

Since the $K_i$ are arbitrary, and since we are seeking solutions which vanish as $t$ approaches infinity, we choose $K_1 = K_2 = K_3 = 0$, and, therefore, also $L_1 = L_2 = L_3 = 0$. Equations (40) then have a particular solution of the form

$$x_0 = K_4 e^{-\rho t}, \quad y_0 = L_1 e^{-\rho t}. \quad (43)$$

Let us impose the initial conditions that $x = b$ at $t = 0$. Since

$$x(t) \underset{\varepsilon}{=} x_0(t) + x_1(t)\varepsilon + x_2(t)\varepsilon^2 + \ldots,$$

then

$$x_0(0) = b, \quad x_j(0) = 0, \quad (j = 1, \ldots \infty). \quad (44)$$

* Moulton, "Periodic Orbits," p. 159, Eqns. (28).

On imposing these initial conditions and using (42), we see that (43) takes the form

$$x_0=be^{-\rho t}, \quad y_0=-mbe^{-\rho t}. \tag{45}$$

In finding successively the values of $x_j$ and $y_j$ $(j=1, \ldots . \infty)$ it is more convenient to use the normal form of equations (26).

As we saw in § 2, the transformations

$$\left.\begin{aligned} x&=u_1+u_2+u_3+u_4,\\ x'&=\sigma\sqrt{-1}\,(u_1-u_2)+\rho(u_3-u_4),\\ y&=n\sqrt{-1}\,(u_1-u_2)+m(u_3-u_4),\\ y'&=-n\sigma(u_1+u_2)+m\rho(u_3+u_4), \end{aligned}\right\} \tag{46}$$

change equations (26) into the normal form

$$\left.\begin{aligned} u_1'-\sigma\sqrt{-1}\,u_1&=\frac{m[\ldots.]\varepsilon}{2(m\sigma-n\rho)\sqrt{-1}}-\frac{\{\ldots.\}\varepsilon}{2(m\rho+n\sigma)},\\ u_2'+\sigma\sqrt{-1}\,u_2&=-\frac{m[\ldots.]\varepsilon}{2(m\sigma-n\rho)\sqrt{-1}}-\frac{\{\ldots.\}\varepsilon}{2(m\rho+n\sigma)},\\ u_3'-\rho u_3&=\frac{-n[\ldots.]\varepsilon}{2(m\sigma-n\rho)}+\frac{\{\ldots.\}\varepsilon}{2(m\rho+n\sigma)},\\ u_4'+\rho u_4&=\frac{n[\ldots.]\varepsilon}{2(m\sigma-n\rho)}+\frac{\{\ldots.\}\varepsilon}{2(m\rho+n\sigma)}, \end{aligned}\right\} \tag{47}$$

where

$$\left.\begin{aligned} [\ldots.]&=\tfrac{3}{2}B[-2x^2+y^2]+2C[2x^3-3xy^2]\varepsilon+\text{ terms in } \varepsilon^2, \varepsilon^3, \ldots.,\\ \{\ldots.\}&=3B\{xy\}+\tfrac{3}{2}Cy\{-4x^2+y^2\}\varepsilon+\text{ terms in } \varepsilon^2, \varepsilon^3, \ldots.. \end{aligned}\right\} \tag{48}$$

These equations have solutions of the form

$$u_i=\sum_{j=0}^{\infty}u_i^{(j)}\varepsilon^j, \qquad (i=1, \ldots., 4).$$

To determine the $u_i^{(j)}$ we substitute (49) in (47) and equate coefficients of like powers of $\varepsilon$ on both sides of the resulting equations. From the terms independent of $\varepsilon$, by applying (46) and (44) we get the value of $x_0$ and $y_0$ as given in (45). Since, by (44),

$$x_0(0)=b, \text{ and } x_j(0)=0, \qquad (j=1, \ldots.\infty), \text{ and } x_j=\sum_{i=1}^{4}u_i^{(j)}, \tag{49}$$

it follows that

$$\left.\begin{aligned} &u_1^{(0)}(0)+u_2^{(0)}(0)+u_3^{(0)}(0)+u_4^{(0)}(0)=b,\\ &u_1^{(j)}(0)+u_2^{(j)}(0)+u_3^{(j)}(0)+u_4^{(j)}(0)=0, \qquad j=(1, \ldots.\infty). \end{aligned}\right\} \tag{50}$$

On equating coefficients of the first power of $\varepsilon$, we have

$$
\left.
\begin{aligned}
u_1^{(1)\prime}-\sigma\sqrt{-1}\,u_1^{(1)} &= \frac{3mB[-2x_0^2+y_0^2]}{4(m\sigma-n\rho)\sqrt{-1}}-\frac{3Bx_0y_0}{2(m\rho+n\sigma)},\\
u_2^{(1)\prime}+\sigma\sqrt{-1}\,u_2^{(1)} &= -\frac{3mB[-2x_0^2+y_0^2]}{4(m\sigma-n\rho)\sqrt{-1}}-\frac{3Bx_0y_0}{2(m\rho+n\sigma)},\\
u_3^{(1)\prime}-\rho u_3^{(1)} &= -\frac{3nB[-2x_0^2+y_0^2]}{4(m\sigma-n\rho)}+\frac{3Bx_0y_0}{2(m\rho+n\sigma)},\\
u_4^{(1)\prime}+\rho u_4^{(1)} &= \frac{3nB[-2x_0^2+y_0^2]}{4(m\sigma-n\rho)}+\frac{3Bx_0y_0}{2(m\rho+n\sigma)}.
\end{aligned}
\right\}\quad(51)
$$

After substituting for $x_0$ and $y_0$ their values from (33) we have

$$
\left.
\begin{aligned}
u_1^{(1)\prime}-\sigma\sqrt{-1}\,u_1^{(1)} &= \left\{\frac{3mB(m^2-2)}{4(m\sigma-n\rho)\sqrt{-1}}+\frac{3mB}{2(m\rho+n\sigma)}\right\}b^2e^{-2\rho t}\equiv M_{12}^{(1)}b^2e^{-2\rho t},\\
u_2^{(1)\prime}+\sigma\sqrt{-1}\,u_2^{(1)} &= \left\{-\frac{3mB(m^2-2)}{4(m\sigma-n\rho)\sqrt{-1}}+\frac{3mB}{2(m\rho+n\sigma)}\right\}b^2e^{-2\rho t}\equiv M_{22}^{(1)}b^2e^{-2\rho t},\\
u_3^{(1)\prime}-\rho u_3^{(1)} &= \left\{-\frac{3nB(m^2-2)}{4(m\sigma-n\rho)}-\frac{3mB}{2(m\rho+n\sigma)}\right\}b^2e^{-2\rho t}\equiv M_{32}^{(1)}b^2e^{-2\rho t},\\
u_4^{(1)\prime}+\rho u_4^{(1)} &= \left\{\frac{3nB(m^2-2)}{4(m\sigma-n\rho)}-\frac{3mB}{2(m\rho+n\sigma)}\right\}b^2e^{-2\rho t}\equiv M_{42}^{(1)}b^2e^{-2\rho t},
\end{aligned}
\right\}\quad(52)
$$

where the $M_{i2}^{(1)}$ $(i=1,\ldots,4)$ denote the expressions in the brackets. The superscript and the first subscript on the $M_{i2}^{(1)}$ are the same as those on the corresponding $u_i^{(1)}$, while the second subscript denotes that they are coefficients of $e^{-2\rho t}$. On integrating (52), we get

$$
\left.
\begin{aligned}
u_1^{(1)} &= c_1^{(1)}e^{\sigma\sqrt{-1}t}-\frac{M_{12}^{(1)}b^2}{2\rho+\sigma\sqrt{-1}}e^{-2\rho t}\equiv c_1^{(1)}e^{\sigma\sqrt{-1}t}-d_{12}^{(1)}b^2e^{-2\rho t},\\
u_2^{(1)} &= c_2^{(1)}e^{-\sigma\sqrt{-1}t}-\frac{M_{22}^{(1)}b^2}{2\rho-\sigma\sqrt{-1}}e^{-2\rho t}\equiv c_2^{(1)}e^{-\sigma\sqrt{-1}t}-d_{22}^{(1)}b^2e^{-2\rho t},\\
u_3^{(1)} &= c_3^{(1)}e^{\rho t}-\frac{M_{32}^{(1)}b^2}{3\rho}e^{-2\rho t}\equiv c_3^{(1)}e^{\rho t}-d_{32}^{(1)}b^2e^{-2\rho t},\\
u_4^{(1)} &= c_4^{(1)}e^{-\rho t}-\frac{M_{42}^{(1)}b^2}{\rho}e^{-2\rho t}\equiv c_4^{(1)}e^{-\rho t}-d_{42}^{(1)}b^2e^{-2\rho t}.
\end{aligned}
\right\}\quad(53)
$$

If then we put $c_1^{(1)}=c_2^{(1)}=c_3^{(1)}=0$ and choose $c_4^{(1)}$ so that the initial conditions (50) are satisfied, we find that

$$c_4^{(1)}=b^2[d_{12}^{(1)}+d_{22}^{(1)}+d_{32}^{(1)}+d_{42}^{(1)}].$$

Since

$$\left.\begin{aligned} x_j&=u_1^{(j)}+u_2^{(j)}+u_3^{(j)}+u_4^{(j)},\\ y_j&=n\sqrt{-1}(u_1^{(j)}-u_2^{(j)})+m(u_3^{(j)}-u_4^{(j)}),\end{aligned}\right\}\quad(54)$$

we see that

$$\left.\begin{aligned} x_1&=[(d_{12}^{(1)}+d_{22}^{(1)}+d_{32}^{(1)}+d_{42}^{(1)})e^{-\rho t}-(d_{12}^{(1)}+d_{22}^{(1)}+d_{32}^{(1)}+d_{42}^{(1)})e^{-2\rho t}]b^2\\ &\qquad\equiv[D_2^{(1)}e^{-\rho t}-D_2^{(1)}e^{-2\rho t}]b^2,\\ y_1&=[-m(d_{12}^{(1)}+d_{22}^{(1)}+d_{32}^{(1)}+d_{42}^{(1)})e^{-\rho t}-\{n\sqrt{-1}(d_{12}^{(1)}-d_{22}^{(1)})\\ &\qquad+m(d_{32}^{(1)}-d_{42}^{(1)})\}e^{-2\rho t}]b^2\equiv[-mD_2^{(1)}e^{-\rho t}-\overline{D}_2^{(1)}e^{-2\rho t}]b^2.\end{aligned}\right\}\quad(55)$$

From the coefficients of $\varepsilon^2$ above, we see that the differential equations which define the $u_i^{(2)}$ $(i=1,\ldots,4)$ are

$$\left.\begin{aligned} u_1^{(2)\prime}-\sigma\sqrt{-1}\,u_1^{(2)}&=\frac{3mB(-2x_0x_1+y_0y_1)}{2(m\sigma-n\rho)\sqrt{-1}}-\frac{3B(x_0y_1+x_1y_0)}{2(m\rho+n\sigma)}\\ &\quad+\frac{mC(2x_0^3-3x_0y_0^2)}{(m\sigma-n\rho)\sqrt{-1}}+\frac{3C(4x_0^2y_0-y_0^3)}{4(m\rho+n\sigma)},\\ u_2^{(2)\prime}+\sigma\sqrt{-1}\,u_2^{(2)}&=-\frac{3mB(-2x_0x_1+y_0y_1)}{2(m\sigma-n\rho)\sqrt{-1}}-\frac{3B(x_0y_1+x_1y_0)}{2(m\rho+n\sigma)}\\ &\quad-\frac{mC(2x_0^3-3x_0y_0^2)}{(m\sigma-n\rho)\sqrt{-1}}+\frac{3C(4x_0^2y_0-y_0^3)}{4(m\rho+n\sigma)},\\ u_3^{(2)\prime}-\rho u_3^{(2)}&=-\frac{3nB(-2x_0y_0+y_0y_1)}{2(m\sigma-n\rho)}+\frac{3B(x_0y_1+x_1y_0)}{2(m\rho+n\sigma)}\\ &\quad-\frac{nC(2x_0^3-3x_0y_0^2)}{m\sigma-n\rho}-\frac{3C(4x_0^2y_0-y_0^3)}{4(m\rho+n\sigma)},\\ u_4^{(2)\prime}+\rho u_4^{(2)}&=+\frac{3nB(-2x_0y_0+y_0y_1)}{2(m\sigma-n\rho)}+\frac{3B(x_0y_1+x_1y_0)}{2(m\rho+n\sigma)}\\ &\quad+\frac{nC(2x_0^3-3x_0y_0^2)}{m\sigma-n\rho}-\frac{3C(4x_0^2y_0-y_0^3)}{4(m\rho+n\sigma)}.\end{aligned}\right\}\quad(56)$$

On substituting for $x_0$, $y_0$, $x_1$, $y_1$ their values from (45) and (55) these equations take the form

$$\left.\begin{aligned} u_1^{(2)\prime}-\sigma\sqrt{-1}\,u_1^{(2)}&=M_{12}^{(2)}b^3e^{-2\rho t}+M_{13}^{(2)}b^3e^{-3\rho t},\\ u_2^{(2)\prime}+\sigma\sqrt{-1}\,u_2^{(2)}&=M_{22}^{(2)}b^3e^{-2\rho t}+M_{23}^{(2)}b^3e^{-3\rho t},\\ u_3^{(2)\prime}-\rho u_3^{(2)}&=M_{32}^{(2)}b^3e^{-2\rho t}+M_{33}^{(2)}b^3e^{-3\rho t},\\ u_4^{(2)\prime}+\rho u_4^{(2)}&=M_{42}^{(2)}b^3e^{-2\rho t}+M_{43}^{(2)}b^3e^{-3\rho t},\end{aligned}\right\}\quad(57)$$

where

$$
\left.\begin{aligned}
M_{12}^{(2)} &= \frac{3mB(m^2-2)D_2^{(1)}}{2(m\sigma-n\rho)\sqrt{-1}}+\frac{6mBD_2^{(1)}}{2(m\rho+n\sigma)}\,;\\
M_{22}^{(2)} &= -\frac{3mB(m^2-2)D_2^{(1)}}{2(m\sigma-n\rho)\sqrt{-1}}+\frac{6mBD_2^{(1)}}{2(m\rho+n\sigma)}\,;\\
M_{32}^{(2)} &= -\frac{3nB(m^2-2)D_2^{(1)}}{2(m\sigma-n\rho)}-\frac{6mBD_2^{(1)}}{2(m\rho+n\sigma)}\,;\\
M_{42}^{(1)} &= \frac{3nB(m^2-2)D_2^{(1)}}{2(m\sigma-n\rho)}-\frac{6mBD_2^{(1)}}{2(m\rho+n\sigma)}\,;\\
M_{13}^{(2)} &= \frac{3mB(2D_2^{(1)}+m\overline{D}_2^{(1)})}{2(m\sigma-n\rho)\sqrt{-1}}-\frac{3B(mD_2^{(1)}-\overline{D}_2^{(1)})}{2(m\rho+n\sigma)}-\frac{mC(3m^2-2)}{(m\sigma-n\rho)\sqrt{-1}}\\
&\qquad+\frac{3mC(m^2-4)}{4(m\rho+n\sigma)}\,;\\
M_{23}^{(2)} &= -\frac{3mB(2D_2^{(1)}+m\overline{D}_2^{(1)})}{2(m\sigma-n\rho)\sqrt{-1}}-\frac{3B(mD_2^{(1)}-\overline{D}_2^{(1)})}{2(m\rho+n\sigma)}+\frac{mC(3m^2-2)}{(m\sigma-n\rho)\sqrt{-1}}\\
&\qquad+\frac{3mC(m^2-4)}{4(m\rho+n\sigma)}\,;\\
M_{33}^{(2)} &= -\frac{3nB(2D_2^{(1)}+m\overline{D}_2^{(1)})}{2(m\sigma-n\rho)}+\frac{3B(mD_2^{(1)}-\overline{D}_2^{(1)})}{2(m\rho+n\sigma)}+\frac{nC(3m^2-2)}{m\sigma-n\rho}\\
&\qquad-\frac{3mC(m^2-4)}{4(m\rho+n\sigma)}\,;\\
M_{43}^{(2)} &= +\frac{3nB(2D_2^{(1)}+m\overline{D}_2^{(1)})}{2(m\sigma-n\rho)}+\frac{3B(mD_2^{(1)}-\overline{D}_2^{(1)})}{2(m\rho+n\sigma)}-\frac{nC(3m^2-2)}{m\sigma-n\rho}\\
&\qquad-\frac{3mC(m^2-4)}{4(m\rho+n\sigma)}\,.
\end{aligned}\right\}\quad(58)
$$

The solutions of (57) have the form

$$
\left.\begin{aligned}
u_1^{(2)} &= c_1^{(2)}e^{\sigma\sqrt{-1}t}-d_{12}^{(2)}b^2e^{-2\rho t}-d_{13}^{(2)}b^3e^{-3\rho t},\\
u_2^{(2)} &= c_2^{(2)}e^{-\sigma\sqrt{-1}t}-d_{22}^{(2)}b^3e^{-2\rho t}-d_{23}^{(2)}b^3e^{-3\rho t},\\
u_3^{(2)} &= c_3^{(2)}e^{\rho t}-d_{32}^{(2)}b^3e^{-2\rho t}-d_{33}^{(2)}b^3e^{-3\rho t},\\
u_4^{(2)} &= c_4^{(2)}e^{-\rho t}-d_{42}^{(2)}b^3e^{-2\rho t}-d_{43}^{(2)}b^3e^{-3\rho t},
\end{aligned}\right\}\quad(59)
$$

where

$$
\left.\begin{aligned}
d_{12}^{(2)} &= \frac{M_{12}^{(2)}}{2\rho+\sigma\sqrt{-1}}\,;\quad d_{22}^{(2)}=\frac{M_{22}^{(2)}}{2\rho-\sigma\sqrt{-1}}\,;\quad d_{32}^{(2)}=\frac{M_{32}^{(2)}}{3\rho}\,;\quad d_{42}^{(2)}=\frac{M_{42}^{(2)}}{\rho}\,;\\
d_{13}^{(2)} &= \frac{M_{13}^{(2)}}{3\rho+\sigma\sqrt{-1}}\,;\quad d_{23}^{(2)}=\frac{M_{23}^{(2)}}{3\rho-\sigma\sqrt{-1}}\,;\quad d_{33}^{(2)}=\frac{M_{33}^{(2)}}{4\rho}\,;\quad d_{43}^{(2)}=\frac{M_{43}^{(2)}}{2\rho}\,.
\end{aligned}\right\}\quad(60)
$$

If we put $c_1^{(2)}=c_2^{(2)}=c_3^{(2)}=0$ and determine $c_4^{(2)}$ to satisfy the initial conditions (50), we have

$$c_4^{(2)}=[\sum_{i=1}^{4} d_{i2}^{(2)}+\sum_{i=1}^{4} d_{i3}^{(2)}]b^3.$$

On substituting these values for $c_i^{(2)}$ $(i=1, \ldots, 4)$ in (59), and writing

$$\left.\begin{aligned}&\sum_{i=1}^{4} d_{i2}^{(2)}=D_2^{(2)}; \quad n\sqrt{-1}(d_{12}^{(2)}-d_{22}^{(2)})+m(d_{32}^{(2)}-d_{42}^{(2)})=\bar{D}_2^{(2)},\\ &\sum_{i=1}^{4} d_{i3}^{(2)}=D_3^{(2)}; \quad n\sqrt{-1}(d_{13}^{(2)}-d_{23}^{(2)})+m(d_{33}^{(2)}-d_{43}^{(2)})=\bar{D}_3^{(2)},\end{aligned}\right\} \quad (61)$$

we have, by (54),

$$\left.\begin{aligned}x_2&=[(D_2^{(2)}+D_3^{(2)})e^{-\rho t}-D_2^{(2)}e^{-2\rho t}-D_3^{(2)}e^{-3\rho t}]b^3,\\ y_2&=[-m(D_2^{(2)}+D_3^{(2)})e^{-\rho t}-\bar{D}_2^{(2)}e^{-2\rho t}-\bar{D}_3^{(2)}e^{-3\rho t}]b^3.\end{aligned}\right\} \quad (62)$$

If we proceed in this way, we can build up in succession the values of $x_j$ and $y_j$ for $j=3, 4, \ldots \infty$. By induction we can get the form of the general term. From (45), (55), (62) we notice that the $x_j$ and $y_j$ are sums of powers of $e^{-\rho t}$, the highest power of $e^{-\rho t}$ occurring being $j+1$. The equations defining $u_i^{(\nu)}$ $(i=1, \ldots, 4)$ have in their right-hand members only sums of powers of $e^{-\rho t}$, the lowest power being $e^{-2\rho t}$ and the highest $e^{-(\nu+1)\rho t}$. When we integrate these equations we will have terms in powers of $e^{-\rho t}$ the lowest power being the first and the highest the $(\nu+1)$-th. These solutions will have the form

$$u_i^{(\nu)}=c_i^{(\nu)}e^{\lambda t}-[\sum_{j=2}^{\nu+1} d_{ij}^{(\nu)}e^{-j\rho t}]b^{\nu+1}, \qquad (i=1, \ldots, 4;$$
$$\lambda=\sigma\sqrt{-1}, -\sigma\sqrt{-1}, \rho, -\rho). \qquad (63).$$

If we put $c_1^{(\nu)}=c_2^{(\nu)}=c_3^{(\nu)}=0$, then to satisfy the initial conditions we must put $c_4^{(\nu)}=b^{\nu+1}\sum_{i=1}^{4}\sum_{j=2}^{\nu+1} d_{ij}^{(\nu)}$. If then we put

$$\left.\begin{aligned}&D_j^{(\nu)}=\sum_{i=1}^{4} d_{ij}^{(\nu)}, \qquad (j=2, \ldots, \nu+1),\\ &\bar{D}_j^{(\nu)}=n\sqrt{-1}(d_{1j}^{(\nu)}-d_{2j}^{(\nu)})+m(d_{3j}^{(\nu)}-d_{4j}^{(\nu)}), \quad (j=2, \ldots, \nu+1),\end{aligned}\right\} \quad (64)$$

it follows that

$$\left.\begin{aligned}x_\nu&=[\sum_{j=2}^{\nu+1} D_j^{(\nu)}e^{-\rho t}-\sum_{k=2}^{\nu+1} D_k^{(\nu)}e^{-k\rho t}]b^{\nu+1},\\ y_\nu&=[-m\sum_{j=2}^{\nu+1} D_j^{(\nu)}e^{-\rho t}-\sum_{k=2}^{\nu+1} \bar{D}_k^{(\nu)}e^{-k\rho t}]b^{\nu+1}.\end{aligned}\right\} \quad (65)$$

When we substitute the results of (45), (55), (62), (65) in (39), we have the values of $x$ and $y$ which are solutions of (38). The $x$ and $y$ belonging to the physical problem as defined by (8) are $\epsilon$ times the $x$ and $y$, respectively, which we have just obtained from (38). On multiplying the values of each $x_j$ and $y_j$ by $\epsilon$ and substituting in (39), we see that the resulting expressions carry $b$ and $\epsilon$

as factors to the same power in each term. Hence $b\epsilon$ is equivalent to a single parameter and may be replaced by $\beta$. Our solutions then may be written

$$\left.\begin{aligned} x&=e^{-\rho t}\beta+[D_2^{(1)}e^{-\rho t}-D_2^{(1)}e^{-2\rho t}]\beta^2+[\sum_{k=2}^{3}D_k^{(2)}e^{-\rho t}-\sum_{k=2}^{3}D_k^{(2)}e^{-k\rho t}]\beta^3+\ldots\\ &\qquad+[\sum_{k=2}^{\nu+1}D_k^{(\nu)}e^{-\rho t}-\sum_{k=2}^{\nu+1}D_k^{(\nu)}e^{-k\rho t}]\beta^{\nu+1}+\ldots,\\ y&=-me^{-\rho t}\beta-[mD_2^{(1)}e^{-\rho t}+\overline{D}_2^{(1)}e^{-2\rho t}]\beta^2-[m\sum_{k=2}^{3}D_k^{(2)}e^{-\rho t}+\sum_{k=2}^{3}\overline{D}_k^{(2)}e^{-k\rho t}]\beta^3\\ &\qquad-\ldots-[m\sum_{k=2}^{\nu+1}D_k^{(\nu)}e^{-\rho t}+\sum_{k=2}^{\nu+1}\overline{D}_k^{(\nu)}e^{-k\rho t}]\beta^{\nu+1}-\ldots \end{aligned}\right\} \quad (66)$$

If, instead of taking $x(0)=b$ as our initial conditions, we had taken

$$x(0)=b-D_2^{(1)}b^2\epsilon-D_3^{(2)}b^3\epsilon^2-D_4^{(3)}b^4\epsilon^3-\ldots,$$

we would have had

$$x_0(0)=b,\ x_j(0)=-D_{j+1}^{(j)}b^{j+1}.$$

Then at the successive steps the constants of integration

$$c_4^{(1)}=c_4^{(2)}=c_4^{3}=\ldots c_4^{(j)}=\ldots=0.$$

The $u_i^{(j)}$ ($i=1,\ldots,4$; $j=0,\ldots\infty$), instead of being expressed as sums of powers in $e^{-\rho t}$, would each be expressed as a single term in $e^{-(j+1)\rho t}$. The solutions (66) would then have the form

$$\left.\begin{aligned} x&=\quad e^{-\rho t}\beta-D_2^{(1)}e^{-2\rho t}\beta^2-D_3^{(2)}e^{-3\rho t}\beta^3-\ldots-D_{\nu+1}^{(\nu)}e^{-(\nu+1)\rho t}\beta^{\nu+1}-\ldots,\\ y&=-me^{-\rho t}\beta-\overline{D}_2^{(1)}e^{-2\rho t}\beta^2-\overline{D}_3^{(2)}e^{-3\rho t}\beta^3-\ldots-\overline{D}_{\nu+1}^{(\nu)}e^{-(\nu+1)\rho t}\beta^{\nu+1}-\ldots \end{aligned}\right\} \quad (68)$$

If we compute the values of the coefficients in these expansions in (68) and write $c$ for $\beta$, we find that solutions (68) are identical with solutions (25) of § IV.

By a method exactly analogous to that just used we could build up a set of solutions arranged in ascending powers of $e^{+\rho t}$, and with a proper choice of initial conditions it could be shown that they were identical with solutions (33) of § IV.

## VII. *Properties of the Orbits.*

When we consider the solutions of the differential equations as given by equations (25) of § IV, which are in the form of power series in $e^{-\rho t}$, we see that the values of $x$ and $y$ continually decrease as $t$ becomes greater and greater, and finally $x$ and $y$ approach the value zero as $t$ becomes infinitely great. Since $x$ and $y$ are the coordinates of the infinitesimal body referred to an equilibrium point as origin, it follows that it will approach nearer and nearer to one of the equilibrium points as $t$ increases. Such orbits are said to be asymptotic to these points.

Similarly, we see that if the infinitesimal body were moving on one of the orbits which are given by equations (33) of § IV, where $x$ and $y$ are expressed in the form of power series in $e^{+\rho t}$, it would approach one of the equilibrium points when $t$ became infinitely large and negative. In other words, if we imagine the infinitesimal body to be placed at one of the equilibrium points, it would gradually leave it on one of these orbits, requiring, however, an infinite time to describe the first small part of the orbit.

(A) *Meaning of the Parameter c.*

When the masses of the finite bodies are given, the only arbitrary in solutions (25) is $c$, where $c=\left(\frac{\partial \zeta}{\partial \omega}\right)_{\omega=0}$. If $c$ is fixed, the $x$ and $y$ are determined uniquely by (25) for all values of $t$ so large that the convergence of the solutions as power series holds. Now the slope of the orbit at any time $t$, for $t$ sufficiently large, is given by

$$\frac{dy}{dx}=\frac{dy}{dt}\Big/\frac{dx}{dt}=\frac{m\rho ce^{-\rho t}+2\rho\bar{D}_2^{(1)}c^2e^{-2\rho t}+3\rho\bar{D}_3^{(2)}c^3e^{-3\rho t}+\ldots\ldots}{-\rho ce^{-\rho t}+2\rho D_2^{(1)}c^2e^{-2\rho t}+3\rho D_3^{(2)}c^3e^{-3\rho t}+\ldots\ldots}. \tag{69}$$

$=-m+$ a power series in $e^{-\rho t}$ beginning with the term in $e^{-\rho t}$. Therefore,

$$\left(\frac{dy}{dx}\right)_{t\doteq\infty}=-m. \tag{70}$$

Hence the direction of approach to the equilibrium point is independent of the value of the arbitrary parameter $c$.

The position of the body in its orbit at the initial time, $t=0$, is given as a power series in $c$ with constant coefficients; so that if $c$ is fixed the position of the body in its orbit at the initial time is determined. It follows therefore that the direction of approach to the equilibrium points is independent of the initial position of the infinitesimal body in its orbit at the initial time. Since $c$ is arbitrary, nothing of generality will be lost in the actual construction of such orbits if we take $c$ equal to unity, and the numerical computation will be simplified. This we do in the numerical computation of an orbit in § VIII.

(B) *The Number and Position of the Asymptotic Orbits.*

In equations (42) of § VI the quantity $m$ is defined

$$m=\frac{\rho^2-1-2A}{2\rho}.$$

It will be shown later* that for each of the equilibrium points ($a$), ($b$), ($c$) the quantity $m$ is negative for all values of $\mu$, where $0 \gtrless \mu \leqq \frac{1}{2}$. Therefore, in the case of the orbits given by (25), it follows that $\left(\frac{dy}{dx}\right)_{t\doteq\infty}=$ a positive quantity. Hence equations (25) represent two orbits, one in the first and one in the third

* See § VII (C).

quadrant in the neighborhood of each of the equilibrium points. For any preassigned value of $\mu$ these two orbits are equally inclined to the $x$-axis in the neighborhood of the origin.

Similarly, it may be shown that equations (33) represent two orbits leaving each of the equilibrium points, one in the second and one in the fourth quadrant. For any given value of $\mu$, these orbits are, in the neighborhood of the origin, the images by reflection on the $x$-axis of the orbits given by (25).

Near the points of equilibrium, that is for $t$ very large, the first terms in the expansion of (25) are the most important and determine the sign of the right-hand members. It follows therefore from (25), since $m$ is negative, that if $c$ is positive $x$ and $y$ are both positive, but if $c$ is negative $x$ and $y$ are both negative. Hence for positive values of $c$ the orbit is in the first quadrant, and for negative values of $c$ the orbit is in the third quadrant in the neighborhood of the equilibrium points.

Similarly, when $t$ is very large and negative, the first terms in the expansion in (20) determine the sign of the right-hand members. For $c$ positive $x$ is positive and $y$ negative, and for $c$ negative $x$ is negative and $y$ positive. Hence for positive values of $c$ the orbit is in the fourth quadrant, and for negative values of $c$ the orbit is in the first quadrant.

The value of $\frac{dy}{dx}$ for a point near the origin is given by (69). If we take the second derivative

$$\frac{d^2y}{dx^2}=\frac{1}{2}\frac{d\left(\frac{dy}{dx}\right)^2}{dt}\Bigg/\frac{dy}{dt}$$

we find that

$$\frac{d^2y}{dx^2}=-2\,[\bar{D}_2^{(1)}+mD_2^{(1)}]+\text{ a power series in } e^{-pt}.$$

Hence the value of $\left(\frac{d^2y}{dx^2}\right)_{t=\infty}$ is independent of $c$, and therefore does not change sign with $c$. Therefore, near the equilibrium points, the orbits in the first and third quadrants lie on the same side of their common tangent line $y+mx=0$. Similarly, it can be shown that the orbits in the second and fourth quadrants lie upon the same side of their common tangent, $y-mx=0$.

(C) *Variation in the Direction of Approach as $\mu$ Varies from Zero to* 1/2.

Now it has been shown* that if $r_1$ and $r_2$ denote the distances of the infinitesimal body from the finite masses $1-\mu$ and $\mu$ respectively, the values of $r_1$ and $r_2$ for the equilibrium points can be expressed, for $\mu$ sufficiently small,

* Moulton, "Introduction to Celestial Mechanics" (New Edition), § 158.

as convergent power series in $\mu^{1/3}$ in the case of the points $(a)$ and $(b)$, and in $\mu$ in the case of the point $(c)$.

For the point $(a)$,

$$r_2=\left(\frac{\mu}{3}\right)^{1/3}+\frac{1}{3}\left(\frac{\mu}{3}\right)^{2/3}-\frac{1}{9}\left(\frac{\mu}{3}\right)^{3/3}\ldots,$$

$$r_1=1+r_2.$$

For the point $(b)$,

$$r_2=\left(\frac{\mu}{3}\right)^{1/3}-\frac{1}{3}\left(\frac{\mu}{3}\right)^{2/3}-\frac{1}{9}\left(\frac{\mu}{3}\right)^{3/3}\ldots,$$

$$r_1=1-r_2.$$

For the point $(c)$,

$$r_2=2-\frac{7}{12}\mu-\frac{23\times 7^2}{12^4}\mu^3-\ldots,$$

$$r_1=r_2-1,$$

$$A=\frac{1-\mu}{r_1^{(0)3}}+\frac{\mu}{r_2^{(0)3}}\ ;\quad m=\frac{\rho^2-1-2A}{2\rho}, \tag{72}$$

we have seen further that $-m$ gives the value of the tangent made by the curve with the positive $x$-axis in the neighborhood of one of the equilibrium points.

The following table has been constructed to show how the elements $r_1$, $r_2$, $A$, and $-m$ vary as the ratio of the finite masses changes from zero to one.

| $\mu$ | $\mu \doteq 0$ | | | $\mu \doteq 0.5$ | | |
|---|---|---|---|---|---|---|
| Point | $(a)$ | $(b)$ | $(c)$ | $(a)$ | $(b)$ | $(c)$ |
| $r_2$ | $\doteq\left(\frac{\mu}{3}\right)^{1/3}$ | $\doteq\left(\frac{\mu}{3}\right)^{1/3}$ | $2-\frac{7}{12}\mu$ | 0.63273 | 0.4308 | 1.700 |
| $r_1$ | $\doteq 1$ | $\doteq 1$ | $1-\frac{7}{12}\mu$ | 1.63273 | 0.5692 | 0.700 |
| $A$ | $\doteq 4$ | $\doteq 4$ | $1+\frac{7}{8}\mu\equiv 1+\varepsilon$ | 2.0886 | 8.9654 | 1.5595 |
| $\rho$ | $\doteq 2.5083$ | $\doteq 2.5083$ | $\sqrt{2\varepsilon}$ | 1.5564 | 4.0305 | 1.1469 |
| $-m$ | $\doteq 0.56224$ | $\doteq 0.56224$ | $\doteq\infty$ | 0.88498 | 0.3324 | 1.2232 |
| $\phi$ | $\doteq 29°\ 20'$ | $\doteq 29°\ 20'$ | $=90°$ | 41° 30′ | 18° 23′ | 50° 44′ |

From the above table we see that as $\mu$ increases from zero to 1/2
for $(a)$, $\phi$ increases from 29° 20′ to 41° 30′ approximately;
for $(b)$, $\phi$ decreases from 29° 20′ to 18° 23′ approximately;
for $(c)$, $\phi$ decreases from 90° 00′ to 50° 44′ approximately.

## VIII. *Continuation of the Orbits beyond the Range of Convergence of the Solutions of the Differential Equations.*

It has been pointed out that the expansions given in (25) and (33) define orbits asymptotic to the equilibrium points for $t$ sufficiently large. It remains to be shown what becomes of the infinitesimal body for smaller values of $t$.

### (A) *Method of Mechanical Quadratures.*

If the $x$- and $y$-coordinates of the moving body be computed for a sufficient number of equidistant values of the time then the coordinates of the body for the next equidistant value of the time may be computed by the method of mechanical quadratures. If, for example, we know the values of $x_1y_1$, $x_2y_2$, $x_3y_3$, $x_4y_4$, $x_5y_5$ at five equidistant values of the time $t_1$, $t_2$, $t_3$, $t_4$, $t_5$ then we shall show how the value of $x_6$ and $y_6$ for the time $t_6$ can be computed, where

$$t_6-t_5=t_5-t_4=\ldots\ldots=t_2-t_1.$$

Form a table of the values of $x$ and their successive differences for the successive values of the time. Form also similar tables for $y$, $x'$, $y'$, $F_2$, $F_4$ where from (1),

$$\left.\begin{aligned}
&\frac{dx}{dt}=x'\\
&\frac{d^2x}{dt^2}=\frac{dx'}{dt}=F_2(x,y,y')=2y'+x-\frac{(1-\mu)(x-x_1)}{r_1^3}-\frac{\mu(x-x_2)}{r_2^3},\\
&\frac{dy}{dt}=y'\\
&\frac{d^2y}{dt^2}=\frac{dy'}{dt}=F_4(x,x',y)=-2x'+y-\frac{(1-\mu)y}{r_1^3}-\frac{\mu y}{r_2^3}.
\end{aligned}\right\} \quad (73)$$

It is necessary to have computed the values of these quantities at sufficient dates that some order of differences obtainable from them will be small. Suppose that such tables of values have been set up for each of the quantities $x$, $y$, $x'$, $y'$, $F_2$, $F_4$ for the dates small $t_1$, $t_2$, $t_3$, ...., $t_n$, and suppose that in these tables the fourth differences are small and approximately constant. In the $F_2$ and $F_4$ tables we assume the next fourth difference about equal to the previous ones, and from the tables compute $F_2^{(n+1)}$ and $F_4^{(n+1)}$. Then the values of the next first differences of $x'$ and $y'$ can be readily computed by means of the formulæ:

$$\left.\begin{aligned}
\Delta_1^{(n+1)}x'&=(t_{n+1}-t_n)\Big[F_2^{(n+1)}-\frac{1}{2}\Delta_1F_2^{(n+1)}-\frac{1}{12}\Delta_2F_2^{(n+1)}-\frac{1}{24}\Delta_3F_2^{(n+1)}\\
&\qquad\qquad-\frac{1}{36}\Delta_4F_2^{(n+1)}\ldots\Big],\\
\Delta_1^{(n+1)}y'&=(t_{n+1}-t_n)\Big[F_4^{(n+1)}-\frac{1}{2}\Delta_1F_4^{(n+1)}-\frac{1}{12}\Delta_2F_4^{(n+1)}-\frac{1}{24}\Delta_3F_4^{(n+1)}\\
&\qquad\qquad-\frac{1}{36}\Delta_4F_4^{(n+1)}\ldots\Big].
\end{aligned}\right\} \quad (74)$$

Having found $\Delta_1^{(n+1)}x'$ and $\Delta_1^{(n+1)}y'$ we can at once compute $x'(t_{n+1})$ and $y'(t_{n+1})$, and complete the table of differences for this date. We can then compute $\Delta_1^{(n+1)}x$ and $\Delta_1^{(n+1)}y$ from $x'(t_{n+1})$, $y'(t_{n+1})$ and their successive differences, by formulæ exactly similar in form to (74), namely,

$$\left.\begin{aligned} \Delta_1^{(n+1)}x &= (t_{n+1}-t_n)\Big[x'^{(n+1)}-\frac{1}{2}\Delta_1 x'^{(n+1)}-\frac{1}{12}\Delta_2 x'^{(n+1)}-\frac{1}{24}\Delta_3 x'^{(n+1)} \\ &\qquad -\frac{1}{36}\Delta_4 x'^{(n+1)}\ldots\Big], \\ \Delta_1^{(n+1)}y &= (t_{n+1}-t_n)\Big[y'^{(n+1)}-\frac{1}{2}\Delta_1 y'^{(n+1)}-\frac{1}{12}\Delta_2 y'^{(n+1)}-\frac{1}{24}\Delta_3 y'^{(n+1)} \\ &\qquad -\frac{1}{36}\Delta_4 y'^{(n+1)}\ldots\Big]; \end{aligned}\right\} \quad (75)$$

whence we obtain at once $x(t_{n+1})$ and $y(t_{n+1})$.

Having found the values of $x$, $y$, $x'$, $y'$ at the date $t_{n+1}$ we compute $F_2^{(n+1)}$ and $F_4^{(n+1)}$ by the second and fourth relations of (73). If the results so obtained agree with the results we already have for $F_2^{(n+1)}$ and $F_4^{(n+1)}$, we assume that the fourth differences which we guessed are correct. But if the computed values of $F_2^{(n+1)}$ and $F_4^{(n+1)}$ differ from the assumed values we replace the latter by the computed values, and repeat the process as before from this point on. We keep on repeating this process, which generally has to be done only once, until the computed value is the same as the value from which it is computed. Having thus found the coordinates of the body for the date $t_{n+1}$, we assume another set of fourth differences for $F_2$ and $F_4$ for the date $t_{n+2}$ and repeat the process. Thus any number of points on the orbit may be determined, and the orbit can thus be continued beyond the range of convergence of the solutions of the differential equations.

### (B) *Jacobi's Constant.*

The question arises how we know that the orbit so continued is really the orbit on which the infinitesimal body would move when forming a part of the system in question.

It has been shown* that in the case of the infinitesimal body

$$V^2=x^2+y^2+\frac{2(1-\mu)}{r_1}+\frac{2\mu}{r_2}-C, \quad (76)$$

or

$$C=x^2+y^2+\frac{2(1-\mu)}{r_1}+\frac{2\mu}{r_2}-x'^2-y'^2. \quad (77)$$

This gives us a relation that must hold between the coordinates and the $x$- and $y$-components of the velocity of the infinitesimal body at all positions of its orbit.

* Moulton, "Introduction to Celestial Mechanics" (New Edition), § 153.

The constant $C$, which is called *Jacobi's Constant*, can be computed from the initial configuration of the system. In the present case the value of $C$ would be the value of the right-hand member of (77) when the body is at rest at the equilibrium point in question. At such points $x'=y'=x=y=0$; $r_2$ and $r_1$ are readily determined from (72).

Formula (77) may then be used as a check on the computation at each step, that is, for each new date. If the integral will not check, when the computation has been accurately made, it is usually necessary to divide the time-interval, and take shorter steps. It then becomes necessary to interpolate values of the coordinates for intermediate values of the time. After obtaining the intermediate values for $F_2$ and $F_4$ by Lagrange's *Interpolation Formula*, the corresponding $x$, $x'$, $y$, $y'$ can be computed out of these by the process described. Thus the orbit of the infinitesimal body can be continued as far as may be desired.

(C) *Specific Form of the Solutions for* $\mu=0.02$.

We proceed now to assign to $\mu$ the arbitrary value $\mu=0.02$, and find the specific form of the solutions and the shape of the orbit. We will consider the orbits in the neighborhood of the equilibrium point ($a$), and consider those given by equations (25), where the expansion is made in powers of $e^{-\rho t}$.

By (72) we have $r_2$ for the point ($a$) given by

$$r_2=\left(\frac{\mu}{3}\right)^{1/3}+\frac{1}{3}\left(\frac{\mu}{3}\right)^{2/3}-\frac{1}{9}\left(\frac{\mu}{3}\right)^{3/3}\cdots\cdot\cdot$$

On neglecting all after the third term in the expansion, we find, for $\mu=0.02$ that $r_2=0.1993$. But because of the neglected terms in (72) this value of $r_2$ does not satisfy with sufficient accuracy the quintic equation of which (72) is the solution, namely,

$$r_2^5+(3-\mu)r_2^4+(3-2\mu)r_2^3-\mu r_2^2-2\mu r_2-\mu=0. \tag{78}$$

If we start with the value $r_2=0.1993$ and apply the method of differential corrections we find $r_2=0.200078$, a value which makes the left member of (78) differ from zero by approximately 0.00000005. Hence, for the point ($a$) we take

$$r_2=0.200078, \quad r_1=1.200078; \tag{79}$$

whence, by (2), (5), and (42) we find in succession

$$\begin{aligned}
A&=\ \ 3.064095; \quad \rho=2.098701; \quad \sigma=\ \ 1.827794;\\
m&=-0.648885; \quad n=2.863898; \quad B=12.952988;\\
m\rho+n\sigma&=\ \ 3.872514; \quad m\sigma-n\rho=\ \ 7.196426;\\
4\rho^2+\sigma^2&=20.958701; \quad m^2-2=-1.578948.
\end{aligned}$$

On substituting these values in equations (25) and putting $c=+1$,* we have

* See § VII (A).

for the equation of the orbit in the first quadrant in the neighborhood of the origin,

$$\left.\begin{aligned} x &= e^{-2.098701t} - 2.944651\, e^{-4.197402t} \ldots, \\ y &= 0.648885\, e^{-2.098701t} + 0.0251562\, e^{-4.197402t} \ldots . \end{aligned}\right\} \qquad (80)$$

For the infinitesimal body at rest at the equilibrium point $(a)$, Jacobi's constant $C$ has the value

$$C = x^2 + y^2 + \frac{2(1-\mu)}{r_1} + \frac{2\mu}{r_2} = 3.225734.$$

Hence, at all points on the orbits asymptotic to $(a)$, Jacobi's constant must have the value 3.225734.

The values of $x$, $y$, $x'$, and $y'$, were computed for the times $t=3$, 2.75, 2.50, 2.25, and 2.00 and the tables of values of $x$, $y$, $x'$, $y'$, $F_2$, $F_4$ and their differences built up. On proceeding as outlined in subsection (A) of this article, the table of values (on p. 248) was obtained. For the first few computations the interval of time taken was one-fourth of the unit of time, and afterwards one-eighth; but in the table are given only the values corresponding to intervals of one-fourth, except in the case of a few dates near which one of the elements changed sign.

A drawing of the orbit represented by the data in the following table is appended.

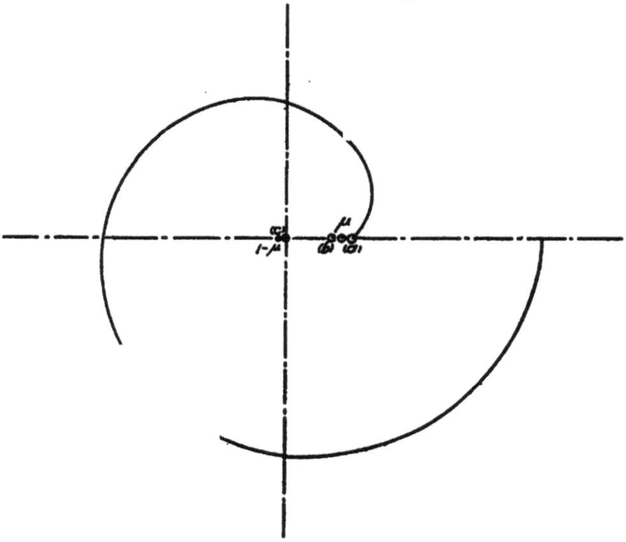

| | | | | |
|---|---|---|---|---|
| 3.00 | 1.18191 | 0.00120 | —0.00383 | —0.00251 |
| 2.75 | 1.18316 | 0.00202 | —0.00642 | —0.00424 |
| 2.50 | 1.18526 | 0.00342 | —0.01071 | —0.00717 |
| 2.25 | 1.18874 | 0.00577 | —0.01773 | —0.01210 |
| 2.00 | 1.19447 | 0.00974 | —0.02896 | —0.02038 |
| 1.75 | 1.20373 | 0.01643 | —0.04633 | —0.03434 |
| 1.50 | 1.21830 | 0.02766 | —0.07175 | —0.05741 |
| 1.25 | 1.24034 | 0.04631 | —0.10604 | —0.09469 |
| 1.00 | 1.27190 | 0.07674 | —0.14716 | —0.15243 |
| 0.75 | 1.31392 | 0.12474 | —0.18816 | —0.23661 |
| 0.50 | 1.36500 | 0.19749 | —0.21733 | —0.35027 |
| 0.25 | 1.42037 | 0.30217 | —0.21993 | —0.49123 |
| 0.00 | 1.47154 | 0.44471 | —0.18149 | —0.65126 |
| —0.25 | 1.50674 | 0.62827 | —0.09068 | —0.81669 |
| —0.50 | 1.51198 | 0.85206 | 0.05873 | —0.96995 |
| —0.75 | 1.47247 | 1.11060 | 0.26679 | —1.09141 |
| —1.00 | 1.3742 | 1.3935 | 0.5269 | —1.1614 |
| —1.25 | 1.2057 | 1.6856 | 0.8262 | —1.1621 |
| —1.50 | 0.9594 | 1.9677 | 1.1459 | —1.0796 |
| —1.75 | 0.6331 | 2.2176 | 1.4628 | —0.9050 |
| —2.00 | 0.2305 | 2.4122 | 1.7508 | —0.6362 |
| —2.125 | 0.0037 | 2.4815 | 1.8756 | —0.4681 |
| —2.25 | —0.2377 | 2.5284 | 1.9830 | —0.2780 |
| —2.375 | —0.4912 | 2.5504 | 2.0699 | —0.0705 |
| —2.50 | —0.7542 | 2.5454 | 2.1333 | 0.1542 |
| —2.75 | —1.2957 | 2.4468 | 2.1794 | 0.6420 |
| —3.00 | —1.8338 | 2.2220 | 2.1041 | 1.1580 |
| —3.25 | —2.3369 | 1.8680 | 1.8979 | 1.6706 |
| —3.50 | —2.7717 | 1.3899 | 1.5593 | 2.1452 |
| —3.75 | —3.1060 | 0.8015 | 1.0961 | 2.5468 |
| —4.00 | —3.3107 | 0.1254 | 0.5255 | 2.8422 |
| —4.125 | —3.3567 | —0.2364 | 0.2077 | 2.9409 |
| —4.25 | —3.3619 | —0.6083 | —0.1265 | 3.0027 |
| —4.50 | —3.2435 | —1.3629 | —0.8263 | 3.0062 |
| —4.75 | —2.9482 | —2.0971 | —1.5344 | 2.8394 |
| —5.00 | —2.4792 | —2.7680 | —2.2081 | 2.4989 |
| —5.25 | —1.8507 | —3.3327 | —2.8035 | 1.9927 |
| —5.50 | —1.0875 | —3.7521 | —3.2789 | 1.3399 |
| —5.75 | —0.2243 | —3.9929 | —3.5975 | 0.5701 |
| —8.875 | 0.2316 | —4.0383 | —3.6886 | 0.1532 |
| —6.00 | 0.6958 | —4.0307 | —3.7303 | —0.2778 |
| —6.25 | 1.6237 | —3.8514 | —3.6580 | —1.1578 |
| —6.50 | 2.5069 | —3.4534 | —3.3729 | —2.0189 |
| —6.75 | 3.2927 | —2.8479 | —2.8803 | —3.9747 |
| —7.00 | 3.9313 | —2.0593 | —2.1988 | —4.2632 |
| —7.25 | 4.3790 | —1.1239 | —1.3597 | |
| —7.50 | 4.6016 | —0.0894 | —0.4070 | |
| —7.625 | 4.4446 | 0.4475 | 0.0947 | |

UNIVERSITY OF MANITOBA, *August* 20, 1913.

# *Some Invariants of the Ternary Quartic.*

By H. Ivah Thomsen.

In his discussion of the ternary quartic, Salmon* has used with advantage the special form $u=ax_1^4+bx_2^4+cx_3^4+6fx_2^2x_3^2+6gx_3^2x_1^2+6hx_1^2x_2^2$. For this form he gives the Hessian

$$h=\bar{a}x_1^6+\bar{b}x_2^6+\bar{c}x_3^6+\bar{a}_2x_1^4x_2^2+\bar{a}_3x_1^4x_3^2+\bar{b}_1x_2^4x_1^2+\bar{b}_3x_2^4x_3^2+\bar{c}_1x_3^4x_1^2+\bar{c}_2x_3^4x_2^2+\overline{m}x_1^2x_2^2x_3^2,$$

where

$$\begin{aligned}&\bar{a}=agh, &&\bar{a}_2=a(bg+hf)-3gh^2, &&\bar{a}_3=a(ch+fg)-3hg^2,\\ &\bar{b}=bhf, &&\bar{b}_3=b(ch+fg)-3hf^2, &&\bar{b}_1=b(af+gh)-3fh^2,\\ &\bar{c}=cfg, &&\bar{c}_1=c(af+gh)-3fg^2, &&\bar{c}_2=c(bg+hf)-3gf^2,\end{aligned}$$
$$\overline{m}=L-3P+18R,$$

and the covariant $S=a_sx_1^4+b_sx_2^4+c_sx_3^4+6f_sx_2^2x_3^2+6g_sx_3^2x_1^2+6h_sx_1^2x_2^2$, where

$$\begin{aligned}&a_s=6g^2h^2, &&f_s=bcgh-f(bg^2+ch^2+R),\\ &b_s=6h^2f^2, &&g_s=cahf-g(ch^2+af^2+R),\\ &c_s=6f^2g^2, &&h_s=abfg-h(af^2+bg^2+R).\end{aligned}$$

Salmon does not give the discriminant of the quartic, though for the special form it may easily be calculated. Thus,

$$u_1=x_1(ax_1^2+3hx_2^2+3gx_3^2),\quad u_2=x_2(3hx_1^2+bx_2^2+3fx_3^2),\quad u_3=x_3(3gx_1^2+3fx_2^2+cx_3^2).$$

By inspection, we see that if $a, b$ or $c=0$, $u$ has one node; if $bc-9f^2$, or $ca-9g^2$ or $ab-9h^2=0$, $u$ has two nodes; if

$$\Delta=\begin{vmatrix}a & 3h & 3g\\ 3h & b & 3f\\ 3g & 3f & c\end{vmatrix}=L-9P+54R=0,$$

$u$ is the product of two conics and has four nodes. Hence, we infer that the discriminant

$$K=L(bc-9f^2)^2(ca-9g^2)^2(ab-9h^2)^2\Delta^4=L(81Q+L^2-9LP-729R^2)^2\Delta^4.$$

We verify this inference by forming $K$ according to the well-known method based on the fact that a double point of $u$ is also a double point of $h$.† This method gives

---

* Salmon, "Treatise on the Higher Plane Curves," 3d Ed., Dublin, 1879, Articles 292–302. References to Salmon are to these articles unless otherwise specified. We use his abbreviations $L=abc$, $P=af^2+bg^2+ch^2$, $R=fgh$ and $Q=bcg^2h^2+cah^2f^2+abf^2g^2$.

† Salmon-Fiedler, "Alg. der lin. Trans.," Art. 90.

$K =$

| $x_1^3x_2x_3$ | $x_2^3x_3x_1$ | $x_3^3x_1x_2$ | $x_1^5$ | $x_2^5$ | $x_3^5$ | $x_2^4x_1$ | $x_3^4x_2$ | $x_1^4x_3$ | $x_3^4x_1$ | $x_1^4x_2$ | $x_2^4x_3$ | $x_1^3x_2^2$ | $x_2^3x_3^2$ | $x_3^3x_1^2$ | $x_1^3x_3^2$ | $x_2^3x_1^2$ | $x_3^3x_2^2$ | $x_2^2x_3^2x_1$ | $x_3^2x_1^2x_2$ | $x_1^2x_2^2x_3$ |
|---|---|---|---|---|---|---|---|---|---|---|---|---|---|---|---|---|---|---|---|---|
| $a$ | $3h$ | $3g$ | 0 | 0 | 0 | 0 | 0 | 0 | 0 | 0 | 0 | 0 | 0 | 0 | 0 | 0 | 0 | 0 | 0 | 0 |
| $3h$ | $b$ | $3f$ | 0 | 0 | 0 | 0 | 0 | 0 | 0 | 0 | 0 | 0 | 0 | 0 | 0 | 0 | 0 | 0 | 0 | 0 |
| $3g$ | $3f$ | $c$ | 0 | 0 | 0 | 0 | 0 | 0 | 0 | 0 | 0 | 0 | 0 | 0 | 0 | 0 | 0 | 0 | 0 | 0 |
| 0 | 0 | 0 | $a$ | 0 | 0 | 0 | 0 | 0 | 0 | 0 | 0 | $3h$ | 0 | 0 | $3g$ | 0 | 0 | 0 | 0 | 0 |
| 0 | 0 | 0 | 0 | $b$ | 0 | 0 | 0 | 0 | 0 | 0 | 0 | 0 | $3f$ | 0 | 0 | $3h$ | 0 | 0 | 0 | 0 |
| 0 | 0 | 0 | 0 | 0 | $c$ | 0 | 0 | 0 | 0 | 0 | 0 | 0 | 0 | $3g$ | 0 | 0 | $3f$ | 0 | 0 | 0 |
| 0 | 0 | 0 | 0 | 0 | 0 | $3h$ | 0 | 0 | 0 | 0 | 0 | $a$ | 0 | 0 | 0 | 0 | 0 | $3g$ | 0 | 0 |
| 0 | 0 | 0 | 0 | 0 | 0 | 0 | $3f$ | 0 | 0 | 0 | 0 | 0 | $b$ | 0 | 0 | 0 | 0 | 0 | $3h$ | 0 |
| 0 | 0 | 0 | 0 | 0 | 0 | 0 | 0 | $3g$ | 0 | 0 | 0 | 0 | 0 | $c$ | 0 | 0 | 0 | 0 | 0 | $3f$ |
| 0 | 0 | 0 | 0 | 0 | 0 | 0 | 0 | 0 | $3g$ | 0 | 0 | 0 | 0 | 0 | $a$ | 0 | 0 | $3h$ | 0 | 0 |
| 0 | 0 | 0 | 0 | 0 | 0 | 0 | 0 | 0 | 0 | $3h$ | 0 | 0 | 0 | 0 | 0 | $b$ | 0 | 0 | $3f$ | 0 |
| 0 | 0 | 0 | 0 | 0 | 0 | 0 | 0 | 0 | 0 | 0 | $3f$ | 0 | 0 | 0 | 0 | 0 | $c$ | 0 | 0 | $3g$ |
| 0 | 0 | 0 | 0 | 0 | 0 | 0 | 0 | $a$ | 0 | 0 | 0 | 0 | 0 | $3g$ | 0 | 0 | 0 | 0 | 0 | $3h$ |
| 0 | 0 | 0 | 0 | 0 | 0 | $b$ | 0 | 0 | 0 | 0 | 0 | $3h$ | 0 | 0 | 0 | 0 | 0 | $3f$ | 0 | 0 |
| 0 | 0 | 0 | 0 | 0 | 0 | 0 | $c$ | 0 | 0 | 0 | 0 | 0 | $3f$ | 0 | 0 | 0 | 0 | 0 | $3g$ | 0 |
| 0 | 0 | 0 | 0 | 0 | 0 | 0 | 0 | 0 | 0 | $a$ | 0 | 0 | 0 | 0 | 0 | $3h$ | 0 | 0 | $3g$ | 0 |
| 0 | 0 | 0 | 0 | 0 | 0 | 0 | 0 | 0 | 0 | 0 | $b$ | 0 | 0 | 0 | 0 | 0 | $3f$ | 0 | 0 | $3h$ |
| 0 | 0 | 0 | 0 | 0 | 0 | 0 | 0 | 0 | $c$ | 0 | 0 | 0 | 0 | 0 | $3g$ | 0 | 0 | $3f$ | 0 | 0 |
| 0 | 0 | 0 | $3\bar{a}$ | 0 | 0 | $\bar{b}_1$ | 0 | 0 | $\bar{c}_1$ | 0 | 0 | $2\bar{a}_2$ | 0 | 0 | $2\bar{a}_3$ | 0 | 0 | $\overline{m}$ | 0 | 0 |
| 0 | 0 | 0 | 0 | $3\bar{b}$ | 0 | 0 | $\bar{c}_2$ | 0 | 0 | $\bar{a}_2$ | 0 | 0 | $2\bar{b}_3$ | 0 | 0 | $2\bar{b}_1$ | 0 | 0 | $\overline{m}$ | 0 |
| 0 | 0 | 0 | 0 | 0 | $3\bar{c}$ | 0 | 0 | $\bar{a}_3$ | 0 | 0 | $\bar{b}_3$ | 0 | 0 | $2\bar{c}_1$ | 0 | 0 | $2\bar{c}_2$ | 0 | 0 | $\overline{m}$ |

Add to row 19 row 4 multiplied by $-3gh$ and perform the obvious similar operations. Then, we see that $K$ is the product of $L$, $\Delta$ and a 15-row determinant. To reduce this determinant, add to row 13 row 1 multiplied by $hf$, row 8 multiplied by $-(af+gh)$, row 4 multiplied by $fg$, row 12 multiplied by $-(af+gh)$; multiply row 1 by $b$ and add row 8 multiplied by $-3h$. Perform the similar operations. Then $K$ is seen to be equal to the product of $L$, $\Delta$, $(ab-9h^2)^2(bc-9f^2)^2(ca-9g^2)^2$ and a 9-row determinant. If in this determinant we add to row 7 row 1 multiplied by $-2g$ and row 4 multiplied by $-2h$, etc., we have

$$K=L\Delta^4(81Q+L^2-9LP-729R^2)^2.$$

Since $s$ is of the same form as $u$, the discriminant of $s$ is

$$K_s=L_s\Delta_s^4(b_sc_s-9f_s^2)^2(c_sa_s-9g_s^2)^2(a_sb_s-9h_s^2)^2.$$

Salmon gives

$$L_s=216R^4,$$
$$P_s=6[Q^2-2PRQ-4R^2Q+2P^2R^2-2PLR^2+4PR^3+6LR^3+3R^4],$$
$$R_s=Q^2-2LRQ-P^2R^2-2PR^3+L^2R^2+4LR^3-R^4.$$

From these values we find

$$\Delta_s=L_s-9P_s+54R_s=-54RM(2Q-R(L+3P)),$$

where

$$M=L-P-2R.$$

As to the remaining factors

$$b_sc_s-9f_s^2=36f^2R^2-9f_s^2=-9(f_s+2fR)(f_s-2fR);$$

and

$$f_s+2fR=(bg-hf)(ch-fg),$$

so that

$$(f_s+2fR)(g_s+2gR)(h_s+2hR)=(Q-R(L+P-R))^2.$$

We will write

$$(f_s-2fR)(g_s-2gR)(h_s-2hR)=V,$$

and find by direct calculation

$$V=Q^2-2(L-P-3R)QR+R^2(L^2-2LP-3P^2+10LR-18PR-27R^2).$$

We have thus shown that

$$K_s=\rho R^8M^4(Q-R(L+P-R))^4(2Q-R(L+3P))^4V^2.^*$$

Referring to the invariants given by Salmon, we find by direct calculation

$$E_1-AB^2=16R^2M(Q-R(L+P-R));$$

---

* By $\rho$ we understand a numerical factor which it is not necessary to specify.

hence it follows that, for the special form we have used, $E_1 - AB^2$ is a factor of $K_s$. Dr. Coble has shown that this is true in general, since if $E_1 - AB^2 = 0$ $s$ consists of two conics.*

We now consider the remaining factor of $K_s$; we call it $S_2$ and have

$$S_2 = \rho(2Q - R(L+3P))^4 V^2.$$

It is three conditions on a conic that it be a repeated line; hence, it is one condition on a ternary $n$-ic that the polar conic of some point in regard to it be a repeated line. If a cubic has this property, it is catalectic.

If the polar conic of a point $y$ as to a quartic $u$ is a repeated line, and $y$ is not on the line, we may take the line as the side $x_2$ of the reference triangle and $y$ as the opposite vertex, $e_2$. Then,

$$\begin{aligned} u = {} & ax_1^4 + bx_2^4 + cx_3^4 + 6gx_3^2x_1^2 + 12lx_1^2x_2x_3 + 12nx_3^2x_1x_2 \\ & + 4a_2x_1^3x_2 + 4a_3x_1^3x_3 + 4c_1x_3^3x_1 + 4c_2x_3^3x_2, \\ u_2 = {} & a_2x_1^3 + bx_2^3 + c_2x_3^3 + 3lx_1^2x_3 + 3nx_3^2x_1. \end{aligned}$$

If we take for $x_1$ and $x_3$ the lines joining $e_2$ to the Hessian points of the points in which $x_2$ meets $u_2$, we shall have $l = n = 0$, and

$$u = ax_1^4 + bx_2^4 + cx_3^4 + 6gx_3^2x_1^2 + 4a_2x_1^3x_2 + 4a_3x_1^3x_3 + 4c_1x_3^3x_1 + 4c_2x_3^3x_2.$$

For this form

$$\begin{aligned} u_1 = {} & ax_1^3 + c_1x_3^3 + 3a_2x_1^2x_2 + 3a_3x_1^2x_3 + 3gx_3^2x_1, \\ u_2 = {} & a_2x_1^3 + bx_2^3 + c_2x_3^3, \\ u_3 = {} & a_3x_1^3 + cx_3^3 + 3gx_1^2x_3 + 3c_1x_3^2x_1 + 3c_2x_3^2x_2, \\ -S = {} & a_2^2c_2^2x_1^2x_2^2 - bx_2[a_2(g^2 - a_3c_1)x_1^3 + (g(a_2c_1 - ac_2) + a_3(a_3c_2 - ca_2))x_1^2x_3 \\ & + (g(c_2a_3 - ca_2) + c_1(a_2c_1 - ac_2))x_1x_3^2 + c_2(g^2 - a_3c_1)x_3^3] \\ & + ba_2c_2x_2^2(a_3x_1^2 + 2gx_1x_3 + c_1x_3^2), \\ h = {} & bx_2^2[(ax_1^2 + 2a_3x_1x_3 + gx_3^2 + 2a_2x_1x_2)(gx_1^2 + 2c_1x_1x_3 + cx_3^2 + 2c_2x_2x_3) \\ & - (a_3x_1^2 + 2gx_1x_3 + c_1x_3^2)^2] + 2a_2c_2x_1^2x_3^2(a_3x_1^2 + 2gx_1x_3 + c_1x_3^2) \\ & - a_2^2x_1^4(gx_1^2 + 2c_1x_1x_3 + cx_3^2 + 2c_2x_2x_3) - c_2^2x_3^4(ax_1^2 + 2a_3x_1x_3 + gx_3^2 + 2a_2x_1x_2). \end{aligned}$$

Hence, we see that the polar cubic of every point on $x_2 = 0$ has a double point at $e_2$; the cuspidal cubics corresponding to $e_1$ and $e_3$; $x_2$ is a factor of the Steinerian of $u$; $S$ has a node at $e_2$, the nodal tangents being $u_{12} = 0$; $h$ has a node at $e_2$, the nodal tangents being $x_1x_3 = 0$.

We are now in a position to write,† in the form of a 15-row determinant, an invariant of a quartic, the vanishing of which expresses the condition that

---

* AMERICAN JOURNAL OF MATHEMATICS, Vol. XXXI, p. 357.

† Assuming that the coefficients of $s$ have been calculated.

the polar conic of some point is a repeated line; we shall prove by using Salmon's special form that this invariant is identical with $S_2$.

For, if the polar conic of $y$ is a repeated line, its line equation vanishes identically. Hence, $y$ satisfies three quartic equations such as $u_{22}u_{33}-u_{23}^2=0$, and three such as $u_{11}u_{23}-u_{31}u_{12}=0$; also $y$ is a node of $s$ and satisfies the three cubic equations $s_1=0$, $s_2=0$, $s_3=0$. From each cubic equation we can get three quartics, making in all fifteen quartic equations which $y$ must satisfy. Eliminating $y$ dialytically from these equations, we have the required determinant. It contains six rows of degree 2 and nine rows of degree 4, and, consequently, when expanded, is of degree 48 in the coefficients of $u$.

For the special form we have

$$u_{22}u_{33}-u_{23}^2=ghy_1^4+bfy_2^4+cfy_3^4+(bc-3f^2)y_2^2y_3^2+(ch+fg)y_3^2y_1^2+(bg+hf)y_1^2y_2^2,$$

$$u_{11}u_{23}-u_{31}u_{12}=-2[(2gh-af)y_1^2y_2y_3-hfy_2^3y_3-fgy_3^3y_2].$$

From $s$ we get three forms such as

$$2a_s y_1^4+g_s y_3^2y_1^2+h_s y_1^2y_2^2$$

and six such as

$$g_s y_1^2y_2y_3+f_s y_2^3y_3+2c_s y_3^3y_2.$$

Hence the 15-row determinant is the product of a 6-row determinant, $\Delta_6$, and a 9-row determinant, $\Delta_9$, where

$$\Delta_6=\left|\begin{array}{cccccc} y_1^4 & y_2^4 & y_3^4 & y_2^2y_3^2 & y_3^2y_1^2 & y_1^2y_2^2 \\ gh & bf & cf & bc-3f^2 & ch+fg & bg+hf \\ ag & hf & cg & ch+fg & ca-3g^2 & af+gh \\ ah & bh & fg & bg+hf & af+gh & ab-3h^2 \\ 2g^2h^2 & 0 & 0 & 0 & g_s & h_s \\ 0 & 2h^2f^2 & 0 & f_s & 0 & h_s \\ 0 & 0 & 2f^2g^2 & f_s & g_s & 0 \end{array}\right|,$$

$$\Delta_9=\left|\begin{array}{ccccccccc} y_1^2y_2y_3 & y_2^2y_3y_1 & y_3^2y_1y_2 & y_1^3y_2 & y_1^3y_3 & y_2^3y_1 & y_2^3y_3 & y_3^3y_1 & y_3^3y_2 \\ 2gh-af & 0 & 0 & 0 & 0 & 0 & -hf & 0 & -fg \\ 0 & 2hf-bg & 0 & 0 & -gh & 0 & 0 & -fg & 0 \\ 0 & 0 & 2fg-ch & -gh & 0 & -hf & 0 & 0 & 0 \\ 0 & 0 & g_s & 2g^2h^2 & 0 & h_s & 0 & 0 & 0 \\ 0 & h_s & 0 & 0 & 2g^2h^2 & 0 & 0 & g_s & 0 \\ h_s & 0 & 0 & 0 & 0 & 0 & 2h^2f^2 & 0 & f_s \\ 0 & 0 & f_s & h_s & 0 & 2h^2f^2 & 0 & 0 & 0 \\ 0 & f_s & 0 & 0 & g_s & 0 & 0 & 2f^2g^2 & 0 \\ g_s & 0 & 0 & 0 & 0 & 0 & f_s & 0 & 2f^2g^2 \end{array}\right|.$$

To expand $\Delta_6$, we multiply column 4 by $gh$ and add to it column 2 multiplied by $-g^2$ and column 3 multiplied by $-h^2$, and perform the obvious similar operations. Then,

$$R^2\Delta_6 = V \begin{vmatrix} gh & bf & cf & 1 & 0 & 0 \\ ag & hf & cg & 0 & 1 & 0 \\ ah & bh & fg & 0 & 0 & 1 \\ 2g^2h^2 & 0 & 0 & 0 & hf & fg \\ 0 & 2h^2f^2 & 0 & gh & 0 & fg \\ 0 & 0 & 2f^2g^2 & gh & hf & 0 \end{vmatrix}.$$

Now, multiply row 1 by $gh$, row 2 by $hf$, row 3 by $fg$, so that $gh$ is a factor of column 1 and of column 4, etc. Taking out these factors, thus getting rid of the factor $R^2$ on the left, we have readily

$$\Delta_6 = -V \begin{vmatrix} 2(af-gh) & bg+hf & ch+fg \\ af+gh & 2(bg-hf) & ch+fg \\ af+gh & bg+hf & 2(ch-fg) \end{vmatrix} = 4V(2Q-R(L+3P)).$$

As to $\Delta_9$, the expression $(2gh-af)(f_s+2fR)+hfg_s+fgh_s$ proves, when expanded, to be symmetric and equal to $2Q-R(L+3P)$. Hence, multiplying row 1 by $f_s+2fR$ and adding to it row 9 multiplied by $hf$ and row 6 multiplied by $fg$, etc., we have

$$\begin{aligned}(f_s+2fR)(g_s+2gR)(h_s+2hR)\Delta_9 \\ =(2Q-R(L+3P))^3\end{aligned} \begin{vmatrix} 2g^2h^2 & 0 & h_s & 0 & 0 & 0 \\ 0 & 2g^2h^2 & 0 & 0 & g_s & 0 \\ 0 & 0 & 0 & 2h^2f^2 & 0 & f_s \\ h_s & 0 & 2h^2f^2 & 0 & 0 & 0 \\ 0 & g_s & 0 & 0 & 2f^2g^2 & 0 \\ 0 & 0 & 0 & f_s & 0 & 2f^2g^2 \end{vmatrix}.$$

Multiplying row 4 by $2g^2h^2$ and subtracting from it row 1 multiplied by $h_s$, etc., we have $\Delta_9 = V(2Q-R(L+3P))^3$. Hence,

$$\Delta_6\Delta_9 = 4(2Q-R(L+3P))^4V^2 = \rho S_2.$$

If we calculate $s$ for a cuspidal quartic, we find that a cusp of $u$ is also a cusp of $s$. Hence, $S_2=0$ if $u$ has a cusp. However, it may happen that both $K=0$ and $S_2=0$, and $u$ is not cuspidal; *i. e.*, the cusp and the point, the polar conic of which is a repeated line, may or may not coincide.*

There is an invariant of lower order which vanishes if $u$ has a cusp; for, if $y$ is a cusp, it satisfies the equations $u_1=0$, $u_2=0$ and $u_3=0$, and we form the invariant by using these equations instead of the three derived from $s$ in

---

* $S_2$ is also a factor of the discriminant of $h$, occurring, probably, to the second degree.

the work given above. This invariant, which we call $G$, is of degree 21 in the coefficients of $u$. For the special form the determinant, as before, consists of two factors $\Delta_6$ and $\Delta_9$, where

$$\Delta_6 = \begin{vmatrix} gh & bf & cf & bc-3f^2 & ch+fg & bg+hf \\ ag & hf & cg & ch+fg & ca-3g^2 & af+gh \\ ah & bh & fg & bg+hf & af+gh & ab-3h^2 \\ a & 0 & 0 & 0 & 3g & 3h \\ 0 & b & 0 & 3f & 0 & 3h \\ 0 & 0 & c & 3f & 3g & 0 \end{vmatrix}$$

and

$$\Delta_9 = \begin{vmatrix} 2gh-af & 0 & 0 & 0 & 0 & 0 & -hf & 0 & -fg \\ 0 & 2hf-bg & 0 & 0 & -gh & 0 & 0 & -fg & 0 \\ 0 & 0 & 2fg-ch & -gh & 0 & -hf & 0 & 0 & 0 \\ 0 & 0 & 3g & a & 0 & 3h & 0 & 0 & 0 \\ 0 & 3h & 0 & 0 & a & 0 & 0 & 3g & 0 \\ 3h & 0 & 0 & 0 & 0 & 0 & b & 0 & 3f \\ 0 & 0 & 3f & 3h & 0 & b & 0 & 0 & 0 \\ 0 & 3f & 0 & 0 & 3g & 0 & 0 & c & 0 \\ 3g & 0 & 0 & 0 & 0 & 0 & 3f & 0 & c \end{vmatrix}.$$

To expand $\Delta_6$ we multiply row 1 by $a$ and subtract from it row 4 multiplied by $gh$, row 5 multiplied by $af$, row 6 multiplied by $af$, etc. Then,

$$\Delta_6 = \begin{vmatrix} a(bc-9f^2), & cah-2afg-3hg^2, & abg-2ahf-3gh^2 \\ bch-2bfg-3hf^2, & b(ca-9g^2), & abf-2bgh-3fh^2 \\ bcg-2chf-3gf^2, & caf-2cgh-3fg^2, & c(ab-9h^2) \end{vmatrix}$$
$$=Q(61L-21P+102R)+L(L^2-10LP+9P^2)$$
$$+LR(14L-46P-565R)+9R^2(5P-6R).$$

To expand $\Delta_9$ we note that if we multiply row 1 by $bc-9f^2$ and add to it row 6 multiplied by $f(ch-3fg)$ and row 9 multiplied by $f(bg-3hf)$, the first element becomes $(bc-9f^2)(3gh-af)-(3hf-bg)(3fg-ch)$, while all other elements of this row are zero.

Hence $(bc-9f^2)(ca-9g^2)(ab-9h^2)\Delta_9$ is equal to the product of three terms such as the first element into the 6-row determinant

$$\begin{vmatrix} a & 0 & 3h & 0 & 0 & 0 \\ 0 & a & 0 & 0 & 3g & 0 \\ 0 & 0 & 0 & b & 0 & 3f \\ 3h & 0 & b & 0 & 0 & 0 \\ 0 & 3g & 0 & 0 & c & 0 \\ 0 & 0 & 0 & 3f & 0 & c \end{vmatrix} = -(bc-9f^2)(ca-9g^2)(ab-9h^2).$$

Hence,

$$\Delta_0 = [(bc-9f^2)(3gh-af)-(3hf-bg)(3fg-ch)]$$
$$[(ca-9g^2)(3hf-bg)-(3fg-ch)(3gh-af)]$$
$$[(ab-9h^2)(3fg-ch)-(3gh-af)(3hf-bg)]$$
$$=8[(LR-3Q)^2+3LR(P-3R)^2]-4[3Q(P+3R)+LR(P-3R)]X$$
$$+2(Q+3PR)X^2-RX^3,$$

where

$$X=L-3P+36R.$$

For the quartic* $u$ the equation

$$\phi=(u_{22}u_{33}-u_{23}^2)\xi_1^2+(u_{33}u_{11}-u_{31}^2)\xi_2^2+(u_{11}u_{22}-u_{12}^2)\xi_3^2$$
$$+2(u_{31}u_{12}-u_{11}u_{23})\xi_2\xi_3+2(u_{12}u_{23}-u_{22}u_{31})\xi_3\xi_1+2(u_{23}u_{31}-u_{33}u_{12})\xi_1\xi_2=0$$

is, for a given $x$ and variable $\xi$, the line equation of the polar conic of $x$. For a given $\xi$ and variable $x$ it is the equation of a point quartic. It is natural to ask the relation of this quartic to the curve $u$ and the line $\xi$.

The polar cubic of $y$ is $y_1u_1+y_2u_2+y_3u_3=0$, and, if $y$ is restricted to the line $\xi$, so that $(y\xi)=0$, $y_1(\xi_2u_1-\xi_1u_2)+y_3(\xi_2u_3-\xi_3u_2)=0$ represents the pencil of polar cubics corresponding to the line.

If, for a given value of the ratio $y_1:y_3$, such a curve has a double point $x$, this point must lie on each of the three curves

$$\frac{\xi_2u_{31}-\xi_3u_{12}}{\xi_2u_{11}-\xi_1u_{12}}=\frac{\xi_2u_{23}-\xi_3u_{22}}{\xi_2u_{12}-\xi_1u_{22}}=\frac{\xi_2u_{33}-\xi_3u_{23}}{\xi_2u_{31}-\xi_1u_{23}}.$$

Reducing these equations, we have three members of the net of quartics determined by the twelve points which are double points of polar cubics of points on the line $\xi$, viz.:

$$\phi_1=\xi_1(u_{22}u_{33}-u_{23}^2)+\xi_2(u_{23}u_{31}-u_{33}u_{12})+\xi_3(u_{12}u_{23}-u_{22}u_{31}),$$
$$\phi_2=\xi_1(u_{23}u_{31}-u_{33}u_{12})+\xi_2(u_{33}u_{11}-u_{31}^2)+\xi_3(u_{31}u_{12}-u_{11}u_{23}),$$
$$\phi_3=\xi_1(u_{12}u_{23}-u_{22}u_{31})+\xi_2(u_{31}u_{12}-u_{11}u_{23})+\xi_3(u_{11}u_{22}-u_{12}^2).$$

Such a set of twelve points lies on $h$; the curve $\eta_1\phi_1+\eta_2\phi_2+\eta_3\phi_3=0$ meets $h$ in the two sets of points corresponding to the line $\xi$ and the line $\eta$. If, in the equation of this curve, we let $\eta_i=\xi_i$, it becomes $\phi=0$.

Hence, we would infer that $\phi$ touches $h$ at the twelve points corresponding to $\xi$. Such, indeed, is the case since for the line $x_1=0$, $\phi=u_{22}u_{33}-u_{23}^2$, and, writing $h$ in the form

$$u_{11}(u_{22}u_{33}-u_{23}^2)+u_{12}(u_{23}u_{31}-u_{33}u_{12})+u_{31}(u_{12}u_{23}-u_{22}u_{31})=0,$$

it is obvious that $h$ reduces to a square if $u_{22}u_{33}-u_{23}^2=0$.

We have already stated, for the special form, the expanded value of $u_{22}u_{33}-u_{23}^2$, the coefficient of $\xi_1^2$ in $\phi$; Salmon gives us the covariant $\sigma$. If we

* This argument may easily be extended to the ternary $n$-ic.

operate on one of these forms with the other, the result is

$$24\,[f(3L+5P+2R)-8af^3+4bcgh],$$

which is, omitting the factor 24, the coefficient of $\xi_1^2$ in the contravariant conic given by Salmon. Hence, this conic may be defined as the envelope of lines such that the curves $\phi$ corresponding to them are apolar to $\sigma$.

Dr. Morley suggested that an invariant of a quartic, $u$, could be written in the form of a 15-row determinant, the vanishing of which would express the condition that it be possible to determine a second quartic, $\bar{u}$, such that the Clebschian of $u$ and $\bar{u}$ vanish identically. We can write down the coefficients of the Clebschian of $u$ and $\bar{u}$ from those of the contravariant $\sigma$ of $u$, which Salmon gives for the general form, by writing for $2bc$, $b\bar{c}+\bar{b}c$, and for $f^2$, $f\bar{f}$, etc.* Equating each of the coefficients of $\sigma$ to zero and eliminating $\bar{a}$, etc., we have the invariant

$$\begin{array}{ccccccccccccccc} \bar a & \bar b & \bar c & \bar f & \bar g & \bar h & \bar l & \bar m & \bar n & \bar a_2 & \bar a_3 & \bar b_1 & \bar b_3 & \bar c_1 & \bar c_2 \end{array}$$

$$E_3=\begin{vmatrix}
0 & c & b & 6f & 0 & 0 & 0 & 0 & 0 & 0 & 0 & 0 & -4c_2 & 0 & -4b_3 \\
c & 0 & a & 0 & 6g & 0 & 0 & 0 & 0 & 0 & -4c_1 & 0 & 0 & -4a_3 & 0 \\
b & a & 0 & 0 & 0 & 6h & 0 & 0 & 0 & -4b_1 & 0 & -4a_2 & 0 & 0 & 0 \\
f & 0 & 0 & a & h & g & 4l & -2a_3 & -2a_2 & -2n & -2m & 0 & 0 & 0 & 0 \\
0 & g & 0 & h & b & f & -2b_3 & 4m & -2b_1 & 0 & 0 & -2n & -2l & 0 & 0 \\
0 & 0 & h & g & f & c & -2c_2 & -2c_1 & 4n & 0 & 0 & 0 & 0 & -2m & -2l \\
0 & 6 & 0 & 2l & -b_3 & -c_2 & 2f & -n & -m & 0 & 0 & c_1 & -g & b_1 & -h \\
0 & 0 & 0 & -a_3 & 2m & -c_1 & -n & 2g & -l & c_2 & -f & 0 & 0 & -h & a_2 \\
0 & 0 & 0 & -a_2 & -b_1 & 2n & -m & -l & 2h & -f & b_3 & -g & a_3 & 0 & 0 \\
0 & 0 & -b_1 & -3n & 0 & 0 & 0 & -3c_2 & -3f & 0 & 0 & -c & c_1 & b_3 & -3m \\
-c_2 & 0 & 0 & 0 & -3l & 0 & -3g & 0 & -3a_2 & c_1 & -3n & 0 & 0 & a_2 & -a \\
0 & -a_3 & 0 & 0 & 0 & -3m & -3b_1 & -3h & 0 & 0 & -b & -3l & a_3 & 0 & b_3 \\
0 & -c_1 & 0 & -3m & 0 & 0 & 0 & -3f & -3b_3 & 0 & 0 & c_2 & -3n & -b & b_1 \\
0 & 0 & -a_2 & 0 & -3n & 0 & -3c_1 & 0 & -3g & -c & c_2 & 0 & 0 & -3l & a_3 \\
-b_3 & 0 & 0 & 0 & 0 & -3l & -3h & -3a_3 & 0 & -3m & b_1 & a_3 & -a & 0 & 0
\end{vmatrix}.$$

For the special form we have $E_3=\Delta_6\Delta_9$, where

$$\Delta_6=\begin{vmatrix}
0 & c & b & 6f & 0 & 0 \\
c & 0 & a & 0 & 6g & 0 \\
b & a & 0 & 0 & 0 & 6h \\
f & 0 & 0 & a & h & g \\
0 & g & 0 & h & b & f \\
0 & 0 & h & g & f & c
\end{vmatrix}
\quad
\Delta_9=\begin{vmatrix}
2f & 0 & 0 & 0 & 0 & 0 & -g & 0 & -h \\
0 & 2g & 0 & 0 & -f & 0 & 0 & -h & 0 \\
0 & 0 & 2h & -f & 0 & -g & 0 & 0 & 0 \\
0 & 0 & -3f & 0 & 0 & -c & 0 & 0 & 0 \\
-3g & 0 & 0 & 0 & 0 & 0 & 0 & 0 & -a \\
0 & -3h & 0 & 0 & -b & 0 & 0 & 0 & 0 \\
0 & -3f & 0 & 0 & 0 & 0 & 0 & -b & 0 \\
0 & 0 & -3g & -c & 0 & 0 & 0 & 0 & 0 \\
-3h & 0 & 0 & 0 & 0 & 0 & -a & 0 & 0
\end{vmatrix}.$$

* This may be verified by calculation; obviously the Clebschian must reduce to a numerical multiple of $\sigma$ if $u\equiv\bar{u}$.

To expand $\Delta_6$ multiply row 1 by $gh$, and subtract from it row 5 multiplied by $ch$ and row 6 multiplied by $bg$. We then have

$$R\Delta_6 = -\begin{vmatrix} P-6R-af^2 & b(fg+ch) & c(hf+bg) \\ a(fg+ch) & P-6R-bg^2 & c(gh+af) \\ a(hf+bg) & b(gh+af) & P-6R-ch^2 \end{vmatrix},$$

and on expansion

$$\Delta_6 = -2[L^2+2LP-3P^2+20LR+36PR-108R^2].$$

To expand $\Delta_9$ multiply row 1 by $a$, and subtract from it row 5 multiplied by $h$ and row 9 multiplied by $g$, when we have immediately

$$\Delta_9 = -8L(af+3gh)(bg+3hf)(ch+3fg)$$
$$= -8L(3Q+R(L+9P+27R)).$$

Referring to the invariants given by Salmon, it will be found that

$$9AD_3+3A^2C_1-27E_1+A^3B+135AB^2=\rho E_3.$$

BALTIMORE, *March* 15, 1915.

# *Functions of Surfaces with Exceptional Points or Curves.*[*]

BY CHARLES A. FISCHER.

In a former paper[†] I have given a definition of the derivative of a function of a surface analogous to Volterra's definition of the derivative of a function of a line, and have proved that if the derivative is continuous and approached uniformly, the first variation of the function is equal to the double integral of the derivative multiplied by the first variation of the dependent variable. The object of the present paper is to extend the theory to the case where there are points or curves where the derivative does not exist. In the first section exceptional points are considered, and an application made to the second variation of a function of a double integral. In the second section exceptional curves are discussed, and in the last an application is made to the variable boundary problem of the calculus of variations.

## § 1. *Exceptional Points.*

It will be assumed that the derivative $L'(S;\, x, y)$ of a function $L(S)$ is continuous and approached uniformly with order $r$[‡] at every point of the surface

$$S: \qquad z = z(x, y),$$

excepting at the point $(x_0, y_0)$. The surfaces

$$S_\epsilon: \qquad z = z(x, y) + \eta(x, y),$$

and

$$S_a: \qquad z = z(x, y) + \omega(x, y, \alpha),$$

will be defined as in the former paper.[§]

Two kinds of exceptional points will be considered. The first kind will be those for which the inequality

$$\lim_{\epsilon=0} \frac{L(S_\epsilon) - L(S)}{\alpha} = 0, \tag{1}$$

is satisfied for every choice of $\eta(x, y)$ and $\alpha$ such that $\eta(x, y)/\alpha$ is bounded everywhere, and at $(x_0, y_0)$

---

* Read before the American Mathematical Society, August 3, 1915.

† Fischer, AMERICAN JOURNAL OF MATHEMATICS, Vol. XXXVI (1914), p. 289.

‡ It can easily be proved that if $L(S)$ is continuous, and $L'(S;\, x, y)$ is approached uniformly, it is also continuous in all arguments. Compare with Evans, *Bulletin of American Mathematical Society*, Vol. XXI (1915), p. 389.

§ Fischer, *loc. cit.*, pp. 290–291.

$$\lim_{\epsilon=0} \frac{1}{\alpha} \cdot \frac{\partial^{i+j}\eta(x_0, y_0)}{\partial x^i \partial y^j}, \qquad (i+j \leqq r),$$

exists. The second kind will be those for which there is a constant $a_0$ such that

$$\lim_{\epsilon=0} \frac{L(S_\epsilon)-L(S)}{\alpha} = a_0 \lim_{\epsilon=0} \frac{\eta(x_0, y_0)}{\alpha},$$

or in a more general case those for which there is a set of constants $a_{ij}$ such that

$$\lim_{\epsilon=0} \frac{L(S_\epsilon)-L(S)}{\alpha} = \sum_{i,j=0}^{n} a_{ij} \lim_{\epsilon=0} \frac{\partial^{i+j}\eta(x_0, y_0)}{\partial x^i \partial y^j}. \tag{2}$$

It will first be proved that the equation

$$\frac{dL(S_a)}{da}\Big|_{a=0} = \iint_R L'(S;\, x, y)\omega_a(x, y, 0)\,dxdy\text{*} \tag{3}$$

is unaffected by an exceptional point of the first kind. A function $\Theta(x, y)$ of class $C^{(r)}$ will be chosen which vanishes excepting in the square

$$x_0-\epsilon < x < x_0+\epsilon, \quad y_0-\epsilon < y < y_0+\epsilon,$$

and is equal to unity in the square

$$x_0-h \leqq x \leqq x_0+h, \quad y_0-h \leqq y \leqq y_0+h,$$

where $h$ is a positive number less than $\epsilon$. The surface $S_{\epsilon a}$ will be defined by the equation

$$S_{\epsilon a}: \qquad z = z(x, y) + \Theta(x, y)\omega(x, y, a).$$

The surfaces $S_a$ and $S_{\epsilon a}$ can then be made to lie in any neighborhood of $S$ of order $r$ by taking $a$ sufficiently small. Since $S_a$ coincides with $S_{\epsilon a}$ in the neighborhood of the exceptional point, the equation

$$\lim_{a=0} \frac{L(S_a)-L(S_{\epsilon a})}{\alpha} = \iint_R L'(S;\, x, y)\,(1-\Theta(x, y))\,\omega_a(x, y, 0)\,dxdy,$$

is satisfied for all values of $\epsilon$. It follows from equation (1) that

$$\lim_{\epsilon, a=0} \frac{L(S_{\epsilon a})-L(S)}{\alpha} = 0.$$

Since the region where $\Theta(x, y) \neq 0$ approaches zero with $\epsilon$, these equations imply equation (3).

If equation (2) is satisfied at the point $(x_0, y_0)$, a new function will be defined by the equation

$$\bar{L}(S) = L(S) - \sum_{i,j=0}^{n} a_{ij} \frac{\partial^{i+j} z(x_0, y_0)}{\partial x^i \partial y^j}.$$

This function will then satisfy equation (1), and consequently equation (3) must be replaced by the equation

* Fischer, *loc. cit.*, p. 291.

$$\frac{dL(S_\alpha)}{d\alpha}\Big|_{\alpha=0}=\iint_R L'(S;\ x,y)\omega_\alpha(x,y,0)\,dxdy+\sum_{i,j=0}^{n} a_{ij}\frac{\partial^{i+j}\omega_\alpha(x_0,y_0,0)}{\partial x^i\partial y^j}. \quad (4)$$

It is easy to prove that if the surface $S$ furnishes a minimum for the function $L(S)$, the constants $a_{ij}$ must all be zero. It has already been proved that $L'(S;\ x,y)=0$ is a necessary condition for a minimum.* Then, if $a_{kl}\neq 0$, the function $\omega(x,y,\alpha)$ can be defined as

$$\omega(x,y,\alpha)=\alpha(x-x_0)^k(y-y_0)^l$$

in the neighborhood of $(x_0,y_0)$, and equation (4) becomes

$$\frac{dL(S_\alpha)}{d\alpha}=a_{kl}k!\,l!\neq 0,$$

and there can be no minimum.

An exceptional point will often occur if a function of a surface is differentiated twice at the same point. For instance, the function $L(S)$ can be defined as $L(S)=\phi(J)$, where $\phi$ is a function of class $C''$, and

$$J=\iint_R f(x,y,z,p_{10},p_{01},\ldots\ldots,p_{mm})\,dxdy.$$

The argument $p_{ij}$ is used to represent $\partial^{i+j}z/\partial x^i\partial y^j$. It can then be easily proved that the derivative

$$L'(S;\ x_0,y_0)=\phi'(J)E(x_0,y_0,z(x_0,y_0),\ldots\ldots)$$

is continuous and approached uniformly with order $4m$, where

$$E(x,y,z,\ldots\ldots)=\sum_{i,j=0}^{m}(-1)^{i+j}\frac{\partial^{i+j}f_{p_{ij}}}{\partial x^i\partial y^j}.\dagger$$

If the second derivative is taken at a different point it is seen to be

$$L''(S;\ x_0,y_0;\ x,y)=\phi''(J)E(x_0,y_0,\ldots\ldots)E(x,y,\ldots\ldots),$$

but $(x_0,y_0)$ is now an exceptional point. At $(x_0,y_0)$ it is found that

$$\lim_{\epsilon=0}\frac{L'(S_\epsilon;\ x_0,y_0)-L'(S;\ x_0,y_0)}{\alpha}=\phi'(J)\sum_{i,j=0}^{2m}\frac{\partial E}{\partial p_{ij}}\lim_{\epsilon=0}\frac{1}{\alpha}\cdot\frac{\partial^{i+j}\eta(x_0,y_0)}{\partial x^i\partial y^j}.$$

This is in the same form as equation (2), and consequently for this case

$$a_{ij}=\phi'(J)\frac{\partial E}{\partial p_{ij}}.$$

If $L(S)$ is a function whose derivatives $L'(S;\ x,y)$ and $L''(S;\ x_0,y_0;\ x,y)$ are continuous and approached uniformly excepting when $x=x_0$ and $y=y_0$, and if, in this case, equation (2) is satisfied, the second derivative of $L(S_\alpha)$ with respect to $\alpha$ is found to be

* Fischer, *loc. cit.*, p. 295. † Compare with Fischer, *loc. cit.*, p. 297.

$$\frac{d^2L(S_a)}{da^2}\Big|_{a=0}=\iint_R\Big\{\iint_R L''(S;\,x_0,\,y_0;\,x,\,y)\omega_a(x_0,\,y_0,\,0)\omega_a(x,\,y,\,0)\,dxdy$$
$$+\sum_{i,\,j=0}^{n} a_{ij}(x_0,\,y_0)\,\frac{\partial^{i+j}\omega_a(x_0,\,y_0,\,0)}{\partial x^i\partial y^j}\,\omega_a(x_0,\,y_0,\,0)$$
$$+L'(S;\,x_0,\,y_0)\omega_{aa}(x_0,\,y_0,\,0)\Big\}dx_0dy_0. \tag{5}$$

The Jacobi condition for a minimum of a double integral can be derived from a special case of equation (5). If $L(S)$ is the double integral $J$, the second derivative $L''(S;\,x_0,y_0;\,x,\,y)$ vanishes. Then, if $\omega(x,\,y,\,a)=a\zeta(x,\,y)$, equation (5) becomes

$$\frac{d^2L(S_a)}{da^2}=\iint_R\sum_{i,\,j=1}^{2m}\frac{\partial E}{\partial p_{ij}}\,\frac{\partial^{i+j}\zeta(x,\,y)}{\partial x^i\partial y^j}\cdot\zeta(x,\,y)\,dxdy.$$

The equation

$$\sum_{i,\,j=0}^{2m}\frac{\partial E}{\partial p_{ij}}\cdot\frac{\partial^{i+j}\zeta}{\partial x^i\partial y^j}=0$$

is the analogue of the Jacobi equation for this problem.* If it has a solution, not identically zero, which vanishes along the boundary of $R$, the second variation of $J$ can be made to vanish. The question whether it can also be made negative or not has not been settled yet, excepting in special cases.

## § 2. *Exceptional Curves.*

Two kinds of exceptional curves will also be considered. The first kind will be those along which

$$\lim_{\epsilon=0}\frac{L(S_\epsilon)-L(S)}{\epsilon a}=0 \tag{6}$$

uniformly in the neighborhood of the given surface. The second kind will be those for which there is a set of continuous functions $a_j(s)$ such that

$$\lim_{\epsilon=0}\frac{L(S_\epsilon)-L(S)}{\epsilon a}=\sum_{j=0}^{n}\lim_{\epsilon=0}\frac{1}{\epsilon a}\int_{s-\epsilon}^{s+\epsilon}a_j(s)\cdot\frac{\partial^j\eta(s,\,0)}{\partial n^j}\,ds, \tag{7}$$

where the variation $\eta(s,\,n)$ is expressed in terms of the length of arc along the exceptional curve and the normal distance to it. This limit is also assumed to be approached uniformly.

If the derivative $L'(S;\,x,\,y)$ is continuous and approached uniformly, excepting along a curve

$$C:\qquad x=x(s),\quad y=y(s),\qquad (a\leq s\leq b),$$

of class $C^{(r)}$, and equation (6) is satisfied along $C$, and if the function $\omega(x,\,y,\,a)$ vanishes at the end-points of $C$, it will be proved that equation (3) is valid. A function $\Theta(x,\,y)$ of class $C^{(r)}$ can be chosen, which is equal to zero at every

*Bolza, "Vorlesungen über Variationsrechnung," p. 675.

point whose distance from $C$ is greater than or equal to $\varepsilon$, and equal to unity at every point whose distance from $C$ is less than or equal to $h$, a positive constant less than $\varepsilon$. The interval $a \leq s \leq b$ will then be divided into $m$ equal parts by the points $s_0 = a, s_1, \ldots, s_m = b$, choosing $m$ in such a way that $\frac{\varepsilon}{2} < s_1 - s_0 < \varepsilon$. The functions $\beta_i(x, y)$, also of class $C^{(r)}$, will be chosen in such a way that the equations

$$\begin{array}{ll} \beta_0(x(s), y(s)) = 0, & (a \leq s \leq b), \\ \beta_i(x(s), y(s)) = 1, & (a \leq s \leq s_i;\ i = 1, 2, \ldots, m), \\ 1 > \beta_i(x(s), y(s)) > 0, & (s_i < s < s_{i+1}), \\ \beta_i(x(s), y(s)) = 0, & (s_{i+1} \leq s \leq b), \end{array}$$

are satisfied. The surfaces $S^{(i)}$ will then be defined by the equations

$$S^{(i)}: \qquad z = z(x, y) + \beta_i(x, y)\Theta(x, y)\omega(x, y, a).$$

Since the surface $S_a$ coincides with $S^{(m)}$ in the neighborhood of $C$ and the hypothesis of the Volterra theorem is satisfied in the rest of the region $R$, the equation

$$\lim_{\varepsilon=0} \frac{L(S_a) - L(S^{(m)})}{a} = \iint_R L'(S;\ x, y)(1 - \Theta(x, y))\omega_a(x, y, 0)\,dxdy \tag{8}$$

must be satisfied. The surfaces $S^{(i+1)}$ and $S^{(i)}$ can now take the place of $S_\varepsilon$ and $S$ in equation (6). Since this limit is approached uniformly, there is a quantity $\xi$, which approaches zero with $\varepsilon$, such that

$$\left| \frac{L(S^{(i+1)}) - L(S^{(i)})}{a} \right| < \xi\varepsilon, \qquad (i = 0, 1, \ldots, m).$$

Adding these inequalities,

$$\left| \frac{L(S^{(m)}) - L(S)}{a} \right| < m\xi\varepsilon < 2\xi(b - a).$$

Consequently,

$$\lim_{\varepsilon=0} \frac{L(S^{(m)}) - L(S)}{a} = 0. \tag{9}$$

Since the region where $\Theta(x, y) \neq 0$ approaches zero with $\varepsilon$, equations (8) and (9) imply equation (3), which was to be proved.

If equation (7) is satisfied along the curve $C$, the new function

$$\bar{L}(S) = L(S) - \sum_{j=0}^{n} \int_a^b a_j(s) \frac{\partial^j z(x(s), y(s))}{\partial n^j} ds$$

will satisfy equation (6), and consequently equation (3) must be replaced by

$$\frac{dL(S_a)}{da}\Big|_{a=0} = \iint_R L'(S;\ x, y)\omega_a(x, y, 0) + \sum_{j=0}^{n} \int_a^b a_j(s) \frac{\partial^j \omega_a}{\partial n^j} ds.$$

A necessary condition for a minimum can be derived from this equation. It follows immediately that, if there is a minimum, the equation

$$\sum_{j=0}^{n} \int_a^b a_j(s) \frac{\partial^j \omega_a}{\partial n^j} ds = 0 \tag{10}$$

must be satisfied for every choice of $\omega(s, n, \alpha)$. This will be proved to be equivalent to the equations

$$a_j(s) = 0, \qquad (j = 0, 1, \ldots, n;\ a \leq s \leq b). \tag{11}$$

Suppose that $a_l(s_1) \neq 0$. Then, since the functions $a_j(s)$ are assumed to be continuous, there is an interval including $s_1$ in which $a_l(s) \neq 0$. Let $\Theta(s)$ be an arbitrary function of class $C^{(r)}$ which agrees in sign with $\omega_l(s)$ in such an interval and vanishes outside of it. Then the function $\omega(s, n, \alpha)$ will be defined in the neighborhood of $C$ by the equation

$$\omega(s, n, \alpha) = \alpha n^l \Theta(s).$$

Consequently,

$$\frac{\partial^j \omega_a(s, n, 0)}{\partial n^j} = 0, \qquad (j \neq l),$$

and

$$\int_a^b a_l(s) \frac{\partial^l \omega(s, n, 0)}{\partial n^l} ds > 0,$$

and equation (10) is not satisfied. Therefore, equations (11) form a necessary condition that the surface $S$ minimize the function $L(S)$.

## § 3. *The Variable Boundary Problem.*

In the work up to this point it has been assumed that the surfaces $S_a$ were all bounded by the same space curve. If this is not the case, as, for instance, if the locus of the bounding curves is a given surface, the surfaces $S_a$ will be supposed to extend beyond this boundary and all intersect along a new space curve. The function $L(S)$ will depend for its value only on the part of $S$ inside the original boundary. Thus, the function $z(x, y)$ will be defined arbitrarily in a region $R'$ surrounding $R$, and the function $\omega(x, y, \alpha)$ defined so as to vanish along the boundary of the region $R+R'$. Under these circumstances the derivative $L'(S;\ x, y)$ will vanish in $R'$ and very likely be infinite along the curve separating $R'$ from $R$. If the function considered is the integral $J$ considered in § 1, the boundary of $R$ is an exceptional curve of the last kind mentioned, and the functions $a_j(s)$ will now be proved to exist. It will be supposed that the surfaces over which the double integral is taken are bounded by a surface

$$\tilde{S}: \qquad z = \tilde{z}(x, y).$$

The curve $C$ will be the projection of the intersection of $\tilde{S}$ and $S$, and the function $\eta(x, y)$ chosen so as to define the derivative $L'(S;\, x_0, y_0)$, where $(x_0, y_0)$ is an arbitrary point of $C$. The distance $n$ will be taken positive inside the region $R$, and the positive direction of $s$ chosen so that $x_s y_n - x_n y_s = 1$. The value of $n$ on the projection of $C_\epsilon$ of the intersection of the surfaces $S_\epsilon$ and $\tilde{S}$ will be called $N(s)$.

Then the equations

$$\left.\begin{aligned}&\tilde{z}(x(s, N(s)), y(s, N(s))) - z(x(s, N(s)), y(s, N(s))) = \eta(x(s, N(s)), y(s, N(s))),\\ &\tilde{z}(x(s, 0), y(s, 0)) - z(x(s, 0), y(s, 0)) = 0,\end{aligned}\right\} \quad (12)$$

must be satisfied. Subtracting and applying Taylor's formula,

$$((\tilde{p}-p)x_n + (\tilde{q}-q)y_n)N(s) = \eta(x(s, N(s)), y(s, N(s))),$$

where the arguments of $\tilde{p}$, $p$, $\tilde{q}$ and $q$ are $x(s, \theta N(s))$, $y(s, \theta N(s))$, $(0<\theta<1)$. Consequently,

$$\lim_{\epsilon=0} \frac{N(s)}{\eta} = \frac{1}{(\tilde{p}-p)x_n + (\tilde{q}-q)y_n}. \quad (13)$$

If the last of equations (12) is differentiated with respect to $s$, it becomes

$$(\tilde{p}-p)x_s + (\tilde{q}-q)y_s = 0.$$

Consequently, equation (13) may be written

$$\lim_{\epsilon=0} \frac{N(s)}{\eta} = \frac{-y_s}{\tilde{p}-p} = \frac{x_s}{\tilde{q}-q}.$$

This expression is finite unless $\tilde{S}$ is tangent to $S$. It follows from the definitions of $S_\epsilon$ and $L(S)$ that

$$L(S_\epsilon) - L(S) = \iint_{\Delta S} \sum_{i,j=0}^{n} f_{ij}(x, y, z+\theta\eta, \ldots\ldots)\eta_{ij}\,dxdy$$
$$-\int_{x(s,0)}^{x(s,N(s))}\int_{y(s,0)}^{y(s,N(s))} f(x, y, z+\eta, \ldots\ldots)\,dxdy, \quad (14)$$

where

$$f_{ij} = f_{p_{ij}},\ \eta_{ij} = \frac{\partial^{i+j}\eta}{\partial x^i \partial y^j} = \frac{\partial^{i+j}\eta(s, n)}{\partial s^{i+j}}\left(\frac{\partial s}{\partial x}\right)^i\left(\frac{\partial s}{\partial y}\right)^j + \ldots\ldots + \frac{\partial \eta}{\partial n}\frac{\partial^{i+j} n}{\partial x^i \partial y^j}, \quad (15)$$

and $\Delta S$ is the part of $R$ where $\eta \neq 0$.

The first integral of equation (14) can be reduced to the expression

$$\iint_{\Delta S} E(x, y, z+\theta\eta, \ldots\ldots)\eta\,dxdy + \int_{s_0-\epsilon}^{s_0+\epsilon}\Big\{y_s \sum_{i=1,\, j=0}^{n} \sum_{k=1}^{i} (-1)^{k-1} \frac{\partial^{k-1} f_{ij}}{\partial x^{k-1}} \eta_{i-kj}$$
$$- x_s \sum_{i=0,\, j=1}^{n} \sum_{l=1}^{j} (-1)^{i+l-1} \frac{\partial^{i+l-1} f_{ij}}{\partial x^i \partial y^{l-1}} \eta_{0j-l}\Big\}ds,$$

by repeated application of Green's theorem.*

The last integral in equation (14) is equal to

$$\int_{s_0-\epsilon}^{s_0+\epsilon} f(x, y, z+\eta, \ldots\ldots)N(s)\,ds,$$

---

* Bolza, *loc. cit.*, p. 654.

where $x=x(s, \theta N(s)), \ldots$. If these expressions are substituted in equation (14), and it is divided by $\epsilon\alpha$, and then the limit taken as $\epsilon \doteq 0$ the double integral vanishes because $\Delta S$ is of the second degree in $\epsilon$ and $\eta/\alpha$ is finite, and the equation becomes

$$\lim_{\epsilon=0} \frac{L(S_\epsilon)-L(S)}{\epsilon\alpha} = \lim_{\epsilon=0} \frac{1}{\epsilon\alpha} \sum_{i,\,j=0}^{n} \int_{s_0-\epsilon}^{s_0+\epsilon} b_{ij}(s)\eta_{ij} ds, \tag{16}$$

where

$$b_{00} = \frac{y_s}{\tilde{p}-p}\left[f+(\tilde{p}-p)\sum_{k=1}^{n}(-1)^{k-1}\frac{\partial^{k-1}}{\partial x^{k-1}}f_{k0}+(\tilde{q}-q)\sum_{i=0,\,l=1}^{n}(-1)^{i+l-1}\frac{\partial^{i+l-1}f_{il}}{\partial x^i \partial y^{l-1}}\right],$$

$$b_{0j} = \frac{y_s}{\tilde{p}-p}\left[(\tilde{p}-p)\sum_{k=1}^{n}(-1)^{k-1}\frac{\partial^{k-1}}{\partial x^{k-1}}f_{kj}+(\tilde{q}-q)\sum_{i=0}^{n}\sum_{l=1}^{n-j}(-1)^{i+l-1}\frac{\partial^{i+l-1}f_{i\,j+l}}{\partial x^i \partial y^{l-1}}\right],$$

$$(j=1, 2, \ldots, n-1),$$

$$b_{ij} = y_s \sum_{k=1}^{n-i}(-1)^{k-1}\frac{\partial^{k-1}f_{i+k\,j}}{\partial x^{k-1}}, \qquad (i=1, 2, \ldots, n-1;\ j=0, 1, \ldots, n).$$

If equations (15) are substituted in equation (16), it takes the form

$$\lim_{\epsilon=0} \frac{L(S_\epsilon)-L(S)}{\epsilon\alpha} = \lim_{\epsilon=0} \frac{1}{\epsilon\alpha} \sum_{i,\,j=0}^{n} \int_{s_0-\epsilon}^{s_0+\epsilon} b'_{ij}(s)\frac{\partial^{i+j}\eta(s, 0)}{\partial s^i \partial n^j}. \tag{17}$$

Since the partial derivatives of $\eta$ vanish at $s_0 \pm \epsilon$, the equations

$$\int_{s_0-\epsilon}^{s_0+\epsilon} b'_{ij}(s)\frac{\partial^{i+j}\eta(s, 0)}{\partial s^i \partial n^j}ds = -\int_{s_0-\epsilon}^{s_0+\epsilon}\frac{db'_{ij}(s)}{ds}\cdot\frac{\partial^{i+j-1}\eta(s, 0)}{\partial s^{i-1}\partial n^j}ds$$

must be satisfied. If this process is repeated often enough equation (17) is reduced to the form of equation (7).

If $f$ is a function of $x$, $y$, $z$, $p$, $q$ only, the only one of the functions $a_j(s)$ which occurs is $a_0(s)$. It is found to be

$$a_0(s) = \frac{y_s}{\tilde{p}-p}(f+(\tilde{p}-p)f_p+(\tilde{q}-q)f_q).$$

The vanishing of this expression constitutes the variable boundary condition for a minimum of such a double integral. If $f$ depends on the partial derivatives of the second order also, the functions $a_0(s)$ and $a_1(s)$ occur. They are found to be

$$\begin{aligned} a_0(s) = \frac{y_s}{\tilde{p}-p} & f+f_{10}y_s-f_{01}x_s-\frac{\partial f_{20}}{\partial x}y_s(1+x_s y_n)-\frac{\partial f_{20}}{\partial y}y_s^2 y_n+\frac{\partial f_{11}}{\partial x}x_s^2 y_n \\ & +\frac{\partial f_{11}}{\partial y}y_s^2 x_n-\frac{\partial f_{02}}{\partial x}x_s^2 x_n+\frac{\partial f_{02}}{\partial y}x_s(1-y_s x_n)-f_{20}(y_{ss}y_n+y_s y_{sn}) \\ & +f_{11}(y_{ss}x_n+y_s x_{sn})-f_{02}(x_{ss}x_n+x_s x_{sn}), \end{aligned}$$

$$a_1(s) = -f_{20}y_s^2+f_{11}x_s y_s-f_{02}x_s^2.$$

If the given double integral depends on derivatives of higher order, the functions $a_j(s)$ can be computed in the same way, but the work is long and will not be given here.

COLUMBIA UNIVERSITY, *July*, 1915.

# *Dupin's Cyclide as a Self-Dual Surface.*

By Mabel M. Young.

## Introduction.

A correlation between the points and the planes of space is determined when it is assumed that by it five arbitrary points are sent into five arbitrary planes. If the correlation sends the points of a surface into its tangent planes, the surface is called self-dual. Its order and class are then the same, and the singularities occurring in the surface considered as an envelop of planes are the dual of those found in the surface taken as a locus of points. The Kummer quartic surface is a well-known example of a surface with this property.* Another self-dual surface, not a special case of the Kummer Surface, is the Cyclide of Dupin. The determination of a group of correlations which transforms this surface into itself is the chief object of this paper.

## The Determination of Four Polarities. §§ 1–3.

### § 1. *Equation of the Surface.*

Dupin's Cyclide is defined by Salmon † as a surface of the fourth order, with four double points and with a double conic in the plane at infinity. To determine its equation from this definition we take a tetrahedron of reference with vertices at the four double points of the surface. A quartic surface having a double conic in the plane at infinity may be put in the form

$$\alpha Q^2+(\beta x)(x)Q+(\gamma x)^2(x)^2=0,$$

where $\alpha$ is a constant, $(\beta x)$ an arbitrary plane, $(\gamma x)^2$ an arbitrary quadric and $(x)$ the plane at infinity. $Q$ is the known quadric cutting out the double conic in $(x)=0$. For simplicity, let $Q$ contain the four lines joining the four double points. Then,

$$Q\equiv x_0x_2+\mu x_1x_3=0.$$

Applying the condition that the four reference points are double points on the surface, we have the relations

$$\gamma_{00}=\gamma_{11}=\gamma_{22}=\gamma_{33}=\gamma_{01}=\gamma_{12}=\gamma_{23}=\gamma_{03}=0,\quad -2\gamma_{02}=\beta_0=\beta_2,\quad -2\gamma_{13}/\mu=\beta_1=\beta_3.$$

---

* Hudson, R. W. H. T., "Kummer's Quartic Surface," § 30.

† Salmon, G., "Geometry of Three Dimensions," 4th edition, §§ 560, 567.

These values of the constants give a final form of the equation with two arbitrary constants, $\mu$ and $(\beta_1-\beta_0)/2\alpha=\lambda$.

$$(x_0x_2+\mu x_1x_3)^2+\lambda(x)\{x_0x_2(x_1+x_3)-\mu x_1x_3(x_0+x_2)\}=0. \qquad (1)$$

## § 2. *The Dual Singularities of the Surface.*

To determine a correlation which is to test the self-duality of this surface, we may make use of the fact that in a self-dual surface the singularities of the surface in points and of the surface in planes are mutually dual. We shall then look first for four double tangent planes and a double quadric cone to correspond to the four double points and the double conic which by definition are present in Dupin's Cyclide.

To find the double tangent planes, we must first consider the nature of the double points to which they are to correspond. If we write the equation of the surface according to the powers of one of its variables, we see that the tangent at a double point is a quadric cone. The double points are then "conical points" at which there are an infinite number of tangent planes enveloping a quadric cone. The dual singularity must then be a plane containing an infinity of points of the surface lying on a conic. As the double points lie by pairs on the lines $\overline{13}$ and $\overline{02}$ of the reference tetrahedron, we examine first the relation to the surface of the planes on these lines. Take the planes on $\overline{13}$, which are of the form

$$kx_0+x_2=0.$$

The tangent cones at $(0,1,0,0)$ and $(0,0,0,1)$ are, respectively,

$$\mu^2x_3^2+\lambda x_0x_2-\mu\lambda x_3(x_0+x_2)=0 \text{ and } \mu^2x_1^2+\lambda x_0x_2-\mu\lambda x_1(x_0+x_2)=0.$$

The line pairs in which a plane of the pencil cuts these cones coincide when

$$\lambda k=(\lambda-2)\pm 2\sqrt{1-\lambda}.$$

There are then in the pencil two planes given by these values of $k$ which are tangent to both of the tangent cones at these double points, and hence to the surface at these points. Moreover, the quartic curve in which each of these planes cuts the surface reduces to a repeated conic given by

$$kx_0+x_2=0, \quad \{2(kx_0^2+\mu x_1x_3)+\lambda(k-1)x_0(x_1+x_3)\}^2=0.$$

At an arbitrary point on this conic there is found to be a single tangent plane, $kx_0+x_2=0$, so that the plane is a trope. Similarly, there are among the planes on $\overline{02}$, $lx_1+x_3=0$, two tropes given by

$$\lambda l=(\lambda+2\mu)\pm 2\sqrt{\mu\lambda+\mu^2}.$$

We have then found four planes which have the properties required for the planes dual to the four double points. These four tropes are

$$k_1x_0+x_2=0,\ k_2x_0+x_2=0,\ l_1x_1+x_3=0,\ l_2x_1+x_3=0, \tag{2}$$

wherf $k_1, k_2, l_1, l_2$ are the two sets of values for $k$ and $l$ already found.

The double conic of the surface is characterized by the fact that at each point the surface has two tangent planes. The dual of the conic must then be a quadric cone, every plane of which meets the surface in two points. Further, the tangent planes of the surface at points of the conic meet the plane $(x)=0$ in lines tangent to the conic. Hence, dually, the two contacts of each plane of the cone lie on a line passing through a fixed point, the vertex of the cone. If we can find such a cone of double tangent lines, we may consider its vertex as a point dually related to the plane of the double conic.

Before finding the equation of the cone and its vertex, it will be useful to notice another locus on its surface. Since the elements are double tangent lines of the surface, the points of contact may be given by a binary quartic with two pairs of equal roots. Let 0 and $\infty$ be the parameters of the points of contact on any element; then the binary quartic is $x_1^2y_2^2=0$. If $z_1+z_2$ is the vertex, its first polar as to the quartic is $(z_1y_2+z_2y_1)x_1y_2=0$; its third $z_1z_2(z_1y_2+x_1z_2)=0$. These have the common point $z_1y_2+x_1z_2$. As this relation occurs on every element, we see that the first and third polar surfaces of the vertex as to (1) meet in a conic on the cone and in a line.

The equation of the cone is most easily found if we rewrite the equation of the surface in terms of the parameter of the points $x_i+\lambda y_i$ in which a line on $x_i$ and $y_i$ meets it. The equation becomes

$$p_4+p_3\gamma+p_2\gamma^2+p_1\gamma^3+p\gamma^4=0, \tag{3}$$

where $p=$equation of surface, and $i!\,p_i=i$-th polar of surface as to a point $z$. If the line cutting the surface is an element of a cone of double tangent lines, it meets the surface in two points only, and (3) has two pairs of equal roots. Take $z$ at the vertex. For simplicity set $p_1\equiv 0$. Then $p_3\equiv 0$ also, and the equation becomes

$$p_4+p_2\gamma^2+p\gamma^4=0. \tag{4}$$

The condition for equal roots is now the vanishing of $4p_4p-p_2^2$. This expression set equal to zero gives an equation in $x$ and $z$ which vanishes when $x$ is a point on a double tangent through $z$. It must then for some special value of $z$ be the equation of the double tangent cone from $z$.

The vertex is the point $(\mu, 1, \mu, 1)$. This fact is most easily established by verifying that for this value of $z$ the discriminant $4p_4p-p_2^2\equiv 0$. The polars, found for this point, are

$$p_1=2[(x)(\mu Q+\lambda R)+\lambda\nu\{x_0x_2(x_1+x_3)-\mu x_1x_3(x_0+x_2)\}],$$
$$p_2=2\nu(\mu Q+\lambda R),\ p_3=2\mu\nu\{\sigma(x_0+x_2)+\mu\rho(x_1+x_3)\},\ p_4=\mu^2\nu^2.$$
$$Q\equiv x_0x_2+\mu x_1x_3;\ R\equiv x_0x_2-\mu^2x_1x_3;\ \mu+1\equiv\nu;\ 1-\lambda\equiv\rho;\ \mu+\lambda\equiv\sigma.$$

Substituting in the discriminant these values of $p_i$ we have, as the equation of the cone

$$4\nu(\mu Q+\lambda R)-\{\sigma(x_0+x_2)+\mu\rho(x_1+x_3)\}^2=0. \tag{5}$$

But a surface on the intersection of $p_1=0$ and $p_2=0$ is easily seen to be $\mu Q+\lambda R=0$; hence $\mu Q+\lambda R$ is known to be identically zero when the polar plane of $(\mu, 1, \mu, 1)$ as to the surface is zero. Therefore the discriminant of (4) vanishes for this point and we see that the cone of double tangent lines, the dual of the double conic in $(x)=0$, is a quadric cone with the vertex $(\mu, 1, \mu, 1)$. This point is accordingly the dual of $(x)=0$.

We have now five pairs of points and planes associated with Dupin's Cyclide, and dually related to each other,—the four double points corresponding to the four tropes, and the vertex of the cone of double tangent lines corresponding to the plane at infinity. These are sufficient to determine the correlation which we wish to consider.

### § 3. *The Equations of the Transformation.*

Let the correlations be given by the transformation

$$\rho u_i=a_{i0}x_0+a_{i1}x_1+a_{i2}x_2+a_{i3}x_3, \tag{6}$$

and assume $a_{ik}=a_{ki}$. This will give rise to a quadric with respect to which the corresponding point and plane will be pole and polar. In pairing the points and planes a point will not then, in general, be on its corresponding plane. Let the points on $\overline{13}$ and $\overline{02}$ be paired with the planes on $\overline{02}$ and $\overline{13}$ respectively:

| *Point.* | *Plane.* |
|---|---|
| (1, 0, 0, 0) | $k_1x_0+x_2=\sigma$ |
| (0, 1, 0, 0) | $l_1x_1+x_3=\sigma$ |
| (0, 0, 1, 0) | $k_2x_0+x_2=\sigma$ |
| (0, 0, 0, 1) | $l_2x_1+x_3=\sigma$ |

If we regard the equation of each plane as the result of applying to the corresponding point the transformation (6), we determine six relations among the coefficients $a_{ik}$:

$$a_{03}=a_{01}=a_{12}=a_{23}=0,\ a_{00}=a_{22},\ a_{11}=a_{33}.$$

The further requirement that by the same transformation $(\mu, 1, \mu, 1)$ goes into $(x)=0$ fixes the ratios

$$\frac{a_{33}}{a_{22}}=\frac{a_{11}}{a_{00}}=\frac{(1+k_1)\mu l_1}{(1+l_1)k_1};\ \frac{a_{13}}{a_{00}}=\frac{\mu(1+k_1)}{k_1(1+l_1)}.$$

But such a transformation is a polarity with respect to a quadric $(ax)^2=0$, the coefficients of which are the constants of the transformation. Hence, setting $k_i+1=n_i$ and $l_i+1=m_i$, we have the quadric

$$Q\equiv(ax)^2\equiv k_1m_1x_0^2+\mu n_1l_1x_1^2+\mu m_1x_2^2+\mu l_1n_1x_3^2+2m_1x_0x_2+2\mu n_1x_1x_3=0.$$

As $k_i$ and $l_i$ have each two values connected by the relation $k_1k_2=l_1l_2=1$, we may immediately derive from $(ax)^2=0$ three other quadrics which meet the conditions, and give rise to three new polarities. Our result may then be stated as a

LEMMA. *There are four transformations, polarites as to four quadrics, which send the point singularities of Dupin's Cyclide into their dual forms.*

The theorem that the entire surface is self-dual under the same polarities will be proved by another method in § 8.

The initial hypothesis about the transformation

$$\rho u_i=\sum_{k=0}^{k=3}a_{ik}x_k$$

was that $a_{ik}=a_{ki}$. If we had assumed $a_{ik}=-a_{ki}$, the transformation would have led to a linear complex, instead of a quadric. But in such a case a point and its corresponding plane are incident. As this is geometrically impossible for $(\mu, 1, \mu, 1)$ and its corresponding plane $(x)=0$, it is evident that we cannot show that these five pairs of points and planes, and much less the points and planes of the surface, belong to a null-system.

## A STUDY OF COVARIANT FORMS. §§ 4–7.

### § 4. *Relations Among the Quadrics $Q_i$.*

Before studying these polarities further, it is perhaps worth while to examine in some detail the geometry of certain covariant forms closely associated with the surface. Among these are the four quadrics $Q_i$ found in § 3, which are closely connected. Because $k_1k_2=l_1l_2=1$, it is possible to write all the quadrics in terms of $k_1$ and $l_1$, and in this form it is readily seen that a linear relation exists among them. The equations* are

$$\left.\begin{aligned}Q_1&\equiv m(kx_0^2+kx_2^2+2x_0x_2)+\mu n(lx_1^2+lx_3^2+2x_1x_3)=0,\\ lQ_2&\equiv m(kx_0^2+kx_2^2+2x_0x_2)+\mu n(x_1^2+x_3^2+2lx_1x_3)=0,\\ kQ_3&\equiv m(x_0^2+x_2^2+2kx_0x_2)+\mu n(lx_1^2+lx_3^2+2x_1x_3)=0,\\ lkQ_4&\equiv m(x_0^2+x_2^2+2kx_0x_2)+\mu n(x_1^2+x_2^2+2lx_1x_3)=0.\end{aligned}\right\}\quad(7)$$

* A transformation of the reference tetrahedron to the self-conjugate tetrahedron of the quadrics gives all forms related to the quadrics more simply. Other forms in frequent use become, however, so complicated and unsymmetrical that this desirable transformation is not made.

That a linear relation exists is known because the Jacobian of the four quadrics vanishes. To determine this relation we combine the equations in pairs in all possible ways.

$$Q-lQ_2=\mu nh(x_1-x_3)^2,\quad kQ_3-klQ_4=\mu nh(x_1-x_3)^2. \tag{8}$$

$$Q_1-kQ_3=mr(x_0-x_2)^2,\quad lQ_2-lkQ_4=mr(x_0-x_2)^2. \tag{9}$$

For each of these pencils of quadrics the invariant $\Phi$* $\equiv 0$,

$$\left.\begin{aligned} lQ_2-kQ_3&=mr\{(x_0-x_2)^2-\mu nh(x_1-x_3)^2\},\\ Q_1-klQ_4&=mr\{(x_0-x_2)^2+\mu nh(x_1-x_3)^2\},\qquad h\equiv l-1;\ r\equiv k-1.\end{aligned}\right\} \tag{10}$$

For these two quadrics the invariants $\Theta$* and $\Theta'\equiv 0$.

$$lQ_2+kQ_3=Q_1+klQ_4=mn\{(x_0+x_2)^2+\mu(x_1+x_3)^2\}. \tag{11}$$

From (8) or (10) we see that the linear relation connecting the quadrics is

$$Q_1-kQ_3-lQ_2+klQ_4\equiv 0.$$

From (8) we see also that the pairs $Q_1$ and $Q_2$, $Q_3$ and $Q_4$ have a repeated conic in the plane $x_1-x_3=0$, and are accordingly tangent along the conic in which they meet it. Equations (9) give the same relation for $Q_1$ and $Q_3$, $Q_2$ and $Q_4$, which have their common curve in $x_0-x_2=0$. Since each quadric contains a fixed conic in $x_0-x_2=0$ and in $x_1-x_3=0$, the intersection of these planes meets all four quadrics at the same two points, and at these points the quadrics have double tangency. From (10) we learn that $Q_1$ and $Q_4$, $Q_2$ and $Q_3$ meet also in other conics in the planes $x_1+x_3=0$, $x_0+x_2=0$, and, accordingly, have double contact when the line

$$x_1+x_3=0,\quad x_0+x_2=0$$

meets each pair. Each quadric has then "ring contact" with two others and double contact with a third. At two points on one line all four quadrics have double contact. On a second line they have double contact by pairs at two sets of points which are readily seen to form a harmonic set.

## § 5. *Certain Tangent Cones.*

The enveloping cone to the surface from a point without is of order 12, since it consists of tangent lines drawn to the intersection of the surface and its first polar. Consider the cone from $(\mu, 1, \mu, 1)$. The first polar of this point as to the surface is

$$(x)(\sigma x_0x_2+\mu^2\rho x_1x_3)+\lambda\nu x_0x_2\overline{x_1+x_3}-\mu x_1x_3\overline{x_0+x_2}=0.$$

This meets the plane at infinity in the conic

$$x_0x_2+\mu x_1x_3=0,\quad (x)=0.$$

* Salmon, G., "Analytic Geometry of Three Dimensions," 5th edition, Vol. I, Chap. 9.

The polar cubic also contains the double points of the surface and the lines joining them, along which it is tangent to the cyclide. The complete cone to the surface must then contain a cone to the double conic, counted twice; the four planes tangent to the surface along its four lines, and the cone of double lines already found. The planes are

$$1)\ x_0-\mu x_1=0;\ 2)\ x_0-\mu x_3=0;\ 3)\ x_2-\mu x_1=0;\ 4)\ x_2-\mu x_3=0. \qquad (12)$$

The cone to the double conic is

$$4\nu(x_0x_2+\mu x_1x_3)-\mu(x)^2=0, \qquad (13)$$

The lines along which planes (12) are tangent to the surface meet the double cone in four points whose coordinates are

$$(0,0,1,-1);\ (0,1,-1,0);\ (1,0,0,-1);\ (1,-1,0,0). \qquad (14)$$

They meet the cone of double lines in points

$$(0,0,\mu\rho,-\sigma);\ (0,-\sigma,\mu\rho,0);\ (\mu\rho,0,0,-\sigma);\ (\mu\rho,-\sigma,0,0).$$

Planes (12) are thus tangent to both cones. Hence, we see that the enveloping cone from $(\mu, 1, \mu, 1)$ to the surface consists of two quadric cones and their four common tangent planes.

At the points (14) the double conic is met also by the four tangent cones on the double points of the surface. These cones are

$$x_2^2+\lambda x_1x_2+\lambda x_2x_3-\mu\lambda x_1x_3=0,\quad \mu^2x_3^2+\lambda x_0x_2-\mu\lambda x_0x_3-\mu\lambda x_2x_3=0,$$
$$x_0^2+\lambda x_0x_1+\lambda x_0x_3-\mu\lambda x_1x_3=0,\quad \mu^2x_1^2+\lambda x_0x_2-\mu\lambda x_0x_1-\mu\lambda x_1x_2=0.$$

Each contains two lines of the surface. The pairs of cones with vertices on the diagonals $\overline{02}$, $\overline{13}$, which are touched by the tropes, meet each other on the surface only at the two remaining double points. Any two cones, except these pairs, have one line in common, along which they are touched by one of the four tangent planes (12). In $(x)=0$ we have thus at each of the four points (14) two cones tangent to each other and to the double conic. Each cone has double contact with the conic, but single contact only with each of the two cones which it meets.

## § 6. *The Four Pinch Points and their Duals.*

In general, at every point $y$ of the double conic the surface has two distinct tangent planes. These are the factors of the polar quadric of $y$,

$$2\{x_0y_2+x_2y_0+\mu(x_1y_3+x_3y_1)\}$$
$$-\{\lambda(y_0+y_2)\pm\sqrt{\lambda^2(y_0+y_2)^2-4\lambda y_0y_2}\}(x)=0. \qquad (15)$$

When, however, the expression under the radical is zero, the two planes coincide. The condition for this is

$$\lambda y_2 = -\{(\lambda-2) \pm 2\sqrt{1-\lambda}\} y_0,$$

with the further necessary relation

$$\lambda y_3 = -\{(\lambda+2\mu) \pm 2\sqrt{\mu^2+\lambda\mu}\} y_1.$$

The constants involved are, respectively, the values of $k_i$ and $l_i$ in the equations of the double tangent planes (2), so that the ratios become

$$y_2/y_0 = -k_i, \quad y_3/y_1 = -l_i.$$

Each of these four points satisfies two tropes, $(x)=0$, and the cone of double tangent lines, as well as the double conic. They are then the vertices of the quadrilateral cut out by the tropes in $(x)=0$ and inscribed in the double conic at its intersections with the cone of double lines. The coordinates of the four points, called "pinch points," are

$$(-h_i,\ r_i,\ k_i h_i,\ -l_i r_i), \qquad (i=1, 2).$$

The tangent planes to the surface at these points are

$$m_i \rho (k_i x_0 + x_2) - n_i \sigma (l_i x_1 + x_3) = 0.$$

The fact that the tropes pass through these points enables us to state the relations between the singular conics in these planes. As we have already seen, the conics in a pair of tropes on a diagonal line meet the double points on this line. At each pinch point two tropes meet which are not on a diagonal, and hence the corresponding singular conics meet also. Each conic accordingly meets one other conic at two double points of the surface and a second at a pinch point. These remarkable points are thus seen to have the following properties. They are the only points of the double conic at which the surface has a single tangent plane and they are each found on the double conic, the cone of double lines, two tropes and two singular conics.

The duals of the pinch points are planes already found. At each point the single tangent plane of the surface cuts $(x)=0$ in a line tangent to the double conic. Dually on the cone of double lines must be an element meeting the surface in two coincident points. The tangent plane on this line meets the surface in four coincident points. Moreover, the pinch points are common to two singular conics not lying in tropes on a diagonal. The corresponding planes must then be common planes of two cones of the surface at double points not on a diagonal. Such planes are the singular planes (12), and the coincident points of contact are the points (14), which are on the conic cut from the cone of double lines by the polar plane of its vertex. The planes of the surface dual to the pinch points are therefore the singular tangent planes of the surface.

## § 7. *A Quartic Curve on the Surface.*

The only space curve on the surface which arises in the preceding discussion is the locus of the points of contact of the cone of double lines and the surface. This curve is also on the polar of the surface as to $(\mu, 1, \mu, 1)$. Since the further intersection of this cubic and the cone is a conic (§ 2), the locus of contacts is a quartic curve* cut out by a pencil of quadrics.†

By combining the equations of the surface, the first polar as to $(\mu, 1, \mu, 1)$ and the cone, we obtain a quartic surface which must contain all points common to the three surfaces. This is

$$[2\nu(x_0x_2+\mu x_1x_3)-(x)\{\sigma(x_0+x_2)+\mu\rho(x_1+x_3)\}]$$
$$[2\nu(x_0x_2+\mu x_1x_3)+(x)\{\sigma(x_0+x_2)+\mu\rho(x_1+x_3)\}]=0. \quad (16)$$

To decide which of these quadrics cuts out the curve under discussion, we may put the equation of the surface in a form which will show its relation to the cone of double lines. Such a form is

$$S=Q_1^2+C_1Q_2,$$

where $S$ is the surface, $C_1$ the cone of double lines, $Q_1$ a quadric through the contacts and $Q_2$ an arbitrary quadric. This equation expresses that $S$ is tangent to $C_1$, where $C_1$ meets $Q_1$. Hence, that one of the two quadrics (16) passes through the curve, the square of which subtracted from the surface gives the cone multiplied by some quadric. $Q_1$ is found to be

$$2\nu(x_0x_2+\mu x_1x_3)-(x)\{\sigma(x_0+x_2)+\mu\rho(x_1+x_3)\}=0. \quad (17)$$

The pencil of quadrics which cuts out the quartic curve of contacts on the cone of double lines is therefore given by linear combinations of (17) and (5).

## MAPPING OF THE SURFACE. §§ 8–11.

### § 8. *Self-duality of Entire Surface.*

By the principles of mapping it is possible to show that the four polarities of § 3 do, in fact, send every point of Dupin's Cyclide into one of its planes. That the surface may be mapped follows at once from the fact that it is rational This is seen by considering the intersections with the surface of a line which moves always touching the double conic and one line of the surface. Three of its four intersections are thus accounted for; the fourth is variable. If the cutting line also meets some arbitrary plane, this point and the variable intersection with the surface are always in 1, 1 correspondence. The surface thus meets the test for rationality and may be mapped.

In mapping a surface of given order, the curves which map plane sections must pass through a sufficient number of base points so that the free inter-

---

* Salmon, G., "Geometry of Three Dimensions," 5th edition, Vol. I, p. 363.

† The breaking up of this curve into two conics is the condition that the surface take the special form of the "Anchor Ring."

sections of any two shall be equal to the order of the surface. To map a quartic, therefore, by means of general cubics, we need five base points. We have in this way $\infty^4$ cubics. Now, as the $\infty^3$ planes of space are linear functions of the four independent planes, they must be mapped by $\infty^3$ cubics on the five base points, which are linear functions of the four fundamental cubics. All such cubics must accordingly meet one further condition. Let this be apolarity to a general line cubic, $(a\xi)^3=0$. Plane sections of the surface are then mapped by a linear system of cubics in the plane.

Before deriving specific transformations for the mapping, we must take account of the fact that the surface has four double points. This leads to a special arrangement of the base points. In general, it is assumed that no three shall lie on a straight line, but when the surface has a double point, this restriction is removed. This is because each of the net of planes which may be passed through a double point on the surface cuts out a rational section, and must be mapped by a cubic of the same genus. There are thus among the $\infty^3$ cubics on five points and apolar to $(a\xi)^3=0$ four nets of rational cubics. Two such nets are: 1) cubics consisting of a line on three points and a net of conics on the two remaining apolar to the polar conic of the line as to $(a\xi)^3=0$; 2) nets of proper rational cubics with the double point at a base point. Two such independent arrangements of three points on a line, as in case 1, may be made. Case 2 arises when two of the points on a line approach coincidence. Hence of the $\infty^3$ cubics through these points, a net will have a double point. Two such nets are possible. We have thus accounted for peculiarities due to the presence of four double points on the surface, if we take the five base points in two coincident pairs on two lines meeting at the fifth point.

Let these points be the vertices of the reference triangle in the mapping plane, and let the coincident pairs be $(0,1,0)$ and a point consecutive to it on $u_2=0$, $(0,0,1)$ and a consecutive point on $u_1=0$. A cubic curve through these points and apolar to $(a\xi)^3=0$ is of the form

$$(au)^3\equiv\alpha_{001}u_0^2u_1+\alpha_{002}u_0^2u_2+\alpha_{112}u_1^2u_2+\alpha_{122}u_1u_2^2 \\ +(a\alpha_{001}+b\alpha_{002}+c\alpha_{112}+d\alpha_{122})u_0u_1u_2=0.$$

Making the first four coefficients zero in turn, we have four cubics meeting the required conditions

$$f_0\equiv u_0u_1u_2+\beta_0u_0^2u_1=px_0,\quad f_1\equiv u_0u_1u_2+\beta_1u_0^2u_2=qx_1,$$
$$f_2\equiv u_0u_1u_2+\beta_2u_1u_2^2=px_2,\quad f_3\equiv u_0u_1u_2+\beta_3u_1^2u_2=qx_3.$$

The elimination of $u$ from these equations gives the surface in form (1), if certain obvious substitutions are made for $p$ and $q$. The transformation then takes the final form

$$\left.\begin{aligned} f_0 &\equiv x_0 = u_0u_1u_2 - \rho u_0^2u_1/\lambda\beta_2, \\ f_1 &\equiv \mu x_1 = u_0u_1u_2 + \sigma u_0^2u_2/\lambda\beta_3, \\ f_2 &\equiv x_2 = u_0u_1u_2 + \beta_2u_1u_2^2, \\ f_3 &\equiv \mu x_3 = u_0u_1u_2 + \beta_3u_1^2u_2. \end{aligned}\right\} \quad (18)$$

We can now show directly that Dupin's Cyclide is self-dual under the polarities of § 3. Any plane cuts the surface in a quartic curve with two double points which maps into a non-rational cubic. Since the point of contact of a tangent plane is a double point on the surface, each tangent plane cuts the quartic surface in a rational quartic which maps into a rational cubic on five points. The $\infty^2$ tangent planes thus give $\infty^2$ rational cubics. Consequently, if a point $y$ on the surface has as polar plane with respect to one of the quadrics $Q_i$ (7) a tangent plane of the surface, the map of the section will be rational. Assume such a tangent plane as the polar plane of $y$. Then,

$$\left.\begin{aligned} P_1(x) &\equiv m(ky_0+y_2)x_0 + \mu n(ly_1+y_3)x_1 + m(y_0+ky_2)x_2 + \mu n(y_1+ly_3)x_3 = 0, \\ P_1(f) &\equiv m(ky_0+y_2)\{u_0u_1u_2 - \rho u_0^2u_1/\lambda\beta_2\} + n(ly_1+y_3)\{u_0u_1u_2 + \sigma u_0^2u_2/\lambda\beta_3\} \\ &\quad + m(y_0+ky_2)\{u_0u_1u_2 + \beta_2u_1u_2^2\} + n(y_1+ly_3)\{u_0u_1u_2 + \beta_3u_1^2u_2\} = 0. \end{aligned}\right\} \quad (19)$$

This cubic is of the form, in Salmon's coefficients,*

$$a_1u_2^2u_1 + u_2(b_2u_1^2 + b_0u_0^2 + 2b_1u_0u_1) + c_1u_0^2u_1 = 0.$$

If this cubic is rational, its discriminant will vanish. The discriminant is

$$\Delta \equiv 27a_1c_1b_0b_2[b_1^4 - 2b_1^2(b_0b_2 + a_1c_1) + (b_0b_2 - a_1c_1)^2].$$

Substituting from (19) the values of the constants and using the relations † which exist among $k$, $l$, $\lambda$, $\mu$, this becomes

$$\begin{aligned} \Delta \equiv\ & (y_0+ky_2)^2(ky_0+y_2)^2(y_1+ly_3)^2(ly_1+y_3)^2 \\ & [(y_0y_2+\mu y_1y_3)^2 + \lambda(y)\{y_0y_2(y_1+y_3) - \mu y_1y_3(y_0+y_2)\}]. \end{aligned}$$

This expresses that the discriminant vanishes identically when 1) $y$ is a point on a singular conic and when 2) $y$ is any point on the original surface (1). As the same result is obtained by using the polar plane of $y$ as to each of the remaining quadrics, we may state the

THEOREM. *Dupin's Cyclide is a surface self-dual under four polarities.*

## § 9. *Plane Equation of Surface.*

We may note here that the discriminant of the map equation of a tangent plane gives the equation of the surface in planes. Substituting for $y$ in the

---

* Salmon, G., "Higher Plane Curves," 3d edition, p. 194.

† $\sqrt{\lambda}\,l_1 = \sqrt{\lambda+\mu} + \sqrt{\mu}$, $\quad \sqrt{\lambda}\,l_2 = \sqrt{\lambda+\mu} - \sqrt{\mu}$

$\sqrt{\lambda}\,k_1 = -\sqrt{1-\lambda} + \sqrt{1}$, $\quad \sqrt{\lambda}\,k_2 = -\sqrt{1-\lambda} - \sqrt{1}$,

$\lambda(k-1)^2 = -4k$, $\quad \lambda(l-1)^2 = 4\mu l$,

$\lambda(k+1)^2 = -4(1-\lambda)k$, $\quad \lambda(l+1)^2 = 4(\mu+\lambda)l$,

$\lambda(k^2+1) = 2k(\lambda-2)$, $\quad \lambda(l^2+1) = 2l(2\mu+\lambda)$.

discriminant its coordinates as the pole of a plane $p$, we have the equation of the surface in the terms of $\eta$, the coordinates of the plane,

$$y_0=\mu h(k\eta_0-\eta_2), \qquad y_1=r(l\eta_1-\eta_3),$$
$$y_2=\mu h(k\eta_2-\eta_0), \qquad y_3=r(l\eta_3-\eta_1),$$
$$S\equiv[\mu^2(4\eta_0\eta_2+\lambda\overline{\eta_0-\eta_2}^2+4\mu\eta_1\eta_3-\lambda\overline{\eta_1-\eta_3}^2]^2+4\mu\lambda(\mu\,\overline{\eta_0+\eta_2}+\eta_1+\eta_3)$$
$$[\mu\overline{\eta_1+\eta_3}\{4\eta_0\eta_2+\lambda(\eta_0-\eta_2)^2\}-(\eta_0+\eta_2)\{4\mu\eta_1\eta_3-\lambda(\eta_1-\eta_3)^2\}]=0.$$

## § 10. *The Mapping of Plane Sections.*

In general, the map of a plane section of this surface is a quartic curve of genus 1, which is in 1, 1 correspondence with a cubic of the same genus in the mapping plane. There are, however, four classes of plane sections, beside those made by the tangent planes, which are rational and accordingly map into rational cubics. A single section also exists which is not in 1, 1 correspondence with its mapping curve. We shall take up separately the maps of these sections.

Two sets of rational sections are those made by planes through the double points and the lines of the surface. Through each double point is a net of planes cutting sections which have this point as a third double point.

| *Double Point.* | *Map of Planes through Point.* |
|---|---|
| $\eta_0$ | $u_2=0$; a net of conics tangent to $u_1=0$ at $(0, 0, 1)$. |
| $\eta_1$ | $u_1=0$; a net of conics tangent to $u_2=0$ at $(0, 1, 0)$. |
| $\eta_2$ | A net of cubics with double point at $(0, 0, 1)$ and tangent to $u_2=0$ at $(0, 1, 0)$. |
| $\eta_3$ | A net of cubics with double point at $(0, 1, 0)$ and tangent to $u_1=0$ at $(0, 0, 1)$. |

By considering the part of the mapping curve common to all curves of the net, we determine the maps of the double points.

| | | | | |
|---|---|---|---|---|
| *Double Point:* | $\eta_0$, | $\eta_1$, | $\eta_2$, | $\eta_3$. |
| *Map:* | $u_2=0$, | $u_1=0$, | $(0, 0, 1)$, | $(0, 1, 0)$. |

Sections of the surface by pencils of planes through a line on the surface break into the line and a pencil of rational cubics with double points on the double conic.

| *Line.* | *Map of Planes through Line.* |
|---|---|
| $\overline{\eta_0\eta_1}$ | $u_1=0$; $u_2=0$; a pencil of lines through $(1, 0, 0)$. |
| $\overline{\eta_2\eta_3}$ | $u_0=0$; a pencil of conics on three reference points. |
| $\overline{\eta_1\eta_2}$ | $u_1=0$; a pencil of conics tangent to $u_2=0$ at $(0, 1, 0)$ and passing through $(0, 0, 1)$. |
| $\overline{\eta_0\eta_3}$ | $u_2=0$; a pencil of conics tangent to $u_1=0$ at $(0, 0, 1)$ and passing through $(0, 1, 0)$. |

(20)

The maps of the lines of the surface are thus known.

$$\begin{array}{lcccc} \textit{Line:} & \overline{\eta_0\eta_1}, & \overline{\eta_2\eta_3}, & \overline{\eta_1\eta_2}, & \overline{\eta_0\eta_3}. \\ \textit{Map:} & u_1u_2=0, & u_0=0, & u_1=0, & u_2=0. \end{array}$$

In each plane pencil is one plane tangent to the surface along the entire line and hence cutting the surface in a conic and the line counted twice.

$$\left.\begin{array}{lcc} \textit{Line.} & \textit{Singular Plane.} & \textit{Map.} \\ \overline{\eta_0\eta_1} & x_2-\mu x_3=0 & u_1u_2[\beta_2u_2-\beta_3u_1]=0 \\ \overline{\eta_2\eta_3} & x_0-\mu x_1=0 & u_0^2[\rho\beta_3u_1+\beta_2\nu u_2]=0 \\ \overline{\eta_1\eta_2} & x_0-\mu x_3=0 & u_1[\rho u_0^2+\beta_3u_1u_2]=0 \\ \overline{\eta_0\eta_3} & x_2-\mu x_1=0 & u_2[\sigma u_0^2-\lambda\beta_2\beta_3u_1u_2]=0 \end{array}\right\} \quad (21)$$

Before considering the remaining classes of rational plane sections of the surface, it is of advantage to map the double conic in the plane at infinity. Its map equation is

$$\mu\lambda\beta_2\beta_3C\equiv\sigma\beta_2u_0^2u_2-\rho\mu\beta_3u_0^2u_1+\mu\lambda\beta_2^2\beta_3u_1u_2^2+\lambda\beta_2\beta_3^2u_1^2u_2+2\nu\lambda\beta_2\beta_3u_0u_1u_2=0. \quad (22)$$

The conic is unique in that it is in 1—2 correspondence with the cubic $C$. The two points of the conic which are double points on every plane section appear in the mapping plane as four common points of the maps of the section and the conic. If a pencil of planes is taken on a secant of the double conic, the corresponding pencil of cubics in the mapping plane will be determined by nine points on $C$,—the five base points, which are on the maps of all plane sections, and the four points which map the intersection of the conic and secant. If, however, the base of the pencil of planes becomes a tangent of the conic, the four variable points on $C$ become two pairs of consecutive points at which each cubic of the pencil has double contact with $C$. Since every fixed point through which a pencil of curves passes with a fixed tangent is the double point of one curve of the pencil, we find in this pencil of non-rational cubics two rational curves with double points, respectively, at one or the other of the two points on $C$. These two cubics map the sections of the surface by the two tangent planes in the plane pencil. At the four points in which the double conic is met by a line on the surface, one of these two tangent planes is a singular plane containing the line. The map of this section accordingly degenerates. At one of the two points on $C$ mapping such a point, we have a double point of a proper rational cubic; at the other a double point of a degenerate cubic. Four exceptional points on the double conic must, moreover, be noted. These are the pinch points, which are mapped by a single point each. The pencil of planes on a tangent to the double conic at one of these points is mapped by a pencil of cubics having contact of the second order with $C$ at the corresponding point. The one rational curve in the pencil has the tangent of $C$ as one tangent at the double point. It maps the single tangent plane of the surface at the pinch point.

If now we consider in relation to $C$ the pencil of planes on a line of the surface (20), the map of each section will contain the two fixed intersections with $C$ which map the common point of line and double conic, while the variable points will map the other double point of the section. The two fixed points will be those mapping the contact of the singular plane in the pencil. In the case of planes on $\overline{\eta_0\eta_1}$, two intersections with $C$ of the variable line must map the variable double point of the section. The third intersection and a fourth point must be fixed, and map the fixed intersection of line and conic. Hence, the variable lines pass through a fixed point and form a pencil. We determine this point from a line in the map of the singular plane in the pencil, $\beta_2 u_2 - \beta_3 u_1 = 0$. This intersects the variable line and touches $C$ at $(-\beta_2\beta_3, \beta_2, \beta_3)$, and meets $C$ again at (1, 0, 0). These two points are accordingly the map of the fixed double point, and $(-\beta_2\beta_3, \beta_2, \beta_3)$ is the vertex of the pencil.

The discussion of the planes through the three other lines gives the same results from a more general point of view. The planes on $\overline{\eta_2\eta_3}$, for example, are mapped by $u_0 = 0$ and conics on three reference points. The map of the singular plane is

$$u_0^2(\rho\beta_3 u_1 + \sigma\beta_2 u_2) = 0.$$

Now $u_0 = 0$ meets $C$ in (0, 1, 0), (0, 0, 1) and $(0, -\mu\beta_2, \beta_3)$. The line

$$\rho\beta_3 u_1 + \sigma\beta_2 u_2 = 0$$

touches $C$ in $(\lambda\beta_2\beta_3, -\beta_2\sigma, \beta_3\rho)$ and has (1, 0, 0) as residual intersection. Of these five points all but $(0, -\mu\beta_2, \beta_3)$ are on all conics of the system. These points, however, are found only on the map of the singular plane, and map its contact with the double conic. The cubics in planes on $\overline{\eta_2\eta_3}$ are then mapped by a pencil of conics on four points of $C$. The two remaining intersections of each conic map the double point of the corresponding section, and these points, by the geometry of the plane cubic, lie on a line through a fixed point of $C$. This is readily seen to be $(-\beta_2\beta_3, \beta_2, \beta_3)$. Exactly similar results are obtained for the planes on the two remaining lines.

We have thus verified for Dupin's Cyclide the properties stated by Clebsch* for the general quartic surface with a double conic; that planes on a line of the surface are mapped by 1) the map of the line and 2) a pencil of curves with base points on the map of the double conic and meeting this curve also in pairs of points which map the single points of the double conic. These point pairs determine a central involution with vertex on the map of the conic.

---

* Borchardt, *Journal für Mathematik*, Vol. LXIX (1868), p. 147.

Sections made by planes on the diagonal lines of the reference tetrahedron are two conics. The map equations give, for planes on $\overline{\eta_1\eta_3}$, a pencil of degenerate conics with double point at $(0,1,0)$, and $u_1=0$; for planes on $\overline{\eta_0\eta_2}$ a similar pencil of conics with double point at $(0,0,1)$, and $u_2=0$. Special cases of these planes are the tropes which touch the surface along conics counted twice.

$$\begin{array}{ll} \textit{Trope.} & \textit{Map.} \\ k_i x_0+x_2=0 & u_1(k_i+1)(u_0+2\beta_2 u_2)^2=0 \\ l_i x_1+x_3=0 & u_2(l_i+1)(u_0+2\beta_3 u_1)^2=0 \end{array} \Bigg\} \quad (23)$$

The points where the tropes meet the double conic are mapped by the coincident point pairs in which the four tangents from $(-\beta_2\beta_3,\ \beta_2,\ \beta_3)$ to $C$ meet the curve. From the coordinates of these points we find that the maps of the tropes pass through them by pairs just as the tropes pass by pairs through the pinch points on the double conic. These four points, which map the pinch points, are

$$\{-2\beta_2\beta_3,\ \ \beta_2(l_i+1),\ \ \beta_3(k_i+1)\} \qquad (i=1,2).$$

A tangent plane to the cone of double lines also cuts the surface in two conics. The map of a section is a cubic for which the invariants* $S$ and $T$ both vanish. Hence, the mapping cubic has a cusp.

We have thus considered in detail the map of the double conic, and the maps of the five sets of planes giving rise to rational sections. These groups of plane sections and the general forms of their maps are as follows:

1. A double infinity of simple tangent planes. These map into $\infty^2$ proper rational cubics.

2. Four nets of planes through the double points. These map into four nets of rational cubics, two of which are degenerate.

3. Four pencils of planes on the lines of the surface. These map into four pencils of degenerate cubics consisting of three lines or a line and a conic.

4. Two pencils of planes on the diagonals of the reference tetrahedron. These map into two pencils of degenerate cubics consisting of three lines.

5. A single infinity of planes tangent to the cone of double lines. These map into $\infty^1$ rational cubics with a cusp.

The section by the plane at infinity, the double conic, has been shown to be in 1, 2 correspondence with a non-rational cubic of the mapping plane, and the relation between the two curves has been found to be the same as that between the corresponding curves on the most general quartic surface with a double conic.

---

* Salmon, G., "Higher Plane Curves", 3d edition, p. 210.

## § 11. *The Map of a Quartic Curve.*

The quartic curve which is the locus of contacts of elements of the double cone is cut out by a quadric (§ 7). Substituting $x_i = f_i$, we have the map equation of the curve. This falls into two factors, one of which is the map of the double conic, the other

$$\beta_3\sigma(\rho u_0^2 u_1 - \lambda\beta_2^2 u_1 u_2^2) - \rho\beta_2(\sigma u_0^2 u_2 - \lambda\beta_3^2 u_1^2 u_2) = \sigma. \tag{24}$$

This shows that the octavic curve in which the quartic and quadric surfaces meet consists of the double conic, and the quartic on the double cone. The cubic (24) is in form similar to the map equations of plane sections, but differs from them in that there is not a linear relation existing between it and the four fundamental cubics.

## GROUP PROPERTIES OF TRANSFORMATIONS ASSOCIATED WITH THE SURFACE.
## §§ 12–13.

## § 12. *The Group of Correlations on the Surface.*

The four polarities (§ 3) and three known collineations $C$ which leave the surface unaltered, constitute a group, $G_8$.

| $C_1$. | $C_2$. | $C_3$. |
|---|---|---|
| $\rho' x_0' = x_2$ | $\rho' x_0' = x_0$ | $\rho' x_0' = x_2$ |
| $\rho' x_1' = x_1$ | $\rho' x_1' = x_3$ | $\rho' x_1' = x_3$ |
| $\rho' x_2' = x_0$ | $\rho' x_2' = x_2$ | $\rho' x_2' = x_0$ |
| $\rho' x_3' = x_3$ | $\rho' x_3' = x_1$ | $\rho' x_3' = x_1$ |

Forming the polars of a point $y$ as to the quadrics $Q$ (7), we have

$$\left.\begin{aligned}
P_1 &\equiv m(ky_0+y_2)x_0+\mu n(ly_1+y_3)x_1+m(y_0+ky_2)x_2+\mu n(y_1+ly_3)x_3,\\
lP_2 &\equiv m(ky_0+y_2)x_0+\mu n(y_1+ly_3)x_1+m(y_0+ky_2)x_2+\mu n(ly_1+y_3)x_3,\\
kP_3 &\equiv m(y_0+ky_2)x_0+\mu n(ly_1+y_3)x_1+m(ky_0+y_2)x_2+\mu n(y_1+ly_3)x_3,\\
lkP_4 &\equiv m(y_0+ky_2)x_0+\mu n(y_1+ly_3)x_1+m(ky_0+y_2)x_2+\mu n(ly_1+y_3)x_3.
\end{aligned}\right\} \tag{25}$$

If we transform any one of these polars by any one of the three collineations, the result is the equation of another polar. For example, $C_1P_1 \equiv P_3$. To operate with one collineation on another gives the third, as $C_1C_3 \equiv C_2$. As the result of carrying out these transformations in all possible ways, we have forms from which we observe the following facts:

1. Any collineation or any polarity operating on itself gives identity.
2. The product of two unlike collineations gives the third, and of two unlike polarities gives a collineation.
3. Each collineation and each polarity is its own inverse.
4. The products are commutative.

All requirements for a group are thus met and we may state our result in a

THEOREM. *The four polarities P and the three collineations C, with identity, form a group $G_8$.*

This group is defined by the equations

$$\begin{array}{ll} C_m^2=1 & P_i^2=1 \\ C_m=C_m^{-1} & P_i=P_i^{-1} \\ C_mC_k=C_l & P_iP_j=C_m \\ C_mC_k=C_kC_m & P_iP_j=P_jP_i \\ (m,k,l=1,2,3;\ m\neq k\neq l) & (i,j=1,2,3,4;\ i\neq j) \end{array}$$

The generating elements are the collineations $C_1$ and $C_2$, and a polarity. The collineations form the only subgroup, which is a "four-group."

The multiplication table of the group is

| | 1 | $C_1$ | $C_2$ | $C_3$ | $P_1$ | $P_2$ | $P_3$ | $P_4$ |
|---|---|---|---|---|---|---|---|---|
| 1 | 1 | $C_1$ | $C_2$ | $C_3$ | $P_1$ | $P_2$ | $P_3$ | $P_4$ |
| $C_1$ | $C_1$ | 1 | $C_3$ | $C_2$ | $P_2$ | $P_1$ | $P_4$ | $P_3$ |
| $C_2$ | $C_2$ | $C_3$ | 1 | $C_1$ | $P_3$ | $P_4$ | $P_1$ | $P_2$ |
| $C_3$ | $C_3$ | $C_2$ | $C_1$ | 1 | $P_4$ | $P_3$ | $P_2$ | $P_1$ |
| $P_1$ | $P_1$ | $P_2$ | $P_3$ | $P_4$ | 1 | $C_1$ | $C_2$ | $C_3$ |
| $P_2$ | $P_2$ | $P_1$ | $P_4$ | $P_3$ | $C_1$ | 1 | $C_3$ | $C_2$ |
| $P_3$ | $P_3$ | $P_4$ | $P_1$ | $P_2$ | $C_2$ | $C_3$ | 1 | $C_1$ |
| $P_4$ | $P_4$ | $P_3$ | $P_2$ | $P_1$ | $C_3$ | $C_2$ | $C_1$ | 1 |

(26)

Any equation symmetric in $x_0$, $x_2$ and $x_1$, $x_3$ is invariant under the subgroup of collineations. Such forms are the equations of the planes $(x_0\pm x_2)+k(x_1\pm x_3)=0$, the quadric $x_0x_2+kx_1x_3=0$ and the surface itself. Invariant lines are given by

$$(1)\ \begin{matrix} x_0=0 \\ x_2=0 \end{matrix} \qquad (2)\ \begin{matrix} x_1=0 \\ x_3=0 \end{matrix} \qquad (3)\ \begin{matrix} x_0+x_2=0 \\ x_1+x_3=0 \end{matrix} \qquad (4)\ \begin{matrix} x_0-x_2=0 \\ x_1-x_3=0 \end{matrix}$$

The first three are invariant in that they are sent into themselves by $G_4$. The points of the last are absolutely invariant. These four lines, moreover, meet in four points and form two pairs of opposite generators on $x_0x_1-x_2x_3=0$. The polar as to the quadrics $Q$ of a point on one generator of a pair is a plane on the other. We may then say that these lines are also invariant under $G_8$.

The 1, 1 correspondence between the points on (4) and the planes on (3) also arises from the operations of the group $G_8$ on points of the surface. By the collineations $C$, a point $(y_0, y_1, y_2, y_3)$ goes into

$$(y_2, y_1, y_0, y_3), \quad (y_0, y_3, y_2, y_1), \quad (y_2, y_3, y_0, y_1).$$

We call these four points a $Y$ set. By the four polarities $Q$ the same point goes into four planes $P$ (25), the points of contact of which with the surface we call an $X$ set. The planes $P$ are found to meet in a point $y'$ and the points $Y$ to lie on a plane $P'$, the polar of $y'$ as to all four quadrics. The point is on the fixed line

$$x_0 - x_2 = 0, \quad x_1 - x_3 = 0;$$

the plane is on the other fixed line

$$x_0 + x_2 = 0, \quad x_1 + x_3 = 0.$$

By the properties of the group it is seen that the eight points $X$ and $Y$ form a closed set of which any one determines the remaining seven, and that the relation between the points $X$ and their polar planes is in all respects the same as that between the points $Y$ and their polar planes. Thus, if we start with a point $X$, we obtain by the polarities four planes touching the surface at points $Y$ and meeting in a point which is the pole as to the quadrics $Q$ of the plane which contains the points $X$. This new point and plane are found to lie respectively on the lines

$$x_0 - x_2 = 0, \quad x_1 - x_3 = 0 \quad \text{and} \quad x_0 + x_2 = 0, \quad x_1 + x_3 = 0.$$

The pairs of points, arising thus on the line

$$x_0 - x_2 = 0, \quad x_1 - x_3 = 0,$$

from every point on the surface, form an involution along the line, the fixed points of which are the two at which the four quadrics $Q$ have double contact. The corresponding pairs of planes lie on the line

$$x_0 + x_2 = 0, \quad x_1 + x_3 = 0$$

also form an involution, the fixed planes of which are the two common tangent planes of the four quadrics at these points of contact.

We have thus the conclusion that the eight points and eight planes arising from the operations of the group on any point of the surface fall into two sets of four. Each set of four planes meets in a point on one line fixed under the group, while each set of four points lies in a plane passing through a second fixed line. The intersection of the planes tangent at one set of four points is the polar point as to the quadrics of the plane containing the four other points.

This group of correlations $G_8$ under which the surface is self-dual, is the only such group. If another group existed, it would be formed by adding to some new collineation group the known correlations. Applying the general collineation

$$\rho x'_i = \sum_{k=1}^{k=4} a_{ik} x_k$$

to the equation of the surface, with the restriction that the surface remain unaltered, we find that the only solutions possible are the four transformations given by the collineation subgroup of $G_8$. The correlation group is, therefore, unique.

## § 13. *The Group of Cremona Transformations in the Mapping Plane.*

We have seen that the transformation of polarity as to the fundamental quadrics $Q$ gives for any point $y$ on the surface a plane tangent to the surface. The point of contact $x$ is a double point on the section by the plane. Hence, the map equation of this plane is a rational cubic with double point $z$. The point $y$ maps into some point $u$ of the mapping plane. Thus, for a pair of points $y$ and $x$, arising from a correlation on the surface, we have a pair $u$ and $z$ in the plane. We now proceed to derive a transformation, which applied to $u$, the map of a given point $y$, gives the point $z$ corresponding to $x$.

The coordinates of the double point $z$ in a curve mapping a tangent plane of the surface may be found by means of the combinant

$$f_0(J_{123})_{u=z}+f_1(J_{023})_{u=z}+f_2(J_{013})_{u=z}+f_3(J_{012})_{u=z}=0. \tag{27}$$

The quantities $f_i$ are the four cubics (18), the quantities $J_i$ are each the Jacobian of three of the cubics. Since the Jacobian of a net passes through the double points of the net, a curve given by (27) has, for a particular value of $z$, a double point at $z$. Let us set this equation identically equal to the map equation of a tangent plane of the surface, and equate coefficients of terms in $u$. We obtain four equations in $y$ and $z$. But since $y$ may be expressed as a function of $u$, our equations give $z$ in terms of $u$. Solving, we have corresponding to the four polar transformations of a point $y$ on the surface, four sets of relations between $z$ and $u$ in the plane.

$$\begin{aligned}
\Pi_1:\ & z_0=+\sqrt{\lambda}\beta_2\beta_3(nmu_0^2+2ln\beta_3u_0u_1+2km\beta_2u_0u_2+4kl\beta_2\beta_3u_1u_2),\\
& z_1=-\sqrt{\sigma}l\beta_2(nmu_0^2+2km\beta_2u_0u_2+2n\beta_3u_0u_1+4k\beta_2\beta_3u_1u_2),\\
& z_2=-\sqrt{-k\rho}\beta_3(nmu_0^2+2ln\beta_3u_0u_1+2m\beta_2u_0u_2+4l\beta_2\beta_3u_1u_2);\\
\Pi_2:\ & z_0=+\sqrt{\lambda l}\,\beta_2\beta_3(nmu_0^2+2km\beta_2u_0u_2+2n\beta_3u_0u_1+4k\beta_2\beta_3u_1u_2),\\
& z_1=-\sqrt{\sigma}\beta_2(nmu_0^2+2km\beta_2u_0u_2+2ln\beta_3u_0u_1+4kl\beta_2\beta_3u_1u_2),\\
& z_2=-\sqrt{-kl\rho}\beta_3(nmu_0^2+2m\beta_2u_0u_2+2n\beta_3u_0u_1+4\beta_2\beta_3u_1u_2);\\
\Pi_3:\ & z_0=+\sqrt{\lambda k}\beta_2\beta_3(nmu_0^2+2\beta_2mu_0u_2+2ln\beta_3u_0u_1+4l\beta_2\beta_3u_1u_2),\\
& z_1=-\sqrt{kl\sigma}\beta_2(nmu_0^2+2m\beta_2u_0u_2+2n\beta_3u_0u_1+4\beta_2\beta_3u_1u_2),\\
& z_2=-\sqrt{-\rho}\beta_3(nmu_0^2+2km\beta_2u_0u_2+2ln\beta_3u_0u_1+4kl\beta_2\beta_3u_1u_2);\\
\Pi_4:\ & z_0=+\sqrt{\lambda kl}\,\beta_2\beta_3(nmu_0^2+2m\beta_2u_0u_2+2n\beta_3u_0u_1+4\beta_2\beta_3u_1u_2),\\
& z_1=-\sqrt{k\sigma}\,\beta_2(nmu_0^2+2m\beta_2u_0u_2+2lm\beta_3u_0u_1+4l\beta_2\beta_3u_1u_2),\\
& z_2=-\sqrt{-l\rho}\,\beta_3(nmu_0^2+2km\beta_2u_0u_2+2n\beta_3u_0u_1+4k\beta_2\beta_3u_1u_2).
\end{aligned}$$

These transformations are involutory. We have then four birational transformations which set up a 1—1 correspondence between points $u$ and $z$. Hence, they are Cremona transformations.

Such a transformation sends the $\infty^2$ lines of the plane into a net of conics on the three singular points of the transformation. Each quadratic expression for $z$ factors into the maps of two tropes on the surface, and from its form is seen to pass through $(0,1,0)$ and $(0,0,1)$. Hence, the singular points for each transformation $\Pi$ are these two reference points and the map of one pinch point. The singular lines of each transformation, given by the Jacobian of the quadratics in $u$, are $u_0=0$ and the maps of the two tropes which pass through this pinch point. The four points on the surface in which it is cut by the line of fixed points

$$x_0-x_2=0, \quad x_1-x_3=0$$

map into four points invariant under all the transformations $\Pi$.

If we apply these transformations to a point $u$, we obtain four points $z$. If we apply the same transformations to the point $z$, we obtain identity and three new points $z'$.

$$\begin{aligned}
T_1:\ & z'_0=\lambda\beta_3^2u_0u_1; && z'_1=\nu u_0^2; && z'_2=\lambda\beta_3^2u_1u_2;\\
T_2:\ & z'_0=\lambda\beta_2^2u_0u_2; && z'_1=\lambda\beta_2^2u_1u_2; && z'_2=-\rho u_0^2;\\
T_3:\ & z'_0=\lambda\beta_2^2\beta_3^2u_1u_2; && z'_1=\sigma\beta_2^2u_0u_2; && z'_2=-\rho\beta_3^2u_0u_1.
\end{aligned}$$

The singular elements are the points and lines of the reference triangle. The fixed elements are the four points invariant under $\Pi$. These transformations T correspond to the collineations on the surface in the same sense that the transformations $\Pi$ correspond to the polarities. The seven elements $\Pi$ and T with identity form an abelian $G_8$ like in all particulars to the correlation group of the surface (26).

We have thus shown that the correspondence on the surface, arising from the correlation group, between a point $y$ and a tangent plane with contact $x$, gives rise in the mapping plane to a group of Cremona transformations which show the same correspondence between the maps of the points $y$ and $x$.

# *Projective Differential Geometry of One-Parameter Families of Space Curves, and Conjugate Nets on a Curved Surface.*

Second Memoir.*

By Gabriel M. Green.

## § 1. *Introduction and Fundamental Equations.*

In a preceding paper,† we have laid the foundations for a purely projective theory of conjugate nets on a surface and for the related theories of a one-parameter family of space curves and the congruence of tangents thereto. In fact, conjugate nets seemed to appear only incidentally, as a convenient avenue of approach to the apparently more general theory of the congruence of tangents to a one-parameter family of space curves. To make this point clearer: the equations

$$y^{(k)}=f^{(k)}(u,v), \qquad (k=1,2,3,4) \tag{1}$$

represent for $v=$const. a one-parameter family of curves $C_u$, if the four $y$'s be interpreted as homogeneous coordinates of a point in space. It is assumed that the surface defined by equations (1) is not developable, and also that the two one-parameter families of curves $C_u$ ($v=$ const.) and $C_v$ ($u=$const.) are not asymptotics on this surface. Under these conditions the four functions $y^{(k)}$ form a fundamental system of solutions of the completely integrable system of partial differential equations

$$\left.\begin{aligned} y_{uu}&=ay_{vv}+by_u+cy_v+dy, \\ y_{uv}&=a'y_{vv}+b'y_u+c'y_v+d'y, \end{aligned}\right\} \tag{2}$$

where $a'^2-a\neq 0$ and $a\neq 0$.‡ If $a'\neq 0$, the parameter net on the integral surface is not conjugate. Suppose $a'\neq 0$; then, in this, the general case, the family of curves conjugate to the family $C_u$ is determined by making the transformation $\bar{u}=U(u,v)$, where $U(u,v)$ satisfies the partial differential equation of the first order, §

$$a'U_u-aU_v=0. \tag{3}$$

* The first three sections were presented to the American Mathematical Society February 28, 1914, and the last four October 30, 1915.

† G. M. Green, "Projective Differential Geometry of One-parameter Families of Space Curves, and Conjugate Nets on a Curved Surface," American Journal of Mathematics, Vol. XXXVII (1915), pp. 215–246. This will be cited as "preceding paper."

‡ Preceding paper, § 1.

§ Preceding paper, p. 221.

The new family of curves $\bar{u}$=const. will then be conjugate to the given family $v$=const., which is undisturbed by the transformation. In fact, the new net is defined by a fundamental system of solutions of a completely integrable system of partial differential equations of the form*

$$\left.\begin{aligned} y_{uu}&=ay_{vv}+by_u+cy_v+dy, \\ y_{uv}&=b'y_u+c'y_v+d'y. \end{aligned}\right\} \tag{4}$$

We have shown, in the memoir cited, that there is no loss of generality whatever in studying a one-parameter family of curves, defined by solutions of the differential equations (2), through the conjugate parameter net defined by system (4), in spite of the fact that the passage from equations (2) to equations (4) requires the integration—in general impossible—of the partial differential equation (3). The second of equations (4) is perfectly symmetric in the parameters $u$ and $v$, but the first is not, and can not easily be made so without sacrifice of simplicity. It appears, then, as we have already said, that the differential equations (4) are more suited—if only on æsthetic grounds—for the study of the congruence of tangents to a one-parameter family of curves, than for that of the conjugate net determined thereby.

It was with these considerations in mind that in the preceding paper we treated the two component families of the conjugate net in an unsymmetric way. In particular, we chose a fundamental system of covariant points as follows. Three of them were a point on the surface and its corresponding first and minus first Laplace transforms; the fourth was the second Laplace transform. We shall in the present paper provide a substitute for this fourth point which will serve equally well in a theory of a one-parameter family of curves and in the theory of a conjugate net.

The introduction of this new covariant point in § 2 follows from a consideration of an important congruence of lines, which, with other congruences associated with the conjugate net, is studied in some detail in the course of the paper. A canonical development in non-homogeneous coordinates of the surface referred to the conjugate net is given, and leads to the consideration of a certain unodal cubic surface. Many questions of interest in the general theory of congruences are discussed in the latter half of the memoir. A conjugate net which is uniquely determined by the given one is introduced; we have ventured to call it the *associate conjugate net*, and lay the foundations for a systematic study thereof in its relation to the given conjugate net. In the last section is given a short treatment of isothermally conjugate nets, in the course of which a purely geometric characterization of these nets is obtained.

* Preceding paper, equations (16).

The completely integrable system (4), upon which we base our theory, has its coefficients restricted by certain integrability conditions. By differentiation, we may obtain from equations (4) the expressions *

$$\left.\begin{aligned} y_{uuu}&=\alpha^{(1)}y_{vv}+\beta^{(1)}y_u+\gamma^{(1)}y_v+\delta^{(1)}y,\\ y_{uuv}&=\alpha^{(2)}y_{vv}+\beta^{(2)}y_u+\gamma^{(2)}y_v+\delta^{(2)}y,\\ y_{uvv}&=\alpha^{(3)}y_{vv}+\beta^{(3)}y_u+\gamma^{(3)}y_v+\delta^{(3)}y,\\ y_{vvv}&=\alpha^{(4)}y_{vv}+\beta^{(4)}y_u+\gamma^{(4)}y_v+\delta^{(4)}y,\end{aligned}\right\} \tag{5}$$

where †

$$\left.\begin{aligned} &\alpha^{(1)}=a(b+c')+a_u,\quad \beta^{(1)}=b_u+ab'_v+d+b^2+b'(c+ab'),\\ &\qquad\gamma^{(1)}=c_u+ac'_v+ad'+bc+c'(c+ab'),\\ &\qquad\delta^{(1)}=d_u+ad'_v+bd+d'(c+ab'),\\ &\alpha^{(2)}=ab',\quad \beta^{(2)}=b'(b+c')+b'_u+d',\quad \gamma^{(2)}=b'c+c'^2+c'_u,\\ &\qquad\delta^{(2)}=b'd+c'd'+d'_u,\\ &\alpha^{(3)}=c',\quad \beta^{(3)}=b'^2+b'_v,\quad \gamma^{(3)}=b'c'+c'_v+d',\quad \delta^{(3)}=b'd'+d'_v,\\ &\alpha^{(4)}=\frac{1}{a}(ab'-c-a_v),\quad \beta^{(4)}=\frac{1}{a}(b'c'+b'_u-b_v+d'),\\ &\gamma^{(4)}=\frac{1}{a}[b'c+c'(c'-b)+c'_u-c_v-d],\\ &\delta^{(4)}=\frac{1}{a}[b'd+d'(c'-b)+d'_u-d'_v].\end{aligned}\right\} \tag{6}$$

It is tacitly assumed, of course, that $a\neq 0$. The conditions of complete integrability are ‡

$$\left.\begin{aligned} \gamma^{(3)}+\alpha^{(3)}_v&=a\beta^{(4)}+\alpha^{(4)}_u,\\ (\alpha^{(3)}-b)\beta^{(4)}+b'\beta^{(3)}+\beta^{(3)}_v&=\alpha^{(4)}\beta^{(3)}+b'\gamma^{(4)}+\beta^{(4)}_u+\delta^{(4)},\\ c'\beta^{(3)}+\gamma^{(3)}_v+\delta^{(3)}&=\alpha^{(4)}\gamma^{(3)}+c\beta^{(4)}+\gamma^{(4)}_u,\\ \alpha^{(3)}\delta^{(4)}+d'\beta^{(3)}+\delta^{(3)}_v&=\alpha^{(4)}\delta^{(3)}+d\beta^{(4)}+d'\gamma^{(4)}+\delta^{(4)}_u,\end{aligned}\right\} \tag{7}$$

the first of which, in virtue of equations (6), being equivalent to the equation

$$\frac{\partial}{\partial v}(b+2c')=\frac{\partial}{\partial u}\left(\frac{2ab'-c-a_v}{a}\right), \tag{8a}$$

or to

$$p_u=b+2c',\quad p_v=\frac{2ab'-c-a_v}{a}. \tag{8b}$$

## § 2. *The Axis Congruence.*

In the preceding paper, we discussed at some length the Laplace transforms of a conjugate net defined by a fundamental system of solutions of equations

* Preceding paper, § 3, equations (20). † Preceding paper, equations (21).
‡ Preceding paper, equations (23).

(4). We found for the minus first and first Laplace transforms the respective expressions

$$\rho = y_u - c'y, \qquad \sigma = y_v - b'y. \tag{9}$$

These two points lie with $y$ in the tangent plane to the surface $S_y$. We wish to obtain a fourth covariant point off this tangent plane, thus determining a covariant tetrahedron of reference. This we did in the preceding paper by choosing the second Laplace transform,*

$$\sigma_1 = \sigma_v - \left(b' + \frac{H_v}{H}\right)\sigma = y_{vv} - \left(2b' + \frac{H_v}{H}\right)y_v + b'\left(b' + \frac{H_v}{H} - \frac{b'_v}{b'}\right)y.$$

We propose to substitute for this point another, in the definition of which the two families $C_u$ and $C_v$, which make up the conjugate net, enter symmetrically. Instead of taking the line $y\sigma_1$ as an edge of the tetrahedron, let us choose the line $yz$ determined by the intersection of the two planes which osculate at the point $y$ the two curves of the net passing through $y$.

It will simplify our subsequent calculations if we write the first of equations (4) in the form

$$y_{vv} = \alpha y_{uu} + \beta y_u + \gamma y_v + \delta y, \tag{10}$$

where

$$\alpha = \frac{1}{a}, \quad \beta = -\frac{b}{a}, \quad \gamma = -\frac{c}{a}, \quad \delta = -\frac{d}{a}. \tag{11}$$

This is always possible, since $a \neq 0$.

We may write equation (10), or, what is the same thing, the first of equations (4), in the form

$$y_{vv} - \gamma y_v = \alpha y_{uu} + \beta y_u + \delta y.$$

The left-hand member represents a point in the osculating plane to the curve $C_v$, and the right-hand member a point in the osculating plane to the curve $C_u$. Consequently, the point

$$z = y_{vv} - \gamma y_v = \alpha y_{uu} + \beta y_u + \delta y \tag{12}$$

lies on the line of intersection of the two osculating planes.

We may therefore associate with each point $y$ of the surface $S_y$ a point $z$, so that through each point of the surface will pass a definite line $yz$. With the totality of points $y$ of the surface will be associated $\infty^2$ lines $yz$, constituting a congruence which Wilczynski has called the *axis congruence*.† The point $z$

* Preceding paper, equations (51).

† E. J. Wilczynski, "The General Theory of Congruences," *Transactions of the American Mathematical Society*, Vol. XVI (1915), pp. 311–327. The present writer made independent use of the axis congruence, in obtaining the canonical development given below, § 3, without a knowledge of Wilczynski's work. This development was, in fact, announced at the February, 1914, meeting of the American Mathematical Society.

is not, however, a covariant point, and therefore can not serve as a fourth vertex for our tetrahedron of reference. This fourth vertex we may determine as follows: The lines of the axis congruence are tangent (in general) to two surfaces, or focal sheets, so that on each line of the congruence are fixed two points of tangency, or focal points. Either of the focal points might serve as a covariant point; but a unique point may be determined by choosing instead the harmonic conjugate of $y$ with respect to the focal points.

In finding the coordinates of the focal points, we make use of a result given in § 8 of the preceding paper. Let a congruence consist of the lines joining corresponding points of two surfaces $S_y$ and $S_z$ defined by the equations

$$y^{(k)}=f^{(k)}(u,v), \quad z^{(k)}=g^{(k)}(u,v), \qquad (k=1,2,3,4).$$

The functions $y$ and $z$ will satisfy a completely integrable system of partial differential equations of the form

$$\left.\begin{aligned} y_v&=a^{(1)}y+b^{(1)}z+c^{(1)}y_u+d^{(1)}z_v, \\ z_u&=a^{(2)}y+b^{(2)}z+c^{(2)}y_u+d^{(2)}z_v, \\ y_{uu}&=\alpha^{(1)}y+\beta^{(1)}z+\gamma^{(1)}y_u+\delta^{(1)}z_v, \\ z_{vv}&=\alpha^{(2)}y+\beta^{(2)}z+\gamma^{(2)}y_u+\delta^{(2)}z_v, \end{aligned}\right\} \tag{13}$$

where the coefficients in the last two equations have nothing to do with the quantities (6).

The focal sheets will be given by the formulas*

$$z_1=z-\lambda_1 y, \quad z_2=z-\lambda_2 y, \tag{14}$$

where $\lambda_1$ and $\lambda_2$ are the roots (which may or may not be distinct) of the quadratic

$$d^{(1)}t^2+(c^{(1)}d^{(2)}-d^{(1)}c^{(2)}-1)t+c^{(2)}=0. \tag{15}$$

We need for our purposes only the first two of equations (11), the coefficients of which we now proceed to find. We have

$$z=y_{vv}-\gamma y_v, \tag{12}$$

$$\left.\begin{aligned} z_u&=y_{uvv}-\gamma y_{uv}-\gamma_u y_v \\ &=\alpha^{(3)}y_{vv}+(\beta^{(3)}-b'\gamma)y_u+(\gamma^{(3)}-c'\gamma-\gamma_u)y_v+(\delta^{(3)}-d'\gamma)y, \\ z_v&=y_{vvv}-\gamma y_{vv}-\gamma_v y_v \\ &=(\alpha^{(4)}-\gamma)y_{vv}+\beta^{(4)}y_u+(\gamma^{(4)}-\gamma_v)y_v+\delta^{(4)}y, \end{aligned}\right\} \tag{16}$$

in which use has been made of the last two of equations (5). Replacing $y_{vv}$ in the second of equations (16) by its value $z+\gamma y_v$, and then solving for $y_v$, the first of equations (13) is obtained. The second of equations (13) may then

* Formulas (14) are obviously equivalent to equations (72) of the preceding paper, since all coordinates are homogeneous.

be obtained from the first of equations (16). Recalling that by (6) $\alpha^{(3)}=c'$, we find without difficulty for four of the coefficients in (13):

$$c^{(1)}=-\beta^{(4)}/(\gamma^{(4)}-\gamma_v+\gamma\alpha^{(4)}-\gamma^2),\quad d^{(1)}=1/(\gamma^{(4)}-\gamma_v+\gamma\alpha^{(4)}-\gamma^2),$$
$$c^{(2)}=\beta^{(3)}-b'\gamma+c^{(1)}(\gamma^{(3)}-\gamma_u),\quad d^{(2)}=d^{(1)}(\gamma^{(3)}-\gamma_u).$$

But by (6),

$$\alpha^{(4)}=b'+\gamma-\frac{a_v}{a},$$

so that

$$\gamma^{(4)}-\gamma_v+\gamma\alpha^{(4)}-\gamma^2=\gamma^{(4)}+b'\gamma+ac_v,$$

and the quadratic (15) becomes

$$t^2-(\beta^{(3)}+\gamma^{(4)}+ac_v)t+(\beta^{(3)}-b'\gamma)(\gamma^{(4)}+b'\gamma+ac_v)-\beta^{(4)}(\gamma^{(3)}-\gamma_u)=0. \quad (17)$$

The discriminant of this quadratic is seen without difficulty to be

$$\Delta=(\beta^{(3)}-\gamma^{(4)}-2b'\gamma-ac_v)^2+4\beta^{(4)}(\gamma^{(3)}-\gamma_u); \quad (18)$$

a somewhat lengthy calculation, in which use is made of equations (6), shows that

$$\beta^{(3)}-\gamma^{(4)}-2b'\gamma-ac_v=\frac{1}{a}(d+ab'^2-c'^2+ab'_v-c'_u+b'c+bc')=\frac{\mathfrak{D}}{a}, \quad (19)$$

where $\mathfrak{D}$ is one of the fundamental invariants of the first memoir.* The discriminant of the quadratic (17) is, therefore,

$$\Delta=\frac{\mathfrak{D}^2+4a^2\beta^{(4)}(\gamma^{(3)}-\gamma_u)}{a^2}, \quad (20)$$

and the two roots are

$$\lambda_1,\lambda_2=\frac{\beta^{(3)}+\gamma^{(4)}+ac_v}{2}\pm\frac{\sqrt{\Delta}}{2a}.$$

The two focal sheets of the axis congruence are therefore given by the formulas

$$z_1=z-\lambda_1 y,\quad z_2=z-\lambda_2 y. \quad (14)$$

The expressions just found for the focal points are of course covariants, and either might serve to fix a fourth vertex for our tetrahedron of reference. A more symmetrical choice, and one which leads to a simpler analytic expression, is made by taking as the fourth vertex the harmonic conjugate of $y$ with respect to the two focal points $z_1$ and $z_2$. This gives the point

$$\tau=z-\frac{\lambda_1+\lambda_2}{2}y=y_{vv}-\gamma y_v-\frac{1}{2}(\beta^{(3)}+\gamma^{(4)}+ac_v)y. \quad (21)$$

If the invariant $\Delta$ given by (20) vanishes, the two focal sheets of the axis congruence coincide with each other and with $\tau$; this can happen only when the

* In the calculation, use is made of equations (24), (29) and (38) of that paper.

lines of the axis congruence are the tangents to a family of asymptotics on a focal sheet.

We shall investigate further the properties of the axis congruence, in its relation to the conjugate net and to associated congruences. First, however, we shall obtain a canonical development for a surface referred to a conjugate net, by making use of the fundamental tetrahedron of reference.

## § 3. *Canonical Development of the Non-homogeneous Coordinates in the Neighborhood of a Point.**

Since the system of differential equations (4) is completely integrable, the derivatives of $y$ to any order are expressible linearly in terms of the fundamental derivatives $y_{uv}$, $y_u$, $y_v$, $y$. Let us write

$$\frac{\partial^{m+n} y}{\partial u^m \partial v^n} = a^{(m,n)} y_{uv} + b^{(m,n)} y_u + c^{(m,n)} y_v + d^{(m,n)} y. \tag{22}$$

Then we have, in particular, for the coefficients of the first order

$$\begin{aligned} a^{(10)} &= 0, & b^{(10)} &= 1, & c^{(10)} &= 0, & d^{(10)} &= 0, \\ a^{(01)} &= 0, & b^{(01)} &= 0, & c^{(01)} &= 1, & d^{(01)} &= 0, \end{aligned}$$

for the coefficients of the second order

$$\begin{aligned} a^{(20)} &= a, & b^{(20)} &= b, & c^{(20)} &= c, & d^{(20)} &= d, \\ a^{(11)} &= 0, & b^{(11)} &= b', & c^{(11)} &= c', & d^{(11)} &= d', \\ a^{(02)} &= 1, & b^{(02)} &= 0, & c^{(02)} &= 0, & d^{(02)} &= 0, \end{aligned}$$

for the coefficients of the third order the expressions (6), and so on.

Let $y$ be any fixed regular point (on the surface), which we may, without loss of generality, suppose to correspond to the parameter values $u=0$, $v=0$. Then, for any point $Y$ in a sufficiently small neighborhood about the point $y$, we have the Taylor expansion

$$Y = y + y_u u + y_v v + \frac{1}{2!}(y_{uu} u^2 + 2 y_{uv} uv + y_{vv} v^2) + \ldots\ldots$$

Substituting for the various derivatives of $y$ their expressions in terms of $y_{uv}$, $y_u$, $y_v$, $y$ as given by (22), and rearranging the series, we find the expansion

$$Y = A y_{uv} + B y_u + C y_v + D y, \tag{23}$$

where

$$\left.\begin{aligned} A &= \tfrac{1}{2} a u^2 + \tfrac{1}{2} v^2 + \tfrac{1}{6} a^{(30)} u^3 + \tfrac{1}{2} a^{(21)} u^2 v + \tfrac{1}{2} a^{(12)} u v^2 + \tfrac{1}{6} a^{(03)} v^3 \\ &\quad + \tfrac{1}{24} a^{(40)} u^4 + \tfrac{1}{6} a^{(31)} u^3 v + \tfrac{1}{4} a^{(22)} u^2 v^2 + \tfrac{1}{6} a^{(13)} u v^3 + \tfrac{1}{24} a^{(04)} v^4 + \ldots, \\ B &= u + \tfrac{1}{2} b u^2 + b' uv + \tfrac{1}{6} b^{(30)} u^3 + \tfrac{1}{2} b^{(21)} u^2 v + \tfrac{1}{2} b^{(12)} u v^2 + \tfrac{1}{6} b^{(03)} v^3 + \ldots, \\ C &= v + \tfrac{1}{2} c u^2 + c' uv + \tfrac{1}{6} c^{(30)} u^3 + \tfrac{1}{2} c^{(21)} u^2 v + \tfrac{1}{2} c^{(12)} u v^2 + \tfrac{1}{6} c^{(03)} v^3 + \ldots, \\ D &= 1 + \tfrac{1}{2} d u^2 + d' uv + \ldots, \end{aligned}\right\} \tag{24}$$

where terms omitted are in each case of higher order than those written down.

* The subsequent sections of the paper are independent of the present one.

Let us choose as three vertices of a new tetrahedron of reference the point $y$ and its minus first and first Laplace transform,

$$\rho=y_u-c'y, \quad \sigma=y_v-b'y.$$

A fourth vertex must be linearly dependent on the points $y_{uv}$, $y_u$, $y_v$, $y$, since these last four are not coplanar. We may therefore take for the fourth vertex any point not in the plane of $y$, $y_u$, $y_v$, given by an expression of the form

$$\tau=y_{uv}-\lambda y_u-\mu y_v-\nu y. \tag{25}$$

We shall seek an expansion having a certain characteristic form, and shall find that the only point $\tau$ which yields this expansion is the covariant point $\tau$ found in § 2.

Taking $\tau$ for the present as defined by (25), we may write the expansion (23) in the form

$$Y=Y_1\tau+Y_2\rho+Y_3\sigma+Y_4 y, \tag{26}$$

where

$$\left.\begin{aligned} &Y_1=A, \quad Y_2=B+A\lambda, \quad Y_3=C+A\mu, \\ &Y_4=D+A(\nu+c'\lambda+b'\mu)+c'B+b'C. \end{aligned}\right\} \tag{27}$$

The coordinates of the point $Y$ referred to the old tetrahedron of reference are of course

$$Y^{(k)}=Y_1\tau^{(k)}+Y_2\rho^{(k)}+Y_3\sigma^{(k)}+Y_4 y^{(k)}, \qquad (k=1,2,3,4).$$

Taking now as a new tetrahedron of reference the points $\tau$, $\rho$, $\sigma$, $y$, we see from this last expression that referred to this new tetrahedron the point $Y$ has the coordinates $Y_1$, $Y_2$, $Y_3$, $Y_4$, given by equations (27), if the unit-point be properly chosen. We have, therefore, for the coordinates of the point $Y$ referred to the tetrahedron $\tau$, $\rho$, $\sigma$, $y$ the expressions

$$\left.\begin{aligned} Y_1={}&A, \\ Y_2={}&u+\tfrac{1}{2}(b+a\lambda)u^2+b'uv+\tfrac{1}{2}\lambda v^2+\tfrac{1}{6}(b^{(30)}+a^{(30)}\lambda)u^3+\tfrac{1}{2}(b^{(21)}+a^{(21)}\lambda)u^2v \\ &+\tfrac{1}{2}(b^{(12)}+a^{(12)}\lambda)uv^2+\tfrac{1}{6}(b^{(03)}+a^{(03)}\lambda)v^3+\ldots, \\ Y_3={}&v+\tfrac{1}{2}(c+a\mu)u^2+c'uv+\tfrac{1}{2}\mu v^2+\tfrac{1}{6}(c^{(30)}+a^{(30)}\mu)u^3+\tfrac{1}{2}(c^{(21)}+a^{(21)}\mu)u^2v \\ &+\tfrac{1}{2}(c^{(12)}+a^{(12)}\mu)uv^2+\tfrac{1}{6}(c^{(03)}+a^{(03)}\mu)v^3+\ldots, \\ Y_4={}&1+c'u+b'v+\tfrac{1}{2}[d+a(\nu+c'\lambda+b'\mu)+bc'+b'c]u^2+(d'+2b'c')uv \\ &+\tfrac{1}{2}(\nu+c'\lambda+b'\mu)v^2+\ldots, \end{aligned}\right\} \tag{28}$$

in which the terms omitted are of higher order in each case than the ones last written.

We now introduce non-homogeneous coordinates by putting

$$\xi=Y_2/Y_4, \quad \eta=Y_3/Y_4, \quad \zeta=Y_1/Y_4. \tag{29}$$

We shall obtain for $\xi$, $\eta$, $\zeta$ power series in $u$ and $v$, and then by eliminating from these three series the variables $u$ and $v$ shall express $\zeta$ as a power series in $\xi$ and $\eta$. We shall ultimately obtain this series up to and including terms of the fourth order; it will, however, simplify our subsequent calculations if we first determine the series up to the third order terms inclusive. We find from the last of equations (28)

$$\frac{1}{Y_4}=1-c'u-b'v+\ldots,$$

so that

$$\left.\begin{aligned}\xi&=Y_2/Y_4=u+\tfrac{1}{2}(b-2c'+a\lambda)u^2+\tfrac{1}{2}\lambda v^2+\ldots,\\ \eta&=Y_3/Y_4=v+\tfrac{1}{2}(c+a\mu)u^2+\tfrac{1}{2}(\mu-2b')v^2+\ldots,\\ \zeta&=Y_1/Y_4=\tfrac{1}{2}au^2+\tfrac{1}{2}v^2+\tfrac{1}{6}(a^{(30)}-3ac')u^3+\tfrac{1}{2}(a^{(21)}-ab')u^2v\\ &\quad+\tfrac{1}{2}(a^{(12)}-c')uv^2+\tfrac{1}{6}(a^{(03)}-3b')v^3+\ldots.\end{aligned}\right\}\qquad(30)$$

From the first two of these we find, correct to terms of the third order inclusive,

$$\begin{aligned}&\xi^2=u^2+(b-2c'+a\lambda)u^3+\lambda uv^2+\ldots, &&\xi^3=u^3+\ldots,\\ &\eta^2=v^2+(c+a\mu)u^2v+(\mu-2b')v^3+\ldots, &&\eta^3=v^3+\ldots,\\ &\xi^2\eta=u^2v+\ldots,\quad \xi\eta^2=uv^2+\ldots.\end{aligned}$$

It is possible, therefore, to express in terms of $\xi$ and $\eta$ those powers of $u$ and $v$ which occur in the expansion for $\zeta$ given by the last of equations (30). We obtain, therefore, the expansion, valid to the terms of the third order,

$$\begin{aligned}\zeta=\tfrac{1}{2}a\xi^2+\tfrac{1}{2}\eta^2+\tfrac{1}{6}(a^{(30)}+3ac'-3ab-3a^2\lambda)\xi^3+\tfrac{1}{2}(a^{(21)}-ab'-c-a\mu)\xi^2\eta\\ +\tfrac{1}{2}(a^{(12)}-c'-a\lambda)\xi\eta^2+\tfrac{1}{6}(a^{(03)}+3b'-3\mu)\eta^3+\ldots.\end{aligned}$$

Remembering that the quantities $a^{(30)}$, $a^{(21)}$, $a^{(12)}$, $a^{(03)}$ are the quantities $\alpha^{(1)}$, $\alpha^{(2)}$, $\alpha^{(3)}$, $\alpha^{(4)}$ given by equations (6), we may reduce the above expansion to

$$\begin{aligned}\zeta=\tfrac{1}{2}a\xi^2+\tfrac{1}{2}\eta^2+\tfrac{1}{6}(4ac'-2ab+a_u-3a^2\lambda)\xi^3-\tfrac{1}{2}(c+a\mu)\xi^2\eta\\ -\tfrac{1}{2}a\lambda\xi\eta^2+\tfrac{1}{6}(4b'+\gamma-aa_v-3\mu)\eta^3+\ldots.\end{aligned}$$

In the expression for $\tau$ given by (25), we have at our disposal the three arbitrary quantities $\lambda$, $\mu$, $\nu$. These we shall make determinate by causing certain terms in the expansion for $\zeta$ to vanish. The most symmetric way of doing this is to suppose the coefficients of $\xi^2\eta$, $\xi\eta^2$, and $\xi^2\eta^2$ to vanish. Our expansion to terms of the third order shows that in this case we have

$$c+a\mu=0,\quad a\lambda=0.$$

But we have supposed that $a\neq0$, so that we choose

$$\lambda=0,\quad \mu=-\frac{c}{a}=\gamma.$$

This choice gives us for (25)

$$\tau=y_{vv}-\gamma y_v-\nu y,\qquad(31)$$

which is a point on the axis of the point $y$, that is, on the line of the axis congruence which passes through the point $y$.

The coefficients of $\xi^3$ and $\eta^3$ become invariants; in fact, it is easy to verify that these are

$$4ac' - 2ab + a_u = 8a\mathfrak{C}',$$
$$4b' - 2\gamma - aa_v = 8\mathfrak{B}',$$

where $\mathfrak{B}'$ and $\mathfrak{C}'$ are fundamental invariants of the first memoir, which may be shown to have the above expressions if use be made of equations (24), (29), and (38) of that paper. Consequently, the expansion for $\zeta$, correct to terms of the third order, is

$$\zeta = \tfrac{1}{2}a\xi^2 + \tfrac{1}{2}\eta^2 + \tfrac{4}{3}a\mathfrak{C}'\xi^3 + \tfrac{4}{3}\mathfrak{B}'\eta^3 + \ldots..$$

We may now find the terms of the fourth order in this expansion. Putting $\lambda=0$, $\mu=\gamma$ in (28), we find

$$Y_1 = A, \quad Y_2 = B,$$
$$Y_3 = v + c'uv + \tfrac{1}{2}\gamma v^2 + \tfrac{1}{6}(c^{(30)} + \gamma a^{(30)})u^3 + \tfrac{1}{2}(c^{(21)} + \gamma a^{(21)})u^2 v$$
$$+ \tfrac{1}{2}(c^{(12)} + \gamma a^{(12)})uv^2 + \tfrac{1}{6}(c^{(03)} + \gamma a^{(03)})v^3 + \ldots.,$$
$$Y_4 = 1 + c'u + b'v + \tfrac{1}{2}(d + bc' + a\nu)u^2 + (d' + 2b'c')uv + \tfrac{1}{2}(\nu + b'\gamma)v^2 + \ldots.,$$

where $A$ and $B$ are given by (24). We find without difficulty that to terms of the third order

$$\xi = u + \tfrac{1}{2}(b - 2c')u^2 + \tfrac{1}{6}(b^{(30)} - 6bc' + 6c'^2 - 3d - 3a\nu)u^3$$
$$+ \tfrac{1}{2}(b^{(21)} - 2d' - 2b'c' - bb')u^2 v + \tfrac{1}{2}(b^{(12)} - b'\gamma - \nu)uv^2 + \tfrac{1}{6}b^{(03)}v^3 + \ldots.,$$
$$\eta = v + \tfrac{1}{2}(\gamma - 2b')v^2 + \tfrac{1}{6}(c^{(30)} + \gamma a^{(30)})u^3 + \tfrac{1}{2}(c^{(21)} + \gamma a^{(21)} - d - bc' - a\nu)u^2 v$$
$$+ \tfrac{1}{2}(c^{(12)} + \gamma a^{(12)} - c'\gamma - 2d' - 2b'c')uv^2$$
$$+ \tfrac{1}{6}(c^{(03)} + \gamma a^{(03)} - 6b'\gamma + 6b'^2 - 3\nu)v^3 + \ldots..$$

If, now, we calculated the expression

$$\zeta - \tfrac{1}{2}a\xi^2 - \tfrac{1}{2}\eta^2 - \tfrac{4}{3}a\mathfrak{C}'\xi^3 - \tfrac{4}{3}\mathfrak{B}'\eta^3, \tag{32}$$

we should find that it begins with terms of the fourth order in $u$ and $v$. We need, therefore, calculate only the terms of the fourth order in the expansions for $\zeta$, $\xi^2$, $\eta^2$, $\xi^3$, and $\eta^3$ in terms of $u$ and $v$. We shall, however, need the explicit expressions for the terms in $u^2v^2$ alone. Only the expansions for $\zeta$, $\xi^2$, and $\eta^2$ yield such terms; they are easily found to be:

$$\text{from } \xi^2\text{:} \quad (b^{(12)} - b'\gamma - \nu)u^2v^2,$$
$$\text{from } \eta^2\text{:} \quad (c^{(21)} + \gamma a^{(21)} - d - bc' - a\nu)u^2v^2,$$
$$\text{from } \zeta\text{:} \quad [\tfrac{1}{4}a^{(22)} - \tfrac{1}{2}c'a^{(12)} - \tfrac{1}{2}b'a^{(21)}$$
$$- \tfrac{1}{4}(d + bc' - 2c'^2 - b'c - 2ab'^2 + 2a\nu)]u^2v^2.$$

Multiplying the first of these three by $-a/2$, the second by $-1/2$, and adding to the last, we obtain the term in $u^2v^2$ which occurs in the expansion of the quantity (32). The coefficient of this term is, therefore,

$$\tfrac{1}{4}a^{(22)}-\tfrac{1}{2}c'a^{(12)}-\tfrac{1}{2}b'a^{(21)}-\tfrac{1}{4}(d+bc'-2c'^2-b'c-2ab'^2+2a\nu)$$
$$-\tfrac{1}{2}a(b^{(12)}-b'\gamma-\nu)-\tfrac{1}{2}(c^{(21)}+\gamma a^{(21)}-d-bc'-a\nu). \qquad (33)$$

From (6) we find that

$$a^{(12)}=c', \quad a^{(21)}=ab', \quad c^{(21)}=b'c+c'^2+c'_u,$$

since $a^{(12)}=\alpha^{(3)}$, $a^{(21)}=\alpha^{(2)}$, $c^{(21)}=\gamma^{(2)}$. Let us write for $b^{(12)}$ its equivalent $\beta^{(3)}$, without substituting its value from (6). Also, since

$$y_{uvv}=\alpha^{(3)}y_{vv}+\beta^{(3)}y_u+\gamma^{(3)}y_v+\delta^{(3)}y,$$

we have by differentiating with respect to $u$

$$y_{uuvv}=\alpha^{(3)}y_{uvv}+\alpha^{(3)}_u y_{vv}+\beta^{(3)}y_{uu}+\ldots,$$

which, with the aid of (4) and (5), becomes

$$y_{uuvv}=a^{(22)}y_{vv}+b^{(22)}y_u+c^{(22)}y_v+d^{(22)}y,$$

where, in particular (since $\alpha^{(3)}=c'$),

$$a^{(22)}=c'_u+c'^2+a\beta^{(3)}.$$

We may now easily reduce the expression (33) to

$$\tfrac{1}{4}(d-c'_u-c'^2+bc'-b'c-a\beta^{(3)}+2a\nu),$$

which by (6) may be written

$$\tfrac{1}{4}(2a\nu-a\gamma^{(4)}-a\beta^{(3)}-c_v).$$

Consequently, in the expansion of the expression (32), the coefficient of $u^2v^2$ may be made to vanish if one chooses

$$\nu=\tfrac{1}{2}(\beta^{(3)}+\gamma^{(4)}+\alpha c_v). \qquad (34)$$

Consequently, the fourth vertex of the tetrahedron of reference, which was given by (25) and then reduced to (31), becomes

$$\tau=y_{vv}-\gamma y_v-\tfrac{1}{2}(\beta^{(3)}+\gamma^{(4)}+\alpha c_v)y,$$

which is the covariant (21) found in § 2.

If we note that to terms of the fourth order $u^4=\xi^4$, $u^3v=\xi^3\eta$, $uv^3=\xi\eta^3$ $v^4=\eta^4$, we find that the expansion for $\zeta$ is of the form

$$\zeta=\tfrac{1}{2}a\xi^2+\tfrac{1}{2}\eta^2+\tfrac{1}{6}a\mathfrak{C}'\xi^3+\tfrac{1}{6}\mathfrak{B}'\eta^3$$
$$+C^{(40)}\xi^4+C^{(31)}\xi^3\eta+\bigstar+C^{(13)}\xi\eta^3+C^{(04)}\eta^4+\ldots, \qquad (35)$$

in which all of the coefficients are relative invariants of the conjugate net. That these coefficients are indeed invariants follows immediately from the fact that the tetrahedron of reference has covariant points as vertices.

The plane $\zeta=0$ cuts the surface in a curve which has a node at the point $\xi=\eta=\zeta=0$. The tangents to the curve at this node are evidently given by the equation

$$a\xi^2+\eta^2=0, \tag{36}$$

or by the equations

$$\eta\pm\sqrt{-a}\,\xi=0.$$

These latter are of course the equations of the tangents to the asymptotic curves which pass through the point on the surface.

We may obtain from (35) a development in which all of the coefficients are *absolute* invariants. To do this we must remove the one arbitrary feature which still remains in our coordinate system, viz., the unit-point of the system. If we put

$$\xi=\lambda x, \quad \eta=\mu y, \quad \zeta=\nu z,$$

the development (35) becomes

$$\begin{aligned} z=\frac{a\lambda^2}{2\nu}x^2+\frac{\mu^2}{2\nu}y^2+\frac{4a\mathfrak{C}'\lambda^3}{3\nu}x^3+\frac{4\mathfrak{B}'\mu^3}{3\nu}y^3+\frac{\lambda^4}{\nu}C^{(40)}x^4 \\ +\frac{\lambda^3\mu}{\nu}C^{(31)}x^3y+\bigstar+\frac{\lambda\mu^3}{\nu}C^{(13)}xy^3+\frac{\mu^4}{\nu}C^{(04)}y^4+\ldots.. \end{aligned}$$

Let us choose $\lambda$, $\mu$ and $\nu$ so that

$$8a\mathfrak{C}'\lambda^3=\nu, \quad 8\mathfrak{B}'\mu^3=\nu, \quad \nu=\sqrt{a}\,\lambda\mu,$$

then

$$8\sqrt{a}\,\mathfrak{C}'\lambda^2=\mu, \quad 8\mathfrak{B}'\mu^2=\sqrt{a}\,\lambda,$$

and

$$\lambda=\frac{1}{8\sqrt[3]{a^{\frac{1}{2}}\mathfrak{B}'\mathfrak{C}'^2}}, \quad \mu=\frac{1}{8\sqrt[3]{\alpha^{\frac{1}{2}}\mathfrak{B}'^2\mathfrak{C}'}}, \quad \nu=\frac{\sqrt{a}}{64\mathfrak{B}'\mathfrak{C}'}, \tag{37}$$

where, as usual, we have written $\alpha=1/a$. The development now takes the final form

$$\begin{aligned} z=\tfrac{1}{2}(I^{(20)}x^2+I^{(02)}y^2)+\tfrac{1}{6}(x^3+y^3) \\ +I^{(40)}x^4+I^{(31)}x^3y+\bigstar+I^{(13)}xy^3+I^{(04)}y^4+\ldots., \end{aligned} \tag{38}$$

where all the coefficients are absolute projective invariants, and, in particular,

$$I^{(20)}=\sqrt[3]{\frac{a^{\frac{1}{2}}\mathfrak{B}'}{\mathfrak{C}'}}, \quad I^{(02)}=\sqrt[3]{\frac{\alpha^{\frac{1}{2}}\mathfrak{C}'}{\mathfrak{B}'}}.$$

That the said coefficients are indeed *absolute* invariants may be verified as follows: As we have already pointed out, they are certainly invariants, since the tetrahedron of reference is covariant. Now, referring to § 5 of the preceding memoir, we see that if we carry out a transformation of the independent variables,

$$\bar{u}=\phi(u), \quad \bar{v}=\psi(u), \tag{39}$$

on the system of differential equations (4), any invariant $\Theta^{(j,k)}$ of system (4) will be transformed into essentially the same function of the coefficients of the new system of differential equations, and that the transformed function is related to the old by an equation of the form

$$\bar{\Theta}^{(j,k)}=\phi_u^j\psi_v^k\Theta^{(j,k)}.$$

We say that the invariant $\Theta^{(j,k)}$ is of weight $(j,k)$. Now, any coefficient $C^{(j,k)}$ of the expansion (35) is of weight $(j, k-2)$, *i. e.*, under the transformation (39) it becomes

$$\bar{C}^{(j,k)}=\phi_u^j\psi_v^{k-2}C^{(j,k)}. \tag{40}$$

Moreover, the quantities (37) undergo the transformations

$$\bar{\lambda}=\phi_u\lambda, \quad \bar{\mu}=\psi_v\mu, \quad \bar{\nu}=\psi_v^2\nu. \tag{41}$$

Consequently, any coefficient

$$I^{(j,k)}=\frac{\lambda^j\mu^k}{\nu}C^{(j,k)}$$

of the expansion (38) is an absolute invariant, since in virtue of (40) and (41) the invariant $I^{(j,k)}$ is of weight $(0,0)$, *i. e.*, undergoes the transformation $\bar{I}^{(j,k)}=I^{(j,k)}$.

It remains for us to describe geometrically the choice of the unit-point of our coordinate system which we have just made. Before doing so, however, we shall write out the transformation of coordinates from the tetrahedron $y_{uv}, y_u, y_v, y$ to the tetrahedron which gives rise to the development (38). A point defined by an expression of the form

$$y_1 y_{uv}+y_2 y_u+y_3 y_v+y_4 y$$

has $y_1, y_2, y_3, y_4$ as its coordinates referred to the tetrahedron $y_{uv}, y_u, y_v, y$. From (27), on putting $\lambda=0$, $\mu=\gamma$, we find for the coordinates of the same point referred to the tetrahedron $\tau, \rho, \sigma, y$ the expressions

$$Y_1=y_1, \quad Y_2=y_2, \quad Y_3=y_3+\gamma y_1, \quad Y_4=y_4+(\nu+b'\gamma)y_1+c'y_2+b'y_3,$$

where

$$\nu=\tfrac{1}{2}(\beta^{(3)}+\gamma^{(4)}+ac_v).$$

Remembering that the coordinates $x$, $y$, $z$ of the development (38) are related to these by the equations

$$\lambda x=Y_2/Y_4, \quad \mu y=Y_3/Y_4, \quad \sqrt{a}\lambda\mu z=Y_1/Y_4,$$

we obtain the transformation from the coordinates $y_1, y_2, y_3, y_4$ referred to the tetrahedron $y_{vv}, y_u, y_v, y$, to the coordinates $x, y, z$ of the development (38):

$$\left.\begin{aligned} \lambda x=\frac{y_2}{(\nu+b'\gamma)y_1+c'y_2+b'y_3+y_4}, \quad \mu y=\frac{y_3+\gamma y_1}{(\nu+b'\gamma)y_1+c'y_2+b'y_3+y_4},\\ \sqrt{a}\lambda\mu z=\frac{y_1}{(\nu+b'\gamma)y_1+c'y_2+b'y_3+y_4}, \end{aligned}\right\} \qquad (42)$$

where

$$\lambda=\frac{1}{8\sqrt[8]{a^4\mathfrak{B}'\mathfrak{C}'^2}}, \quad \mu=\frac{1}{8\sqrt[8]{a^4\mathfrak{B}'^2\mathfrak{C}'}}, \quad \nu=\frac{\beta^{(3)}+\gamma^{(4)}+ac_v}{2}.$$

Let $x_1, x_2, x_3, x_4$ be a set of homogeneous coordinates corresponding to the non-homogeneous coordinates $x, y, z$. Then we may write

$$\sqrt{a}\lambda\mu x_1=y_1, \quad \lambda x_2=y_2, \quad \mu x_3=y_3+\gamma y_1, \quad x_4=(\nu+b'\gamma)y_1+c'y_2+b'y_3+y_4. \qquad (43)$$

We shall need a covariant point, different from the point $\tau$, which does not lie in the tangent plane to the surface. The line joining the first and second Laplace transforms of a point $y$ of the surface is determined by the two points

$$\sigma=y_v-b'y, \quad \sigma_v=y_{vv}-b'y_v-b'_v y,$$

and therefore lies in the osculating plane to the curve $C_v$ at $y$. The point

$$\sigma_v+(b'-\gamma)\sigma=y_{vv}-\gamma y_v+(b'\gamma-b'^2-b'_v)y=y_{vv}-\gamma y_v+(b'\gamma-\beta^{(3)})y$$

evidently lies on the line $y\tau$. Referred to the tetrahedron $y_{vv}, y_u, y_v, y$ it has coordinates $(1, 0, -\gamma, b'\gamma, -\beta^{(3)})$; using (43), we find that, *referred to the fundamental tetrahedron of reference, the point in which the line joining the first and second Laplace transforms of a point $y$ meets the corresponding axis $y\tau$ has coordinates proportional to*

$$x_1=\frac{64\mathfrak{B}'\mathfrak{C}'}{\sqrt{a}}, \quad x_2=0, \quad x_3=0, \quad x_4=\tfrac{1}{2}(\gamma^{(4)}-\beta^{(3)}+ac_v+2b'\gamma)=-\frac{\mathfrak{D}}{2a}. \qquad (44)$$

The final expression for $x_4$ follows from equation (19).

Let us now consider the cubic surface

$$z=\tfrac{1}{2}(I^{(20)}x^2+I^{(02)}y^2)+\tfrac{1}{6}(x^3+y^3), \qquad (45)$$

or, in homogeneous coordinates (with reference to the covariant tetrahedron),

$$F\equiv 6x_1x_4^2-3I^{(20)}x_2^2x_4-3I^{(02)}x_3^2x_4-x_2^3-x_3^3=0.$$

Denoting by $F_{x_1}$ the partial derivative $\partial F/\partial x_1$, etc., we find

$$F_{x_1}=6x_4^2, \quad F_{x_2}=-6I^{(20)}x_2x_4-3x_2^2, \quad F_{x_3}=-6I^{(02)}x_3x_4-3x_3^2,$$
$$F_{x_4}=12x_1x_4-3I^{(20)}x_2^2-3I^{(02)}x_3^2. \qquad (46)$$

All of these vanish simultaneously if and only if $x_2=x_3=x_4=0$, *i. e.*, the vertex

(1, 0, 0, 0) is the only singular point of the surface. The only one of the second derivatives which does not vanish at this point is $F_{x_4 x_4}$, so that the cone of the second order which is tangent to the cubic surface at the singular point degenerates into the two coincident planes $x_4^2=0$. The surface is, therefore, a *unodal* cubic, the vertex $\tau$ (1, 0, 0, 0) of the tetrahedron of reference is its *unode*, and the face $x_4=0$ of the tetrahedron is its *uniplane*.

It is not difficult to see that any unodal cubic surface having as its unode the point (1, 0, 0, 0) and as its uniplane the plane $x_4=0$ must have an equation of the form

$$F \equiv x_1 x_4^2 + \phi(x_2, x_3, x_4) = 0,$$

where $\phi$ is a homogeneous polynomial of the third degree in $x_2, x_3, x_4$. In non-homogeneous coordinates this equation may by written

$$z + \phi(x, y, 1) = 0. \tag{47}$$

The arbitrary function $\phi$ may be determined uniquely by subjecting the cubic surface to a further condition. In fact, let us suppose that the unodal cubic has contact of the third order with our original surface, which is given by the canonical development,

$$z = \tfrac{1}{2}(I^{(20)}x^2 + I^{(02)}y^2) + \tfrac{1}{6}(x^3 + y^3) + \ldots\ldots$$

Then the value of $z$ obtained from (47) must agree with this expansion to terms of the third order inclusive. Consequently,

$$\phi(x, y, 1) \equiv \tfrac{1}{2}(I^{(20)}x^2 + I^{(02)}y^2) + \tfrac{1}{6}(x^3 + y^3),$$

which makes (47) coincide with (45). We have then the result:

*The cubic surface*

$$z = \tfrac{1}{2}(I^{(20)}x^2 + I^{(02)}y^2) + \tfrac{1}{6}(x^3 + y^3), \tag{45}$$

*is completely characterized by the following properties:* (1) *It has a unode at the vertex $\tau$ of the covariant tetrahedron,* (2) *its uniplane is the face of this tetrahedron opposite the point $y$, and* (3) *it has contact of the third order with the surface $S_y$ at the point $y$.**

The tangent plane to the surface $S_y$ at $y$ cuts the unodal cubic in the cubic curve

*The unodal cubic just characterized closely resembles the *canonical cubic* introduced by Wilczynski in his study of surfaces referred to their asymptotic curves. Cf. his "Projective Differential Geometry of Curved Surfaces," second memoir, *Transactions of the American Mathematical Society*, Vol. IX (1908), pp. 104 *et seq.* A unodal cubic may, of course, always be uniquely determined as follows, in relation to a point $y$ of a curved surface. In the tangent plane to the surface at $y$ fix a line $l$ not passing through $y$, and off the tangent plane fix a point $P$. Then just one unodal cubic may be determined having contact of the third order with the surface at $y$, and having $P$ as its unode and the plane passing through $P$ and $l$ as its uniplane.

$$z=0, \quad 3I^{(20)}x^2+3I^{(02)}y^2+2x^3+2y^3=0, \tag{48a}$$

or in homogeneous coordinates

$$x_1=0, \quad 3I^{(20)}x_2^2x_4+3I^{(02)}x_3^2x_4+2x_2^3+2x_3^3=0. \tag{48b}$$

The line $x_1=0$, $x_4=0$, *i. e.*, the edge $\rho\sigma$ of the covariant tetrahedron, cuts this cubic in the three points given by the equation $x_2^3+x_3^3=0$ in conjunction with $x_1=0$, $x_4=0$, *i. e.*, the three points

$$(0, -1, 1, 0), \quad (0, -\varepsilon, 1, 0), \quad (0, -\varepsilon^2, 1, 0), \tag{49}$$

where $\varepsilon$ is an imaginary cube root of unity. Consider the point $(0, -1, 1, 0)$. The tangent plane to the unodal cubic surface at this point has the equation

$$x_2+x_3+(I^{(20)}+I^{(02)})x_4=0, \tag{50}$$

and the tangent to the cubic curve (48) at the same point is obtained by adjoining to equation (50) the equation $x_1=0$. Unfortunately, the plane (50) passes through the point $\tau$ (1, 0, 0, 0). The point given by equations (44) does not lie in the tangent plane to the surface at $y$; the plane passing through it and through the tangent to the cubic curve (48) at the point $(0, -1, 1, 0)$ has the equation

$$(I^{(20)}+I^{(02)})(\mathfrak{D}x_1+128\sqrt[3]{a\mathfrak{B}'\mathfrak{C}'}x_4)+128\sqrt[3]{a\mathfrak{B}'\mathfrak{C}'}(x_2+x_3)=0. \tag{51}$$

*The unit point of the coordinate system which gives rise to the development* (38) *is such that equation* (51), *or, in non-homogeneous coordinates, the equation*

$$(I^{(20)}+I^{(02)})(\mathfrak{D}z+128\sqrt[3]{a\mathfrak{B}'\mathfrak{C}'})+128\sqrt[3]{a\mathfrak{B}'\mathfrak{C}'}(x+y)=0, \tag{52}$$

*represents a plane, which passes through the point of intersection of the line joining the first and second Laplace transforms of the point $y$ with the corresponding axis $y\tau$, and through the line which is tangent to the cubic curve in which the unodal cubic surface is cut by the tangent plane at $y$, the point of tangency of this line being one of the three points in which the line $\rho\sigma$ intersects the said cubic curve.** *This last fact explains the presence, in the development* (38), *of the cube roots of the invariants*

$$\sqrt[3]{a\mathfrak{B}'\mathfrak{C}'^2}, \quad \sqrt[3]{a\mathfrak{B}'^2\mathfrak{C}'}.$$

We recall that the development (38) was obtained by expanding the non-homogeneous coordinates $\xi$, $\eta$, $\zeta$ of a point of the surface in power series in $u$ and $v$, and then expressing $\zeta$ as a power series in $\xi$ and $\eta$ by eliminating

* Of course, if $\mathfrak{D}=0$, the point (44) coincides with the point $\tau$, and, therefore, the $z$ coordinate will be absent from equation (52). The unit-point must then be interpreted through the use of some other covariant point off the tangent plane.

between the expansions (30) the parameters $u$ and $v$. Instead of doing this, we might have eliminated $u$ between the expressions for $\xi$ and $\eta$ on the one hand, and $\xi$ and $\zeta$ on the other; thus obtaining $\eta$ and $\zeta$ as power series

$$\eta=\mathfrak{P}_1(\xi, v), \quad \zeta=\mathfrak{P}_2(\xi, v)$$

in $\xi$ and $v$. Such expansions would properly be regarded as characteristic of the one-parameter family of curves $C_u(v=\text{const.})$. In the form thus obtained, however, the two power series would not have all of their coefficients invariants —even relative—because the arbitrariness in the choice of the parameter $v$ must still be removed. How this may be done in a case like the present one, so as to yield expansions whose coefficients are all invariants, has been discussed by the author in connection with a one-parameter family of curves in the plane.* The developments for a one-parameter family of space curves were announced by the writer, together with the expansion (38), at a meeting of the American Mathematical Society, to the report of which we refer for the actual forms of the series involved.†

## § 4. *The Parametric Ruled Surfaces and the Developables of the Axis Congruence.*

Let $R^{(u)}$ be the ruled surface generated by the axes $yz$ corresponding to the points of a curve $C_u(v=\text{const.})$ on the surface $S_y$, and $R^{(v)}$ the ruled surface generated by the axes corresponding to the points of a curve $C_v(u=\text{const.})$ on $S_y$. The osculating plane to the curve $C_u$ at the point $y$ contains in it the axis of the point $y$, and also the tangent to the curve $C_u$. Consequently, this plane is tangent to the ruled surface $R^{(u)}$ at $y$, since the axis and the tangent to $C_u$ can not coincide. *A parametric curve on the surface $S_y$ is an asymptotic curve on the corresponding ruled surface of the axis congruence.*

It may happen, however, that a parametric ruled surface, say $R^{(u)}$, be developable. Since the generators of a curved developable are its only asymptotics, it follows that the curve $C_u$ can be asymptotic on $R^{(u)}$ only if $R^{(u)}$ be a plane. We examine this point more closely. If the line $yz$ is to generate a developable, then the four points $y$, $z$, $y_u$, $z_u$ must be coplanar. We have

$$\begin{aligned} z &= y_{vv} - \gamma y_v, \\ z_u &= y_{uvv} - \gamma y_{uv} - \gamma_u y_v \\ &= \alpha^{(3)} y_{vv} + (\beta^{(3)} - b'\gamma) y_u + (\gamma^{(3)} - c'\gamma - \gamma_u) y_v + (\delta^{(3)} - d'\gamma) y. \end{aligned}$$

If $y$, $z$, $y_u$, $z_u$ are coplanar, their expressions must be linearly dependent, that is, we must have

* Cf. G. M. Green, "One-parameter Families of Curves in the Plane," *Transactions of the American Mathematical Society*, Vol. XV (1914), pp. 277–290. See § 2.

† *Bulletin of the American Mathematical Society*, Series 2, Vol. XX (1913–14), p. 397.

$$\begin{vmatrix} 1 & -\gamma \\ \alpha^{(3)} & \gamma^{(3)}-c'\gamma-\gamma_u \end{vmatrix}=0.$$

Since by (6) $\alpha^{(3)}=c'$, this reduces to

$$\gamma^{(3)}-\gamma_u=0. \tag{53}$$

This is easily seen to be the condition that the curves $C_u$ be plane curves, as follows: The curves $C_u$ are plane if and only if the points $y$, $y_u$, $y_{uu}$, $y_{uuu}$ be coplanar. But

$$\begin{aligned} y_{uu} &= ay_{vv} + by_u + cy_v + dy, \\ y_{uuu} &= ay_{uvv} + a_u y_{vv} + by_{uu} + b_u y_u + cy_{uv} + c_u y_v + dy_u + d_u y \\ &= (a\alpha^{(3)} + ab + a_u) y_{vv} + (\,) y_u + (a\gamma^{(3)} + bc + c'c + c_u) y_v + (\,) y, \end{aligned}$$

in which the coefficients of $y_u$ and $y$ do not concern us. The points $y$, $y_u$, $y_{uu}$, $y_{uuu}$ are thefore linearly dependent if and only if

$$\begin{vmatrix} a & c \\ ac'+ab+a_u & a\gamma^{(3)}+bc+c'c+c_u \end{vmatrix}=0,$$

which, remembering that $\gamma=-c/a$, we may easily reduce to (53). Consequently, *if a parametric ruled surface $R^{(u)}$ of the axis congruence be developable, the corresponding parametric curve $C_u$ is a plane curve and vice versa; the surface $R^{(u)}$ is, in fact, the plane of this curve.*

Unless the invariant (20) vanishes, the lines of the axis congruence may be assembled into two families of developable surfaces. We shall determine these developables presently, but may now state the theorem: *A necessary and sufficient condition that a conjugate net on a surface $S_y$ consist of two families of plane curves, is that the developables of the axis congruence cut the surface $S_y$ in the said conjugate net. The two families of developables will then consist entirely of planes.**

Let us now fix our attention upon a point $y$ of the surface $S_y$, and the two curves $C_u$ and $C_v$ of the conjugate net which pass through $y$. Consider the ruled surface $R^{(u)}$, which we suppose non-developable. Any plane through a generator of a skew ruled surface is tangent to that surface at some point of the generator; consequently, the osculating plane to the curve $C_v$ at $y$, since it passes through the axis at $y$, must be tangent to $R^{(u)}$ somewhere along that axis. Now, any point on the axis is given by the expression

$$R=z+\nu y,$$

for a suitable value of $\nu$. If we suppose $\nu$ to be a function of $u$, $v$, then, as the

---

* The necessity of this condition is of course obvious.

point $y$ traces the curve $C_u$ on $S_y$, the line $yz$ generates the ruled surface $R^{(u)}$, and the point $R$ traces a curve on $R^{(u)}$. A point on the tangent to this curve at $R$ is found by differentiating the expression for $R$ with respect to $u$. We have

$$\begin{aligned} R_u &= z_u + \nu y_u + \nu_u y \\ &= c' y_{vv} + (\beta^{(3)} - b'\gamma + \nu) y_u + (\gamma^{(3)} - c'\gamma - \gamma_u) y_v + (\delta^{(3)} - d'\gamma + \nu_u) y. \end{aligned} \qquad (54)$$

We wish to determine the function $\nu$ so that for the point $y$ the osculating plane to $C_v$ will be tangent to the surface $R^{(u)}$ at $R$, and will therefore contain in it the point $R^{(u)}$. Consequently, the points $y$, $y_v$, $y_{vv}$, $R_u$ must be linearly dependent, which by (54) can be the case if and only if

$$\nu = b'\gamma - \beta^{(3)}.$$

*The osculating plane to the curve $C_v$ ($u=const.$) at the point $y$ is tangent to the ruled surface $R^{(u)}$ at the point given by the expression*

$$R = y_{vv} - \gamma y_v - (\beta^{(3)} - b'\gamma) y. \qquad (55)$$

Similarly, writing $R' = z + \nu' y$, we have

$$\begin{aligned} R'_v &= z_v + \nu' y_v + \nu'_v y \\ &= (\alpha^{(4)} - \gamma) y_{vv} + \beta^{(4)} y_u + (\gamma^{(4)} - \gamma_v + \nu' y_v) + (\delta^{(4)} + \nu'_v) y, \end{aligned} \qquad (56)$$

which lies in the same plane with $y$, $y_u$, and $z = y_{vv} - \gamma y_v$ if and only if

$$\begin{vmatrix} \alpha^{(4)} - \gamma & \gamma^{(4)} - \gamma_v + \nu' \\ 1 & -\gamma \end{vmatrix} = 0,$$

that is, by (6),

$$\nu' = -\gamma\alpha^{(4)} - \gamma^{(4)} + \gamma^2 + \gamma_v = b'\gamma - \gamma^{(4)} - \frac{c_v}{a}.$$

*The osculating plane to the curve $C_u$ ($v=const.$) at the point $y$ is tangent to the ruled surface $R^{(v)}$ at the point given by the covariant*

$$R' = y_{vv} - \gamma y_v - (\gamma^{(4)} + b'\gamma + ac_v) y. \qquad (57)$$

In § 3 we found an expression which gives the point of intersection of the axis with the line joining the first and second Laplace transforms.* This expression coincides with (55). By symmetry we infer also that (57) gives the point in which the axis is cut by the line joining the minus first and minus second Laplace transforms. It is not difficult to see geometrically why the point $R$ is the intersection of the axis with the line $\sigma\sigma_1$ ($\sigma_1$ denoting the first Laplace transform of $\sigma$, *i. e.*, the second Laplace transform of $y$), if we observe that for $v$=const. the lines $\sigma\sigma_1$ generate a developable surface, which is pre-

---

* See page 300.

cisely the developable enveloped by the osculating planes to the curves $u=\text{const.}$ at the points where these curves meet the same fixed curve $v=\text{const.}$†

The harmonic conjugate of the point $y$ with respect to the points $R$ and $R'$ is

$$\tau=y_{vv}-\gamma y_v-\tfrac{1}{2}(\beta^{(3)}+\gamma^{(4)}+ac_v)y,$$

which is precisely the covariant (21). Consequently, *the harmonic conjugate of the point y with respect to the two focal points of the axis congruence coincides with the harmonic conjugate of y with respect to the two points in which the axis is met by the line joining the first and second Laplace transforms and the line joining the minus first and minus second Laplace transforms.*

The above theorem would be trivial if the points $R$ and $R'$ were the focal points of the axis congruence. Let us determine when this can occur. If the surface $S_R$, for example, be a focal sheet of the axis congruence, then the axis $yR$ is tangent to the surface $S_R$. In other words, since the points $R_u$ and $R_v$ lie in the tangent plane to the surface $S_R$, we must have the four points $y$, $R$, $R_u$, $R_v$ coplanar. Now, putting $\nu=b'\gamma-\beta^{(3)}$ in (54), we find

$$R_u=c'y_{vv}+(\gamma^{(3)}-c'\gamma-\gamma_u)y_v+(\,)y.$$

Also,

$$\begin{aligned} R_v&=y_{vvv}-\gamma y_{vv}-\gamma_v y_v-(\beta^{(3)}-b'\gamma)y_v+(\,)y \\ &=(\alpha^{(4)}-\gamma)y_{vv}+\beta^{(4)}y_u+(\gamma^{(4)}-\beta^{(3)}+b'\gamma-\gamma_v)y_v+(\,)y, \\ R&=y_{vv}-\gamma y_v-(\beta^{(3)}-b'\gamma)y. \end{aligned}$$

If $y$, $R$, $R_u$, $R_v$ are to be linearly dependent, we must have

$$\begin{vmatrix} c' & 0 & \gamma^{(3)}-c'\gamma-\gamma_u \\ \alpha^{(4)}-\gamma & \beta^{(4)} & \gamma^{(4)}-\beta^{(3)}+b'\gamma-\gamma_v \\ 1 & 0 & -\gamma \end{vmatrix}=0,$$

*i. e.*,

$$\beta^{(4)}(\gamma^{(3)}-\gamma_u)=0.$$

† Most of the above considerations, and some of those which are to follow, have been extended by the author to the case of general nets—not conjugate—on a curved surface. The axis congruence may be defined for such a net in exactly the same way, as consisting of the lines of intersection of the osculating planes to the curves of the net. The minus first and first Laplace transforms are replaced by the second focal sheets of the two congruences formed by the tangents to the curves $C_u$ and $C_v$ of the net. Calling these covariants $\rho$ and $\sigma$, we find that on the surface $S_\sigma$ the tangent to a curve $u=\text{const.}$ intersects the axis $yz$ in a point $P$. But for $v=\text{const.}$ the tangents to the curves $u=\text{const.}$ do not form a developable, since the parameter net can not be conjugate on $S_\sigma$ if it is not on $S_y$. Nevertheless, the ruled surface $R^{(u)}$ is touched at a point $R$ by the osculating plane to the curve $C_v$ at $y$, or, as we may say, the ruled surface $R^{(u)}$ is cut along a curve $C_R$ by the developable generated by the planes which osculate the curves $C_v$ at the points where these curves meet a fixed curve $C_u$. Now, as we have just seen, there is no developable to coincide with this which at the same time takes the place of the one formed by lines joining the first and second Laplace transforms in the case of a conjugate net. We may infer that *the point R is the point of intersection of the axis with the tangent to the curve $u=\text{const.}$ on $S_\sigma$ if and only if the parameter net on $S_y$ is conjugate.*

We have already seen that if $\gamma^{(3)}-\gamma_u=0$, the curves $C_u$ are plane. The equation

$$y_{vvv}=\alpha^{(4)}y_{vv}+\beta^{(4)}y_u+\gamma^{(4)}y_v+\delta^{(4)}y$$

shows that if $\beta^{(4)}=0$ the curves $C_v$ are plane. Consequently, *the surface $S_R$ is a focal sheet of the axis congruence if and only if at least one of the families of the conjugate net consists of plane curves. By symmetry, the surface $S_{R'}$ will then be the other focal sheet.*

We now proceed to determine the two families of developables of the axis congruence. Let the point $y$ take the position $y+\delta y$, where, of course, $\delta y=y_u\delta u+y_v\delta v$; then the corresponding point $z$ will move to $z+\delta z$, where $\delta z=z_u\delta u+z_v\delta v$. If $y$ is to move so that the corresponding axis $yz$ generates a developable, the four points $y$, $z$, $\delta y$, $\delta z$ must be coplanar. We have

$$\begin{aligned}
\delta y&=y_u\delta u+y_v\delta v,\\
z&=y_{vv}-\gamma y_v,\\
\delta z&=z_u\delta u+z_v\delta v\\
&=[c'\delta u+(\alpha^{(4)}-\gamma)\delta v]y_{vv}+[(\beta^{(3)}-b'\gamma)\delta u+\beta^{(4)}\delta v]y_u\\
&\qquad+[(\gamma^{(3)}-c'\gamma-\gamma_u)\delta u+(\gamma^{(4)}-\gamma_v)\delta v]y_v+(\ )y,
\end{aligned}$$

on putting $\nu=0$, $\nu'=0$ in (54) and (56). Let us suppose $y$, $z$, $\delta y$, $\delta z$ coplanar, then their expressions must be linearly dependent, and the determinant of the coefficients of $y_{vv}$, $y_u$, $y_v$, $y$ therein must vanish. We thus obtain

$$\begin{vmatrix}
0 & \delta u & \delta v\\
1 & 0 & -\gamma\\
c'\delta u+(\alpha^{(4)}-\gamma)\delta v & (\beta^{(3)}-b'\gamma)\delta u+\beta^{(4)}\delta v & (\gamma^{(3)}-c'\gamma-\gamma_u)\delta u+(\gamma^{(4)}-\gamma_v)\delta v
\end{vmatrix}=0,$$

which is easily reduced to the quadratic

$$a(\gamma^{(3)}-\gamma_u)\delta u^2-\mathfrak{D}\delta u\delta v-a\beta^{(4)}\delta v^2=0, \tag{58}$$

where the invariant $\mathfrak{D}$ is given by (19). This may be regarded as a differential equation defining a net of curves on the surface $S_y$. Wilczynski has called these the *axis curves.* If a point $y$ of the surface $S_y$ moves along an axis curve, the corresponding axis generates a developable surface of the axis congruence. Through a point $y$ of $S_y$ pass two axis curves; we call the two tangents to these curves at $y$ the *axis tangents* of the point $y$.

In § 2 we determined the two sheets of the focal surface, and found it necessary to solve the quadratic (17). The form of this quadratic differs from that of (58), though the two have, of course, the same discriminant, given by (20).

If we make use of equations (6), we obtain

$$a\beta^{(4)}=H+2b'_u-b_v, \quad \gamma^{(3)}-\gamma_u=K+2c'_v-\gamma_u,$$

where

$$H=d'+b'c'-b'_u, \quad K=d'+b'c'-c'_v \tag{59}$$

are the Laplace-Darboux invariants of our original conjugate net. We may write (8a) in the form

$$b_v+2c'_v=2b'_u+\gamma_u-\frac{\partial^2}{\partial u\,\partial v}\log a,$$

so that

$$\gamma^{(3)}-\gamma_u=K+2b'_u-b_v-\frac{\partial^2}{\partial u\,\partial v}\log a.$$

We may therefore take the differential equation of the axis curves in the form

$$a\left(K+2b'_u-b_v-\frac{\partial^2}{\partial u\,\partial v}\log a\right)\delta u^2-\mathfrak{D}\delta u\delta v-(H+2b'_u-b_v)\delta v^2=0. \tag{60}$$

The theorems concerning the developables of the axis congruence given at the beginning of this paragraph may of course be read off from (58). We point out one further fact, concerning the relation of the developables of the axis congruence to the parametric ruled surfaces $R^{(u)}$ and $R^{(v)}$. Considering (58) as a quadratic in $\delta v/\delta u$, let us denote its two roots by $a_1$ and $a_2$. Then the expression $y_u+a_1y_v$ represents a point on one of the axis tangents corresponding to the point $y$. Remembering this, we may without difficulty show that *the plane determined by the axis and either of the axis tangents is tangent to each of the parametric ruled surfaces $R^{(u)}$ and $R^{(v)}$ at the two focal points.*

## § 5. *The Ray Congruence, Ray, Anti-Ray and Anti-Axis Curves.*

The quantities

$$\rho=y_u-c'y, \quad \sigma=y_v-b'y \tag{61}$$

are respectively the minus first and first Laplace transforms of $y$. They both lie in the tangent plane to the surface $S_y$ at $y$. The surface $S_\rho$ is the second focal sheet of the congruence of tangents to the curves $C_u$ on $S_y$, and the surface $S_\sigma$ the second focal sheet of the congruence of tangents to the curves $C_v$ on $S_y$. Wilczynski has called the line joining the points $\rho$, $\sigma$ corresponding to a point $y$ the *ray* of the point $y$, and the totality of rays, which constitute a congruence, the *ray congruence*. He has also pointed out a dualistic correspondence between the axis and ray congruences.*

---

* E. J. Wilczynski, "The General Theory of Congruences," *Transactions of the American Mathematical Society*, Vol. XVI (1915), pp. 311–327. Cf., in particular, § 3.

In our preceding memoir, we determined the focal sheets of the ray congruence. These are given by the formulas

$$R=\rho+r_1\sigma, \quad S=\rho+r_2\sigma, \tag{62}$$

where $r_1$ and $r_2$ are the roots of the quadratic *

$$Hr^2+\mathfrak{D}r-aK=0. \tag{63}$$

We may find the developables of the ray congruence just as we did those of the axis congruence. We determine the curves along which $y$ must move, in order that the corresponding *rays* may generate developables. If $y$ moves to $y+\delta y$, the corresponding $\rho$ and $\sigma$ move to $\rho+\delta\rho$ and $\sigma+\delta\sigma$, where $\delta\rho=\rho_u\delta u+\rho_v\delta v$, $\delta\sigma=\sigma_u\delta u+\sigma_v\delta v$. We have

$$\begin{aligned}\rho_u&=y_{uu}-c'y_u-c'_uy\\&=ay_{vv}+(b-c')y_u+cy_v+(d-c'_u)y,\\\rho_v&=y_{uv}-c'y_v-c'_vy\\&=b'y_u+(d'-c'_v)y,\\\sigma_u&=c'y_v+(d'-b'_u)y,\\\sigma_v&=y_{vv}-b'y_v-b'_vy,\end{aligned}$$

so that we may write

$$\left.\begin{aligned}&\rho=y_u-c'y, \quad \sigma=y_v-b'y,\\&\delta\rho=a\delta u\cdot y_{vv}+[(b-c')\delta u+b'\delta v]y_u+c\delta u\cdot y_v+[(d-c'_u)\delta u+(d'-c'_v)\delta v]y,\\&\delta\sigma=\delta v\cdot y_{vv}+(c'\delta u-b'\delta v)y_v+[(d'-b'_u)\delta u-b'_v\delta v]y.\end{aligned}\right\} \tag{64}$$

If in its motion the line $\rho\sigma$ is to describe a developable, the four points $\rho$, $\sigma$, $\rho+\delta\rho$, $\sigma+\delta\sigma$ must lie in a plane, *i. e.*, the expressions for $\rho$, $\sigma$, $\delta\rho$, $\delta\sigma$ must be linearly dependent. Equating to zero the determinant of the coefficients of $y_{vv}$, $y_u$, $y_v$, $y$ in (64), and expanding, we obtain as the differential equation which determines the developables of the ray congruence

$$aH\delta u^2-\mathfrak{D}\delta u\delta v-K\delta v^2=0, \tag{65}$$

where $\mathfrak{D}$ is the invariant (19), and $H$ and $K$ are the Laplace-Darboux invariants (59). The differential equation defines a net of curves, the *ray curves*, on the surface $S_y$. The tangents at $y$ to the two ray curves which pass through $y$ we call the *ray tangents* of the point $y$.

A net of curves closely related to the ray curves is defined by the differential equation

$$aH\delta u^2+\mathfrak{D}\delta u\delta v-K\delta v^2=0, \tag{66}$$

which differs from that of the ray curves only in the sign of the middle term.

---

* We have changed our notation slightly, so that the quadratic, which follows equations (74) of the preceding paper, differs in form from the present one.

Its roots are therefore the negatives of the roots of (65). Let $R_1$, $R_2$ be the ray tangents of the point $y$, and $C_1$, $C_2$ the conjugate tangents, *i. e.*, the tangents at $y$ to the curves of our conjugate parameter net. Let $R_1'$ be the harmonic conjugate of $R_1$ with respect to $C_1$ and $C_2$, and $R_2'$ the harmonic conjugate of $R_2$ with respect to $C_1$ and $C_2$. Then $R_1'$ and $R_2'$ are the tangents at $y$ to the curves of the net defined by (66). This net is therefore defined geometrically in terms of the conjugate net and the net of ray curves. We shall call it the *anti-ray* net, and the tangents at $y$ to the two curves of the net passing through $y$ the *anti-ray tangents. The anti-ray tangents are the harmonic conjugates of the ray tangents with respect to the conjugate tangents.*

In the same way we may define the *anti-axis* curves by means of the differential equation

$$a\left(K+2b_u'-b_v-\frac{\partial^2}{\partial u\partial v}\log a\right)\delta u^2+\mathfrak{D}\,\delta u\,\delta v-(H+2b_u'-b_v)\delta v^2=0. \qquad (67)$$

The anti-axis tangents are the harmonic conjugates of the axis tangents with respect to the conjugate tangents.

In the preceding paper,* we found that by a transformation of the independent variables,

$$\bar{u}=\phi(u, v), \quad \bar{v}=\psi(u, v)$$

the conjugate net is replaced by a new parameter net, which is asymptotic if and only if $\phi$ and $\psi$ satisfy the same quadratic partial differential equation of the first order,

$$\phi_u^2+a\phi_v^2=0, \quad \psi_u^2+a\psi_v^2=0.$$

We may throw this into a form analogous to (60) and (65). *The differential equation of the asymptotic curves on the surface $S_y$ is*

$$a\,\delta u^2+\delta v^2=0. \qquad (68)$$

Let us now regard the differential equations, of the various nets of curves which we have defined, as binary quadratics in $\delta u$, $\delta v$. The two roots of any one of the quadratics will give the directions of the two tangents at any point $y$ to the two curves of that particular net which pass through $y$. The simultaneous invariant of the two quadratics (65) and (68) is $a(H-K)$; if this is zero, the ray tangents separate the asymptotic tangents harmonically. Since $a\neq 0$, we have the theorem of Wilczynski: *A conjugate net has equal Laplace-Darboux invariants if and only if its ray curves form a conjugate net.*

---

* Cf. § 7 thereof.

Taking the differential equation of the axis curves in the form (60), and calculating the simultaneous invariant of this and (68), we find that *the axis curves form a conjugate net if and only if*

$$H-K+\frac{\partial^2}{\partial u\,\partial v}\log a=0.^{*} \tag{69}$$

In the preceding memoir,† we found a necessary and sufficient condition that the congruence of tangents to the curves $C_u$ form a $W$-congruence. This condition is the vanishing of an invariant which we denoted by $W$, but the negative of which we now denote by $W^{(u)}$. By making use of equations (24) and (29) of the memoir cited, we find without difficulty that

$$W^{(u)}=2b'_u-b_v-\frac{\partial^2}{\partial u\,\partial v}\log a. \tag{70}$$

The condition was obtained incidentally in relating the formulas of that paper to those of Wilczynski's general theory of congruences. We may, however, obtain it independently, by a procedure similar to that which we now follow in deriving the condition that the congruence of tangents to the family of curves $C_v$ form a $W$-congruence. This congruence of tangents has the surfaces $S_y$ and $S_\sigma$ as focal sheets, and will be a $W$-congruence if and only if the asymptotics on these two surfaces correspond. We have

$$\begin{gathered}\sigma=y_v-b'y,\quad \sigma_u=c'y_v+(d'-b'_u)y,\quad \sigma_v=y_{vv}-b'y_v-b'_vy,\\ \sigma_{uu}=Hy_u+(c'^2+c'_u)y_v+(c'd'+d'_u-b'_{uu})y,\\ \sigma_{vv}=(\alpha^{(4)}-b')y_{vv}+\beta^{(4)}y_u+(\gamma^{(4)}-2b'_v)y_v+(\delta^{(4)}-b'_{vv})y,\end{gathered}$$

from which five equations we may eliminate $y_{vv}$, $y_u$, $y_v$, $y$ and obtain an equation of the form

$$\sigma_{uu}=a_1\sigma_{vv}+b_1\sigma_u+c_1\sigma_v+d_1\sigma,$$

where, in particular,

$$a_1=H/\beta^{(4)}.$$

This differential equation for $\sigma$ is of the form of the first of equations (4); consequently, the asymptotics on the surface $S_\sigma$ are given by the differential equation $a_1\delta u^2+\delta v^2=0$. If the congruence of lines $y\sigma$ is to be a $W$-congruence,

* The configuration which we are studying is self-dual—concerning which remark see the paper by Wilczynski cited at the beginning of this section. The dualistic correspondence between the axis and ray of a point $y$ is there exhibited in detail. It is therefore to be expected that equation (69) is equivalent to the statement that the Laplace-Darboux invariants of the system of differential equations adjoint to system (4) are equal. This is in fact the case; however, by the adjoint system we do not mean the Lagrange adjoint, but the system of form (4) which is satisfied by the coordinates of the tangent plane to the surface $S_y$.

† Equations (68).

this differential equation must coincide with equation (68). Therefore, $a_1=a$, so that

$$H-a\beta^{(4)}=0.$$

But from (6), $a\beta^{(4)}=H+2b'_u-b_v$, and we have the result: *A necessary and sufficient condition that the congruence of tangents to the curves $C_v$ on the surface $S_y$ be a W-congruence is the vanishing of the invariant*

$$W^{(v)}=2b'_u-b_v. \tag{71}$$

We may therefore write the differential equation (60) of the axis curves in the form

$$a(K+W^{(u)})\delta u^2-\mathfrak{D}\delta u\delta v-(H+W^{(v)})\delta v^2=0. \tag{72}$$

The method which we have just followed in obtaining the invariant $W^{(v)}$ leads at once to the definition of two important nets on the surface $S_y$. It is easily verified that $\rho$ satisfies an equation of the form

$$\rho_{uu}=a_{-1}\rho_{vv}+b_{-1}\rho_u+c_{-1}\rho_v+d_{-1}\rho,$$

where

$$a_{-1}=a(\gamma^{(3)}-\gamma_u)/K,$$

so that the asymptotics on the surface $S_\rho$ are given by the differential equation $a_{-1}\delta u^2+\delta v^2=0$. This, however, defines the net of curves on the surface $S_y$ which corresponds to the asymptotic net on $S_\rho$. *The congruence of tangents to the curves $C_u$ on $S_y$ sets up a point-to-point correspondence between its focal sheets $S_y$ and $S_\rho$, and the net of curves on $S_y$ which corresponds to the asymptotic net on $S_\rho$ is defined by the differential equation*

$$a(\gamma^{(3)}-\gamma_u)\delta u^2+K\delta v^2=0. \tag{73}$$

*Similarly, the differential equation*

$$H\delta u^2+\beta^{(4)}\delta v^2=0 \tag{74}$$

*defines the net of curves on $S_y$ which corresponds to the asymptotic net on $S_\sigma$.*

The tangents at a point $y$ to the curves of either of these nets are separated harmonically by the parametric conjugate tangents at $y$. They coincide with the asymptotic tangents only when the corresponding congruence is a $W$-congruence.

An interesting special case, to which we shall return later, is that in which the congruences of tangents to the curves $C_u$ and $C_v$ are both of them $W$-congruences. If, in the differential equation (72) of the axis curves, we put $W^{(u)}=0$, $W^{(v)}=0$, we obtain for this case

$$aK\delta u^2-\mathfrak{D}\delta u\delta v-H\delta v^2=0.$$

But the focal points of the ray corresponding to a point $y$ are given by the formulas

$$R=\rho+r_1\sigma, \quad S=\rho+r_2\sigma, \tag{62}$$

where $r_1$ and $r_2$ are the roots of the quadratic

$$Hr^2+\mathfrak{D}r-aK=0. \tag{63}$$

These points are evidently the same as those in which the two axis tangents meet the ray. Moreover, it is easily seen that this can happen only when both $W^{(u)}$ and $W^{(v)}$ vanish. Consequently, *the axis tangents of a point $y$ meet the corresponding ray in the focal points of the ray if and only if both of the congruences of tangents to the curves $C_u$ and $C_v$ on $S_y$ are W-congruences, i. e.,* $W^{(u)}=0$, $W^{(v)}=0$.

A glance at equation (65) will show that *the ray tangents meet the ray in the focal points of the ray if and only if the conjugate net on $S_y$ has equal Laplace-Darboux invariants, i. e., $H-K=0$.* This affords a new geometric characterization of a conjugate net with equal Laplace-Darboux invariants.

Many interesting questions present themselves in connection with the various nets which we have defined in this section. Aside from the applications to be made in a later paragraph, we refrain from further consideration of these matters, although an exhaustive study of the interrelations of these nets, and their bearing on the general theory of conjugate nets and congruences, is greatly to be desired.

## § 6. *The Associate Conjugate Net.*

We noted in the preceding section that the asymptotic curves of the surface $S_y$ are given by the differential equation

$$a\delta u^2+\delta v^2=0. \tag{68}$$

We shall now define a new net of curves, which will be useful later. The differential equation

$$a\delta u^2-\delta v^2=0 \tag{75}$$

determines a net of curves, which for reasons which will appear presently we shall call the *associate conjugate net.* The tangents to the two curves of the net at $y$ we shall call the *associate conjugate tangents.* From (68) and (75), we see that the associate conjugate tangents separate the asymptotic tangents harmonically. Consequently, the net defined by (75) is actually a conjugate net. Moreover, the form of equation (75) shows that the associate conjugate tangents separate harmonically the original conjugate tangents. Now, there is evidently one and only one pair of lines which separates harmonically each of two given pairs. *The associate conjugate net, which is defined by the differential equation*

$$a\delta u^2-\delta v^2, \tag{75}$$

*is uniquely characterized by the property, that the tangents to the two curves of the net at a point y are harmonically separated both by the pair of asymptotic tangents and by the pair of original conjugate tangents.*

The relation between a conjugate net and its associate conjugate net is of course a reciprocal one. In fact, it may be convenient to speak of two conjugate nets as *associated conjugate nets*, if this relation subsists between them, without distinguishing the nets one from the other.

We shall now set up the completely integrable system of partial differential equations, of form (4), for the surface $S_y$ referred to the associate conjugate net. By the transformation

$$\bar{u}=\phi(u,v), \quad \bar{v}=\psi(u,v), \tag{76}$$

where

$$\phi_u^2-a\phi_v^2=0, \quad \psi_u^2-a\psi_v^2=0, \tag{77}$$

the associate conjugate net is made parametric. We may choose for $\phi$ and $\psi$ functions satisfying the equations

$$\phi_u+\sqrt{a}\,\phi_v=0, \quad \psi_u-\sqrt{a}\,\psi_v=0. \tag{78}$$

The derivatives of any function $y(u, v)$ undergo the transformations

$$y_u=\bar{y}_u\phi_u+\bar{y}_v\psi_u, \quad y_v=\bar{y}_u\phi_v+\bar{y}_v\psi_v, \tag{79}$$

$$\left.\begin{aligned} y_{uu} &= \bar{y}_{uu}\phi_u^2+2\bar{y}_{uv}\phi_u\psi_u+\bar{y}_{vv}\psi_u^2+\bar{y}_u\phi_{uu}+\bar{y}_v\psi_{uu}, \\ y_{uv} &= \bar{y}_{uu}\phi_u\phi_v+\bar{y}_{uv}(\phi_u\psi_v+\phi_v\psi_u)+\bar{y}_{vv}\psi_u\psi_v+\bar{y}_u\phi_{uv}+\bar{y}_v\psi_{uv}, \\ y_{vv} &= \bar{y}_{uu}\phi_v^2+2\bar{y}_{uv}\phi_v\psi_v+\bar{y}_{vv}\psi_v^2+\bar{y}_u\phi_{vv}+\bar{y}_v\psi_{vv}, \end{aligned}\right\} \tag{80}$$

where $\bar{y}_u$, $\bar{y}_v$, etc., stand for $\partial y/\partial\bar{u}$, $\partial y/\partial\bar{v}$, etc. These formulas hold for any transformation of the form (76). We now impose on $\phi$ and $\psi$ the conditions (78), from which we have

$$\phi_u=-\sqrt{a}\,\phi_v, \quad \psi_u=\sqrt{a}\,\psi_v, \tag{81}$$

and on differentiation

$$\phi_{uu}=-\sqrt{a}\,\phi_{uv}-\frac{a_u}{2\sqrt{a}}\phi_v, \quad \psi_{uu}=\sqrt{a}\,\psi_{uv}+\frac{a_u}{2\sqrt{a}}\psi_v,$$

$$\phi_{uv}=-\sqrt{a}\,\phi_{vv}-\frac{a_v}{2\sqrt{a}}\phi_v, \quad \psi_{uv}=\sqrt{a}\,\psi_{vv}+\frac{a_v}{2\sqrt{a}}\psi_v.$$

From these we find

$$\left.\begin{aligned}\phi_{uu}&=a\phi_{vv}+\tfrac{1}{2}\left(a_v-\frac{a_u}{\sqrt{a}}\right)\phi_v, \quad \phi_{uv}=-\sqrt{a}\,\phi_{vv}-\frac{a_v}{2\sqrt{a}}\phi_v,\\ \psi_{uu}&=a\psi_{vv}+\tfrac{1}{2}\left(a_v+\frac{a_u}{\sqrt{a}}\right)\psi_v, \quad \psi_{uv}=\sqrt{a}\,\psi_{vv}+\frac{a_v}{2\sqrt{a}}\psi_v.\end{aligned}\right\} \tag{82}$$

In fact, any derivative of $\phi$, for instance, is expressible entirely in terms of derivatives of $\phi$ taken with respect to $v$ alone; the corresponding expression for the same derivative of $\psi$ is then obtained by changing $\phi$ into $\psi$ and $\sqrt{a}$ into $-\sqrt{a}$. The only derivatives of the third order which we shall need are

$$\left.\begin{aligned}\phi_{uvv}&=-\sqrt{a}\,\phi_{vvv}-\frac{a_v}{\sqrt{a}}\phi_{vv}+\left(\frac{a_v^2}{4a\sqrt{a}}-\frac{a_{vv}}{2\sqrt{a}}\right)\phi_v,\\ \psi_{uvv}&=\sqrt{a}\,\psi_{vvv}+\frac{a_v}{\sqrt{a}}\psi_{vv}+\left(\frac{a_{vv}}{2\sqrt{a}}-\frac{a_v^2}{4a\sqrt{a}}\right)\psi_v.\end{aligned}\right\} \tag{83}$$

Transformation (79) becomes, by (81),

$$y_u=-\sqrt{a}\,\bar{y}_u\phi_v+\sqrt{a}\,\bar{y}_v\psi_v, \quad y_v=\bar{y}_u\phi_v+\bar{y}_v\psi_v, \tag{84}$$

the inverse of which is

$$\bar{y}_u=\frac{1}{2\sqrt{a}\,\phi_v}(-y_u+\sqrt{a}\,y_v), \quad \bar{y}_v=\frac{1}{2\sqrt{a}\,\psi_v}(y_u+\sqrt{a}\,y_v). \tag{85}$$

Multiplying the last of equations (80) by $a$, subtracting from the first, and using (78) and (82), we find

$$y_{uu}-a y_{vv}=-4a\bar{y}_{uv}\phi_v\psi_v+\tfrac{1}{2}\left(a_v-\frac{a_u}{\sqrt{a}}\right)\bar{y}_u\phi_v+\tfrac{1}{2}\left(a_v+\frac{a_u}{\sqrt{a}}\right)\bar{y}_v\psi_v, \tag{86}$$

whence from (85)

$$4a\bar{y}_{uv}\phi_v\psi_v=-y_{uu}+a y_{vv}+\frac{a_u}{2a}y_u+\frac{a_v}{2\sqrt{a}}y_v. \tag{87}$$

Also, from the second of (80),

$$y_{uv}=-\sqrt{a}\,\bar{y}_{uu}\phi_v^2+\sqrt{a}\,\bar{y}_{vv}\psi_v^2+\bar{y}_u\phi_{uv}+\bar{y}_v\psi_{uv}; \tag{88}$$

using this, and the equation found by multiplying the last of (80) by $a$ and adding to the first, we obtain without difficulty

$$\left.\begin{aligned}4a\bar{y}_{uu}\phi_v^2&=y_{uu}-2\sqrt{a}\,y_{uv}+a y_{vv}+y_u\left(2\sqrt{a}\frac{\phi_{vv}}{\phi_v}+\frac{a_v}{\sqrt{a}}-\frac{a_u}{2a}\right)-y_v\left(2a\frac{\phi_{vv}}{\phi_v}+\frac{a_v}{2}\right),\\ 4a\bar{y}_{vv}\psi_v^2&=y_{uu}+2\sqrt{a}\,y_{uv}+a y_{vv}-y_u\left(2\sqrt{a}\frac{\psi_{vv}}{\psi_v}+\frac{a_v}{\sqrt{a}}+\frac{a_u}{2a}\right)-y_v\left(2a\frac{\psi_{vv}}{\psi_v}+\frac{a_v}{2}\right).\end{aligned}\right\} \tag{89}$$

The formulas which we have thus far written hold for any function $y$ whatever. We shall now suppose that $y$ is the dependent variable in system (4), and shall obtain the analogous completely integrable system for $y$ as a function of the variables $\bar{u}$, $\bar{v}$. The second of equations (4) is

$$y_{uv}=b'y_u+c'y_v+d'y,$$

so that by (84) we have

$$y_{uv}=\phi_v(-\sqrt{a}\,b'+c')\bar{y}_u+\psi_v(\sqrt{a}\,b'+c')\bar{y}_v+d'\,\bar{y}, \tag{90}$$

and consequently by substituting in the left-hand member of (88) we may express $\bar{y}_{uu}$ linearly in terms of $\bar{y}_{vv}$, $\bar{y}_u$, $\bar{y}_v$, $\bar{y}$. We shall write the result presently. From the first of equations (4) we have

$$\begin{aligned} y_{uu}-ay_{vv}&=by_u+cy_v+dy\\ &=\phi_v(-\sqrt{a}\,b+c)\bar{y}_u+\psi_v(\sqrt{a}\,b+c)\bar{y}_v+d\bar{y}, \end{aligned}$$

which when substituted in the left-hand member of (86) gives an equation of the Laplace type for $\bar{y}$. We thus find the required system of differential equations.

*The completely integrable system of partial differential equations*

$$\bar{y}_{uu}=\bar{a}\,\bar{y}_{vv}+\bar{b}\,\bar{y}_u+\bar{c}\,\bar{y}_v+\bar{d}\,\bar{y},\quad \bar{y}_{uv}=\bar{b}'\,\bar{y}_u+\bar{c}'\,\bar{y}_v+\bar{d}'\,\bar{y}, \tag{91}$$

*where*

$$\left.\begin{aligned} &\bar{a}=\frac{\psi_v^2}{\phi_v^2},\quad \bar{b}=\frac{1}{\phi_v}\left(b'-\frac{c'}{\sqrt{a}}-\frac{a_v}{2a}-\frac{\phi_{vv}}{\phi_v}\right),\quad \bar{c}=\frac{\psi_v}{\phi_v^2}\left(\frac{a_v}{2a}-b'-\frac{c'}{\sqrt{a}}+\frac{\psi_{vv}}{\psi_v}\right),\\ &\bar{d}=-\frac{1}{\phi_v^2}\frac{d'}{\sqrt{a}},\quad \bar{b}'=\frac{1}{8a\psi_v}\left(a_v-\frac{a_u}{\sqrt{a}}+2\sqrt{a}\,b-2c\right),\\ &\bar{c}'=\frac{1}{8a\phi_v}\left(a_v+\frac{a_u}{\sqrt{a}}-2\sqrt{a}\,b-2c\right),\quad \bar{d}'=-\frac{1}{4a\phi_v\psi_v}d, \end{aligned}\right\} \tag{92}$$

*has as its integral surfaces the integral surfaces of system* (4), *but referred to the associate conjugate net as parameter curves.*

We may define for the associate conjugate net the axis congruence, ray congruence, etc. Let us call these the *associate axis congruence, associate ray congruence*, etc. The associate axis of a point $y$ on the surface is the line joining the point $y$ to the point

$$\bar{z}=\bar{y}_{vv}+\frac{\bar{c}}{\bar{a}}\bar{y}_v.$$

We wish to express this in terms of the coefficients and variables of system (4). Equations (89) hold for any function $y$; if $y$ satisfy (4), however, we may express $\bar{y}_{vv}$ linearly in terms of $y_{vv}$, $y_u$, $y_v$, $y$:

$$\begin{aligned} 4a\bar{y}_{vv}\psi_v^2=2ay_{vv}&+\left(b+2\sqrt{a}b'-2\sqrt{a}\frac{\psi_{vv}}{\psi_v}-\frac{a_v}{\sqrt{a}}-\frac{a_u}{2a}\right)y_u\\ &+\left(c+2\sqrt{a}c'-2a\frac{\psi_{vv}}{\psi_v}-\frac{a_v}{2}\right)y_v+(d+2\sqrt{a}d')y. \end{aligned} \tag{93}$$

We find, therefore, using this, (85), and (92),

$$4a\psi_v^2\bar{z}=2ay_{vv}+\left(b-2c'-\frac{a_u}{2a}\right)y_u+\left(c-2ab'+\frac{a_v}{2}\right)y_v+()y,$$

in which the coefficient of $y$ is immaterial. The coefficient of $y_u$ is an invariant; in fact, if use be made of equations (24), (29), and (38) of the previous memoir, we find without difficulty that

$$\mathfrak{C}'=\tfrac{1}{4}\left(2c'-b+\frac{a_u}{2a}\right),\quad \mathfrak{B}'=\tfrac{1}{4}\left(2b'+\frac{c}{a}-\frac{a_v}{2a}\right), \tag{94}$$

consequently,

$$2\psi_v^2\bar{z}=y_{vv}-\frac{2}{a}\mathfrak{C}'y_u+\left(\frac{c}{a}-2\mathfrak{B}'\right)y_v+()y,$$

or

$$2\psi_v^2\bar{z}=z-\frac{2}{a}\mathfrak{C}'y_u-2\mathfrak{B}'y_v+()y. \tag{95}$$

This equation affords a means for interpreting geometrically the invariants $\mathfrak{B}'$ and $\mathfrak{C}'$; this will complete the geometric interpretation of the set of fundamental invariants of the preceding memoir. Remembering that the points $z$, $y$, $y_u$ determine the osculating plane to the curve $C_u$ on $S_y$, and the points $z$, $y$, $y_v$ the osculating plane to the curve $C_v$, we may state the theorem:

*The associate axis of a point $y$ of the surface $S_y$ lies in the osculating plane to the curve $C_u$ ($v$=const.) on $S_y$ if and only if the invariant $\mathfrak{B}'$ vanishes; it lies in the osculating plane to the curve $C_v$ ($u$=const.) if and only if the invariant $\mathfrak{C}'$ vanishes.*

We recall that if both $\mathfrak{B}'$ and $\mathfrak{C}'$ are identically zero, the surface $S_y$ is a quadric.* Consequently, *the axis congruence of a conjugate net on a surface coincides with the associate axis congruence if and only if the surface is a quadric.*

For the minus first and first Laplace transforms of the associate conjugate net we find without difficulty, using the first of formulas (85),

$$\begin{aligned}8a\phi_v\bar{\rho}&=8a\phi_v(\bar{y}_u-\bar{c}'\bar{y})\\&=4\sqrt{a}(-y_u+\sqrt{a}y_v)-\left(a_v+\frac{a_u}{\sqrt{a}}-2\sqrt{a}b-2c\right)y\\&=4\sqrt{a}(-\rho+\sqrt{a}\sigma)-\left(a_v+\frac{a_u}{\sqrt{a}}-2\sqrt{a}b-2c+4\sqrt{a}c'-4ab'\right)y,\end{aligned}$$

so that

$$2\sqrt{a}\phi_v\bar{\rho}=-\rho+\sqrt{a}\sigma+2(\sqrt{a}\mathfrak{B}'-\mathfrak{C}')y. \tag{96}$$

---

* Preceding paper, end of § 7.

Similarly,

$$2\sqrt{\bar{a}}\psi_v\bar{\sigma}=\rho+\sqrt{a}\sigma+2(\sqrt{a}\mathfrak{B}'+\mathfrak{C}')y. \qquad (97)$$

The point of intersection of the ray $\rho\sigma$ of a point $y$ with its associate ray $\bar{\rho}\bar{\sigma}$ is given by

$$(\rho\sigma \# \bar{\rho}\bar{\sigma})=\mathfrak{B}'\rho-\mathfrak{C}'\sigma.$$

If $\mathfrak{B}'=0$, this point coincides with the first Laplace transform $\sigma$, and if $\mathfrak{C}'=0$ it coincides with the minus first Laplace transform $\rho$. Combining these with the results just found in connection with the axis and associated axis, we may state that *if at a point $y$ on the surface $S_y$ the associate axis lies in the osculating plane to the curve $C_u$ (or $C_v$) then the corresponding ray meets the associate ray in the first Laplace transform (or minus first Laplace transform) of $y$.*

It is also easily seen that *the quadrics are the only ruled surfaces for which either of the above cases may arise. Moreover, the ray congruence coincides with the associate ray congruence if and only if the axis congruence coincides with the associate axis congruence, in which case the surface $S_y$ is a quadric.*

It is not our purpose here to give an extended discussion of the associate conjugate net in its relation to the original conjugate net. Analytically, a complete discussion would be very complicated, since we should have to express the invariants of the associate net in terms of those of the original net. We may predict the form of some of these expressions. For example, the condition that the surface $S_y$ be ruled is $a\mathfrak{B}'^2+\mathfrak{C}'^2=0$ in terms of the original parameters, and $\bar{a}\bar{\mathfrak{B}}'^2+\bar{\mathfrak{C}}'^2=0$ in terms of the new ones. Consequently, there must be a relation of the form

$$\bar{a}\bar{\mathfrak{B}}'^2+\bar{\mathfrak{C}}'^2=\lambda(a\mathfrak{B}'^2+\mathfrak{C}'^2),$$

in which the factor $\lambda$ is easily found if use be made of (92) and (82). It should be noted that in this invariant the curves $C_u$ and $C_v$ enter symmetrically; if for an invariant of this kind its expression in the new variables $\bar{u}$, $\bar{v}$ were essentially the same as its expression in the old variables $u$, $v$, there would be little interest in studying the associate conjugate net. However, the following example will show that this is not the case.

Let us calculate the invariant $\bar{H}-\bar{K}$. The vanishing thereof is a necessary and sufficient condition that the associate ray curves on $S_y$ form a conjugate net. If we put

$$m=a_v-2c, \quad n=2\sqrt{a}b-\frac{a_u}{\sqrt{a}},$$

we have from (92)

$$b'=\frac{1}{8a\psi_v}(m+n),\quad \bar{c}'=\frac{1}{8a\phi_v}(m-n).$$

Consequently, if we denote $\partial\bar{b}'/\partial\bar{u}$ by $\bar{b}'_u$, we have by (85)

$$\bar{b}'_u=\frac{1}{2\sqrt{a}\phi_v}\left(-\frac{\partial\bar{b}'}{\partial u}+\sqrt{a}\,\frac{\partial\bar{b}'}{\partial v}\right),$$

so that

$$16\bar{b}'_u=\frac{1}{\sqrt{a}\phi_v\psi_v}\Big[-\frac{1}{a}(m_u+n_u)+\frac{1}{\sqrt{a}}(m_v+n_v)$$
$$+(m+n)\left\{\frac{a_u}{a^2}-\frac{a_v}{a\sqrt{a}}+\frac{\psi_{uv}}{a\psi_v}-\frac{\psi_{vv}}{\sqrt{a}\psi_v}\right\}\Big],$$

which by (82) becomes

$$16\bar{b}'_u=\frac{1}{\sqrt{a}\phi_v\psi_v}\left[-\frac{1}{a}(m_u+n_u)+\frac{1}{\sqrt{a}}(m_v+n_v)+(m+n)\left\{\frac{a_u}{a^2}-\frac{a_v}{2a\sqrt{a}}\right\}\right].\quad (98)$$

Similarly,

$$16\bar{c}'_v=\frac{1}{\sqrt{a}\phi_v\psi_v}\left[\frac{1}{a}(m_u-n_u)+\frac{1}{\sqrt{a}}(m_v-n_u)+(m-n)\left\{-\frac{a_u}{a^2}-\frac{a_v}{2a\sqrt{a}}\right\}\right],\quad (99)$$

so that

$$16\sqrt{a}\phi_v\psi_v(\bar{H}-\bar{K})=16\sqrt{a}\phi_v\psi_v(\bar{c}'_v-\bar{b}'_u)$$
$$=\frac{2}{a}m_u-\frac{2}{\sqrt{a}}n_v-\frac{2a_u}{a^2}m+\frac{a_v}{a\sqrt{a}}n.$$

Substituting the values for $m$ and $n$, we have

$$16\sqrt{a}\phi_v\psi_v(\bar{H}-\bar{K})=\frac{2}{a}(a_{uv}-2c_u)-\frac{2}{\sqrt{a}}\left(2\sqrt{a}b_v+\frac{ba_v}{\sqrt{a}}-\frac{a_{uv}}{\sqrt{a}}+\frac{a_ua_v}{2a\sqrt{a}}\right)$$
$$-\frac{2a_u}{a^2}(a_v-2c)+\frac{a_v}{a\sqrt{a}}\left(2\sqrt{a}b-\frac{a_u}{\sqrt{a}}\right)$$
$$=\frac{4a_{uv}}{a}-\frac{4a_ua_v}{a^2}-\frac{4c_u}{a}-4b_v+\frac{4ca_u}{a^2}.$$

But from the integrability condition (8a), we have

$$b_v+2c'_v=2b'_u-\frac{c_u}{a}+\frac{ca_u}{a^2}-\frac{a_{uv}}{a}+\frac{a_ua_v}{a^2},$$

so that

$$4\sqrt{a}\phi_v\psi_v(\bar{H}-\bar{K})=2c'_v-2b'_u+\frac{2a_{uv}}{a}-\frac{2a_ua_v}{a^2},$$

and finally
$$2\sqrt{a}\phi_v\psi_v(\overline{H}-\overline{K})=c'_v-b'_u+\frac{\partial^2}{\partial u\partial v}\log a$$
$$=H-K+\frac{\partial^2}{\partial u\partial v}\log a. \qquad (100)$$

The right-hand member is the invariant appearing in equation (69), the geometric interpretation of which was given in the preceding section. Recalling also Wilczynski's theorem, that for a net with equal Laplace-Darboux invariants the ray curves form a conjugate net, we may state the theorem:

*If the axis curves corresponding to a conjugate net themselves form a conjugate net, then the associate ray curves also form a conjugate net, and conversely. If the ray curves form a conjugate net, then the associate axis curves also form a conjugate net, and conversely.*

This theorem is sufficient to show that the consideration of the associate conjugate net will not lead to trivial results. Some properties of a conjugate net are enjoyed also by the associate net; we have already seen an obvious instance of this, and in the next section shall find another which is of greater interest, and not at all self-evident. Other properties, however, are not common to the two nets, even when the two component families of each net are concerned in a symmetric way. It would therefore seem desirable *to determine all the properties which hold for both of two associated conjugate nets.* We shall not pursue this study any further, although in the next section we shall make an important application of the associate conjugate net. The subject undoubtedly deserves closer investigation; in fact, it would appear that all properties of a conjugate net might well be described in connection with its associate conjugate net. For instance, the two families of developables of a congruence touch the two focal sheets in a conjugate net on either; if the associate conjugate nets on the two sheets also correspond, the congruence is a $W$-congruence. The equations of the present paragraph constitute an analytic starting-point for the theory, whose more systematic development we must leave for a future occasion.

## § 7. *Isothermally Conjugate Nets.*

We shall in the present section apply some of the concepts introduced in previous sections to the study of isothermally conjugate nets. If in the second fundamental form of a surface, viz., $D\,\delta u^2+2D'\,\delta u\,\delta v+D''\,\delta v^2$, the coefficient $D'=0$, the parametric net is conjugate, and Bianchi calls it *isothermally conjugate* if in addition $D=D''$, or can be made so by a transformation $\overline{u}=U(u)$, $\overline{v}=V(v)$. This is equivalent to demanding that by such a transformation the coefficient $a$ in the first of equations (4) be reducible to unity. A necessary and sufficient condition for this is without difficulty seen to be

$$\frac{\partial^2}{\partial u\,\partial v}\log a=0. \tag{101}$$

*The conjugate nets on the integral surfaces of system* (4) *are isothermally conjugate if and only if*

$$\frac{\partial^2}{\partial u\,\partial v}\log a=0.$$

Isothermally conjugate nets have received increased attention of late. However, until quite recently their only characterization was analytic, until Wilczynski* gave a geometric interpretation of the condition (101). His interpretation consists in the determination of an algebraic relation between three absolute projective invariants which have themselves been previously characterized geometrically. We shall presently give a new geometric interpretation of condition (101) consisting entirely of descriptive geometric properties.

That the left-hand member of (101) is actually a projective invariant of the conjugate net may be seen from equations (70) and (71), from which we have

$$W^{(v)}-W^{(u)}=\frac{\partial^2}{\partial u\,\partial v}\log a. \tag{102}$$

This leads at once to the theorem of Demoulin and Tzitzéica: *If the tangents to the curves $C_u$ and the tangents to the curves $C_v$ on $S_y$ both form W-congruences, the conjugate net $C_u$, $C_v$ is isothermally conjugate.* In § 5 we gave a geometric characterization of a conjugate net having this property; we found that for such a net the axis tangents at any point of the surface meet the corresponding ray in the focal points of the ray.

We found also in § 5 that the axis curves form a conjugate net if and only if

$$H-K+\frac{\partial^2}{\partial u\,\partial v}\log a=0. \tag{69}$$

Moreover, if $H-K=0$, the ray curves form a conjugate net. Consequently, *if the axis curves and ray curves both form conjugate nets, the parametric conjugate net is isothermally conjugate. An isothermally conjugate net has equal Laplace-Darboux invariants if and only if its axis curves form a conjugate net.*

---

* "The General Theory of Congruences," *Transactions of the American Mathematical Society*, Vol. XVI (July, 1915), pp. 311–327.

A consideration of the differential equations of the axis curves and ray curves shows that *the only isothermally conjugate nets for which the axis curves coincide with the ray curves are those having equal Laplace-Darboux invariants in addition to being subject to the condition of Demoulin and Tzitzéica: the tangents to the curves $C_u$, and hence also the tangents to the curves $C_v$, form a W-congruence.*

In § 4 we found that a conjugate net for which both families of curves are plane is completely characterized by the fact that the axis curves coincide with the conjugate net. Since we have

$$a\beta^{(4)}=H+2b'_u-b_v,\quad \gamma^{(3)}-\gamma_u=K+2b'_u-b_v-\frac{\partial^2}{\partial u\,\partial v}\log a,$$

and since $\beta^{(4)}=0$ and $\gamma^{(3)}-\gamma_u=0$ are respectively the conditions that the curves $C_v$ and the curves $C_u$ be plane, *a conjugate net consisting of two families of plane curves is isothermally conjugate if and only if it has equal Laplace-Darboux invariants.*

Surfaces on which one or both families of lines of curvature are plane have been extensively studied; it would seem well worth while to consider also conjugate nets, either general or of some particular kind, for which one or both families are plane. We are not aware of any systematic treatment of the subject, however, in spite of its apparent promise.

Let us return now to the general isothermally conjugate net. In order to give a geometric characterization thereof, we shall use the differential equations of the axis curves, of the anti-ray curves, and of the associate conjugate net. These are, respectively,

$$a\left(K+2b'_u-b_v-\frac{\partial^2}{\partial u\partial v}\log a\right)\delta u^2-\mathfrak{D}\delta u\delta v-(H+2b'_u-b_v)\delta v^2=0, \tag{60}$$

$$aH\delta u^2+\mathfrak{D}\delta u\delta v-K\delta v^2=0, \tag{66}$$

$$a\delta u^2-\ \delta v^2=0. \tag{75}$$

We recall that the anti-ray tangents are the harmonic conjugates of the ray tangents with respect to the parametric conjugate tangents. Let us regard the above three equations as binary quadratics. The Jacobian of the two binary quadratics

$$a_0x_1^2+2a_1x_1x_2+a_2x_2^2=0,\quad b_0x_1^2+2b_1x_1x_2+b_2x_2^2=0,$$

is

$$(a_0b_1-a_1b_0)x_1^2+(a_0b_2-a_2b_0)x_1x_2+(a_1b_2-a_2b_1)x_2^2=0,$$

and its roots give the pair which separates harmonically each of the pairs defined by the two quadratics. The Jacobian of the two quadratics (60) and (75) is

$$a\mathfrak{D}\delta u^2+2a\left(H-K+\frac{\partial^2}{\partial u\partial v}\log a\right)\delta u\delta v+\mathfrak{D}\delta v^2=0, \tag{103}$$

and defines the pair of lines which separates harmonically both the pair of axis tangents and the pair of associate conjugate tangents. The Jacobian of the two quadratics (66) and (75) is

$$a\mathfrak{D}\delta u^2+2a(H-K)\delta u\delta v+\mathfrak{D}\delta v^2=0, \tag{104}$$

and defines the pair of lines which separates harmonically both the pair of anti-ray tangents and the pair of associate conjugate tangents.

The two Jacobians (103) and (104) coincide if and only if $\partial^2 \log a/\partial u\partial v=0$, *i. e.*, the parametric conjugate net is isothermally conjugate. This means that the double points of the involution determined by the pair of axis tangents and the pair of associate conjugate tangents coincide in this case with the double points of the involution determined by the pair of anti-ray tangents and the pair of associate conjugate tangents. In other words, the three pairs defined by the quadratics (60), (66), and (75) belong to the same involution. We have then the theorem:

*A necessary and sufficient condition that a conjugate net on a surface be isothermally conjugate is that for each point of the surface the pair of axis tangents, the pair of anti-ray tangents, and the pair of associate conjugate tangents form pairs of the same involution.*

In the geometric characterization just given, the axis tangents may be replaced by the anti-axis tangents (defined by equation (67)), and the anti-ray tangents at the same time by the ray tangents.

We shall now investigate the nature of the original conjugate net, if its associate conjugate net be isothermally conjugate. We have by (92) $\bar{a}=\psi_v^2/\phi_v^2$, and we must calculate the expression $\partial^2 \log \bar{a}/\partial\bar{u}\partial\bar{v}$ in terms of the coefficients and variables of system (4). This may be done most expeditiously as follows:

$$\begin{aligned}\frac{\partial^2}{\partial\bar{u}\partial\bar{v}}\log\bar{a}&=2\frac{\partial^2}{\partial\bar{u}\partial\bar{v}}\log\psi_v-2\frac{\partial^2}{\partial\bar{u}\partial\bar{v}}\log\phi_v\\&=2\frac{\partial}{\partial\bar{v}}\left(\frac{\partial}{\partial\bar{u}}\log\psi_v\right)-2\frac{\partial}{\partial\bar{u}}\left(\frac{\partial}{\partial\bar{v}}\log\phi_v\right)\\&=2\frac{\partial}{\partial\bar{v}}\left[\frac{1}{2\sqrt{a}\phi_v}\left(-\frac{\partial}{\partial u}\log\psi_v+\sqrt{a}\frac{\partial}{\partial v}\log\psi_v\right)\right]\\&\qquad-2\frac{\partial}{\partial\bar{u}}\left[\frac{1}{2\sqrt{a}\psi_v}\left(\frac{\partial}{\partial u}\log\phi_v+\sqrt{a}\frac{\partial}{\partial v}\log\phi_v\right)\right],\end{aligned}$$

where we have made use of the formula of differentiation (85). We thus obtain

$$\begin{aligned}\frac{\partial^2}{\partial u\partial \bar v}\log \bar a &= \frac{\partial}{\partial \bar v}\left[\frac{1}{\sqrt{a}\phi_v}\left(-\frac{\psi_{uv}}{\psi_v}+\frac{\sqrt{a}\psi_{vv}}{\psi_v}\right)\right]-\frac{\partial}{\partial \bar u}\left[\frac{1}{\sqrt{a}\psi_v}\left(\frac{\phi_{uv}}{\phi_v}+\frac{\sqrt{a}\phi_{vv}}{\phi_v}\right)\right]\\ &= -\frac{\partial}{\partial \bar v}\left[\frac{1}{\sqrt{a}\phi_v}\cdot\frac{a_v}{2\sqrt{a}}\right]+\frac{\partial}{\partial \bar u}\left[\frac{1}{\sqrt{a}\psi_v}\cdot\frac{a_v}{2\sqrt{a}}\right]\\ &= -\frac{1}{2}\frac{\partial}{\partial \bar v}\left[\frac{1}{\phi_v}\frac{\partial}{\partial v}\log a\right]+\frac{1}{2}\frac{\partial}{\partial \bar u}\left[\frac{1}{\psi_v}\frac{\partial}{\partial v}\log a\right]\\ &= -\frac{1}{2\sqrt{a}\phi_v\psi_v}\frac{\partial^2}{\partial u\partial v}\log a. \end{aligned}\tag{105}$$

In the final reduction, use has again been made of (85) and (82).

Equation (105) leads to the important result: *If either of two associated conjugate nets is isothermally conjugate, the other is also.* Isothermal conjugacy is therefore one of the properties of a conjugate net which is also a property of the associate conjugate net. It is in a sense independent of the nature of the surface, since there exist an infinite number of isothermally conjugate nets on any curved surface. We have here, then, a property which is common to both of two associated conjugate nets, or subsists for neither. We have not succeeded in finding any others,* so that it may be interesting to determine just how many of these properties there are. We shall leave this question open for the present.

In the course of this and the preceding memoir, we have studied a single conjugate net, with its related configurations, but have not touched upon the larger questions concerning conjugate nets in general on a surface, nor the properties of conjugate nets which are preserved under certain transformations. We hope soon to give a general theory of the transformation of conjugate nets from a purely projective point of view.

HARVARD UNIVERSITY, *August* 11, 1915.

* We mean, of course, properties which do not depend entirely on the nature of the surface; for instance, if a surface is ruled for one conjugate net thereon, it is ruled for every other net.

# *The Asymptotic Equation and Satellite Conic of the Plane Quartic.*

By Teresa Cohen.

The four tangents at the intersections of a line $(\xi x)$ with a quartic curve $(ax)^4$ can theoretically be obtained as follows:

Let $y$ be one of these intersections; then the four tangents are given by eliminating $y$ from

$$(ay)^4=0, \text{ the condition } y \text{ be on the quartic,}$$
$$(ax)(ay)^3=0, \text{ the tangent to the quartic at } y,$$
$$(\xi y)=0, \text{ the condition that } y \text{ be on the line.}$$

The eliminant is of seventh degree in the coefficients of the quartic, fourth degree in $x$ and twelfth degree in $\xi$; all of which may be summed up by saying that it is an $A^7x^4\xi^{12}$. For a given $\xi$ it gives the four tangents at the intersections of the line $(\xi x)$ with the quartic; hence, it is called the *asymptotic equation* of the quartic. But, equally well, for a given $x$, it gives the twelve points of contact of tangents from $x$ to the curve.

The four tangents may be thought of as making up a second quartic, and on them and $(ax)^4$ can be built up a whole pencil of quartics, each member of which cuts the quartic twice at the points where $(\xi x)$ cuts it. Therefore in the pencil must be a member made up of the line $(\xi x)$ counted twice and a conic on the remaining eight points where the four tangents cut the quartic, the so-called *satellite conic*. If $S=0$ be this conic and $T=0$ the four tangents,

$$T+\lambda(ax)^4=(x\xi)^2S; \tag{1}$$

or, more symmetrically, since $T$ bears the same relation to $(a\xi)^{12}$, the quartic in lines, as to $(ax)^4$, the quartic in points,

$$T+\lambda(ax)^4(a\xi)^{12}=(x\xi)^2S. \tag{1'}$$

Since $T$ is an $A^7x^4\xi^{12}$, $S$ must be an $A^7x^2\xi^{10}$. The object of this paper is to express $S$, and therefore $T$, in terms of known comitants of the quartic $(ax)^4$.

The $A^7x^2\xi^{10}$, considered as an equation in $x$, is the satellite conic of a line $\xi$; that is, it is the conic on the eight points where the tangents to the quartic at its intersections with the line $\xi$ again meet the quartic. But considered as an equation in $\xi$, it is a curve of class 10 on the ten tangents to the quartic from each point of contact of the twelve tangents from a point $x$. Each double point of the quartic absorbs two of the tangents; therefore, the degree of the conic in $\xi$ decreases by two for each double point, so that the existence of six double points causes the satellite to vanish.

To express this satellite in terms of known forms of the quartic, it is best to take up special cases where the satellite is known. Three such have been found sufficient.

## *Special Cases.*

Case I. $$4x_0(x_1^3+x_2^3)=0.$$

This represents four lines, three of which are on a point. Owing to the presence of six double points, $S\equiv 0$.

Case II. $$x_0^4+x_1^4+x_2^4=0.$$

To obtain the four tangents it is necessary to eliminate $y$ from

$$y_0^4+y_1^4+y_2^4=0,\quad x_0y_0^3+x_1y_1^3+x_2y_2^3=0,\quad \xi_0y_0+\xi_1y_1+\xi_2y_2=0.$$

This may be accomplished by means of the eliminant of

$$\alpha_0y_0^4+\alpha_1y_1^4+\alpha_2y_2^4=0,\quad \beta_0y_0^3+\beta_1y_1^3+\beta_2y_2^3=0,\quad y_0+y_1+y_2=0,$$

which is found to be *

$$\overset{3}{\Sigma}\alpha_0^3(\beta_1-\beta_2)^4+3\overset{6}{\Sigma}\alpha_1^2\alpha_2\beta_0\{\beta_0^3+16\beta_0^2\beta_2+10\beta_0\beta_2^2-4\beta_1(\beta_0-\beta_2)^2\}$$
$$+18\alpha_0\alpha_1\alpha_2\{4(\beta)\beta_0\beta_1\beta_2-(\beta_1^2\beta_2^2)\}.$$

Let $y_i=\xi_iy_i'$, $\alpha_i=\dfrac{x_i}{\xi_i^3}$, $\beta_i=\dfrac{1}{\xi_i^4}$. Then the second set of equations reduces to the first set and the eliminant, after clearing of fractions, becomes

$$\overset{3}{\Sigma}x_0^4(\xi_1^{12}+3\xi_1^8\xi_2^4+3\xi_1^4\xi_2^8+\xi_2^{12})+\overset{6}{\Sigma}4x_0^3x_1\xi_0^3\xi_1(-\xi_1^8+12\xi_1^4\xi_2^4-3\xi_2^8)$$
$$+\overset{3}{\Sigma}6x_0^2x_1^2\xi_0^2\xi_1^2(\xi_0^4\xi_1^4+5\xi_0^4\xi_2^4+5\xi_1^4\xi_2^4-3\xi_2^8)+\overset{3}{\Sigma}12x_0^2x_1x_2\xi_0^2\xi_1\xi_2(2\xi_0^4\xi_1^4+2\xi_0^4\xi_2^4-\xi_1^8$$
$$+6\xi_1^4\xi_2^4-\xi_2^8).$$

This, by equation (1), must be equal to

$$\lambda(x_0^4+x_1^4+x_2^4)+(x\xi)^2[\overset{3}{\Sigma}x_0^2\xi_0^2\{A\xi_0^8+B(\xi_0^4\xi_1^4+\xi_0^4\xi_2^4)+C(\xi_1^8+\xi_2^8)+D\xi_1^4\xi_2^4\}$$
$$+\overset{3}{\Sigma}2x_1x_2\xi_1\xi_2\{E\xi_0^8+F(\xi_0^4\xi_1^4+\xi_0^4\xi_2^4)+G(\xi_1^8+\xi_2^8)+H\xi_1^4\xi_2^4\}],$$

---

* This is from a remark by Professor Morley.

where the expression in brackets is the satellite conic. Since $\lambda$ enters into the coefficient of $x_i^4$ only, $A$, $B$, etc., may be obtained from the other coefficients.

$$A=-1,\quad B=-3,\quad C=-3,\quad D=21,\quad E=-3,\quad F=3,\quad G=1,\quad H=3.$$

$$\therefore\ S=\overset{3}{\Sigma}x_0^2\xi_0^2[-\xi_0^8-3(\xi_0^4\xi_1^4+\xi_0^4\xi_2^4)-3(\xi_1^8+\xi_2^8)+21\xi_1^4\xi_2^4]$$
$$+\overset{3}{\Sigma}x_1x_2\xi_1\xi_2\cdot 2[-3\xi_0^8+3(\xi_0^4\xi_1^4+\xi_0^4\xi_2^4)+(\xi_1^8+\xi_2^8)+3\xi_1^4\xi_2^4].$$

Case III. *The Lemniscate:* $6(x_1^2x_2^2+x_2^2x_0^2+x_0^2x_1^2)=0.$*

It is a property of the lemniscate that a tangent at a point $y$ meets the curve again in two points that lie on the conic $(y/x)$. Now, suppose $y$ eliminated between

$$\frac{y_0}{x_0}+\frac{y_1}{x_1}+\frac{y_2}{x_2}=0,\quad y_1^2y_2^2+y_2^2y_0^2+y_0^2y_1^2=0,\quad y_0\xi_0+y_1\xi_1+y_2\xi_2=0,$$

the operation being a simple one because two of the equations are linear. The eliminant is an octavic, made up of the four conics $(y/x)$ for the four points $y$ in which the line $\xi$ cuts the lemniscate. Since each conic passes through the reference points, which are double points of the lemniscate, the thirty-two intersections of octavic and quartic are made up of the three reference points, each taken eight times, and the eight intersections of the curve with the four tangents at its intersections with the line $\xi$. Therefore, it must be possible to find a quartic $(A_0x_1^2x_2^2)+(B_0x_0^2x_1x_2)$ such that

$$\overset{3}{\Sigma}x_1^4x_2^4\xi_1^2\xi_2^2+\overset{6}{\Sigma}x_0^4x_1^3x_2\cdot-2\xi_0^2\xi_1\xi_2+\overset{3}{\Sigma}x_0^4x_1^2x_2^2\xi_0^2(\xi_0^2+\xi_1^2+\xi_2^2)$$
$$+\overset{3}{\Sigma}x_0^3x_1^3x_2^2\cdot 2\xi_0\xi_1(-\xi_0^2-\xi_1^2+2\xi_2^2)$$
$$+(x_0^2x_1^2+x_0^2x_2^2+x_1^2x_2^2)[(A_0x_1^2x_2^2)+(B_0x_0^2x_1x_2)]=x_0^2x_1^2x_2^2S,$$

$$A_0=-\xi_1^2\xi_2^2,\quad B_0=2\xi_0^2\xi_1\xi_2.$$

Therefore, multiplying by $\xi_0^2\xi_1^2\xi_2^2$, the square of the double points,

$$S=\overset{3}{\Sigma}x_0^2\xi_0^6\xi_1^2\xi_2^2+\overset{3}{\Sigma}x_1x_2\xi_1\xi_2\cdot 2[3\xi_0^4\xi_1^2\xi_2^2-(\xi_0^2\xi_1^4\xi_2^2+\xi_0^2\xi_1^2\xi_2^4)].$$

## *The Expression of the Satellite in Terms of Known Comitants of the Quartic.*

The satellite, an $A^7x^2\xi^{10}$, may be written as

$$S\equiv C_{7,2,10}+(x\xi)C_{7,1,9}+(x\xi)^2C_{7,0,8}.$$

Therefore it is not necessary to deal with all possible terms at once, since we

---

* See R. A. Roberts, "Examples on Conics and Higher Plane Curves."

must have that $S-C_{7,2,10}=(x\xi)S'$ (*i. e.*, contains $(x\xi)$ as a factor), and, again, $S'-C_{7,1,9}=(x\xi)S''$.

Let us, then, first consider the possibilities for the $C_{7,2,10}$. If the two simplest contravariants of the quartic be taken as

$$(s\xi)^4=s\equiv\tfrac{1}{2}\,|\alpha\beta\xi|^4 \text{ and } (t\xi)^6=t\equiv\tfrac{1}{6}\,|\alpha\beta\xi|^2|\beta\gamma\xi|^2|\gamma\alpha\xi|^2,$$

and if it be understood that

$$s_is_js_k\equiv(s\xi)^{4-i}(s\alpha)^i(s'\xi)^{4-j}(s'\alpha)^j(s''\xi)^{4-k}(s''\alpha)^k(\alpha x)^{4-i-j-k}$$

and

$$t_it_j=(t\xi)^{6-i}(t\alpha)^i(t'\xi)^{6-j}(t'\alpha)^j(\alpha x)^{4-i-j},$$

then

$$C_{7,2,10}=M_{011}ss_1^2+M_{002}s^2s_2+N_{11}t_1^2+N_{02}tt_2,$$

where the $M$'s and $N$'s are coefficients to be determined.

A relation on these coefficients is most easily obtained from Case I, where the satellite vanishes and, therefore, $C_{7,2,10}$ must itself contain $(x\xi)$.

$$\begin{aligned}
C_{7,2,10}&=\sum^{2}x_0x_1[\xi_0^6\xi_1^4(8N_{11}+\tfrac{32}{5}N_{02})+\xi_0^6\xi_1\xi_2^3(216M_{011}+\tfrac{32}{5}N_{02})]\\
&\quad+\sum^{2}x_1^2[\xi_0^5\xi_1^5(8N_{11}+\tfrac{48}{5}N_{02})+\xi_0^5\xi_1^2\xi_2^3(432M_{011}+576M_{002}+8N_{11}+\tfrac{48}{5}N_{02})]\\
&=(x_0\xi_0+x_1\xi_1+x_2\xi_2)\,[x_1(A\xi_0^5\xi_1^4+B\xi_0^5\xi_1\xi_2^3)+x_2(A\xi_0^5\xi_2^4+B\xi_0^5\xi_1^3\xi_2)]\\
&=\sum^{2}(x_0x_1\xi_0+x_1^2\xi_1)\,(A\xi_0^5\xi_1^4+B\xi_0^5\xi_1\xi_2^3)+x_1x_2(\xi_0^5\xi_1^4\xi_2+\xi_0^5\xi_1\xi_2^4)\,(A+B).
\end{aligned}$$

$$A=8N_{11}+\tfrac{32}{5}N_{02}=8N_{11}+\tfrac{48}{5}N_{02}\quad\therefore\ N_{02}=0.$$

$$B=216M_{011}+\tfrac{32}{5}N_{02}=432M_{011}+576M_{002}+8N_{11}+\tfrac{48}{5}N_{02}$$

$$\therefore\ 216M_{011}+576M_{002}+8N_{11}=0.$$

$$A+B=8N_{11}+216M_{011}=0\quad\therefore\ M_{002}=0,\quad M_{011}:N_{11}=1:-27=-6:162.^*$$

$$\therefore\ S=-6(ss_1^2-27t_1^2)+(x\xi)S'.$$

For this case

$$S'=\sum^{2}x_1(-1296\xi_0^5\xi_1^4+1296\xi_0^5\xi_1\xi_2^3).$$

The correctness of this determination of $C_{7,2,10}$ may be tested on the other cases. For Case II,

$$\begin{aligned}
S-C_{7,2,10}&=\lambda[\sum^{3}x_0^2\xi_0^2\{-\xi_0^8-3(\xi_0^4\xi_1^4+\xi_0^4\xi_2^4)-3(\xi_1^8+\xi_2^8)+21\xi_1^4\xi_2^4\}\\
&\quad+\sum^{3}x_1x_2\xi_1\xi_2\{-6\xi_0^8+6(\xi_0^4\xi_1^4+\xi_0^4\xi_2^4)+2(\xi_1^8+\xi_2^8)+6\xi_1^4\xi_2^4\}]\\
&\quad+6\sum^{3}x_0^2\xi_0^2\{\xi_0^8+(\xi_0^4\xi_1^4+\xi_0^4\xi_2^4)-3\xi_1^4\xi_2^4\}\\
&=(x\xi)\sum^{3}x_0\xi_0[a\xi_0^8+b(\xi_0^4\xi_1^4+\xi_0^4\xi_2^4)+c(\xi_1^8+\xi_2^8)+d\xi_1^4\xi_2^4].
\end{aligned}$$

$$\lambda=1,\ a=5,\ b=3,\ c=-3,\ d=3.$$

$$\therefore\ S'=\sum^{3}x_0\xi_0[5\xi_0^8+3(\xi_0^4\xi_1^4+\xi_0^4\xi_2^4)-3(\xi_1^8+\xi_2^8)+3\xi_1^4\xi_2^4].$$

---

* The factor $-6$ is introdñced because it simplifies subsequent work. It is introduced here to avoid confusion.

For Case III,

$$S-C_{7,2,10}=\lambda[\overset{3}{\Sigma}x_0^2\xi_0^2\cdot\xi_0^4\xi_1^2\xi_2^2+\overset{3}{\Sigma}x_1x_2\xi_1\xi_2\{6\xi_0^4\xi_1^2\xi_2^2-2(\xi_0^2\xi_1^4\xi_2^2+\xi_0^2\xi_1^2\xi_2^4)\}]$$
$$+27\cdot216[\overset{3}{\Sigma}x_0^2\xi_0^2\{2\xi_0^4\xi_1^2\xi_2^2-5(\xi_0^2\xi_1^4\xi_2^2+\xi_0^2\xi_1^2\xi_2^4)+2(\xi_1^6\xi_2^2+\xi_1^2\xi_2^6)+4\xi_1^4\xi_2^4\}$$
$$+\overset{3}{\Sigma}x_1x_2\xi_1\xi_2\{2(\xi_0^6\xi_1^2+\xi_0^6\xi_2^2)+4(\xi_0^4\xi_1^4+\xi_0^4\xi_2^4)-28\xi_0^4\xi_1^2\xi_2^2$$
$$+2(\xi_0^2\xi_1^6+\xi_0^2\xi_2^6)+6(\xi_0^2\xi_1^4\xi_2^2+\xi_0^2\xi_1^2\xi_2^4)\}]$$
$$=(x\xi)\overset{3}{\Sigma}x_0\xi_0\{a\xi_0^4\xi_1^2\xi_2^2+b(\xi_0^2\xi_1^4\xi_2^2+\xi_0^2\xi_1^2\xi_2^4)+c(\xi_1^6\xi_2^2+\xi_1^2\xi_2^6)+d\xi_1^4\xi_2^4\}.$$

$\lambda=3\cdot27\cdot216$, $a=5\cdot27\cdot216$, $b=-5\cdot27\cdot216$, $c=2\cdot27\cdot216$, $d=4\cdot27\cdot216$.

$$\therefore\ S'=27\cdot216\overset{3}{\Sigma}x_0\xi_0[5\xi_0^4\xi_1^2\xi_2^2-5(\xi_0^2\xi_1^4\xi_2^2+\xi_0^2\xi_1^2\xi_2^4)+2(\xi_1^6\xi_2^2+\xi_1^2\xi_2^6)+4\xi_1^4\xi_2^4].$$

Turning now to the $C_{7,1,9}$, we have as possible terms

$$C_{7,1,9}=M_{111}s_1^3+M_{012}ss_1s_2+N_{12}t_1t_2+N_{03}tt_3.^{*}$$

Here it is easiest to work first with Case II.

$$S'-C_{7,1,9}=\overset{3}{\Sigma}x_0\xi_0[\xi_0^8(5-M_{111}-M_{012})+(\xi_0^4\xi_1^4+\xi_0^4\xi_2^4)(3-M_{012})$$
$$-3(\xi_1^8+\xi_2^8)+\xi_1^4\xi_2^4(3-N\tfrac{12}{45})].$$

If $(x\xi)$ is to be a factor, the coefficient of $x_i\xi_i$ must be the same for $i=0, 1, 2$; *i. e.*, must be symmetrical in $\xi_i$. Therefore,

$$5-M_{111}-M_{012}=-3, \qquad 3-M_{012}=3-N\tfrac{12}{45},$$
$$\therefore\ M_{111}=8-M_{012}, \qquad N_{12}=45M_{012}. \tag{2}$$

In Case I,

$$S'-C_{7,1,9}=x_0\xi_0(\xi_0^5\xi_1^3+\xi_0^5\xi_2^3)(-27M_{111}-\tfrac{8}{5}N_{12}-\tfrac{8}{5}N_{03})$$
$$+\overset{2}{\Sigma}x_1\xi_1[\xi_0^5\xi_1^3(-1296-\tfrac{32}{5}N_{12}-\tfrac{36}{5}N_{03})$$
$$+\xi_0^5\xi_2^3(1296-162M_{111}-144M_{012}-\tfrac{8}{5}N_{12}-\tfrac{36}{5}N_{03})],$$

$$27M_{111}+\tfrac{8}{5}N_{12}+\tfrac{8}{5}N_{03}=1296+\tfrac{32}{5}N_{12}+\tfrac{36}{5}N_{03}=-1296+162M_{111}$$
$$+144M_{012}+\tfrac{8}{5}N_{12}+\tfrac{36}{5}N_{03}.$$

By use of relations (2), these may be solved.

$$M_{111}=\tfrac{176}{13},\quad M_{012}=-\tfrac{72}{13},\quad N_{12}=-\tfrac{3240}{13},\quad N_{03}=\tfrac{4320}{91},$$
$$\therefore\ S'=\tfrac{8}{13}(22s_1^3-9ss_1s_2-405t_1t_2+\tfrac{540}{7}tt_3)+(x\xi)S''.$$

For Case II,

$$S''=-3(\xi^8)+\tfrac{111}{3}(\xi_1^4\xi_2^4).$$

For Case I,

$$S''=-\tfrac{3888}{91}\xi_0^5(\xi_1^3+\xi_2^3).$$

---

* $s^2s_2$ is a numerical multiple of $(x\xi)s^2s_4$.

For Case III,

$$\begin{aligned}S'-C_{7,1,9}=\sum^{3} x_0\xi_0[&\xi_0^8(-18M_{012}-\tfrac{2}{5}N_{12}-\tfrac{6}{5}N_{03})\\ &+(\xi_0^6\xi_1^2+\xi_0^6\xi_2^2)(-81M_{111}-126M_{012}-\tfrac{2}{5}N_{12}-\tfrac{3}{5}N_{03})\\ &+(\xi_0^4\xi_1^4+\xi_0^4\xi_2^4)(-243M_{111}-270M_{012}-\tfrac{12}{5}N_{12}+\tfrac{27}{5}N_{03})\\ &+\xi_0^4\xi_1^2\xi_2^2(135\cdot216-486M_{111}-540M_{012}+\tfrac{444}{5}N_{12}+\tfrac{378}{5}N_{03})\\ &+(\xi_0^2\xi_1^6+\xi_0^2\xi_2^6)(-243M_{111}-234M_{012}-\tfrac{26}{5}N_{12}+\tfrac{39}{5}N_{03})\\ &+(\xi_0^2\xi_1^4\xi_2^2+\xi_0^2\xi_1^2\xi_2^4)(-135\cdot216-729M_{111}-702M_{012}-\tfrac{978}{5}N_{12}-\tfrac{693}{5}N_{03})\\ &+(\xi_1^8+\xi_2^8)(-81M_{111}-72M_{012}-\tfrac{14}{5}N_{12}+3N_{03})\\ &+(\xi_1^6\xi_2^2+\xi_1^2\xi_2^6)(54\cdot216-324M_{111}-288M_{012}+\tfrac{196}{5}N_{12}+12N_{03})\\ &+\xi_1^4\xi_2^4(108\cdot216-486M_{111}-432M_{012}+84N_{12}+18N_{03}).\end{aligned}$$

$$S''=\tfrac{648}{91}[20(\xi^8)-46(\xi_0^6\xi_1^2)-132(\xi_1^4\xi_2^4)+987(\xi_0^4\xi_1^2\xi_2^2)].$$

Finally, for the $C_{7,0,8}$ the possibilities are

$$C_{7,0,8}=M_{004}s^2s_4+M_{022}ss_2^2+M_{112}s_1^2s_2+N_{04}tt_4+N_{13}t_1t_3+N_{22}t_2^2.^*$$

Instead of dealing with all of these terms at once it is simpler to take

$$C'_{7,0,8}=M_{112}s_1^2s_2+N_{04}tt_4+N_{13}t_1t_3+N_{22}t_2^2.$$

Then, $S''-C'_{7,0,8}=sS'''$; *i. e.*, contains $s$ as a factor.

For Case I,

$$S''-C'_{7,0,8}=\xi_0^5(\xi_1^3+\xi_2^3)(-\tfrac{3888}{91}-36M_{112}-\tfrac{32}{5}N_{04}-\tfrac{22}{5}N_{13}-\tfrac{96}{25}N_{22})=0,$$

$$\therefore \tfrac{1944}{91}+18M_{112}+\tfrac{16}{5}N_{04}+\tfrac{11}{5}N_{13}+\tfrac{48}{25}N_{22}=0. \qquad (3)$$

For Case II,

$$S''-C'_{7,0,8}=\sum^{3}\xi_0^8(-3-M_{112})+\sum^{3}\xi_1^4\xi_2^4(\tfrac{111}{13}-\tfrac{1}{225}N_{22})=(\xi^4)\cdot A(\xi^4)$$

$$\therefore A=-3-M_{112}=\tfrac{111}{26}-\tfrac{1}{450}N_{22} \text{ or } -\tfrac{189}{13}-2M_{112}+\tfrac{1}{225}N_{22}=0. \qquad (4)$$

For Case III,

$$\begin{aligned}S''-C'_{7,0,8}=&\sum^{3}\xi_0^8(\tfrac{648}{91}\cdot20-18M_{112}+\tfrac{4}{5}N_{04}-\tfrac{6}{5}N_{13}-\tfrac{22}{25}N_{22})\\ &+\sum^{6}\xi_0^6\xi_1^2(-\tfrac{648}{91}\cdot46-180M_{112}+\tfrac{16}{5}N_{04}+\tfrac{18}{5}N_{13}+\tfrac{32}{25}N_{22})\\ &+\sum^{3}\xi_0^4\xi_1^4(-\tfrac{648}{91}\cdot132-324M_{112}+\tfrac{24}{5}N_{04}+\tfrac{48}{5}N_{13}+\tfrac{108}{25}N_{22})\\ &+\sum^{3}\xi_0^4\xi_1^2\xi_2^2(\tfrac{648}{91}\cdot987-756M_{112}-\tfrac{168}{5}N_{04}-\tfrac{294}{5}N_{13}-\tfrac{2592}{25}N_{22})\\ =&3[(\xi^4)+2(\xi_1^2\xi_2^2)][A(\xi^4)+B(\xi_1^2\xi_2^2)]\\ =&3[A(\xi^8)+(B+2A)(\xi_0^6\xi_1^2)+(2A+2B)(\xi_1^4\xi_2^4)\\ &\qquad+(2A+5B)(\xi_0^4\xi_1^2\xi_2^2)].\end{aligned}$$

---

* $ss_1s_3$ is a numerical multiple of $s^2s_4$.

From this comes the single relation

$$-\tfrac{27}{91}\cdot 1377+\tfrac{9}{5}N_{04}+\tfrac{18}{5}N_{13}+\tfrac{122}{25}N_{22}=0. \tag{5}$$

(3), (4), (5) give only three equations to solve for four unknowns. They may be solved for three in terms of the fourth, say $M_{112}$.

$$N_{04}=5\left(\tfrac{9}{91}\cdot 3085+\tfrac{920}{21}M_{112}\right),\quad N_{13}=5\left(\tfrac{9}{91}\cdot -10280-\tfrac{3022}{21}M_{112}\right),$$
$$N_{22}=25\left(\tfrac{9}{91}\cdot 1323+\tfrac{378}{21}M_{112}\right).$$

Using these relations to obtain the coefficients of $S'''$ in terms of $M_{112}$, it remains only to equate $S'''$ to $M_{004}ss_4+M_{022}s_2^2$.

For Case II,

$$S'''-M_{004}ss_4-M_{022}s_2^2=(\xi^4)(-3-M_{112}-3M_{004}-M_{022})=0.$$
$$\therefore 3+3M_{004}+M_{022}+M_{112}=0. \tag{6}$$

For Case III,

$$S'''-M_{004}ss_4-M_{022}s_2^2=\sum^3\xi_0^4\left(\tfrac{3}{13}\cdot 6622+\tfrac{1874}{9}M_{112}-54M_{004}-14M_{022}\right)$$
$$+\sum^3\xi_1^2\xi_2^2\left(-\tfrac{3}{13}\cdot 27052-\tfrac{8228}{9}M_{112}-108M_{004}-52M_{022}\right).$$
$$\therefore \tfrac{3}{13}\cdot 6622+\tfrac{1874}{9}M_{112}-54M_{004}-14M_{022}=0. \tag{7}$$
$$-\tfrac{3}{13}\cdot 27052-\tfrac{8228}{9}M_{112}-108M_{004}-54M_{022}=0. \tag{8}$$

These two relations, together with (6), can be solved for the three unknowns.

$M_{004}=-\frac{49}{13}$, $M_{022}=\frac{9}{26}\cdot 45$, $M_{112}=-\frac{9}{26}\cdot 21$, $N_{04}=\frac{-9}{91}\cdot 675$, $N_{13}=\frac{9}{91}\cdot 1485$, $N_{22}=0$.

Therefore,

$$S=-6(ss_1^2-27t_1^2)+(x\xi)\cdot\tfrac{8}{13}\left(22s_1^3-9ss_1s_2-405t_1t_2+\tfrac{540}{7}tt_3\right)$$
$$+(x\xi)^2\cdot\tfrac{9}{13}\left(-\tfrac{49}{9}s^2s_4+\tfrac{45}{2}ss_2^2-\tfrac{21}{2}s_1^2s_2-\tfrac{675}{7}tt_4+\tfrac{1485}{7}t_1t_3\right).$$

## *The Satellite for the General Quartic.*

This expression for the satellite conic may be verified by use of the general quartic

$$ax_0^4+bx_1^4+cx_2^4+4a_1x_0^3x_1+4a_2x_0^3x_2+4b_0x_0x_1^3+4c_0x_0x_2^3+4b_2x_1^3x_2+4c_1x_1x_2^3$$
$$+6fx_1^2x_2^2+6gx_2^2x_0^2+6hx_0^2x_1^2+12lx_0^2x_1x_2+12mx_0x_1^2x_2+12nx_0x_1x_2^2=0,$$

where the conic can be found for a special reference triangle and line $\xi$. Let $(x\xi)\equiv x_0$, and let two of the tangents at its intersections with the quartic be $x_1=0$ and $x_2=0$. Then, in the equation of the quartic,

$$b=c=b_0=c_0=0.$$

Beside the two reference points $x_0=0$ cuts the quartic in two points given by

$$4b_2x_1^2+6fx_1x_2+4c_1x_2^2=0. \tag{9}$$

Let these two points be $(0, A, 1)$ and $(0, B, 1)$. The tangent at the first of these points is

$$x_0(3mA^2+3nA)+x_1(3b_2A^2+3fA+c_1)+x_2(b_2A^3+3fA^2+3cA)=0,$$

or, making use of the fact that $A$ satisfies (9) and dividing by $A$,

$$x_0(3mA+3n)+x_1(2b_2A+\tfrac{3}{2}f)+x_2(\tfrac{3}{2}fA+2c_1)=0.$$

Similarly, the tangent at $(0, B, 1)$ is

$$x_0(3mB+3n)+x_1(2b_2B+\tfrac{3}{2}f)+x_2(\tfrac{3}{2}fB+2c_1)=0.$$

Multiplying these two together and making use of the symmetric functions of the roots of a quadratic, we have the equation of the two tangents in terms of the coefficients of the quartic. Then by (1),

$$\begin{aligned}2x_1x_2[x_0^2(72b_2n^2+72c_1m^2-108fmn)&+x_0x_1(96b_2c_1m-54f^2m)+x_0x_2(96b_2c_1n-54f^2n)\\ &+x_1^2(32b_2^2c_1-18b_2f^2)+x_1x_2(48b_2c_1f-27f^3)+x_2^2(32b_2c_1^2-18c_1f^2)]\\ &+\lambda(ax_0^4+4a_1x_0^3x_1+4a_2x_0^3x_2+6hx_0^2x_1^2+12lx_0^2x_1x_2+6gx_0^2x_2^2\\ &+12mx_0x_1^2x_2+12nx_0x_1x_2^2+4b_2x_1^3x_2+6fx_1^2x_2^2+4c_1x_1x_2^3)=x_0^2S.\end{aligned}$$

$$\lambda=(-16b_2c_1+9f^2).$$

$$\begin{aligned}S=x_0^2(-16ab_2c_1+9af^2)&+2x_0x_1(-32a_1b_2c_1+18a_1f^2)+2x_0x_2(-32a_2b_2c_1+18a_2f^2)\\ &+x_1^2(-96b_2c_1h+54f^2h)+2x_1x_2(-96b_2c_1l+72b_2n^2+72c_1m^2\\ &+54f^2l-108fmn)+x_2^2(-96b_2c_1g+54fg^2).\end{aligned}$$

The formula to be tested is

$$\begin{aligned}S=M_{011}ss_1^2+N_{11}t_1^2+(x\xi)(M_{111}s_1^3+M_{012}ss_1s_2+N_{12}t_1t_2+N_{03}tt_3)\\ +(x\xi)^2(M_{004}s^2s_4+M_{022}ss_2^2+M_{112}s_1^2s_2+N_{04}tt_4+N_{13}t_1t_3),\end{aligned}$$

where

$$M_{011}=-6,\quad N_{11}=162,\quad M_{111}=\tfrac{176}{13},\quad M_{012}=-\tfrac{72}{13},\quad N_{12}=\tfrac{-3240}{13},\quad N_{03}=\tfrac{4320}{91},$$
$$M_{004}=\tfrac{-49}{13},\quad M_{022}=\tfrac{9}{26}\cdot 45,\quad M_{112}=\tfrac{-9}{26}\cdot 21,\quad N_{04}=\tfrac{-9}{91}\cdot 675,\quad N_{13}=\tfrac{9}{91}\cdot 1485.$$

In this case it becomes

$$\begin{aligned}S=x_0^2[ab_2^3c_1^3(&-64M_{011}-64M_{111}-64M_{012}-192M_{004}-64M_{022}-64M_{112}\equiv 64)\\ +ab_2^2c_1^2f^2(&144M_{011}+4N_{11}+144M_{111}+144M_{012}+4N_{12}+4N_{03}+432M_{004}\\ &+144M_{022}+144M_{112}+\tfrac{24}{5}N_{04}+4N_{13}\equiv -36)\\ +ab_2c_1f^4(&-108M_{011}-4N_{11}-108M_{111}-108M_{012}-4N_{12}-4N_{03}-324M_{004}\\ &-108M_{022}-108M_{112}-\tfrac{24}{5}N_{04}-4N_{13}\equiv 0)\\ +af^6(&27M_{011}+N_{11}+27M_{111}+27M_{012}+N_{12}+N_{03}-81M_{004}+27M_{002}\\ &+27M_{112}+\tfrac{6}{5}N_{04}+N_{13}\equiv 0)\end{aligned}$$

$$+(a_1b_2^2c_1^3m+a_2b_2^3c_1^2n)(96M_{011}+144M_{111}+144M_{012}+576M_{004}+192M_{022}$$
$$+192M_{112}=0)$$
$$+(a_1b_2^2c_1^2fn+a_2b_2^2c_1^2fm)(-96M_{011}-\tfrac{8}{3}N_{11}-144M_{111}-144M_{012}-4N_{12}$$
$$-\tfrac{22}{5}N_{03}-576M_{004}-192M_{022}-192M_{112}-\tfrac{32}{5}N_{04}-\tfrac{16}{3}N_{13}=0)$$
$$+(b_2^3c_1^2gl+b_2^2c_1^3hl)(-64M_{012}-576M_{004}-192M_{022}-64M_{112}=0)$$
$$+(b_2^3c_1fg^2+b_2c_1^3fh^2)(-\tfrac{4}{5}N_{12}-\tfrac{12}{5}N_{03}-4N_{04}-\tfrac{12}{5}N_{13}=0)$$
$$+b_2^2c_1^2fgh(32M_{012}+\tfrac{8}{15}N_{12}+\tfrac{8}{5}N_{03}+288M_{004}+64M_{022}+32M_{112}--\tfrac{24}{5}N_{04}+\tfrac{8}{5}N_{13}=0)$$
$$+(a_1b_2c_1^2f^2m+a_2b_2c_1f^2n)(-144M_{011}-4N_{11}-216M_{111}-216M_{012}-6N_{12}$$
$$-\tfrac{33}{5}N_{03}-864M_{004}-288M_{022}-288M_{112}-\tfrac{48}{5}N_{04}--8N_{13}=0)$$
$$+b_2^2c_1^2fl^2(64M_{012}+\tfrac{16}{15}N_{12}+\tfrac{16}{5}N_{03}+576M_{004}+224M_{022}+64M_{112}+\tfrac{64}{5}N_{04}+\tfrac{16}{5}N_{13}=0)$$
$$+(b_2^3c_1gn^2+b_2c_1^3hm^2)(-36M_{011}-108M_{111}-48M_{012}+\tfrac{4}{15}N_{12}-16M_{022}$$
$$-150M_{112}+\tfrac{7}{5}N_{13}=0)$$
$$+(b_2^2c_1^3gm^2+b_2^2c_1^2hn^2)(\tfrac{4}{9}N_{11}+56M_{012}+\tfrac{4}{5}N_{12}+576M_{004}+192M_{022}+62M_{112}$$
$$+\tfrac{7}{15}N_{13}=0)$$
$$+b_2^2c_1^2lmn(-72M_{011}+\tfrac{8}{9}N_{11}-216M_{111}-128M_{012}+\tfrac{16}{15}N_{12}-576M_{004}$$
$$-224M_{022}-368M_{112}+\tfrac{56}{15}N_{13}=0)$$
$$+(a_1b_2c_1f^3n+a_2b_2c_1f^3m)(144M_{011}+\tfrac{16}{3}N_{11}+216M_{111}+216M_{012}+8N_{12}$$
$$+\tfrac{44}{5}N_{03}+864M_{004}+288M_{022}+288M_{112}+\tfrac{64}{5}N_{04}+\tfrac{32}{3}N_{13}=0)$$
$$+(b_2^2c_1f^2gl+b_2c_1^2f^2hl)(96M_{012}+\tfrac{8}{5}N_{12}+\tfrac{24}{5}N_{03}+864M_{004}+288M_{022}+96M_{112}$$
$$+8N_{04}+\tfrac{24}{5}N_{13}=0)$$
$$+(b_2^2c_1fgmn+b_2c_1^2fhmn)(72M_{011}+\tfrac{4}{3}N_{11}+216M_{111}+84M_{012}+4N_{12}$$
$$+\tfrac{27}{5}N_{03}+32M_{022}+294M_{112}+16N_{04}+\tfrac{41}{5}N_{13}=0)$$
$$+(b_2^2c_1fln^2+b_2c_1^2flm^2)(72M_{011}+\tfrac{4}{3}N_{11}+216M_{111}+72M_{012}+\tfrac{52}{15}N_{12}+\tfrac{6}{5}N_{03}$$
$$-32M_{022}+306M_{112}-\tfrac{32}{5}N_{04}+\tfrac{27}{5}N_{13}=0)$$
$$+(b_2^2c_1mn^3+b_2c_1^2m^3n)(81M_{111}-\tfrac{8}{5}N_{12}+32M_{022}+180M_{112}-\tfrac{14}{3}N_{13}=0)$$
$$+(b_2^2f^3g^2+c_1^2f^3h^2)(\tfrac{2}{5}N_{12}+\tfrac{6}{5}N_{03}+2N_{04}+\tfrac{6}{5}N_{13}=0)$$
$$+b_2c_1f^3gh(-48M_{012}--\tfrac{16}{15}N_{12}-\tfrac{16}{5}N_{03}-432M_{004}-104M_{022}-48M_{112}$$
$$+4N_{04}-\tfrac{16}{5}N_{13}=0)$$
$$+(a_1c_1f^4m+a_2b_2f^4n)(54M_{011}+2N_{11}+81M_{111}+81M_{012}+3N_{12}+\tfrac{33}{10}N_{03}$$
$$+324M_{004}+108M_{022}+108M_{112}+\tfrac{24}{5}N_{04}+4N_{13}=0)$$
$$+b_2c_1f^3l^2(-96M_{012}-\tfrac{32}{15}N_{12}-\tfrac{32}{5}N_{03}-864M_{004}-328M_{022}-96M_{112}-20N_{04}$$
$$-\tfrac{32}{5}N_{13}=0)$$
$$+(b_2^2f^2gn^2+c_1^2f^2hm^2)(27M_{011}+N_{11}+81M_{111}+36M_{012}+\tfrac{6}{5}N_{12}+12M_{022}$$
$$+108M_{112}=0)$$

$$+(b_2c_1f^3gm^2+b_2c_1f^3hn^2)(-36M_{011}-\tfrac{4}{3}N_{11}-108M_{111}-126M_{012}-\tfrac{22}{5}N_{12}$$
$$-\tfrac{27}{5}N_{08}-864M_{004}-304M_{022}-240M_{112}-16N_{04}-\tfrac{48}{5}N_{13}=0)$$
$$+b_2c_1f^3lmn(-18M_{011}-\tfrac{2}{3}N_{11}-54M_{111}+72M_{012}-\tfrac{4}{5}N_{12}+\tfrac{18}{5}N_{08}+864M_{004}$$
$$+376M_{022}+6M_{112}+\tfrac{128}{5}N_{04}-\tfrac{8}{5}N_{13}=0)$$
$$+(b_2^3fn^4+c_1^3fm^4)(-81M_{111}-\tfrac{4}{5}N_{12}-180M_{112}-\tfrac{14}{5}N_{13}=0)$$
$$+b_2c_1fm^2n^2(-324M_{111}-\tfrac{8}{5}N_{12}-\tfrac{12}{5}N_{08}-48M_{022}-684M_{112}-\tfrac{48}{5}N_{04}-\tfrac{26}{5}N_{13}=0)$$
$$+(a_1f^5n+a_2f^5m)(-54M_{011}-2N_{11}-81M_{111}-81M_{012}-3N_{12}-\tfrac{33}{10}N_{08}$$
$$-324M_{004}-108M_{022}-108M_{112}-\tfrac{24}{5}N_{04}-4N_{13}=0)$$
$$+(b_2f^4gl+c_1f^4hl)(-36M_{012}-\tfrac{4}{5}N_{12}-\tfrac{12}{5}N_{08}-324M_{004}-108M_{022}-36M_{112}$$
$$-4N_{04}-\tfrac{12}{5}N_{13}=0)$$
$$+(b_2f^3gmn+c_1f^3hmn)(-54M_{011}-2N_{11}-162M_{111}-63M_{012}-\tfrac{19}{5}N_{12}-\tfrac{27}{10}N_{08}$$
$$-24M_{022}-216M_{112}-8N_{04}-\tfrac{24}{5}N_{13}=0)$$
$$+(b_2f^3ln^2+c_1f^3lm^2)(-54M_{011}-2N_{11}-162M_{111}-54M_{012}-\tfrac{18}{5}N_{12}-\tfrac{3}{5}N_{08}$$
$$+24M_{022}-216M_{112}+\tfrac{16}{5}N_{04}-\tfrac{24}{5}N_{13}=0)$$
$$+(b_2f^2mn^3+c_1f^2m^3n)(243M_{111}+\tfrac{18}{5}N_{12}-24M_{022}+504M_{112}+\tfrac{56}{5}N_{13}=0)$$
$$+f^5gh(18M_{012}+\tfrac{2}{5}N_{12}+\tfrac{6}{5}N_{08}+162M_{004}+42M_{022}+18M_{112}-\tfrac{4}{5}N_{04}+\tfrac{6}{5}N_{13}=0)$$
$$+f^5l^2(36M_{012}+\tfrac{4}{5}N_{12}+\tfrac{12}{5}N_{08}+324M_{004}+120M_{022}+36M_{112}+\tfrac{34}{5}N_{04}+\tfrac{12}{5}N_{13}=0)$$
$$+(f^4gm^2+f^4hn^2)(27M_{011}+N_{11}+81M_{111}+63M_{012}+\tfrac{13}{5}N_{12}+\tfrac{27}{10}N_{08}+324M_{004}$$
$$+120M_{022}+144M_{112}+8N_{04}+\tfrac{24}{5}N_{13}=0)$$
$$+f^4lmn(54M_{011}+2N_{11}+162M_{111}+18M_{012}+\tfrac{14}{5}N_{12}-\tfrac{9}{5}N_{08}-324M_{004}$$
$$-156M_{022}+180M_{112}-\tfrac{64}{5}N_{04}+\tfrac{12}{5}N_{13}=0)$$
$$+f^3m^2n^2(-162M_{111}-\tfrac{12}{5}N_{12}+\tfrac{6}{5}N_{08}+36M_{022}-324M_{112}+\tfrac{24}{5}N_{04}-\tfrac{36}{5}N_{13}=0)]$$
$$+\sum^{3}2x_0x_1[a_1b_2^3c_1^3(-64M_{011}-32M_{111}-32M_{012}=128)$$
$$+a_1b_2^2c_1^2f^2(144M_{011}+4N_{11}+72M_{111}+72M_{012}+2N_{12}+\tfrac{7}{5}N_{08}=-72)$$
$$+b_2^3c_1^2gm(-4M_{012}+\tfrac{8}{90}N_{12}=0)$$
$$+b_2^2c_1^3hm(96M_{011}+72M_{111}+68M_{012}+\tfrac{8}{90}N_{12}=0)$$
$$+b_2^3c_1^2ln(96M_{011}+72M_{111}+80M_{012}-\tfrac{16}{90}N_{12}=0)$$
$$+b_2^3c_1fgn(-6M_{012}-\tfrac{8}{80}N_{12}-\tfrac{21}{10}N_{08}=0)$$
$$+b_2^2c_1^2fhn(-96M_{011}-\tfrac{8}{3}N_{11}-72M_{111}-70M_{012}-\tfrac{172}{90}N_{12}-\tfrac{7}{10}N_{08}=0)$$
$$+b_2^2c_1^2flm(-96M_{011}-\tfrac{8}{3}N_{11}-72M_{111}-76M_{012}-\tfrac{196}{90}N_{12}-\tfrac{14}{5}N_{08}=0)$$
$$+b_2^3c_1n^3(-36M_{011}-\tfrac{81}{2}M_{111}-48M_{012}-\tfrac{8}{30}N_{12}=0)$$
$$+b_2^2c_1^2m^2n(-72M_{011}+\tfrac{4}{3}N_{11}-\tfrac{135}{2}M_{111}-72M_{012}+\tfrac{16}{30}N_{12}=0)$$
$$+a_1b_2c_1f^4(-108M_{011}-4N_{11}-54M_{111}-54M_{012}-2N_{12}-\tfrac{7}{5}N_{08}=0)$$

$$+b_2^2c_1f^2gm\,(9M_{012}+\tfrac{1}{5}N_{12}+\tfrac{21}{10}N_{03}\equiv 0)$$
$$+b_2c_1^2f^2hm\,(-144M_{011}-4N_{11}-108M_{111}-105M_{012}-\tfrac{92}{30}N_{12}-\tfrac{21}{10}N_{03}\equiv 0)$$
$$+b_2^2c_1f^2ln\,(-144M_{011}-4N_{11}-108M_{111}-114M_{012}-\tfrac{86}{30}N_{12}-\tfrac{21}{10}N_{03}\equiv 0)$$
$$+b_2^2c_1fmn^2\,(144M_{011}+\tfrac{8}{3}N_{11}+135M_{111}+156M_{012}+\tfrac{68}{30}N_{12}+\tfrac{21}{5}N_{03}\equiv 0)$$
$$+b_2c_1^2fm^3\,(72M_{011}+\tfrac{4}{3}N_{11}+\tfrac{135}{2}M_{111}+72M_{012}+\tfrac{52}{30}N_{12}+\tfrac{14}{5}N_{03}\equiv 0)$$
$$+b_2^2f^3gn\,(\tfrac{9}{2}M_{012}+\tfrac{1}{10}N_{12}+\tfrac{21}{20}N_{03}\equiv 0)$$
$$+b_2c_1f^3hn\,(144M_{011}+\tfrac{16}{3}N_{11}+108M_{111}+\tfrac{213}{2}M_{012}+\tfrac{119}{30}N_{12}+\tfrac{49}{20}N_{03}\equiv 0)$$
$$+b_2c_1f^3lm\,(144M_{011}+\tfrac{16}{3}N_{11}+108M_{111}+111M_{012}+\tfrac{122}{30}N_{12}+\tfrac{35}{10}N_{03}\equiv 0)$$
$$+b_2^2f^2n^3\,(27M_{011}+N_{11}+\tfrac{81}{2}M_{111}+36M_{012}+\tfrac{14}{10}N_{12}\equiv 0)$$
$$+b_2c_1f^2m^2n\,(-54M_{011}-2N_{11}-\tfrac{81}{2}M_{111}-54M_{012}-\tfrac{18}{10}N_{12}-\tfrac{21}{5}N_{03}\equiv 0)$$
$$+a_1f^6\,(27M_{011}+N_{11}+\tfrac{27}{2}M_{111}+\tfrac{27}{2}M_{012}+\tfrac{1}{2}N_{12}+\tfrac{7}{20}N_{03}\equiv 0)$$
$$+b_2f^4gm\,(-\tfrac{9}{2}M_{012}-\tfrac{1}{10}N_{12}-\tfrac{21}{20}N_{03}\equiv 0)$$
$$+c_1f^4hm\,(54M_{011}+2N_{11}+\tfrac{81}{2}M_{111}+\tfrac{81}{2}M_{012}+\tfrac{15}{10}N_{12}+\tfrac{21}{20}N_{03}\equiv 0)$$
$$+b_2f^4ln\,(54M_{011}+2N_{11}+\tfrac{81}{2}M_{111}+\tfrac{81}{2}M_{012}+\tfrac{15}{10}N_{12}+\tfrac{21}{20}N_{03}\equiv 0)$$
$$+b_2f^3mn^2\,(-108M_{011}-4N_{11}-\tfrac{243}{2}M_{111}-117M_{012}+\tfrac{44}{10}N_{12}-\tfrac{21}{10}N_{03}\equiv 0)$$
$$+c_1f^3m^3\,(-54M_{011}-2N_{11}-54M_{111}-54M_{012}-2N_{12}-\tfrac{7}{5}N_{03}\equiv 0)$$
$$+f^5hn\,(-54M_{011}-2N_{11}-\tfrac{81}{2}M_{111}-\tfrac{81}{2}M_{012}-\tfrac{15}{10}N_{12}-\tfrac{21}{20}N_{03}\equiv 0)$$
$$+f^5lm\,(-54M_{011}-2N_{11}-\tfrac{81}{2}M_{111}-\tfrac{81}{2}M_{012}-\tfrac{15}{10}N_{12}-\tfrac{21}{20}N_{03}\equiv 0)$$
$$+f^4m^2n\,(81M_{011}+3N_{11}+81M_{111}+81M_{012}+3N_{12}+\tfrac{21}{10}N_{03}\equiv 0)\,]$$

$$+\sum^{2} x_1^2\,[\,b_2^3c_1^3h\,(-64M_{011}\equiv 384)$$
$$+b_2^3c_1^2mn\,(24M_{011}+\tfrac{8}{9}N_{11}\equiv 0)$$
$$+b_2^2c_1^2f^2h\,(144M_{011}+4N_{11}\equiv -216)$$
$$+b_2^3c_1fn^2\,(36M_{011}+\tfrac{4}{3}N_{11}\equiv 0)$$
$$+b_2^2c_1^2fm^2\,(-24M_{011}-\tfrac{8}{9}N_{11}\equiv 0)$$
$$+b_2^2c_1f^2mn\,(-90M_{011}-\tfrac{10}{3}N_{11}\equiv 0)$$
$$+b_2c_1f^4h\,(-108M_{011}-4N_{11}\equiv 0)$$
$$+b_2c_1f^3m^2\,(54M_{011}+2N_{11}\equiv 0)$$
$$+b_2^2f^2n^2\,(-27M_{011}-N_{11}\equiv 0)$$
$$+b_2f^4mn\,(54M_{011}+2N_{11}\equiv 0)$$
$$+f^6h\,(27M_{011}+N_{11}\equiv 0)$$
$$+f^5m^2\,(-27M_{011}-N_{11}\equiv 0)\,]$$

$$+2x_1x_2\,[\,b_2^3c_1^3l\,(-64M_{011}\equiv 384)$$
$$+\,(b_2^3c_1^2n^2+b_2^2c_1^3m^2)\,(60M_{011}+\tfrac{4}{9}N_{11}\equiv -288)$$
$$+b_2^2c_1^2f^2l\,(144M_{011}+4N_{11}\equiv -216)$$

$$
\begin{aligned}
&+b_2^2c_1^2fmn(-120M_{011}-\tfrac{16}{9}N_{11}=432)\\
&+(b_2^2c_1f^2n^2+b_2c_1^2f^2m^2)(-81M_{011}-3N_{11}=0)\\
&+b_2c_1f^4l(-108M_{011}-4N_{11}=0)\\
&+b_2c_1f^3mn(162M_{011}+6N_{11}=0)\\
&+(b_2f^4n^2+c_1f^4m^2)(27M_{011}+N_{11}=0)\\
&+f^6l(27M_{011}+N_{11}=0)\\
&+f^5mn(-54M_{011}-2N_{11}=0)]\\
=-4b_2^2c_1^2[&x_0^2(-16ab_2c_1+9af^2)+2x_0x_1(-32a_1b_2c_1+18a_1f^2)\\
&+2x_0x_2(-32a_2b_2c_1+18a_2f^2)+x_1^2(-96b_2c_1h+54f^2h)\\
&+2x_1x_2(-96b_2c_1l+72b_2n^2+c_1m^2+54f^2l-108fmn)\\
&\qquad\qquad+x_2^2(-96b_2c_1g+54f^2g).
\end{aligned}
$$

Therefore the expression for the satellite is correct.

It follows directly from (1′), when $(a\xi)^{12}$ is taken as $s^3-27t^2$, which in this case becomes $4b_2^3c_1^3(-16b_2c_1+9f^2)$, that

$$T-(ax)^4(a\xi)^4=(x\xi)^2S.$$

*Therefore the asymptotic equation of the quartic is*

$$
\begin{aligned}
(s^3-27t^2)(ax)^4-6(ss_1^2-27t_1^2)(x\xi)^2&\\
+\tfrac{8}{13}(22s_1^3-9ss_1s_2-405t_1t_2+\tfrac{540}{7}tt_3)(x\xi)^3&\\
+\tfrac{9}{13}(-\tfrac{49}{9}s^2s_4+\tfrac{45}{2}ss_2^2-\tfrac{21}{2}s_1^2s_2-\tfrac{675}{7}tt_4+\tfrac{1485}{7}t_1t_3)(x\xi)^4&=0.
\end{aligned}
$$

JOHNS HOPKINS UNIVERSITY, *December* 20, 1915.

# AMERICAN

# Journal of Mathematics

EDITED BY

**FRANK MORLEY**

WITH THE COÖPERATION OF

A. COHEN, CHARLOTTE A. SCOTT

AND OTHER MATHEMATICIANS

PUBLISHED UNDER THE AUSPICES OF THE JOHNS HOPKINS UNIVERSITY

*Πραγμάτων ἔλεγχος οὐ βλεπομένων*

VOLUME XXXVIII, NUMBER 4

BALTIMORE: THE JOHNS HOPKINS PRESS

LEMCKE & BUECHNER, *New York.*
G. E. STECHERT & CO., *New York.*
E. STEIGER & CO., *New York.*
ARTHUR F. BIRD, *London.*
WILLIAM WESLEY & SON, *London.*
A. HERMANN, *Paris.*
MAYER & MÜLLER, *Berlin.*

OCTOBER, 1916

Entered as Second-Class Matter at the Baltimore, Maryland, Postoffice.

# *Conditions for the Complete Reducibility of Groups of Linear Substitutions.*

By Henry Taber.

## § 1. *Conditions Sufficient for the Complete Reducibility of the Group G.*

Let $A_1, A_2, \ldots, A_m$ be any system of linearly independent linear homogeneous transformations in $n$ variables ($m \leq n^2$) constituting a hypercomplex number system so that

$$A_i A_j = \Sigma_{k=1}^{m} \gamma_{ijk} A_k, \qquad (i, j = 1, 2, \ldots, m). \tag{1}$$

Let $A_i$, for $i$ equal successively to $1, 2, \ldots, m$, be defined by the system of equations

$$\xi'_\mu = \Sigma_{\nu=1}^{n} \alpha_{\mu\nu}^{(i)} \xi_\nu, \qquad (\mu = 1, 2, \ldots, n), \tag{2}$$

or

$$(\xi'_1, \xi'_2, \ldots, \xi'_n) = \begin{pmatrix} \alpha_{11}^{(i)}, & \alpha_{12}^{(i)}, & \ldots, & \alpha_{1n}^{(i)} \\ \alpha_{21}^{(i)}, & \alpha_{22}^{(i)}, & \ldots, & \alpha_{2n}^{(i)} \\ \ldots & \ldots & \ldots & \ldots \\ \alpha_{n1}^{(i)}, & \alpha_{n2}^{(i)}, & \ldots, & \alpha_{nn}^{(i)} \end{pmatrix} (\xi_1, \xi_2, \ldots, \xi_n) \tag{3}$$

Then, for any two integers $i, j$ from 1 to $m$,

$$\Sigma_{\lambda=1}^{n} \alpha_{\mu\lambda}^{(i)} \alpha_{\lambda\nu}^{(j)} = \Sigma_{k=1}^{m} \gamma_{ijk} \alpha_{\mu\nu}^{(k)}, \qquad (\mu, \nu = 1, 2, \ldots, n). \tag{4}$$

By equation (1),

$$\begin{aligned} \Sigma_{i=1}^{m} x_i A_i \cdot \Sigma_{j=1}^{m} y_j A_j &= \Sigma_{i=1}^{m} \Sigma_{j=1}^{m} x_i y_j A_i A_j \\ &= \Sigma_{i=1}^{m} \Sigma_{j=1}^{m} (x_i y_j \Sigma_{k=1}^{m} \gamma_{ijk} A_k) \\ &= \Sigma_{k=1}^{m} (\Sigma_{i=1}^{m} \Sigma_{j=1}^{m} x_i y_j \gamma_{ijk}) A_k . \end{aligned} \tag{5}$$

Whence it follows that the aggregate of numbers of the hypercomplex number system $(A_1, A_2, \ldots, A_m)$, that is, the aggregate of transformations

$$X = x_1 A_1 + x_2 A_2 + \ldots + x_m A_m , \tag{6}$$

for every system of scalars $x_1, x_2, \ldots, x_m$, constitutes a group $G$ with $m$ essential parameters.* Therefore, *any linear function of transformations of G*

* For, if $\Sigma_{i=1}^{m} x_i A_i = \Sigma_{i=1}^{m} y_i A_i$, then $\Sigma_{i=1}^{m} (x_i - y_i) A_i = 0$; and, therefore, $x_i - y_i = 0$ for $i = 1, 2, \ldots, m$, since, otherwise, the $A$'s are not linearly independent, which is contrary to supposition.

*is also a transformation of this group.* The general transformation $X=\Sigma_{i=1}^{m}x_iA_i$ of $G$ is defined by the system of equations

$$\xi'_\mu=\Sigma_{i=1}^{m}\Sigma_{\nu=1}^{n}x_i\alpha_{\mu\nu}^{(i)}\xi_\nu, \qquad (\mu=1, 2, \ldots, n), \tag{7}$$

or

$$(\xi'_1, \xi'_2, \ldots, \xi'_n)=\begin{pmatrix}\Sigma_{i=1}^{m}x_i\alpha_{11}^{(i)}, & \Sigma_{i=1}^{m}x_i\alpha_{12}^{(i)}, & \ldots, & \Sigma_{i=1}^{m}x_i\alpha_{1n}^{(i)} \\ \Sigma_{i=1}^{m}x_i\alpha_{21}^{(i)}, & \Sigma_{i=1}^{m}x_i\alpha_{22}^{(i)}, & \ldots, & \Sigma_{i=1}^{m}x_i\alpha_{2n}^{(i)} \\ \ldots & \ldots & \ldots & \ldots \\ \Sigma_{i=1}^{m}x_i\alpha_{n1}^{(i)}, & \Sigma_{i=1}^{m}x_i\alpha_{n2}^{(i)}, & \ldots, & \Sigma_{i=1}^{m}x_i\alpha_{nn}^{(i)}\end{pmatrix}(\xi_1, \xi_2, \ldots, \xi_n) \tag{8}$$

I shall denote the determinant of $X$ by $|X|$; and I shall assume that $|X|\neq 0$ for every transformation of $G$. Whence it follows that the group $G$, that is, the aggregate of the numbers of the system $(A_1, A_2, \ldots, A_m)$ contains the identical transformation, which will be denoted simply by 1;* and, if $|X|\neq 0$, that $G$ contains $X^{-1}$. For, if

$$A=a_1A_1+a_2A_2+\ldots+a_mA_m \tag{9}$$

is any transformation of $G$, we have, by Cayley's theorem,

$$\phi(A)\equiv A^n-p_1A^{n-1}+\ldots\mp p_{n-1}A\pm p_n=0, \tag{10}$$

where

$$\phi(\rho)\equiv|\rho-A|\equiv\begin{vmatrix}\rho-\Sigma_{i=1}^{m}a_i\alpha_{11}^{(i)}, & -\Sigma_{i=1}^{m}a_i\alpha_{12}^{(i)}, & \ldots, & -\Sigma_{i=1}^{m}a_i\alpha_{1n}^{(i)} \\ -\Sigma_{i=1}^{m}a_i\alpha_{21}^{(i)}, & \rho-\Sigma_{i=1}^{m}a_i\alpha_{22}^{(i)}, & \ldots, & -\Sigma_{i=1}^{m}a_i\alpha_{2n}^{(i)} \\ \ldots & \ldots & \ldots & \ldots \\ -\Sigma_{i=1}^{m}a_i\alpha_{n1}^{(i)}, & -\Sigma_{i=1}^{m}a_i\alpha_{n2}^{(i)}, & \ldots, & \rho-\Sigma_{i=1}^{m}a_i\alpha_{nn}^{(i)}\end{vmatrix},\dagger \tag{11}$$

and thus $p_n=|A|$. Therefore, if $|A|\neq 0$,

$$1=\frac{\mp 1}{p_n}(A^n-p_1A^{n-1}+\ldots\mp p_{n-1}A),$$

$$A^{-1}=\frac{\mp 1}{p_n}(A^{n-1}-p_1A^{n-2}+\ldots\mp p_{n-1});$$

and thus 1 and $A^{-1}$, being linear in powers of $A$, are transformations of $G$.

Let $\rho_1, \rho_2, \ldots, \rho_r$ respectively of multiplicity $\eta_1, \eta_2, \ldots, \eta_r$, be the distinct roots of the characteristic equation, $\phi(\rho)\equiv|\rho-A|=0$, of $A=\Sigma_{i=1}^{m}a_iA_i$; in which case,

$$\phi(\rho)\equiv(\rho-\rho_1)^{\eta_1}(\rho-\rho_2)^{\eta_2}\ldots(\rho-\rho_r)^{\eta_r}. \tag{12}$$

I shall denote by $\bar{r}$ the greatest number of distinct roots of $\phi(\rho)=0$ for any transformation $A$ of $G$. Since $G$ contains the identical transformation, there is at least one transformation of $G$ whose characteristic equation has $\bar{r}$ distinct

*The transformation $\rho 1$, defined by the system of equations

$$\xi'_\mu=\rho\xi_\mu \qquad (\mu=1, 2, \ldots, n),$$

will, following Cayley, be denoted simply by $\rho$, and will be identified with the scalar $\rho$.

† *Phil. Trans.*, Vol. CXLVIII (1858), p. 24.

non-zero roots. For, let $A$ be so chosen that $r=\bar{r}$. Since 1 is a transformation of $G$, so also is $A+\lambda$ for any scalar $\lambda$. But the roots of the characteristic equation of $A+\lambda$ are given by $\rho_h+\lambda$ for $h=1, 2, \ldots, \bar{r}$; and these $\bar{r}$ roots will be distinct and other than zero provided $\lambda$ is properly chosen. Wherefore, *$\bar{r}$ is equal to the greatest number of distinct non-zero roots of $\phi(\rho)=0$ for any transformation $A$ of $G$.*

We may transform the hypercomplex number system $(A_1, A_2, \ldots, A_m)$ by the substitution of new units, and express the general transformation of $G$ in terms of the latter. Thus, if

$$A'_i=\tau_{i1}A_1+\tau_{i2}A_2+\ldots+\tau_{im}A_m, \qquad (i=1, 2, \ldots, m), \tag{13}$$

the determinant of transformation not being zero, then

$$\begin{aligned} X&=x_1A_1+x_2A_2+\ldots+x_mA_m \\ &=x'_1A'_1+x'_2A'_2+\ldots+x'_mA'_m, \end{aligned} \tag{14}$$

the $x'$'s being determined by the system of equations

$$x'_1\tau_{1j}+x'_2\tau_{2j}+\ldots+x'_m\tau_{mj}=x_j, \qquad (j=1, 2, \ldots, m). \tag{15}$$

We then have

$$A'_iA'_j=\Sigma_{k=1}^{m}\gamma'_{ijk}A'_k, \qquad (i, j=1, 2, \ldots, m), \tag{16}$$

the $\gamma'$'s being linear functions of the $\gamma$'s. Any linear function of the $A'$'s is a transformation of $G$.

Let $A=\Sigma_{i=1}^{m}a_iA_i$ be any given transformation of $G$, or number of the hypercomplex number system $(A_1, A_2, \ldots, A_m)$. I shall employ $S_1A$, $S_2A$, in designation the *first* and *second scalar* of $A$ respectively, to denote those functions of the coefficients $a_1, a_2, \ldots, a_m$ and the constants $\gamma_{ijk}$ of multiplication of the system defined as follows:

$$S_1A=\frac{1}{m}\Sigma_{i=1}^{m}\Sigma_{j=1}^{m}a_i\gamma_{ijj}, \quad S_2A=\frac{1}{m}\Sigma_{i=1}^{m}\Sigma_{j=1}^{m}a_i\gamma_{jij}. \tag{17}$$

These functions constitute generalizations of the scalar function of quaternions, to which they reduce, becoming identical, when $m=4$ and when the system $(A_1, A_2, A_3, A_4)$ is equivalent to that constituted by the four units of quaternions. They conform to the following theorem:

THEOREM I. *Let $A_1, A_2, \ldots, A_m$ be any system of linearly independent linear homogeneous transformations in $n$ variables constituting a hypercomplex number system, so that*

$$A_iA_j=\Sigma_{k=1}^{m}\gamma_{ijk}A_k, \qquad (i, j=1, 2, \ldots, m);$$

*and let $G$ denote the group constituted by the aggregate of transformations*

$$A=a_1A_1+a_2A_2+\ldots+a_mA_m$$

*for every system of scalars* $a_1, a_2, \ldots, a_m$. *Let*

$$S_1 A = \frac{1}{m} \Sigma_{i=1}^{m} \Sigma_{j=1}^{m} a_i \gamma_{ijj}, \quad S_2 A = \frac{1}{m} \Sigma_{i=1}^{m} \Sigma_{j=1}^{m} a_i \gamma_{jij}.$$

*Then both* $S_1 A$ *and* $S_2 A$ *are invariant to any linear transformation of the number system* $(A_1, A_2, \ldots, A_m)$; *that is to say, if*

$$A'_i = \tau_{i1} A_1 + \tau_{i2} A_2 + \ldots + \tau_{im} A_m, \qquad (i = 1, 2, \ldots, m),$$

(*the determinant of the transformation not being zero*),

$$A'_i A'_j = \Sigma_{k=1}^{m} \gamma'_{ijk} A'_k, \qquad (i, j = 1, 2, \ldots, m),$$

*and if*

$$A = a_1 A_1 + a_2 A_2 + \ldots + a_m A_m = a'_1 A'_1 + a'_2 A_2 + \ldots + a'_m A_m,$$

*then*

$$S_1 A = \frac{1}{m} \Sigma_{i=1}^{m} \Sigma_{j=1}^{m} a_i \gamma_{ijj} = \frac{1}{m} \Sigma_{i=1}^{m} \Sigma_{j=1}^{m} a'_i \gamma'_{ijj},$$

$$S_2 A = \frac{1}{m} \Sigma_{i=1}^{m} \Sigma_{j=1}^{m} a_i \gamma_{jij} = \frac{1}{m} \Sigma_{i=1}^{m} \Sigma_{j=1}^{m} a'_i \gamma'_{jij}.$$

*If* $\rho$ *is any scalar, and*

$$B = b_1 A_1 + b_2 A_2 + \ldots + b_m A_m$$

*is any second transformation of* $G$, *then*

$$\begin{aligned} S_1 \rho A &= \rho S_1 A, & S_2 \rho A &= \rho S_2 A, \\ S_1(A \pm B) &= S_1 A \pm S_1 B, & S_2(A \pm B) &= S_2 A \pm S_2 B, \\ S_1(AB) &= S_1(BA), & S_2(AB) &= S_2(BA); \end{aligned}$$

*and, if* $\varepsilon$ *is a modulus of the system,*

$$S_1 \varepsilon = 1, \quad S_2 \varepsilon = 1.$$

*Let* $\bar{r}$ *denote the maximum number of distinct non-zero roots possessed by the characteristic equation of any transformation of* $G$. *If, for any positive integer* $p$, *either*

$$S_1 A^{p+h} = 0, \qquad (h = 0, 1, 2, \ldots, \bar{r} - 1),$$

*or*

$$S_2 A^{p+h} = 0, \qquad (h = 0, 1, 2, \ldots, \bar{r} - 1),$$

$A$ *is nilpotent; conversely, if* $A$ *is nilpotent,*

$$S_1 A^p = 0, \quad S_2 A^p = 0,$$

*for every positive integer* $p$. *If* $A$ *is idempotent, there is a system of just* $mS_1 A > 0$ *linearly independent transformations of* $G$ *satisfying the equation*

$$AX = X,$$

*in terms of which every transformation of* $G$ *satisfying this equation can be expressed linearly, and a system of just* $mS_2A>0$ *linearly independent transformations of G satisfying the equation*

$$XA=X,$$

*in terms of which every transformation of G satisfying this equation can be expressed linearly.**

I shall denote by $T_{\mu\nu}$, for $\mu, \nu=1, 2, \ldots, n$, the transformation

$$\xi_1'=0, \ldots, \xi_{\mu-1}'=0, \quad \xi_\mu'=\xi_\nu, \quad \xi_{\mu+1}'=0, \ldots, \xi_n'=0, \tag{18}$$

the coefficients of whose matrix are all zero except that in the $\mu$-th row and $\nu$-th column, which is equal to unity. The $n^2$ transformation which we obtain by giving $\mu$ and $\nu$ all integer values from 1 to $n$ are linearly independent; and in terms of these $n^2$ transformations any linear homogeneous transformation in $n$ variables can be expressed linearly. In particular, we have

$$1=T_{11}+T_{22}+\ldots+T_{nn}, \tag{19}$$

$$A_i=\Sigma_{\mu=1}^{n}\Sigma_{\nu=1}^{n}a_{\mu\nu}^{(i)}T_{\mu\nu}, \qquad (i=1, 2, \ldots, m), \tag{20}$$

$$A=\Sigma_{i=1}^{m}a_iA_i=\Sigma_{i=1}^{m}\Sigma_{\mu=1}^{n}\Sigma_{\nu=1}^{n}a_ia_{\mu\nu}^{(i)}T_{\mu\nu}. \tag{21}$$

Moreover,

$$T_{\mu\lambda}T_{\lambda\nu}=T_{\mu\nu}, \quad T_{\mu\lambda}T_{\lambda'\nu}=0, \qquad (\mu, \lambda, \lambda', \nu=1, 2, \ldots, n;\ \lambda'\neq\lambda); \tag{22}$$

and thus these $n^2$ transformations constitute a *quadrate*, that is, one of a special class of hypercomplex number systems. The first and second scalar functions for a number system constituting a quadrate are equal for any given number of the quadrate; and, therefore, but a single symbol is requisite for the scalar functions in the case of a quadrate. I shall write $\bar{S}T$ to denote the scalar of any number

$$T=\Sigma_{\mu=1}^{n}\Sigma_{\nu=1}^{n}c_{\mu\nu}T_{\mu\nu} \tag{23}$$

---

* See papers by the author in the *Trans. Am. Math. Soc.*, Vol. V (1904), p. 522, and *Proc. Am. Acad. Arts and Scs.*, Vol. XLI (1905), p. 61, Vol. XLVIII (1913), p. 628. In the first two of these papers the propositions in the above theorem are established, except that it is not there shown that $A=\Sigma_{i=1}^{m}a_iA_i$ is nilpotent if $S_1A^{p+h}=0$ or $S_2A^{p+h}=0$ for $h=0, 1, \ldots, \bar{r}-1$. This may be proved as follows: Let $\rho_1, \rho_2, \ldots, \rho_s$, respectively of multiplicity $\eta_1, \eta_2, \ldots, \eta_s$, be the distinct non-zero roots of $\phi(\rho)\equiv|\rho-A|=0$, the characteristic equation of the matrix $A$; and let

$$\psi(\rho)\equiv\rho^\nu-q_1\rho^{\nu-1}+\ldots\pm q_\nu=0$$

be the reduced characteristic equation of $A$. Then

$$\psi(\rho)\equiv\rho^{\kappa_0}(\rho-\rho_1)^{\kappa_1}\ldots(\rho-\rho_s)^{\kappa_s},$$

where $1\leqq\kappa_i\leqq\eta_i\,(i=1, 2, \ldots, s)$ and $0\leqq\kappa_0\leqq n-\Sigma_{i=1}^{s}\eta_i$. The syzygy of lowest order between powers of $A$, including $A^0=1$, is $\psi(A)=0$. Therefore, the syzygy of lowest order between powers of $A$, excluding the zero-th power, or 1, is $\Omega(A)\equiv\psi(A)=0$ if $\kappa_0\neq0$, otherwise, is $\Omega(A)\equiv A\psi(A)=0$. If now $r$ denotes the maximum number of distinct non-zero roots of $\phi(\rho)=0$ for any number $A$ of the number system $(A_1, A_2, \ldots, A_m)$, it is then the maximum number of distinct non-zero roots of the equation $\psi(\rho)=0$, and, therefore, the maximum number of distinct non-zero roots of $\Omega(\rho)=0$ for any number $A$ of the system. But in the third of the papers cited above, pp. 635–637, it is shown that, if $\bar{r}$ is the maximum number of distinct non-zero roots of $\Omega(\rho)=0$ for any number $A$ of the system, then $A$ is nilpotent if $S_1A^{p+h}=0$ or $S_2A^{p+h}=0$ for $h=0, 1, \ldots, \bar{r}-1$.

A number $A$ of a hypercomplex number system is *nilpotent* if $A\neq0$, but $A^p=0$ for some positive number $p$; $A$ is *idempotent* if $A^2=A\neq0$.

of the quadrate. By (22), it follows from *Theorem I* that

$$\bar{S}T_{\mu\mu}=\frac{1}{n}, \quad \bar{S}T_{\mu\nu}=0, \qquad (\mu,\nu=1,2,\ldots,n;\ \nu\neq\mu).^* \tag{24}$$

Therefore,

$$\bar{S}T=\Sigma_{\mu=1}^{n}\Sigma_{\nu=1}^{n}c_{\mu\nu}\bar{S}T_{\mu\nu}=\frac{1}{n}\Sigma_{\mu=1}^{n}c_{\mu\mu}. \tag{25}$$

Since $n\bar{S}T$ is the sum of the constituents in the principal diagonal of the matrix of $T$, it follows that $n\bar{S}T$ is equal to the sum of the roots of the characteristic equation, $|\rho-T|=0$ of $T$.† If $T$ is idempotent, the roots of its characteristic equation are 0 and 1. Wherefore, *if $T$ is idempotent, $n\bar{S}T$ is equal to the multiplicity of the root 1 of its characteristic equation.*

The symbol $\bar{S}$ is significant when applied to any transformation of $G$, as this group is a subgroup of the general linear homogeneous group; and we have

$$\bar{S}A_i=\frac{1}{n}\Sigma_{\mu=1}^{n}\alpha_{\mu\mu}^{(i)}, \qquad (i=1,2,\ldots,m), \tag{26}$$

$$\bar{S}A=\frac{1}{n}\Sigma_{i=1}^{m}\Sigma_{\mu=1}^{n}a_i\alpha_{\mu\mu}^{(i)}. \tag{27}$$

Let now

$$\left.\begin{aligned}\Delta_1&=\begin{vmatrix}S_1A_1A_1, & S_1A_1A_2, & \ldots, & S_1A_1A_m\\ S_1A_2A_1, & S_1A_2A_2, & \ldots, & S_1A_2A_m\\ \ldots & \ldots & \ldots & \ldots\\ S_1A_mA_1, & S_1A_mA_2, & \ldots, & S_1A_mA_m\end{vmatrix},\\ \Delta_2&=\begin{vmatrix}S_2A_1A_1, & S_2A_1A_2, & \ldots, & S_2A_1A_m\\ S_2A_2A_1, & S_2A_2A_2, & \ldots, & S_2A_2A_m\\ \ldots & \ldots & \ldots & \ldots\\ S_2A_mA_1, & S_2A_mA_2, & \ldots, & S_2A_mA_m\end{vmatrix},\end{aligned}\right\} \tag{28}$$

* Let $\bar{S}_1T$ and $\bar{S}_2T$ denote the first and second scalar with respect to the quadrate of any number $T$ of the quadrate. For any two distinct integers $\mu$, $\nu$ from 1 to $n$, $T_{\mu\nu}$ is nilpotent by (22); and, therefore,

$$\bar{S}_1T_{\mu\nu}=0=\bar{S}_2T_{\mu\nu}, \qquad (\mu\neq\nu)$$

by *Theorem I*. Moreover, for $1\leqq\mu\leqq n$, $T_{\mu\mu}$ is idempotent; and there is a system of just $n$ linearly independent numbers of the quadrate, namely, $T_{\mu1}, T_{\mu2}, \ldots, T_{\mu n}$, satisfying the equation $T_{\mu\mu}T=T$, and in terms of which every number satisfying this equation can be expressed linearly. Similarly, there are just $n$ linearly independent numbers of the quadrate satisfying the equation $TT_{\mu\mu}=T$, and in terms of which every number satisfying this equation can be expressed linearly. Therefore, by *Theorem I*,

$$n^2\bar{S}_1T_{\mu\mu}=n=n^2\bar{S}_2T_{\mu\mu}.$$

Wherefore, if

$$T=\Sigma_{\mu=1}^{n}\Sigma_{\nu=1}^{n}c_{\mu\nu}T_{\mu\nu},$$

then

$$\bar{S}_1T=\frac{1}{n}\Sigma_{\mu=1}^{n}c_{\mu\mu}=\bar{S}_2T.$$

† See p. 338.

denote, respectively, the resultants of the systems of equations

$$S_1XA_i = x_1S_1A_1A_i + x_2S_1A_2A_i + \ldots + x_mS_1A_mA_i = 0, \quad (i=1, 2, \ldots, m), \tag{29}$$

and

$$S_2XA_i = x_1S_2A_1A_i + x_2S_2A_2A_i + \ldots + x_mS_2A_mA_i = 0, \quad (i=1, 2, \ldots, m); \tag{30}$$

and let

$$\nabla = \begin{vmatrix} \bar{S}A_1A_1, & \bar{S}A_1A_2, & \ldots, & \bar{S}A_1A_m \\ \bar{S}A_2A_1, & \bar{S}A_2A_2, & \ldots, & \bar{S}A_2A_m \\ \ldots & \ldots & \ldots & \ldots \\ \bar{S}A_mA_1, & \bar{S}A_mA_2, & \ldots, & \bar{S}A_mA_m \end{vmatrix} \tag{31}$$

denote the resultant of the system of equations

$$\bar{S}XA_i = x_1\bar{S}A_1A_i + x_2\bar{S}A_2A_i + \ldots + x_m\bar{S}A_mA_i = 0, \quad (i=1, 2, \ldots, m). \tag{32}$$

In the *Proceedings of the American Academy of Arts and Sciences*, Vol. XLVIII (1913), p. 627 *et seq.*, I have shown that $\Delta_1 = \Delta_2$, and that any number $X = \Sigma_{i=1}^{m} x_i A_i$ of the system $(A_1, A_2, \ldots, A_m)$, that is, any transformation $X$ of $G$ satisfying either one of the three systems of equations (29), (30), or (32), is also a solution of the other two; whence it follows that the nullities of $\Delta_1$, $\Delta_2$ and $\nabla$ are equal. I have also shown that the aggregate of numbers of the number system $(A_1, A_2, \ldots, A_m)$ satisfying either one of these three systems of equations constitute a maximum invariant nilpotent subsystem of $(A_1, A_2, \ldots, A_m)$; that the condition necessary and sufficient that the number system shall contain no invariant nilpotent subsystem, in which case the system is either *simple* or *semi-simple* (Cartan), is $\Delta_1 = \Delta_2 \neq 0$ or $\nabla \neq 0$; and that, when this condition is satisfied, $S_1A = S_2A$ for any transformation $A$ of $G$.*

I shall now assume that $\Delta_1 = \Delta_2 \neq 0$, in which case the number system $(A_1, A_2, \ldots, A_m)$ containing no invariant nilpotent subsystem, is constituted

---

* See *Theorems II, IV*, and *V* of the paper cited above, also pp. 650, 655–656. It is also there shown that the equations $\Delta_1 = 0$, $\Delta_2 = 0$, $\nabla = 0$ are invariant to any transformation of the units of the number system $(A_1, A_2, \ldots, A_m)$. Thus, if this number system is transformed by the introduction of new units defined by equations (13), and, if we denote by $\Delta'_1$, $\Delta'_2$, $\nabla'$ respectively the results of replacing $A_1, A_2, \ldots, A_m$ in these determinants by $A'_1, A'_2, \ldots, A'_m$ respectively, then, if $\tau$ denotes the determinant of the substitution, we have

$$\Delta'_1 = \tau^2\Delta_1, \quad \Delta'_2 = \tau^2\Delta_2, \quad \nabla' = \tau^2\nabla.$$

The *nullity* of $\Delta_1$ is zero if $\Delta_1 \neq 0$; and the nullity of $\Delta_1$ is equal to $p$ if the minors of $\Delta_1$ of order $n-p+1$ are all zero, but not all the minors of order $n-p$.

A subsystem $(B_1, B_2, \ldots, B_{m'})$ of the system $(A_1, A_2, \ldots, A_m)$ is *invariant* to the latter if

$$A_iB_j = c'_{ij1}B_1 + \ldots c'_{ijm'}B_{m'}$$
$$B_jA_i = c''_{ij1}B_1 + \ldots c''_{ijm'}B_{m'} \qquad (i=1, 2, \ldots, m;\ j=1, 2, \ldots, m').$$

The subsystem $(B_1, B_2, \ldots, B_m)$ is *nilpotent* if it contains no idempotent number, in which case, every number of the system is nilpotent. A *maximal invariant nilpotent subsystem* of $(A_1, A_2, \ldots, A_m)$ contains every invariant nilpotent subsystem of $(A_1, A_2, \ldots, A_m)$.

A system termed by Cartan is *simple* if it contains no invariant subsystem, and *semi-simple* if its invariant subsystems are simple. See following note.

by $\varkappa(1\leq\varkappa\leq\bar{r})$ mutually nilfactorial quadrates of $s_1^2, s_2^2, \ldots, s_\varkappa^2$ units respectively, where

$$m=\Sigma_{p=1}^{\varkappa} s_p^2, \quad \bar{r}=\Sigma_{p=1}^{\varkappa} s_p. \tag{33}$$

That is to say, we can now find $m$ linearly independent numbers

$$I_{uv}^{(p)}=\Sigma_{i=1}^{m}\tau_{uvi}^{(p)}A_i, \qquad (p=1, 2, \ldots, \varkappa;\ u, v=1, 2, \ldots, s_p) \tag{34}$$

of the number system $(A_1, A_2, \ldots, A_m)$ such that

$$I_{uw}^{(p)} I_{wv}^{(p)}=I_{uv}^{(p)}, \quad I_{uw}^{(p)} I_{w'v}^{(p)}=0,$$
$$(p=1, 2, \ldots, \varkappa;\ u, w, w', v=1, 2, \ldots, s_p;\ w'\neq w), \tag{35}$$

$$I_{uv}^{(p)} I_{u'v'}^{(q)}=0,$$
$$(p, q=1, 2, \ldots, \varkappa;\ q\neq p;\ u, v=1, 2, \ldots, s_p;\ u', v'=1, 2, \ldots, s_q); \tag{36}$$

and these $m$ numbers we can now take as a new system of units.* We shall then have

$$A_i=\Sigma_{p=1}^{\varkappa}\Sigma_{u=1}^{s_p}\Sigma_{v=1}^{s_p} g_{puv}^{(i)} I_{uv}^{(p)}, \qquad (i=1, 2, \ldots, m). \tag{37}$$

Further, for any transformation $X=\Sigma_{i=1}^{m}x_i A_i$ of $G$,

$$X=\Sigma_{p=1}^{\varkappa}\Sigma_{u=1}^{s_p}\Sigma_{v=1}^{s_p} x_{uv}^{(p)} I_{uv}^{(p)}, \tag{38}$$

where

$$x_{uv}^{(p)}=\Sigma_{i=1}^{m} x_i g_{puv}^{(i)}, \qquad (p=1, 2, \ldots, \varkappa;\ u, v=1, 2, \ldots, s_p); \tag{39}$$

and, conversely, the transformation $X$ determined by equation (38), for any system of scalars

$$x_{uv}^{(p)}, \qquad (p=1, 2, \ldots, \varkappa;\ u, v=1, 2, \ldots, s_p),$$

belongs to $G$ in consequence of (34) and what was stated on p. 337. Finally, since 1 is a transformation of $G$, we have

$$1=\Sigma_{p=1}^{\varkappa}\Sigma_{u=1}^{s_p}\Sigma_{v=1}^{s_p} g_{puv}^{(0)} I_{uv}^{(p)};$$

and, therefore, by (35) and (36),

$$I_{u'u'}^{(p')}=I_{u'u'}^{(p')}\,1\,I_{u'u'}^{(p')}=g_{p'u'u'}^{(0)} I_{u'u'}^{(p')}$$
$$0=I_{u'u'}^{(p')}\,1\,I_{v'v'}^{(p')}=g_{p'u'v'}^{(0)} I_{u'v'}^{(p')}, \quad (p'=1, 2, \ldots, \varkappa;\ u', v'=1, 2, \ldots, s_{p'};\ v'\neq u'):$$

whence follows

$$g_{p'u'u'}^{(0)}=1, \quad g_{p'u'v'}^{(0)}=0, \qquad (p'=1, 2, \ldots, \varkappa;\ u', v'=1, 2, \ldots, s_{p'};\ v'\neq u');$$

and thus

$$1=\Sigma_{p=1}^{\varkappa}\Sigma_{u=1}^{s_p} I_{uu}^{(p)}. \tag{40}$$

---

* See paper referred to above, pp. 647–649. See also note to p. 341 of the present paper where it is shown that $\bar{r}$ is equal to the number of distinct roots of $\Omega(\rho)=0$. Cartan has shown that a hypercomplex number system containing no invariant nilpotent subsystem is constituted by simple systems (quadrates) mutually nilfactorial, that is, such that the product is zero of any number of any one constituent simple system by any number of any other of the constituent simple systems. See *Comptes Rendus*, Vol. CXXIV (1897), p. 1218.

From equations (35) and (36), it now follows, by *Theorem I*, that

$$mS_1 I_{uu}^{(p)} = s_p = mS_2 I_{uu}^{(p)}, \qquad (p=1, 2, \ldots, \varkappa;\ u=1, 2, \ldots, s_p).^* \tag{41}$$

From equation (35), we derive

$$\bar{S} I_{uu}^{(p)} = \bar{S} I_{uv}^{(p)} I_{vu}^{(p)} = \bar{S} I_{vu}^{(p)} I_{uv}^{(p)} = \bar{S} I_{vv}^{(p)}, \qquad (p=1, 2, \ldots, \varkappa;\ u, v=1, 2, \ldots, s_p). \tag{42}$$

Whence, if we write

$$\sigma_p = n\bar{S} I_{uu}^{(p)}, \qquad (p=1, 2, \ldots, \varkappa;\ u=1, 2, \ldots, s_p) \tag{43}$$

and

$$n_p = s_p \sigma_p, \qquad (p=1, 2, \ldots, \varkappa), \tag{44}$$

we shall have

$$n = n\bar{S}1 = \Sigma_{p=1}^{\varkappa} \Sigma_{u=1}^{s_p} n\bar{S} I_{uu}^{(p)} = \Sigma_{p=1}^{\varkappa} s_p \sigma_p = \Sigma_{p=1}^{\varkappa} n_p, \tag{45}$$

in consequence of (40).

By (35), $I_{uu}^{(p)}$ is idempotent for $1 \leqq p \leqq \varkappa$ and $1 \leqq u = s_p$; and thus it follows from what was stated on p. 342 that $\sigma_p = n\bar{S} I_{uu}^{(p)}$ is the multiplicity of the root 1 of the characteristic equation of $I_{uu}^{(p)}$. Whence, if

$$J_{uu}^{(p)} = \Sigma_{\mu=N_{p-1}+\overline{u-1}\sigma_p+1}^{N_{p-1}+u\sigma_p} T_{\mu\mu}, \qquad (p=1, 2, \ldots, \varkappa;\ u=1, 2, \ldots, s_p), \tag{46}$$

where

$$N_0 = 0,\ N_p = s_1\sigma_1 + s_2\sigma_2 + \ldots + s_p\sigma_p = n_1 + n_2 + \ldots + n_p, \qquad (p=1, 2, \ldots, \varkappa), \tag{47}$$

it follows, by (35), (36), and (40), that a linear homogeneous transformation $\omega$ of non-zero determinant in the variables $\xi_1, \xi_2, \ldots, \xi_n$ and $\xi_1', \xi_2', \ldots, \xi_n'$ can be found such that

$$I_{uu}^{(p)} = \omega J_{uu}^{(p)} \omega^{-1}, \qquad (p=1, 2, \ldots, \varkappa;\ u=1, 2, \ldots, s_p).\dagger \tag{48}$$

---

* For if $X = \Sigma_{q=1}^{\kappa} \Sigma_{u'=1}^{s_q} \Sigma_{v'=1}^{s_q} x_{u'v'}^{(q)} I_{u'v'}^{(q)}$ is any transformation of $G$ satisfying the equation $I_{uu}^{(p)} X = X$, then $x_{u'v'}^{(q)} = 0$ for $q \neq p$, and $x_{u'v'}^{(p)} = 0$ for $u' \neq u$. Therefore, there is a system of just $s_p$ linearly independent transformations of $G$ satisfying this equation, namely, $I_{uv'}^{(p)}$ for $v'=1, 2, \ldots, s_p$; and, in terms of these transformations, every transformation of $G$ satisfying the above equation can be expressed linearly. Similarly, $I_{u'u}^{(p)}$ for $u'=1, 2, \ldots, s_p$ constitute a system of linearly independent transformations of $G$ satisfying the equation $XI_{uu}^{(p)} = X$, and any transformation of $G$ satisfying this equation can be expressed linearly in terms of these transformations.

† Thus, let

$$A_0 = \Sigma_{p=1}^{\kappa} \Sigma_{u=1}^{s_p} \lambda_u^{(p)} I_{uu}^{(p)},$$

where the $\lambda$'s are distinct. The $\lambda$'s are then the roots of the characteristic equation of $A_0$; $\lambda_u^{(p)}$, for $1 \leqq p \leqq \kappa$ and $1 \leqq u \leqq s_p$, being of multiplicity $\sigma_p$. For, by (35) and (36),

$$A_0^{\mu} = \Sigma_{p=1}^{\kappa} \Sigma_{u=1}^{s_p} (\lambda_u^{(p)})^{\mu} I_{uu}^{(p)}$$

for any integer $\mu$; and, therefore, by (43),

$$nSA_0^{\mu} = \Sigma_{p=1}^{\kappa} \Sigma_{u=1}^{s_p} (\lambda_u^{(p)})^{\mu} nSI_{uu}^{(p)} = \Sigma_{p=1}^{\kappa} \Sigma_{u=1}^{s_p} \sigma_p (\lambda_u^{(p)})^{\mu}.$$

But $nSA_0^{\mu}$, see p. 342, is the sum of the roots of the characteristic equation of $A_0^{\mu}$; and, therefore, is the

Let now

$$J_{uv}^{(p)}=\omega^{-1}I_{uv}^{(p)}\omega, \qquad (p=1,2,\ldots,\varkappa;\ u,v=1,2,\ldots,s_p;\ v\neq u). \tag{49}$$

Then, by (35), (36), and (48),

$$J_{uw}^{(p)}J_{wv}^{(p)}=\omega^{-1}I_{uw}^{(p)}\omega\cdot\omega^{-1}I_{wv}^{(p)}\omega=\omega^{-1}I_{uv}^{(p)}\omega=J_{uv}^{(p)},$$

$$J_{uw}^{(p)}J_{w'v}^{(p)}=\omega^{-1}I_{uw}^{(p)}\omega\cdot\omega^{-1}I_{w'v}^{(p)}\omega=0,$$

$$(p=1,2,\ldots,\varkappa;\ u,w,w',v=1,2,\ldots,s_p;\ w'\neq w), \tag{50}$$

$$J_{uv}^{(p)}J_{u'v'}^{(q)}=\omega^{-1}I_{uv}^{(p)}\omega\cdot\omega^{-1}I_{u'v'}^{(q)}\omega=0,$$

$$(p,q=1,2,\ldots,\varkappa;\ q\neq p;\ u,v=1,2,\ldots,s_p;\ u'\,v'=1,2,\ldots,s_q); \tag{51}$$

and, therefore, if

$$J^{(p)}=\Sigma_{u=1}^{s_p}J_{uu}^{(p)}=\Sigma_{\mu=1}^{n_p}T_{N_{p-1}+\mu,\,N_{p-1}+\mu}, \qquad (p=1,2,\ldots,\varkappa), \tag{52}$$

we then have

$$J_{uv}^{(p)}J^{(q)}=0=J^{(q)}J_{uv}^{(p)}, \quad (p,q=1,2,\ldots,\varkappa;\ q\neq p;\ u,v=1,2,\ldots,s_p). \tag{53}$$

Whence follows

$$J_{uv}^{(p)}=\Sigma_{\mu=1}^{n_p}\Sigma_{\nu=1}^{n_p}\theta_{uv\mu\nu}^{(p)}T_{N_{p-1}+\mu,\,N_{p-1}+\nu}, \quad (p=1,2,\ldots,\varkappa;\ u,v=1,2,\ldots,s_p), \tag{54}$$

where, for $p=1,2,\ldots,\varkappa$ and $u=1,2,\ldots,s_p$,

$$\theta_{u,u,\overline{u-1}\,\sigma_p+\mu,\,\overline{u-1}\,\sigma_p+\mu}^{(p)}=1,\ \theta_{u,u,w\sigma_p+\mu,\,w\sigma_p+\mu}^{(p)}=0,$$

$$(w=0,1,\ldots,\overline{u-2},u,\ldots,s_p-1;\ \mu=1,2,\ldots,\sigma_p) \tag{55}$$

$$\theta_{uu\mu\nu}^{(p)}=0, \qquad (\mu,\nu=1,2,\ldots,n_p;\ \nu\neq\mu) \tag{56}$$

by (46). Wherefore, if, for any system of $m$ scalars $x_{uv}^{(p)}$, $(p=1,2,\ldots,\varkappa;$ $u,v=1,2,\ldots,s_p)$,

$$\Theta_p=\Sigma_{u=1}^{s_p}\Sigma_{v=1}^{s_p}x_{uv}^{(p)}\omega^{-1}I_{uv}^{(p)}\omega=\Sigma_{u=1}^{s_p}\Sigma_{v=1}^{s_p}x_{uv}^{(p)}J_{uv}^{(p)}, \qquad (p=1,2,\ldots,\varkappa), \tag{57}$$

---

sum of the $\mu$-th powers of the roots of the characteristic equation of $A_0$. Further, by (40),

$$A_0-\lambda_u^{(p)}\,(1\leqq p\leqq\kappa;\ 1\leqq u\leqq s_p)$$

is linear in the $I$'s, the coefficient of $I_{uu}^{(p)}$ being zero; and thus

$$\psi(A_0)=(A_0-\lambda_1^{(1)})\ldots.(A_0-\lambda_{s_1}^{(1)})\ldots.(A_0-\lambda_1^{(\kappa)})\ldots.(A_0-\lambda_{s_\kappa}^{(\kappa)})=0$$

by (35) and (36). Therefore, the nullity of $A_0-\lambda_u^{(p)}$ is equal to $\sigma_p$, the multiplicity of $\lambda_u^{(p)}$. Whence it follows that a matrix $\omega$ of non-zero determinant can be found such that

$$A_0=\omega J\omega^{-1},$$

where

$$J=\Sigma_{p=1}^{\kappa}\Sigma_{u=1}^{s_p}\Sigma_{\mu=N_{p-1}+\overline{u-1}\sigma_p+1}^{N_{p-1}+u\sigma_p}\lambda_u^{(p)}T_{\mu\mu},$$

and $N_p$ is given by (47). If now

$$\chi_u^{(p)}(\rho)\equiv\frac{\psi(\rho)}{\rho-\lambda_u^{(p)}},$$

$$f_u^{(p)}(\rho)\equiv\chi_u^{(p)}(\rho)/\chi_u^{(p)}(\lambda_u^{(p)}),$$

then

$$I_{uu}^{(p)}=f_u^{(p)}(A_0)=f_u^{(p)}(\omega J\omega^{-1})=\omega f_u^{(p)}(J)\omega^{-1}=\omega J_{uu}^{(p)}\omega^{-1},$$

for $p=1,2,\ldots,\kappa$ and $u=1,2,\ldots,s_p$, where $J_{uu}$ is given by (46).

it then follows that

$$\Theta_p = \Sigma_{\mu=1}^{n_p} \Sigma_{\nu=1}^{n_p} \vartheta_{\mu\nu}^{(p)} T_{N_{p-1}+\mu,\, N_{p-1}+\nu}, \qquad (p=1, 2, \ldots, \varkappa), \tag{58}$$

where

$$\vartheta_{\mu\nu}^{(p)} = \Sigma_{u=1}^{s_p} \Sigma_{v=1}^{s_p} x_{uv}^{(p)} \theta_{uv\mu\nu}^{(p)}, \qquad (p=1, 2, \ldots, \varkappa;\ \mu, \nu = 1, 2, \ldots, n_p). \tag{59}$$

Whence, if

$$X = \Sigma_{p=1}^{\varkappa} \Sigma_{u=1}^{s_p} \Sigma_{v=1}^{s_p} x_{uv}^{(p)} I_{uv}^{(p)} \tag{60}$$

is the general transformation of $G$, and

$$\Theta = \omega^{-1} X \omega, \tag{61}$$

it follows that

$$\Theta = \Sigma_{p=1}^{\varkappa} \Theta_p = \Sigma_{p=1}^{\varkappa} \Sigma_{\mu=1}^{n_p} \Sigma_{\nu=1}^{n_p} \vartheta_{\mu\nu}^{(p)} T_{N_{p-1}+\mu,\, N_{p-1}+\nu}; \tag{62}$$

that is to say, $\Theta = \omega^{-1} X \omega$ is represented by the matrix

$$\begin{vmatrix} \Theta_1, & 0, & \ldots, & 0 \\ 0, & \Theta_2, & \ldots, & 0 \\ \ldots & \ldots & \ldots & \ldots \\ 0, & 0, & \ldots, & \Theta_\varkappa \end{vmatrix}$$

in which $\Theta_p$ $(p=1, 2, \ldots, \varkappa)$ represents the square array

$$\begin{vmatrix} \vartheta_{11}^{(p)}, & \vartheta_{12}^{(p)}, & \ldots, & \vartheta_{1n_p}^{(p)} \\ \vartheta_{21}^{(p)}, & \vartheta_{22}^{(p)}, & \ldots, & \vartheta_{2n_p}^{(p)} \\ \ldots & \ldots & \ldots & \ldots \\ \vartheta_{n_p1}^{(p)}, & \vartheta_{n_p2}^{(p)}, & \ldots, & \vartheta_{n_p n_p}^{(p)} \end{vmatrix}$$

with $n_p^2$ constituents, and the zero's represent rectangular arrays whose constituents are all zero. In other words, if

$$(\xi_1', \xi_2', \ldots, \xi_n') = \Theta(\xi_1, \xi_2, \ldots, \xi_n), \tag{63}$$

then

$$(\xi'_{N_{p-1}+1}, \xi'_{N_{p-1}+2}, \ldots, \xi'_{N_p}) = \begin{pmatrix} \vartheta_{11}^{(p)}, & \vartheta_{12}^{(p)}, & \ldots, & \vartheta_{1n_p}^{(p)} \\ \vartheta_{21}^{(p)}, & \vartheta_{22}^{(p)}, & \ldots, & \vartheta_{2n_p}^{(p)} \\ \ldots & \ldots & \ldots & \ldots \\ \vartheta_{n_p1}^{(p)}, & \vartheta_{n_p2}^{(p)}, & \ldots, & \vartheta_{n_p n_p}^{(p)} \end{pmatrix} (\xi_{N_{p-1}+1}, \xi_{N_{p-1}+2}, \ldots, \xi_{N_p}) \qquad (p=1, 2, \ldots, \varkappa). \tag{64}$$

For the further determination of the matrices $\Theta_1, \Theta_2, \ldots, \Theta_\varkappa$, we may proceed as follows: Let $\varkappa=1$, in which case

$$m = s_1^2, \quad n = s_1 \sigma_1 = n_1 = N_1,$$

by (33), (44), (45), and (47). In the AMERICAN JOURNAL OF MATHEMATICS, Vol. XII (1899), p. 391, I have shown that a quadrate of order $n=n'n''$ is the product of two quadrates of order $n'$ and $n''$ respectively, or, what is the same thing, that a matrix of order $n=n'n''$ may be regarded as a matrix of order $n'$

whose constituents are matrices of order $n''$.* Whence, if $T$ denotes the general linear homogeneous transformation in $n_1=s_1\sigma_1$ variables, it is represented by the matrix

$$\begin{vmatrix} M_{11}, & M_{12}, & \ldots, & M_{1s_1} \\ M_{21}, & M_{22}, & \ldots, & M_{2s_1} \\ \ldots & \ldots & \ldots & \ldots \\ M_{s_1 1}, & M_{s_1 2}, & \ldots, & M_{s_1 s_1} \end{vmatrix},$$

where the $M$'s are matrices of order $\sigma_1$. Or, otherwise expressed, let $\bar{T}_{uv}$, for $u, v=1, 2, \ldots, s_1$, denote the linear transformation

$$\bar{\xi}'_1=0, \ldots, \bar{\xi}'_{u-1}=0,\ \bar{\xi}'_u=\bar{\xi}_v,\ \bar{\xi}'_{u+1}=0, \ldots, \bar{\xi}'_{s_1}=0 \tag{65}$$

in $s_1$ variables; and let $\bar{\bar{T}}_{\mu\nu}$, for $\mu, \nu=1, 2, \ldots, \sigma_1$, denote the linear transformation

$$\bar{\bar{\xi}}'_1=0, \ldots, \bar{\bar{\xi}}'_{\mu-1}=0,\ \bar{\bar{\xi}}'_\mu=\bar{\bar{\xi}}_\nu,\ \bar{\bar{\xi}}'_{\mu+1}=0, \ldots, \bar{\bar{\xi}}'_{\sigma_1}=0 \tag{66}$$

in $\sigma_1$ variables. Then

$$\bar{T}_{uw}\bar{T}_{wv}=\bar{T}_{uv},\quad \bar{T}_{uw}\bar{T}_{w'v}=0, \qquad (u, w, w', v=1, 2, \ldots, s_1;\ w'\neq w), \tag{67}$$

$$1=\bar{T}_{11}+\bar{T}_{22}+\ldots+\bar{T}_{s_1 s_1}, \tag{68}$$

$$\bar{\bar{T}}_{\mu\lambda}\bar{\bar{T}}_{\lambda\nu}=\bar{\bar{T}}_{\mu\nu},\quad \bar{\bar{T}}_{\mu\lambda}\bar{\bar{T}}_{\lambda'\nu}=0, \qquad (\mu, \lambda, \lambda', \nu=1, 2, \ldots, \sigma_1;\ \lambda'\neq\lambda), \tag{69}$$

$$1=\bar{\bar{T}}_{11}+\bar{\bar{T}}_{22}+\ldots+\bar{\bar{T}}_{\sigma_1\sigma_1}, \tag{70}$$

corresponding to (22) and (19). If now we assume that

$$\bar{T}_{uv}\bar{\bar{T}}_{\mu\nu}=\bar{\bar{T}}_{\mu\nu}\bar{T}_{uv}, \qquad (u, v=1, 2, \ldots, s_1;\ \mu, \nu=1, 2, \ldots, \sigma_1), \tag{71}$$

then, for any two integers $\mu'$, $\nu'$ from 1 to $n_1=s_1\sigma_1$, we may put

$$T_{\mu'\nu'}=\bar{T}_{uv}\bar{\bar{T}}_{\mu\nu}, \tag{72}$$

where $\mu$, $\nu$ are the smallest positive residues of $\mu'$, $\nu'$ respectively *modulo* $\sigma_1$, and $u$, $v$ are determined by the equations

$$\mu'=(u-1)\sigma_1+\mu, \quad \nu'=(v-1)\sigma_1+\nu. \tag{73}$$

For the units $T_{\mu'\nu'}$ will then have the required multiplication table of the quadrate of order $n=n_1=s_1\sigma_1$ given by (22). The general number $T$ of this quadrate, that is, the general linear homogeneous transformation in $n_1=s_1\sigma_1$ variables, is now given by the equation

$$\begin{aligned} T=\Sigma_{\mu=1}^{n_1}\Sigma_{\nu=1}^{n_1} c_{\mu\nu}T_{\mu\nu} &= \Sigma_{u,v=1}^{s_1}\Sigma_{\mu,\nu=1}^{\sigma_1} c_{\overline{u-1}\sigma_1+\mu,\ \overline{v-1}\sigma_1+\nu}\bar{T}_{uv}\bar{\bar{T}}_{\mu\nu} \\ &= \Sigma_{u=1}^{s_1}\Sigma_{v=1}^{s_1} M_{uv}\bar{T}_{uv}=\Sigma_{u=1}^{s_1}\Sigma_{v=1}^{s_1}\bar{T}_{uv}M_{uv}, \end{aligned} \tag{74}$$

where

$$M_{uv}=\Sigma_{\mu=1}^{\sigma_1}\Sigma_{\nu=1}^{\sigma_1} c_{\overline{u-1}\sigma_1+\mu,\ \overline{v-1}\sigma_1+\nu}\bar{\bar{T}}_{\mu\nu}, \qquad (u, v=1, 2, \ldots, s_1).\dagger \tag{75}$$

* This theorem was obtained independently by E. Study. See "Math. papers read at the Internat. Math. Congress, Chicago, 1893", p. 378.

† Two matrices $\Sigma_{u=1}^{s_1}\Sigma_{v=1}^{s_1} M_{uv}\bar{T}_{uv}$ and $\Sigma_{u=1}^{s_1}\Sigma_{v=1}^{s_1} M_{uv}^{(1)}\bar{T}_{uv}$ are equal if, and only if, $M_{uv}^{(1)}=M_{uv}$ for $u, v=1, 2, \ldots, s_1$.

In particular, by (46) and (70),

$$J^{(1)}_{uu}=\Sigma^{\sigma_1}_{\mu=1} T_{\overline{u-1}\,\sigma_1+\mu,\,\overline{u-1}\,\sigma_1+\mu}=\Sigma^{\sigma_1}_{\mu=1} \overline{T}_{uu}\,\overline{\overline{T}}_{\mu\mu}=\overline{T}_{uu}, \quad (u=1, 2, \ldots, s_1). \quad (76)$$

Let

$$J^{(1)}_{uv}=\Sigma^{s_1}_{u'=1}\Sigma^{s_1}_{v'=1} M^{(u,v)}_{u'v'}\,\overline{T}_{u'v'}, \qquad (u, v=1, 2, \ldots, s_1;\ v\neq u).$$

From (50), (67), (71), and (76), it follows that

$$J^{(1)}_{uv}=J^{(1)}_{uu}\,J^{(1)}_{uv}\,J^{(1)}_{vv}=\overline{T}_{uu}\,\Sigma^{s_1}_{u'=1}\Sigma^{s_1}_{v'=1} M^{(u,v)}_{u'v'}\,\overline{T}_{u'v'}\,\overline{T}_{vv}=M^{(u,v)}_{uv}\,\overline{T}_{uv}, \quad (u, v=1, 2, \ldots, s_1);$$

and thus

$$M^{(u,v)}_{u'v'}=0,^{*} \qquad (u, v, u', v'=1, 2, \ldots, s_1;\ u'\neq u;\ \text{or } v'\neq v).$$

We may, therefore, delete the indices and write simply

$$J^{(1)}_{uv}=M'_{uv}\,\overline{T}_{uv}, \qquad (u, v=1, 2, \ldots, s_1), \qquad (77)$$

where $M'_{uv}$ is a matrix of order $\sigma_1$; when, by (76),

$$M'_{uu}=1, \qquad (u=1, 2, \ldots, s_1). \qquad (78)$$

From (77), it follows, by (50), (67), and (71), that

$$\begin{aligned} M'_{uw}\,M'_{wv}\,\overline{T}_{uv}&=M'_{uw}\,\overline{T}_{uw}\cdot M'_{wv}\,\overline{T}_{wv}\\ &=J^{(1)}_{uw}\,J^{(1)}_{wv}\\ &=J^{(1)}_{uv}=M'_{uv}\,\overline{T}_{uv},\\ &\qquad (u, v, w=1, 2, \ldots, s_1); \end{aligned}$$

and, therefore,

$$M'_{uw}\,M'_{wv}=M'_{uv}, \qquad (u, v, w=1, 2, \ldots, s_1). \qquad (79)$$

In particular, by (78),

$$M'_{uv}\,M'_{vu}=M'_{uu}=1, \qquad (u, v=1, 2 \ldots, s_1). \qquad (80)$$

Let

$$\overline{\omega}=M'_{11}\,\overline{T}_{11}+M'_{12}\,\overline{T}_{22}+\ldots+M'_{1s_1}\,\overline{T}_{s_1s_1}. \qquad (81)$$

Then, by (67), (68), (71), and (80),

$$\overline{\omega}^{-1}=M'_{11}\,\overline{T}_{11}+M'_{21}\,\overline{T}_{22}+\ldots+M'_{s_11}\,\overline{T}_{s_1s_1};\dagger \qquad (82)$$

and, therefore, by (67), (71), and (76),

$$\overline{\omega}\,J^{(1)}_{uu}=\overline{\omega}\,\overline{T}_{uu}=M'_{1u}\,\overline{T}_{uu},\quad J^{(1)}_{uu}\,\overline{\omega}^{-1}=\overline{T}_{uu}\,\overline{\omega}^{-1}=M'_{u1}\,\overline{T}_{uu}, \quad (u=1, 2, \ldots, s_1). \quad (83)$$

Whence follows

$$\begin{aligned} \overline{\omega}\,J^{(1)}_{uv}\,\overline{\omega}^{-1}&=\overline{\omega}\,J^{(1)}_{uu}\cdot J^{(1)}_{uv}\cdot J^{(1)}_{vv}\,\overline{\omega}^{-1}\\ &=M'_{1u}\,\overline{T}_{uu}\cdot M'_{uv}\,\overline{T}_{uv}\cdot M'_{v1}\,\overline{T}_{vv}\\ &=M'_{1u}\,M'_{uv}\,M'_{v1}\,\overline{T}_{uu}\,\overline{T}_{uv}\,\overline{T}_{vv}=\overline{T}_{uv},\\ &\qquad (u, v=1, 2, \ldots, s_1), \end{aligned} \qquad (84)$$

---

* See preceding note.

† That is, if $\overline{\omega}$ is defined by (81), and $\overline{\omega}^{(1)}=\Sigma^{s_1}_{u=1}M'_{u1}\overline{T}_{uu}$, then $\overline{\omega}\,\overline{\omega}^{(1)}=1$; and therefore, $\overline{\omega}^{(1)}=\overline{\omega}^{-1}$.

by (50), (67), (71), (77), (78), and (79). Wherefore, if

$$\Theta_1 = \Sigma_{u=1}^{s_1} \Sigma_{v=1}^{s_1} x_{uv}^{(1)} J_{uv}^{(1)}, \tag{85}$$

then, by (70) and (72),

$$\begin{aligned}\bar{\omega}\,\Theta_1\,\bar{\omega}^{-1} &= \Sigma_{u=1}^{s_1} \Sigma_{v=1}^{s_1} x_{uv}^{(1)}\,\bar{\omega}\,J_{uv}^{(1)}\,\bar{\omega}^{-1}\\ &= \Sigma_{u=1}^{s_1} \Sigma_{v=1}^{s_1} x_{uv}^{(1)}\,\bar{T}_{uv}\\ &= \Sigma_{u=1}^{s_1} \Sigma_{v=1}^{s_1} \Sigma_{\mu=1}^{\sigma_1} x_{uv}^{(1)}\,\bar{T}_{uv}\,\bar{\bar{T}}_{\mu\mu}\\ &= \Sigma_{u=1}^{s_1} \Sigma_{v=1}^{s_1} x_{uv}^{(1)} \Sigma_{\mu=1}^{\sigma_1} T_{\overline{u-1}\sigma_1+\mu,\ \overline{v-1}\sigma_1+\mu}\,;\end{aligned} \tag{86}$$

that is to say, $\bar{\omega}\Theta_1\bar{\omega}^{-1}$ is represented by the matrix

$$\begin{vmatrix} x_{11}^{(1)} L, & x_{12}^{(1)} L, & \dots, & x_{1s_1}^{(1)} L\\ x_{21}^{(1)} L, & x_{22}^{(1)} L, & \dots, & x_{2s_1}^{(1)} L\\ \dots & \dots & \dots & \dots\\ x_{s_1 1}^{(1)} L, & x_{s_1 2}^{(1)} L, & \dots, & x_{s_1 s_1}^{(1)} L \end{vmatrix},$$

where $L$ is a matrix of order $\sigma_1$ whose constituents are all zero, except those in the principal diagonal which are all equal to unity.

By the composition with $\bar{\omega}$ of an interchange of the subscripts of the $N_1 = n_1 = s_1\sigma_1$ variables $\xi_1, \xi_2, \dots, \xi_{N_1}$, and, therefore, by a properly chosen linear substitution $\bar{\omega}_1$ with constant coefficients in these variables, we shall have

$$\bar{\omega}_1\,\Theta_1\,\bar{\omega}_1^{-1} = \Sigma_{u=1}^{s_1} \Sigma_{v=1}^{s_1} \Sigma_{\mu=1}^{\sigma_1} x_{uv}^{(1)}\, T_{\overline{\mu-1}s_1+u,\ (\mu-1)s_1+v}\,, \tag{87}$$

in which case $\bar{\omega}_1\,\Theta\,\bar{\omega}_1^{-1}$ is represented by the matrix

$$\Psi^{(1)} = \begin{vmatrix} \Psi_1^{(1)}, & 0, & \dots, & 0\\ 0, & \Psi_2^{(1)}, & \dots, & 0\\ \dots & \dots & \dots & \dots\\ 0, & 0, & \dots, & \Psi_{\sigma_1}^{(1)} \end{vmatrix}, \tag{88}$$

where the zeros represent rectangular arrays whose constituents are all zero, and

$$\Psi_1^{(1)} = \Psi_2^{(1)} = \dots = \Psi_{\sigma_1}^{(1)} = \begin{vmatrix} x_{11}^{(1)}, & \dots, & x_{1s_1}^{(1)}\\ \dots & \dots & \dots\\ x_{s_1 1}^{(1)}, & \dots, & x_{s_1 s_1}^{(1)} \end{vmatrix}. \tag{89}$$

That is to say, if

$$(\xi_1', \xi_2', \dots, \xi_{N_1}') = \bar{\omega}_1\,\Theta_1\,\bar{\omega}_1^{-1}(\xi_1, \xi_2, \dots, \xi_{N_1}), \tag{90}$$

then

$$\begin{aligned}&(\xi'_{(\mu-1)s_1+1},\ \xi'_{(\mu-1)s_1+2},\ \dots,\ \xi'_{\mu s_1})\\ &= \left(\begin{matrix} x_{11}^{(1)}, & x_{12}^{(1)}, & \dots, & x_{1s_1}^{(1)}\\ x_{21}^{(1)}, & x_{22}^{(1)}, & \dots, & x_{2s_1}^{(1)}\\ \dots & \dots & \dots & \dots\\ x_{s_1 1}^{(1)}, & x_{s_1 2}^{(1)}, & \dots, & x_{s_1 s_1}^{(1)} \end{matrix}\right)(\xi_{(\mu-1)s_1+1},\ \xi_{(\mu-1)s_1+2},\ \dots,\ \xi_{\mu s_1}),\\ &\qquad (\mu = 1, 2, \dots, \sigma_1).\end{aligned} \tag{91}$$

In particular, when $\varkappa=1$ (and, as assumed above $\Delta_1=\Delta_2\neq 0$), if $\sigma_1=1$, then

$$m=s_1^2=n_1^2, \quad n=s_1=n_1=N_1.$$

In this case,

$$\overline{\omega}_1\,\Theta_1\,\overline{\omega}_1^{-1}(\xi_1, \xi_2, \ldots, \xi_n)=\left(\begin{array}{l} x_{11}^{(1)}, x_{12}^{(1)}, \ldots, x_{1n}^{(1)} \\ x_{21}^{(1)}, x_{22}^{(1)}, \ldots, x_{2n}^{(1)} \\ \ldots\ldots\ldots\ldots\ldots \\ x_{n1}^{(1)}, x_{n2}^{(1)}, \ldots, x_{nn}^{(1)} \end{array}\right)(\xi_1, \xi_2, \ldots, \xi_n).$$

Wherefore, by (60), (61), and (62), if

$$\omega_0=\omega\,\overline{\omega}_1^{-1},$$

the general transformation $X$ of $G$ is given by the equation

$$X=\omega\,\Theta\,\omega^{-1}=\omega\,\Theta_1\,\omega^{-1}=\omega_0\left|\begin{array}{l} x_{11}^{(1)}, x_{12}^{(1)}, \ldots, x_{1n}^{(1)} \\ x_{21}^{(1)}, x_{22}^{(1)}, \ldots, x_{2n}^{(1)} \\ \ldots\ldots\ldots\ldots\ldots \\ x_{n1}^{(1)}, x_{n2}^{(1)}, \ldots, x_{nn}^{(1)} \end{array}\right|\omega_0^{-1};$$

and, since the $x$'s are arbitrary, there is no flat through the origin invariant to $G$, which is, therefore, irreducible.* Conversely, if $G$ is irreducible, then $\Delta_1=\Delta_2\neq 0$. For, in this case, by Burnside's theorem,† $G$ contains $n^2$ linearly independent transformations; and thus $m=n^2$. But since $A_1, A_2, \ldots, A_m$, the maximum number of linearly independent transformations of $G$ are, by (20), expressible linearly in terms of the $m=n^2$ matrices $T_{\mu\nu}(\mu, \nu=1, 2, \ldots, n)$, it then follows that the latter are linearly expressible in terms of the $A$'s: therefore, the number system $(A_1, A_2, \ldots, A_m)$, being equivalent to a quadrate, can contain no invariant nilpotent subsytem; and thus $\Delta_1=\Delta_2\neq 0$.

Let now $\varkappa>1$; and for $1\leqq p\leqq\varkappa$, see (57), let

$$\Theta_p=\Sigma_{u=1}^{s_p}\Sigma_{v=1}^{s_p}x_{uv}^{(p)}\,\omega^{-1}I_{uv}^{(p)}\,\omega=\Sigma_{u=1}^{s_p}\Sigma_{v=1}^{s_p}x_{uv}^{(p)}\,J_{uv}^{(p)}. \tag{92}$$

Then, by (58), $\Theta_p$ may be regarded as a transformation in the $n_p$ variables $\xi_{N_{p-1}+1}, \xi_{N_{p-1}+2}, \ldots, \xi_{N_p}$; and, by what has just been shown (pp. 347–350), it follows that a linear substitution $\overline{\omega}^p$ in these variables with constant coefficients can be found such, if

$$(\xi'_{N_{p-1}+1}, \xi'_{N_{p-1}+2}, \ldots, \xi'_{N_p})=\overline{\omega}_p\,\Theta_p\,\overline{\omega}_p^{-1}(\xi_{N_{p-1}+1}, \xi_{N_{p-1}+2}, \ldots, \xi_{N_p}), \tag{93}$$

---

* See p. 358. A subgroup of the general linear homogeneous group in $n$ variables is irreducible if, and only if, there is a $p$-flat $(1\leqq p<n)$ through the origin invariant to every transformation of the subgroup.

† *Proc. Lond. Math. Soc.*, 2d Ser. Vol. III (1905), p. 433.

that then

$$
\begin{aligned}
&(\xi'_{N_{p-1}+\overline{\nu-1}s_p+1},\ \xi'_{N_{p-1}+\overline{\nu-1}s_p+2},\ \ldots,\ \xi'_{N_{p-1}+\nu s_p}) \\
&\quad = \left( \begin{array}{l} x^{(p)}_{11},\ x^{(p)}_{12},\ \ldots,\ x^{(p)}_{1s_p} \\ x^{(p)}_{21},\ x^{(p)}_{22},\ \ldots,\ x^{(p)}_{2s_p} \\ \ldots\ldots\ldots\ldots\ldots \\ x^{(p)}_{s_p1},\ x^{(p)}_{s_p2},\ \ldots,\ x^{(p)}_{s_ps_p} \end{array} \right)(\xi_{N_{p-1}+\overline{\nu-1}s_p+1},\ \xi_{N_{p-1}+\overline{\nu-1}s_p+2},\ \ldots,\ \xi_{N_{p-1}+\nu s_p}) \\
&\qquad (\nu=1,2,\ldots,\sigma_p); \qquad (94)
\end{aligned}
$$

that is,

$$\overline{\omega}_p\,\Theta_p\,\overline{\omega}_p^{-1}=\Sigma^{s_p}_{u=1}\Sigma^{s_p}_{v=1}\Sigma^{\sigma_p}_{\nu=1}x^{(p)}_{uv}\,T_{N_{p-1}+\overline{\nu-1}s_p+u,\ N_{p-1}+\overline{\nu-1}s_p+v}. \qquad (95)$$

The $\varkappa$ sets of variables $\xi_{N_{p-1}+1}, \xi_{N_{p-1}+2}, \ldots, \xi_{N_p}$, for $p=1, 2, \ldots, \varkappa$, are mutually exclusive and comprise all the $\xi$'s. Whence, if

$$\Theta=\Sigma^{\kappa}_{p=1}\Sigma^{s_p}_{u=1}\Sigma^{s_p}_{v=1}x^{(p)}_{uv}\,J^{(p)}_{uv}\doteq\Sigma^{\kappa}_{p=1}\Theta_p, \qquad (96)$$

and if

$$\Psi=\Sigma^{\kappa}_{p=1}\Sigma^{s_p}_{u=1}\Sigma^{s_p}_{v=1}\Sigma^{\sigma_p}_{\mu=1}x^{(p)}_{uv}\,T'_{N_{p-1}+\overline{\mu-1}s_p+u,\ N_{p-1}+\overline{\mu-1}s_p+v}, \qquad (97)$$

in which case

$$\Psi=\left| \begin{array}{cccc} \Psi^{(1)}, & 0, & \ldots, & 0 \\ 0, & \Psi^{(2)}, & \ldots, & 0 \\ \ldots & \ldots & \ldots & \ldots \\ 0, & 0, & \ldots, & \Psi^{(\kappa)} \end{array} \right|, \qquad (98)$$

where

$$\Psi^{(p)}=\left| \begin{array}{cccc} \Psi^{(p)}_1, & 0, & \ldots, & 0 \\ 0, & \Psi^{(p)}_2, & \ldots, & 0 \\ \ldots & \ldots & \ldots & \ldots \\ 0, & 0, & \ldots, & \Psi^{(p)}_{\sigma_p} \end{array} \right| \qquad (p=1,2,\ldots,\varkappa), \qquad (99)$$

the zeros in this matrix (also the zeros in the preceding matrix) representing rectangular arrays whose constituents are all zero, and

$$\Psi^{(p)}_1=\Psi^{(p)}_2=\ldots\Psi^{(p)}_{\sigma_p}=\left| \begin{array}{l} x^{(p)}_{11},\ x^{(p)}_{12},\ \ldots,\ x^{(p)}_{1s_p} \\ x^{(p)}_{21},\ x^{(p)}_{22},\ \ldots,\ x^{(p)}_{2s_p} \\ \ldots\ldots\ldots\ldots\ldots \\ x^{(p)}_{s_p1},\ x^{(p)}_{s_p2},\ \ldots,\ x^{(p)}_{s_ps_p} \end{array} \right| \qquad (p=1,2,\ldots,\varkappa), \qquad (100)$$

it then follows that a linear substitution $\overline{\omega}_0$ in the $\xi$'s with constant coefficients can be found such that

$$\overline{\omega}_0\,\Theta\,\overline{\omega}_0^{-1}=\Psi. \qquad (101)$$

Wherefore, if

$$X=\Sigma^{\kappa}_{p=1}\Sigma^{s_p}_{u=1}\Sigma^{s_p}_{v=1}x^{(p)}_{uv}\,I^{(p)}_{uv}, \qquad (102)$$

where the $x$'s are any set of $m=\Sigma^{\kappa}_{p=1}s_p^2$ scalars, is any transformation of $G$, and

$$\omega'=\omega\,\overline{\omega}_0^{-1}, \qquad (103)$$

then, by (61),

$$X=\omega\,\Theta\,\omega^{-1}=\omega'\,\Psi\,\omega'^{-1}, \qquad (104)$$

where the coefficients of $\omega'$ are constant. That is to say, if

$$(\xi'_1, \xi'_2, \dots, \xi'_n) = \omega'^{-1} X \omega' (\xi_1, \xi_2, \dots, \xi_n), \qquad (105)$$

then

$$\begin{aligned} &(\xi'_{N_{p-1}+\overline{\mu-1}s_p+1}, \xi'_{N_{p-1}+\overline{\mu-1}s_p+2}, \dots, \xi'_{N_{p-1}+\mu s_p}) \\ &= \begin{pmatrix} x_{11}^{(p)}, & x_{12}^{(p)}, & \dots, & x_{1s_p}^{(p)} \\ x_{21}^{(p)}, & x_{22}^{(p)}, & \dots, & x_{2s_p}^{(p)} \\ \dots & \dots & \dots & \dots \\ x_{s_p1}^{(p)}, & x_{s_p2}^{(p)}, & \dots, & x_{s_ps_p}^{(p)} \end{pmatrix} (\xi_{N_{p-1}+\overline{\mu-1}s_p+1}, \xi_{N_{p-1}+\overline{\mu-1}s_p+2}, \dots, \xi_{N_{p-1}+\mu s_p}) \\ &\qquad (p=1, 2, \dots, \varkappa;\ \mu=1, 2, \dots, \sigma_p). \end{aligned} \qquad (106)$$

Since the $m$ scalars $x_{uv}^{(p)}$ $(p=1, 2, \dots, \varkappa;\ u, v=1, 2, \dots, s_p)$ are arbitrary, it follows that the group $G_\mu^{(p)}$ defined, for $1 \leqq p \leqq \varkappa$ and $1 \leqq \mu \leqq \sigma_p$, by equations (106) is irreducible. Whence it follows that $G$ is completely reducible.* We have, therefore, the following theorem.

THEOREM II. *If $A_1, A_2, \dots, A_m$ is any system of $m$ linearly independent linear homogeneous transformations in $n$ variables $(m \leqq n^2)$ constituting a hypercomplex number system, so that*

$$A_i A_j = \Sigma_{k=1}^m \gamma_{ijk} A_k, \qquad (i, j=1, 2, \dots, m),$$

*the aggregate of numbers $\Sigma_{i=1}^m x_i A_i$ of this number system constitute a group which is completely reducible if the discriminant of either of the quadratic forms $S_1(\Sigma_{i=1}^m x_i A_i)^2$, $S_2(\Sigma_{i=1}^m x_i A_i)^2$, or $\bar{S}(\Sigma_{i=1}^m x_i A_i)^2$, is not zero (in which case the discriminant of neither of the other two forms is zero), provided the determinant is not zero of every transformation of the group.*†

## § 2. *Conditions Sufficient for Similarity of Two Subgroups of the General Linear Homogeneous Group the Coefficients of whose Transformations are Linear Functions of Certain Parameters.*

Let $A_1, A_2, \dots, A_m$ and $B_1, B_2, \dots, B_m$ be two systems of linearly independent linear homogeneous transformations in $n$ variables, each constituting a hypercomplex number system, so that

$$A_i A_j = \Sigma_{k=1}^m \gamma_{ijk} A_k, \qquad (i, j=1, 2, \dots, m), \qquad (107)$$

$$B_i B_j = \Sigma_{k=1}^m \gamma'_{ijk} B_k, \qquad (i, j=1, 2, \dots, m); \qquad (108)$$

---

* The group $G$ is irreducible if $\varkappa=1$, $\sigma=1$. I follow Loewy in regarding the case of irreducibility as included in that of complete reducibility. See p. 359.

† Since, for $1 \leqq i \leqq m$, $1 \leqq j \leqq m$,

$$\frac{\partial^2}{\partial x_i \partial x_j} S_1 (\Sigma_{k=1}^m x_k A_k)^2 = 2 S_1 A_i A_j,$$

it follows that $\Delta_1$ is the discriminant of $S_1(\Sigma_{k=1}^m x_k A_k)^2$. Similarly, $\Delta_2$ and $\nabla$ are, respectively, the discriminants of $S_2(\Sigma_{k=1}^m x_k A_k)^2$ and $\bar{S}(\Sigma_{k=1}^m x_k A_k)^2$.

and let $G$ and $H$, respectively, be the groups constituted by the aggregates of transformations

$$X=\Sigma_{i=1}^{m} x_i A_i, \tag{109}$$

$$Y=\Sigma_{i=1}^{m} y_i B_i, \tag{110}$$

for all systems of values of the $x$'s and the $y$'s.

The condition necessary and sufficient that the groups $G$ and $H$ shall be *similar* is that a linear substitution $\Omega$ with constant coefficients can be found such that

$$Y=\Sigma_{i=1}^{m} y_i B_i=\Sigma_{i=1}^{m} x_i \Omega A_i \Omega^{-1}=\Omega X \Omega^{-1}, \tag{111}$$

where the $x$'s are determined as linear functions of the $y$'s by the equations

$$y_i=\tau_{1i} x_1+\tau_{2i} x_2+\ldots\ldots+\tau_{mi} x_m, \qquad (i=1, 2, \ldots\ldots, m). \tag{112}$$

Let, first, $G$ and $H$ be similar. Then the number systems $(A_1, A_2, \ldots\ldots, A_m)$ and $(B_1, B_2, \ldots\ldots, B_m)$ are equivalent. For, if

$$B'_i=\Sigma_{h=1}^{m} \tau_{ih} B_h=\Omega A_i \Omega^{-1}, \qquad (i=1, 2, \ldots\ldots, m), \tag{113}$$

by (111) and (112), it follows that the $B'$'s are linearly independent and that

$$B'_i B'_j=\Sigma_{k=1}^{m} \gamma_{ijk} B'_k, \qquad (i, j=1, 2, \ldots\ldots, m). \tag{114}$$

Moreover,

$$\bar{S} B'_i=\bar{S}\Omega A_i \Omega^{-1}=\bar{S} A_i, \qquad (i=1, 2, \ldots\ldots, m). \tag{115}$$

Conversely, let now the number systems $(A_1, A_2, \ldots\ldots, A_m)$ and $(B_1, B_2, \ldots\ldots, B_m)$ be equivalent; and let $B'_1, B'_2, \ldots\ldots, B'_m$, where

$$B'_i=\tau_{i1} B_1+\tau_{i2} B_2+\ldots\ldots+\tau_{im} B_m, \qquad (i=1, 2, \ldots\ldots, m), \tag{116}$$

be $m$ linearly independent functions of the $B$'s such that

$$B'_i B'_j=\Sigma_{i=1}^{m} \gamma_{ijk} B_k, \qquad (i, j=1, 2, \ldots\ldots, m). \tag{117}$$

Moreover, let

$$\bar{S} B'_i=\bar{S} A_i, \qquad (i=1, 2, \ldots\ldots, m). \tag{118}$$

Then, as will be shown below, $G$ and $H$ are similar, provided the discriminant of the quadratic form $\Sigma_{i=1}^{m}\Sigma_{j=1}^{m} x_i x_j S A_i A_j$ is not zero; but not, in general, otherwise. For, let

$$\begin{vmatrix} \bar{S} B'_1 B'_1, \ldots\ldots, \bar{S} B'_1 B'_m \\ \ldots\ldots\ldots\ldots\ldots\ldots \\ \bar{S} B'_m B'_1, \ldots\ldots, \bar{S} B'_m B'_m \end{vmatrix} = \begin{vmatrix} \bar{S} A_1 A_1, \ldots\ldots, \bar{S} A_1 A_m \\ \ldots\ldots\ldots\ldots\ldots\ldots \\ \bar{S} A_m A_1, \ldots\ldots, \bar{S} A_m A_m \end{vmatrix} \neq 0. \tag{119}$$

Then the number system $(A_1, A_2, \ldots\ldots, A_m)$ is constituted by $\kappa$ mutually nil-factorial quadrates $I_{uv}^{(p)}$ $(u, v=1, 2, \ldots\ldots, s_p)$ for $p=1, 2, \ldots\ldots, \kappa$; that is, we can find

$$m=s_1^2+s_2^2+\ldots\ldots+s_\kappa^2, \qquad (\kappa \geqq 1), \tag{120}$$

linearly independent numbers of this system, namely,

$$I_{uv}^{(p)}=\Sigma_{h=1}^{m}\tau_{uvh}^{(p)}A_h, \qquad (p=1,2,\ldots,\varkappa;\ u,v=1,2,\ldots,s_p), \tag{121}$$

(which we may take as a new system of units) such that

$$I_{uw}^{(p)}I_{wv}^{(p)}=I_{uv}^{(p)},\ I_{uw}^{(p)}I_{w'v}^{(p)}=0,$$
$$(p=1,2,\ldots,\varkappa;\ u,v,w,w'=1,2,\ldots,s;\ w'\neq w), \tag{122}$$

$$I_{uv}^{(p)}I_{u'v'}^{(q)}=0,$$
$$(p,q=1,2,\ldots,\varkappa;\ q\neq p;\ u,v=1,2,\ldots,s_p;\ u',v'=1,2,\ldots,s_q); \tag{123}$$

and we shall then have

$$m\,S_1 I_{uu}^{(p)}=s_p=m\,S_2 I_{uu}^{(p)}, \qquad (p=1,2,\ldots,\varkappa;\ u=1,2,\ldots,s_p), \tag{124}$$

and

$$\bar{S}I_{11}^{(p)}=\bar{S}I_{22}^{(p)}=\ldots.=\bar{S}I_{s_p s_p}^{(p)}, \qquad (p=1,2,\ldots,\varkappa).^{*} \tag{125}$$

For any system of scalars $x_1, x_2, \ldots, x_m$,

$$X=\Sigma_{i=1}^{m}x_i A_i=\Sigma_{p=1}^{\varkappa}\Sigma_{u=1}^{s_p}\Sigma_{v=1}^{s_p}x_{uv}^{(p)}I_{uv}^{(p)}, \tag{126}$$

where $x_{uv}^{(p)}$, for $p=1,2,\ldots,\varkappa$ and $u,v=1,2,\ldots,s_p$ are determined by the equations

$$x_i=\Sigma_{p=1}^{\varkappa}\Sigma_{u=1}^{s_p}\Sigma_{v=1}^{s_p}\tau_{uvi}^{(p)}x_{uv}^{(p)}, \qquad (i=1,2,\ldots,m). \tag{127}$$

If now

$$n\,\bar{S}I_{uu}^{(p)}=\sigma_p, \qquad (p=1,2,\ldots,\varkappa;\ u,v=1,2,\ldots,s_p), \tag{128}$$

and

$$N_0=0,\ N_p=s_1\sigma_1+s_2\sigma_2+\ldots.+s_p\sigma_p, \qquad (p=1,2,\ldots,\varkappa), \tag{129}$$

it follows, by what has been shown in § 1, that a linear substitution $\omega'$ with constant coefficients can be found such that

$$X=\omega'\Psi\omega'^{-1}, \tag{130}$$

where $\Psi$ is defined by equations (98); that is, if

$$(\xi_1',\xi_2',\ldots,\xi_n')=\omega'^{-1}X\omega'(\xi_1,\xi_2,\ldots,\xi_n), \tag{131}$$

then

$$(\xi'_{N_{p-1}+\overline{\mu-1}s_p+1}\ \xi'_{N_{p-1}+\overline{\mu-1}s_p+2}\ \ldots.\ \xi'_{N_{p-1}+\mu s_p})$$
$$=\begin{pmatrix} x_{11}, & x_{12}, & \ldots, & x_{1s_p} \\ x_{21}, & x_{22}, & \ldots, & x_{2s_p} \\ \ldots & \ldots & \ldots & \ldots \\ x_{s_p1}, & x_{s_p2}, & \ldots, & x_{s_ps_p}\end{pmatrix}(\xi_{N_{p-1}+\overline{\mu-1}s_p+1}\ \xi_{N_{p-1}+\overline{\mu-1}+2}\ \ldots.\ \xi_{N_{p-1}+\overline{\mu-1}s_p})$$
$$(p=1,2,\ldots,\varkappa;\ \mu=1,2,\ldots,\sigma_p). \tag{132}$$

Let

$$K_{uv}^{(p)}=\Sigma_{h=1}^{m}\tau_{uvh}^{(p)}B_h', \qquad (p=1,2,\ldots,\varkappa;\ u,v=1,2,\ldots,s_p). \tag{133}$$

---

* Cf. pp. 343, 344 and 345.

Then the $m=\Sigma_{p=1}^{\varkappa} s_p^2$ $K$'s are linearly independent; and it follows from equation (117) that

$$K_{uw}^{(p)} K_{wv}^{(p)} = K_{uv}^{(p)}, \quad K_{uw}^{(p)} K_{w'v}^{(p)} = 0,$$
$$(p=1, 2, \ldots, \varkappa;\ u, v, w, w'=1, 2, \ldots, s_p;\ w' \neq w), \tag{134}$$
$$K_{uv}^{(p)} K_{u'v'}^{(q)} = 0,$$
$$(p, q=1, 2, \ldots, \varkappa;\ q \neq p;\ u, v=1, 2, \ldots, s_p;\ u', v'=1, 2, \ldots, s_q); \tag{135}$$

and, therefore,

$$m\, S_1 K_{uu}^{(p)} = s_p, \qquad (p=1, 2, \ldots, \varkappa;\ u, v=1, 2, \ldots, s_p). \tag{136}$$

For any system of scalars $y_1, y_2, \ldots, y_m$,

$$Y = \Sigma_{i=1}^{m} y_i B_i = \Sigma_{i=1}^{m} x_i B_i' = \Sigma_{p=1}^{\varkappa} \Sigma_{u=1}^{s_p} \Sigma_{v=1}^{s_p} x_{uv}^{(p)} K_{uv}^{(p)} \tag{137}$$

provided $x_1, x_2, \ldots, x_m$ are determined by the equations

$$y_i = \tau_{1i} x_1 + \tau_{2i} x_2 + \ldots + \tau_{mi} x_m, \qquad (i=1, 2, \ldots, m), \tag{138}$$

and $x_{uv}^{(p)}$ $(p=1, 2, \ldots, \varkappa;\ u, v=1, 2, \ldots, s_p)$ by (127). Moreover,

$$n\, \bar{S} K_{uu}^{(p)} = n\, \Sigma_{h=1}^{m} \tau_{uuh}^{(p)} \bar{S} B_h' = n\, \Sigma_{h=1}^{m} \tau_{uuh}^{(p)} \bar{S} A_h = n\, \bar{S} I_{uu}^{(p)} = \sigma_p,$$
$$(p=1, 2, \ldots, \varkappa;\ u=1, 2, \ldots, s_p). \tag{139}$$

Therefore, by what was shown in § 1, a linear substitution $\omega''$ with constant coefficients can be found such that

$$Y = \omega'' \Psi \omega''^{-1}, \tag{140}$$

where $\Psi$ is defined by equations (98); that is, if

$$(\xi_1', \xi_2', \ldots, \xi_n') = \omega''^{-1} Y \omega'' (\xi_1, \xi_2, \ldots, \xi_n), \tag{141}$$

the $\xi'$'s are defined by equations (132).

Whence if

$$\Omega = \omega'' \omega'^{-1}, \tag{142}$$

it follows that

$$Y = \omega'' \Psi \omega''^{-1} = \Omega \omega' \Psi \omega'^{-1} \Omega^{-1} = \Omega X \Omega^{-1}; \tag{143}$$

that is, $G$ and $H$ are similar. On the other hand, if

$$\begin{vmatrix} \bar{S} A_1 A_1, \ldots, \bar{S} A_1 A_m \\ \ldots\ldots\ldots\ldots \\ \bar{S} A_m A_1, \ldots, \bar{S} A_m A_m \end{vmatrix} = \begin{vmatrix} \bar{S} B_1' B_1', \ldots, \bar{S} B_1' B_m' \\ \ldots\ldots\ldots\ldots \\ \bar{S} B_m' B_1', \ldots, \bar{S} B_m' B_m' \end{vmatrix} = 0,$$

the groups $G$ and $H$ are not, in general, similar. Thus, let $n=4$; and let $C$ denote the linear homogeneous transformation

$$C(\xi_1, \xi_2, \xi_3, \xi_4) = \begin{pmatrix} 0, 1, 0, 0 \\ 0, 0, 1, 0 \\ 0, 0, 0, 1 \\ 0, 0, 0, 0 \end{pmatrix} (\xi_1, \xi_2, \xi_3, \xi_4).$$

Let

$$A_1 = 1, \; A_2 = C^2; \quad B_1 = 1, \; B_2 = C^3.$$

Then $A_1$ and $A_2$ are linearly independent, as are $B_1$ and $B_2$; and

$$A_1^2 = A_1, \; A_1 A_2 = A_2 = A_2 A_1, \; A_2^2 = 0,$$
$$B_1^2 = B_1, \; B_1 B_2 = B_2 = B_2 B_1, \; B_2^2 = 0.$$

Therefore, the number systems $(A_1, A_2)$ and $(B_1, B_2)$ have the same multiplication table. Moreover,

$$\bar{S} A_1 = 1 = \bar{S} B_1, \; \bar{S} A_2 = 0 = \bar{S} B_2.$$

But the group

$$(\xi_1', \xi_2', \xi_3', \xi_4') = \left( \begin{array}{cccc} x_1, & 0, & x_2, & 0 \\ 0, & x_1, & 0, & x_2 \\ 0, & 0, & x_1, & 0 \\ 0, & 0, & 0, & x_1 \end{array} \right)(\xi, \xi, \xi, \xi),$$

constituted by the aggregate of transformations $X = x_1 A_1 + x_2 A_2$, is not similar to the group

$$(\xi_1', \xi_2', \xi_3', \xi_4') = \left( \begin{array}{cccc} y_1, & 0, & 0, & y_2 \\ 0, & y_1, & 0, & 0 \\ 0, & 0, & y_1, & 0 \\ 0, & 0, & 0, & y_1 \end{array} \right)(\xi_1, \xi_2, \xi_3, \xi_4),$$

constituted by the aggregate of transformations $Y = y_1 B_1 + y_2 B_2$.

From what has been shown above we have now the following theorem:

THEOREM III. *If $A_1, A_2, \ldots, A_m$ and $B_1, B_2, \ldots, B_m$ are two systems of linearly independent linear homogeneous transformations in $n$ variables constituting the same hypercomplex number system, so that*

$$A_i A_j = \Sigma_{k=1}^{m} \gamma_{ijk} A_k, \; B_i B_j = \Sigma_{k=1}^{m} \gamma_{ijk} B_k, \qquad (i, j = 1, 2, \ldots, m),$$

*and if, moreover,*

$$\bar{S} A_i = \bar{S} B_i, \qquad (i = 1, 2, \ldots, m),$$

*then the groups $G$ and $H$, constituted by the aggregates of transformations $\Sigma_{i=1}^{m} x_i A_i$ and $\Sigma_{i=1}^{m} y_i B_i$, respectively, for all values of the $x$'s and the $y$'s are similar, provided*

$$\left| \begin{array}{cccc} \bar{S} A_1 A_1, & \bar{S} A_1 A_2, & \ldots, & \bar{S} A_1 A_m \\ \bar{S} A_2 A_1, & \bar{S} A_2 A_2, & \ldots, & \bar{S} A_2 A_m \\ \cdots & \cdots & \cdots & \cdots \\ \bar{S} A_m A_1, & \bar{S} A_m A_2, & \ldots, & \bar{S} A_m A_m \end{array} \right| = \left| \begin{array}{cccc} \bar{S} B_1 B_1, & \bar{S} B_1 B_2, & \ldots, & \bar{S} B_1 B_m \\ \bar{S} B_2 B_1, & \bar{S} B_2 B_2, & \ldots, & \bar{S} B_2 B_m \\ \cdots & \cdots & \cdots & \cdots \\ \bar{S} B_m B_1, & \bar{S} B_m B_2, & \ldots, & \bar{S} B_m B_m \end{array} \right| \neq 0,$$

*but not, in general, otherwise.*

## § 3. *Conditions Sufficient for Complete Reducibility with Respect to a Given Domain of Rationality.*

In this and the following sections of this paper, $R$ will denote a domain of rationality determined by a finite or infinite number of scalars; *e. g.*, $R$ may denote the domain of rational numbers, or the domain of all real numbers, or even the domain of all scalars real and imaginary. A linear homogeneous transformation,

$$(\xi_1', \xi_2', \ldots, \xi_n') = \Sigma_{\mu=1}^{n} \Sigma_{\nu=1}^{n} c_{\mu\nu} T_{\mu\nu}(\xi_1, \xi_2, \ldots, \xi_n)$$
$$= \left( \begin{vmatrix} c_{11}, & c_{12}, & \ldots, & c_{1n} \\ c_{21}, & c_{22}, & \ldots, & c_{2n} \\ \ldots & \ldots & \ldots & \ldots \\ c_{n1}, & c_{n2}, & \ldots, & c_{nn} \end{vmatrix} \xi_1, \xi_2, \ldots, \xi_n \right),$$

will be said *to belong to R* or to be *rational with respect to R* if its coefficients are rational with respect to $R$. I shall assume that the linearly independent linear homogeneous transformations $A_1, A_2, \ldots, A_m$ constituting a hypercomplex number system belong to $R$. Then the constants $\gamma_{ijk}$ $(i, j, k = 1, 2, \ldots, m)$ of multiplication of the number system $(A_1, A_2, \ldots, A_m)$ are rational with respect to $R$.* A number $X = \Sigma_{i=1}^{m} x_i A_i$ of this system is a transformation belonging to $R$ if, and only if, the $x$'s are rational with respect to $R$.† It follows from (5) that the aggregate of numbers of the system $(A_1, A_2, \ldots, A_m)$ that are transformations belonging to $R$ constitute a group. This group will be denoted in what follows by $\bar{G}$; and $G$ will, as in § 1 and § 2, denote the aggregate, constituting a group, of all numbers $\Sigma_{i=1}^{m} x_i A_i$ whatever of this system. The group $\bar{G}$ is obviously a subgroup of $G$. By what was stated, p. 337, it follows that the sum of two transformations of $\bar{G}$ is also a transformation of $\bar{G}$.

The group $\bar{G}$ is said to be *reducible with respect to R* if a linear substitution $\bar{\omega}$, one and the same for every transformation $X = \Sigma_{i=1}^{m} x_i A_i$ of $\bar{G}$ (that is, with constant coefficients), can be found such that

$$X = \Sigma_{i=1}^{m} x_i A_i = \bar{\omega} \begin{vmatrix} X_{11}, & X_{12} \\ 0, & X_{22} \end{vmatrix} \bar{\omega}^{-1},$$

where $X_{11}$, $X_{22}$ are square matrices of order $p$ and $n-p$ respectively $(1 < p < n)$,

---

* For the coefficients of the linear equations (4) are rational with respect to $R$; and these equations determine the $\gamma$'s uniquely, since, otherwise, the $A$'s are not linearly independent.

† Since the $A$'s belong to $R$, it follows, see equation (3), that $a_{\mu\nu}^{(i)}$ is rational with respect to $R$ for $i = 1, 2, \ldots, m$ and $\mu, \nu = 1, 2, \ldots, n$. Therefore, $\Sigma_{i=1}^{m} x_i a_{\mu\nu}^{(i)}$, for $\mu, \nu = 1, 2, \ldots, n$, is rational with respect to $R$, and thus $\Sigma_{i=1}^{m} x_i A_i$ belongs to $R$, if the $x$'s are rational with respect to $R$. Conversely, if $\Sigma_{i=1}^{m} x_i A_i$ belongs to $R$, in which case $\Sigma_{i=1}^{m} x_i a_{\mu\nu}^{(i)}$ is rational with respect to $R$ for $\mu, \nu = 1, 2, \ldots, n$, then the $x$'s are rational with respect to $R$. For the $x$'s are then determined by $n^2$ linear equations whose coefficients are rational with respect to $R$; and these equations determine the $x$'s uniquely, since, otherwise, the $A$'s are not linearly independent.

$X_{12}$ is a rectangular matrix with $p$ rows and $n-p$ columns, 0 denotes a rectangular matrix with $n-p$ rows and $p$ columns whose constituents are all zero, and the constituents of $X_{11}$, $X_{12}$ and $X_{22}$ are rational with respect to $R$. Otherwise, $\overline{G}$ is *irreducible with respect to R*. If $\overline{G}$ is reducible with respect to $R$, the constituents of $\overline{\omega}$ can be taken rational with respect to $R$, since the equations determining these constituents are linear with coefficients belonging to $R$.

The aggregate of transformations $X_{11}$ and the aggregate of transformations $X_{22}$, for all transformations $X$ of $\overline{G}$, constitute groups $\overline{G}^{(1)}$, $\overline{G}^{(2)}$, respectively, of linear homogeneous transformations belonging to $R$ in $p$ and $n-p$ variables, respectively. Either $\overline{G}^{(1)}$ or $\overline{G}^{(2)}$, or both, may be reducible with respect to $R$. Ultimately, if $\overline{G}$ is reducible with respect to $R$, a linear substitution $\omega$ with constant coefficients can be found (which can be so chosen that its coefficients shall lie in $R$) such that

$$X=\Sigma_{i=1}^{m} x_i A_i=\omega \begin{vmatrix} X_{11}, & X_{12}, & \ldots, & X_{1\kappa} \\ 0, & X_{22}, & \ldots, & X_{2\kappa} \\ \cdots & \cdots & \cdots & \cdots \\ 0, & 0, & \ldots, & X_{\kappa\kappa} \end{vmatrix} \omega^{-1} \tag{144}$$

for any transformation $X$ of $\overline{G}$, where the zeros denote rectangular matrices whose constituents are all zero, $X_{11}, X_{22}, \ldots, X_{\kappa\kappa}$ are square matrices, $X_{12}$, $\ldots, X_{\kappa-1,\kappa}$ rectangular matrices, whose constituents are rational with respect to $R$, and the aggregate of transformations $X_{uu}$ $(u=1, 2, \ldots, \varkappa)$ for all transformations $X$ of $\overline{G}$, constituting a group $\overline{G}_u$ irreducible with respect to $R$.

The group $\overline{G}$, if reducible with respect to $R$, is said to be *completely reducible with respect to R*, if $\omega$ can be so chosen that

$$X_{uv}=0, \qquad (u=1, 2, \ldots, \varkappa-1;\ v=u+1, u+2, \ldots, \varkappa);$$

otherwise *not completely reducible with respect to R*. The case in which $\overline{G}$ is irreducible with respect to $R$ may be regarded as comprised in the case of complete reducibility, namely, when $\varkappa=1$; and following Loewy, I shall regard complete reducibility as embracing irreducibility.*

Since the coefficients of the transformations $A_1, A_2, \ldots, A_m$ of $\overline{G}$ are rational with respect to $R$, it follows that the constituents of the determinant

$$\nabla \equiv \begin{vmatrix} \overline{S} A_1 A_1, & \overline{S} A_1 A_2, & \ldots, & \overline{S} A_1 A_m \\ \overline{S} A_2 A_1, & \overline{S} A_2 A_2, & \ldots, & \overline{S} A_2 A_m \\ \cdots & \cdots & \cdots & \cdots \\ \overline{S} A_m A_1, & \overline{S} A_m A_2, & \ldots, & \overline{S} A_m A_m \end{vmatrix} \tag{145}$$

are rational with respect to $R$; and, since the constants $\gamma_{ijk}$ $(i, j, k=1, 2, \ldots, m)$

* *Trans. Am. Math. Soc.*, Vol. VI (1905), p. 506.

of multiplication of the number system $(A_1, A_2, \ldots, A_m)$ are rational with respect to $R$, the constituents $S_1 A_i A_j$ and $S_2 A_i A_j$ $(i, j=1, 2, \ldots, m)$ of the determinants $\Delta_1$ and $\Delta_2$ are rational with respect to $R$. I shall now show that the group $\overline{G}$ is completely reducible with respect to $R$ if $\nabla \neq 0$, in which case $\Delta_1 = \Delta_2 \neq 0$.

For, let $\nabla \neq 0$, in which case the number system $(A_1, A_2, \ldots, A_m)$ contains no invariant nilpotent subsystem. Either $\overline{G}$ is irreducible with respect to $R$, in which case the theorem is true, or $\overline{G}$ is reducible with respect to $R$. Let it be assumed that $\overline{G}$ is reducible with respect to $R$; and let

$$A_i = \omega \begin{vmatrix} A_{11}^{(i)}, & A_{12}^{(i)}, & \ldots, & A_{1\kappa}^{(i)} \\ 0, & A_{22}^{(i)}, & \ldots, & A_{2\kappa}^{(i)} \\ \cdots & \cdots & \cdots & \cdots \\ 0, & 0, & \ldots, & A_{\kappa\kappa}^{(i)} \end{vmatrix} \omega^{-1}, \qquad (i=1, 2, \ldots, m), \tag{146}$$

where $\omega$ and $A_{uv}^{(i)}$ $(u=1, 2, \ldots, \kappa;\ v=u, u+1, \ldots, \kappa)$ belong to $R$. Then

$$X_{uv} = \Sigma_{i=1}^{m} x_i A_{uv}, \qquad (u=1, 2, \ldots, \kappa;\ v=u, u+1, \ldots, \kappa), \tag{147}$$

as will be seen on comparing with (144).

Let the order of the matrices $A_{uu}^{(1)}, A_{uu}^{(2)}, \ldots, A_{uu}^{(m)}$ $(u=1, 2, \ldots, \kappa)$ be $n_u$. Then

$$n = n_1 + n_2 + \ldots + n_\kappa. \tag{148}$$

If now, we denote by $\bar{S}^{(u)} A_{uu}^{(i)}$ for $1 \leq i \leq m$ and $1 \leq u \leq \kappa$ the sum of the constituents in the principal diagonal of $A_{uu}^{(i)}$ divided by $n_u$, then, by (25),

$$\bar{S} A_i = \bar{S} \omega^{-1} A_i \omega = \frac{1}{n} [n_1 \bar{S}^{(1)} A_{11}^{(i)} + n_2 \bar{S}^{(2)} A_{22}^{(i)} + \ldots + n_\kappa \bar{S}^{(\kappa)} A_{\kappa\kappa}^{(i)}], \qquad (i=1, 2, \ldots, m). \tag{149}$$

From (146) and (1), it follows that

$$\begin{vmatrix} A_{11}^{(i)} A_{11}^{(j)}, & \Sigma_{w=1}^{2} A_{1w}^{(i)} A_{w2}^{(j)}, & \ldots, & \Sigma_{w=1}^{\kappa} A_{1w}^{(i)} A_{w\kappa}^{(j)} \\ 0, & A_{22}^{(i)} A_{22}^{(j)}, & \ldots, & \Sigma_{w=2}^{\kappa} A_{2w}^{(i)} A_{w\kappa}^{(j)} \\ \cdots & \cdots & \cdots & \cdots \\ 0, & 0, & \ldots, & A_{\kappa\kappa}^{(i)} A_{\kappa\kappa}^{(j)} \end{vmatrix}$$

$$= \omega^{-1} A_i \omega \cdot \omega^{-1} A_j \omega = \omega^{-1} A_i A_j \omega = \Sigma_{k=1}^{m} \gamma_{ijk} \omega^{-1} A_k \omega$$

$$= \begin{vmatrix} \Sigma_{k=1}^{m} \gamma_{ijk} A_{11}^{(k)}, & \Sigma_{k=1}^{m} \gamma_{ijk} A_{12}^{(k)}, & \ldots, & \Sigma_{k=1}^{m} \gamma_{ijk} A_{1\kappa}^{(k)} \\ 0, & \Sigma_{k=1}^{m} \gamma_{ijk} A_{22}^{(k)}, & \ldots, & \Sigma_{k=1}^{m} \gamma_{ijk} A_{2\kappa}^{(k)} \\ \cdots & \cdots & \cdots & \cdots \\ 0, & 0, & \ldots, & \Sigma_{k=1}^{m} \gamma_{ijk} A_{\kappa\kappa}^{(k)} \end{vmatrix} \qquad (i, j=1, 2, \ldots, m); \tag{150}$$

and, therefore,

$$\Sigma_{w=u}^{v} A_{uw}^{(i)} A_{wv}^{(j)} = \Sigma_{k=1}^{m} \gamma_{ijk} A_{uv}^{(k)}, \quad (u=1, 2, \ldots, \kappa;\ v=u, u+1, \ldots, \kappa). \tag{151}$$

In particular,

$$A^{(i)}_{uu} A^{(j)}_{uu} = \Sigma^m_{k=1} \gamma_{ijk} A^{(k)}_{uu}, \qquad (u=1, 2, \ldots, \varkappa). \tag{152}$$

Let now

$$B_i = \omega \begin{vmatrix} A^{(i)}_{11}, & 0, & \ldots, & 0 \\ 0, & A^{(i)}_{22}, & \ldots, & 0 \\ \ldots & \ldots & \ldots & \ldots \\ 0, & 0, & \ldots, & A^{(i)}_{\varkappa\varkappa} \end{vmatrix} \omega^{-1}, \qquad (i=1, 2, \ldots, m). \tag{153}$$

The linear homogeneous transformations $B_1, B_2, \ldots, B_m$ then belong to $R$; and, by (152),

$$B_i B_j = \omega \begin{vmatrix} A^{(i)}_{11} A^{(j)}_{11}, & 0, & \ldots, & 0 \\ 0, & A^{(i)}_{22} A^{(j)}_{22}, & \ldots, & 0 \\ \ldots & \ldots & \ldots & \ldots \\ 0, & 0, & \ldots, & A^{(i)}_{\varkappa\varkappa} A^{(j)}_{\varkappa\varkappa} \end{vmatrix} \omega^{-1}$$

$$= \omega \begin{vmatrix} \Sigma^m_{i=1} \gamma_{ijk} A^{(k)}_{11}, & 0, & \ldots, & 0 \\ 0, & \Sigma^m_{i=1} \gamma_{ijk} A^{(k)}_{22}, & \ldots, & 0 \\ \ldots & \ldots & \ldots & \ldots \\ 0, & 0, & \ldots, & \Sigma^m_{i=1} \gamma_{ijk} A^{(k)}_{\varkappa\varkappa} \end{vmatrix} \omega^{-1}$$

$$= \Sigma^m_{k=1} \gamma_{ijk} B_k, \qquad (i, j=1, 2, \ldots, m). \tag{154}$$

Further,

$$\bar{S} B_i = \bar{S} \omega^{-1} B_i \omega = \frac{1}{n} [n_1 \bar{S}^{(1)} A^{(i)}_{11} + n_2 \bar{S}^{(2)} A^{(i)}_{22} + \ldots + n_\varkappa \bar{S}^{(\varkappa)} A^{(i)}_{\varkappa\varkappa}] = \bar{S} A_i,$$
$$(i=1, 2, \ldots, m); \tag{155}$$

and, therefore,

$$\bar{S} B_i B_j = \Sigma^m_{i=1} \gamma_{ijk} \bar{S} B_k = \Sigma^m_{i=1} \gamma_{ijk} \bar{S} A_k = \bar{S} A_i A_j, \qquad (i=1, 2, \ldots, m). \tag{156}$$

Whence it follows that

$$\begin{vmatrix} \bar{S} B_1 B_1, & \ldots, & \bar{S} B_1 B_m \\ \ldots & \ldots & \ldots \\ \bar{S} B_m B_1, & \ldots, & \bar{S} B_m B_m \end{vmatrix} = \begin{vmatrix} \bar{S} A_1 A_1, & \ldots, & \bar{S} A_1 A_m \\ \ldots & \ldots & \ldots \\ \bar{S} A_m A_1, & \ldots, & \bar{S} A_m A_m \end{vmatrix} \neq 0. \tag{157}$$

Finally, the matrices $B_1, B_2, \ldots, B_m$ are linearly independent. For, otherwise

$$c_1 B_1 + c_2 B_2 + \ldots + c_m B_m = 0,$$

where the $c$'s are not all zero, in which case,

$$(c_1 B_1 + c_2 B_2 + \ldots + c_m B_m) B_i = 0, \qquad (i=1, 2, \ldots, m).$$

Therefore,

$$c_1 \bar{S} A_1 A_i + c_2 \bar{S} A_2 A_i + \ldots + c_m \bar{S} A_m A_i$$
$$= c_1 \bar{S} B_1 B_i + c_2 \bar{S} B_2 B_i + \ldots + c_m \bar{S} B_m B_i$$
$$= \bar{S}(c_1 B_1 + c_2 B_2 + \ldots + c_m B_m) B_i = 0, \qquad (i=1, 2, \ldots, m);$$

and thus $\nabla = 0$, which is contrary to supposition.

We have now shown that the two systems of linearly independent linear homogeneous transformations $A_1, A_2, \ldots, A_m$ and $B_1, B_2, \ldots, B_m$ in $n$ variables constitute number systems having the same constants of multiplication and containing no invariant nilpotent subsystems; and that $\bar{S} A_i = \bar{S} B_i$ for $i = 1, 2, \ldots, m$. Therefore, by *Theorem III*, a linear substitution $\omega'$ with constant coefficients can be found such that

$$\Sigma_{i=1}^{m} x_i A_i = \omega' \Sigma_{i=1}^{m} x_i B_i \omega'^{-1} \tag{158}$$

for every system of values of the $m$ scalars $x_1, x_2, \ldots, x_m$. Whence, if

$$\Omega = \omega' \omega, \tag{159}$$

it follows that

$$\Sigma_{i=1}^{m} x_i A = \Omega \begin{vmatrix} \Sigma_{i=1}^{m} x_i A_{11}^{(i)}, & 0, & \ldots, & 0 \\ 0, & \Sigma_{i=1}^{m} x_i A_{22}^{(i)}, & \ldots, & 0 \\ \ldots & \ldots & \ldots & \ldots \\ 0, & 0, & \ldots, & \Sigma_{i=1}^{m} x_i A_{\kappa\kappa}^{(i)} \end{vmatrix} \Omega^{-1}. \tag{160}$$

*A fortiori*, this is true for all transformations $\Sigma_{i=1}^{m} x_i A_i$ of $G$ for which the $x$'s are rational with respect to $R$, that is for all transformations of the subgroup $\bar{G}$ of $G$. Whence it follows that $\bar{G}$ is completely reducible with respect to $R$. We have, therefore, the following theorem:

THEOREM IV. *If $A_1, A_2, \ldots, A_m$ are $m$ linearly independent linear homogeneous transformations in $n$ variables whose coefficients are rational with respect to an arbitrarily given domain of rationality $R$, and if $A_1, A_2, \ldots, A_m$ constitute a hypercomplex number system, so that*

$$A_i A_j = \Sigma_{i=1}^{m} \gamma_{ijk} A_k, \qquad (i, j = 1, 2, \ldots, m),$$

*the aggregate of transformations $\Sigma_{i=1}^{m} x_i A_i$, for all values of the $x$'s rational with respect to $R$, constitute a group which is completely reducible with respect to $R$ provided*

$$\begin{vmatrix} \bar{S} A_1 A_1, & \bar{S} A_1 A_2, & \ldots, & \bar{S} A_1 A_m \\ \bar{S} A_2 A_1, & \bar{S} A_2 A_2, & \ldots, & \bar{S} A_2 A_m \\ \ldots & \ldots & \ldots & \ldots \\ \bar{S} A_m A_1, & \bar{S}_m A_2 A, & \ldots, & \bar{S} A_m A_m \end{vmatrix} \neq 0,$$

*or, what is the same thing, provided*

$$\begin{vmatrix} S_1 A_1 A_1, & S_1 A_1 A_2, & \ldots, & S_1 A_1 A_m \\ S_1 A_2 A_1, & S_1 A_2 A_2, & \ldots, & S_1 A_2 A_m \\ \ldots & \ldots & \ldots & \ldots \\ S_1 A_m A_1, & S_1 A_m A_2, & \ldots, & S_1 A_m A_m \end{vmatrix} = \begin{vmatrix} S_2 A_1 A_1, & S_2 A_1 A_2, & \ldots, & S_2 A_1 A_m \\ S_2 A_2 A_1, & S_2 A_2 A_2, & \ldots, & S_2 A_2 A_m \\ \ldots & \ldots & \ldots & \ldots \\ S_2 A_m A_1, & S_2 A_m A_2, & \ldots, & S_2 A_m A_m \end{vmatrix} \neq 0,$$

## § 4. *Conditions Necessary and Sufficient for Complete Reducibility with Respect to a Given Domain of Rationality.*

Employing the notation of the preceding section, let the $m$ linearly independent linear homogeneous transformations $A_1, A_2, \ldots, A_m$, constituting a hypercomplex number system, belong to $R$. I shall now assume that $\nabla=0$; and I shall show that then $\overline{G}$ is reducible, but not completely reducible with respect to $R$.

Let the nullity of $\nabla$ be $m_0$ $(1\leqq m_0\leqq m)$. There is then a system of just $m_0$ linearly independent numbers,

$$N_u=\alpha_{u1}A_1+\alpha_{u2}A_2+\ldots+\alpha_{um}A_m, \qquad (u=1, 2, \ldots, m_0), \tag{161}$$

of the number system $(A_1, A_2, \ldots, A_m)$ satisfying the equations

$$\begin{aligned}&\overline{S}(x_1A_1+x_2A_2+\ldots+x_mA_m)A_i\\ &=x_1\overline{S}A_1A_i+x_2\overline{S}A_2A_i+\ldots+x_m\overline{S}A_mA_i=0, \qquad (i=1, 2, \ldots, m),\end{aligned} \tag{162}$$

and constituting a maximal invariant nilpotent subsystem of $(A_1, A_2, \ldots, A_m)$.* Since, as stated above, p. 359, the coefficients, $SA_iA_j$ $(i, j=1, 2, \ldots, m)$, of the linear equations determining the $N$'s are rational with respect to $R$, we may so choose $N_1, N_2, \ldots, N_{m_0}$ that they shall belong to $R$. Let the $N$'s be so chosen. Since they constitute an invariant subsystem of $(A_1, A_2, \ldots, A_m)$ it follows that

$$N_uN_v=\Sigma_{w=1}^{m_0}\gamma_{uvw}^{(0)}N_w, \qquad (u, v=1, 2, \ldots, m_0), \tag{163}$$

$$A_iN_u=\Sigma_{v=1}^{m_0}\gamma_{iuv}^{(1)}N_v \qquad (i=1, 2, \ldots, m;\ u=1, 2, \ldots, m_0), \tag{164}$$

where, since $A_1, A_2, \ldots, A_m$ and $N_1, N_2, \ldots, N_{m_0}$ belong to $R$, the $\gamma$'s are rational with respect to $R$.

The linear homogeneous transformations $1, N_1, N_2, \ldots, N_{m_0}$ are linearly independent † and by (163) constitute a number system (of which 1 is the modulus) whose constants of multiplication are rational with respect to $R$; and thus, since $N_1, N_2, \ldots, N_{m_0}$ belong to $R$, the aggregate of transformations

$$Y=y_0+\Sigma_{u=1}^{m_0}y_uN_u, \tag{165}$$

for any systems of scalars $y_0, y_1, \ldots, y_{m_0}$ rational with respect to $R$, constitute a group $\overline{H}$ of transformations belonging to $R$. Since $N_1, N_2, \ldots, N_{m_0}$

---

* See p. 343.

† For if

$$c_01+\Sigma_i^{m_0}c_iN_i=0,$$

then

$$c_0=c_0\overline{S}1+\Sigma_{i=1}^{m_0}c_i\overline{S}N_i=\overline{S}(c_01+\Sigma_{i=1}^{m_0}c_iN_i)=0.$$

Therefore, $\Sigma_{i=1}^{m_0}c_iN_i=0$; and thus

$$c_1=c_2=\ldots c_{m_0}=0.$$

are transformations of $\overline{G}$, the group $\overline{H}$ is a subgroup of $\overline{G}$. From *Theorem VI*, proved in the next section, it follows that a linear function,

$$N = c_1 N_1 + c_2 N_2 + \ldots . c_{m_0} N_{m_0} \neq 0, \tag{166}$$

of the $N$'s with coefficients rational with respect to $R$, can be found such that

$$N_u N = 0, \qquad (u = 1, 2, \ldots, m_0); \tag{167}$$

and that a linear substitution $\omega$ can be found whose coefficients are rational with respect to $R$ such that

$$N_u = \omega \begin{vmatrix} 0, & b_{12}^{(u)}, & \ldots, & b_{1n}^{(u)} \\ 0, & 0, & \ldots, & b_{2n}^{(u)} \\ \ldots & \ldots & \ldots & \ldots \\ 0, & 0, & \ldots, & 0 \end{vmatrix} \omega^{-1}, \qquad (u = 1, 2, \ldots, m_0), \tag{168}$$

where the $b$'s are all rational with respect to $R$, in which case

$$N = \Sigma_{u=1}^{m_0} c_u N_u = \omega \begin{vmatrix} 0, & b_{12}, & \ldots, & b_{1n} \\ 0, & 0, & \ldots, & b_{2n} \\ \ldots & \ldots & \ldots & \ldots \\ 0, & 0, & \ldots, & 0 \end{vmatrix} \omega^{-1}, \tag{169}$$

where $b_{\mu\nu}$ $(\mu = 1, 2, \ldots, n-1;\ \nu = \mu+1, \mu+2, \ldots, n)$ is rational with respect to $R$. By the aid of this theorem, we may now show that $\overline{G}$ is reducible with respect to $R$.

Thus let $X = \Sigma_{i=1}^{m} x_i A_i$, where $x_1, x_2, \ldots, x_m$ are rational with respect to $R$, be any transformation of $\overline{G}$; and let

$$\omega^{-1} X \omega = \begin{vmatrix} \beta_{11}(x), & \beta_{12}(x), & \ldots, & \beta_{1n}(x) \\ \beta_{21}(x), & \beta_{22}(x), & \ldots, & \beta_{2n}(x) \\ \ldots & \ldots & \ldots & \ldots \\ \beta_{n1}(x), & \beta_{n2}(x), & \ldots, & \beta_{nn}(x) \end{vmatrix}. \tag{170}$$

Then $\beta_{11}(x), \beta_{12}(x), \ldots, \beta_{nn}(x)$ are rational with respect to $R$.

By equation (164),

$$\begin{aligned} XN &= \Sigma_{i=1}^{m} x_i A_i \cdot \Sigma_{u=1}^{m_0} c_u N_u \\ &= \Sigma_{i=1}^{m} \Sigma_{u=1}^{m_0} x_i c_u A_i N_u \\ &= \Sigma_{i=1}^{m} \Sigma_{u=1}^{m_0} \Sigma_{v=1}^{m_0} x_i c_u \gamma_{iuv} N_v = \Sigma_{v=1}^{m_0} x'_v N_v, \end{aligned} \tag{171}$$

provided

$$x'_v = \Sigma_{i=1}^{m} \Sigma_{u=1}^{m_0} x_i c_u \gamma_{iuv}, \qquad (v = 1, 2, \ldots, m_0); \tag{172}$$

and, therefore,

$$\begin{aligned} \omega^{-1} X \omega \cdot \omega^{-1} N \omega &= \omega^{-1} XN \omega \\ &= \omega^{-1} \Sigma_{v=1}^{m_0} x'_v N_v \omega \\ &= \Sigma_{v=1}^{m_0} x'_v \omega^{-1} N_v \omega. \end{aligned} \tag{173}$$

Whence it follows that every constituent of $\omega^{-1} X \omega \cdot \omega^{-1} N \omega$ in and below the principal diagonal is equal to zero, since this is true for $\Sigma_{v=1}^{m_0} x'_v \omega^{-1} N_v \omega$. That is,

$$\beta_{\mu 1}(x)\, b_{1\nu} + \ldots + \beta_{\mu\lambda}(x)\, b_{\lambda\nu} + \ldots + \beta_{\mu,\nu-1}(x)\, b_{\nu-1,\nu} = 0,$$
$$(\mu = 2, 3, \ldots, n;\ \nu = 2, 3, \ldots, \mu). \qquad (174)$$

Not every constituent of $\omega^{-1} N \omega$ is zero, since otherwise $N=0$. We may, therefore, assume that $b_{\lambda\nu'} \neq 0$ for $1 < \nu' \leqq n$ and $\lambda < \nu'$.

Therefore, not every coefficient is zero in the linear relation we obtain from (174) by putting $\nu = \nu'$. That is to say, the same linear relation, rational with respect to $R$, obtains between the first $\nu' - 1$ constituents of the $\mu$-th row ($\mu \geqq \nu'$) of every transformation of the group of linear homogeneous transformations $\omega^{-1} X \omega$ belonging to $R$. But then by Maschke's theorem, this group is reducible with respect to $R$.* Whence it follows that $\overline{G}$ is reducible with respect to $R$. That is to say, *if $A_1, A_2, \ldots, A_m$ are $m$ linearly independent linear homogeneous transformations which belong to $R$ and constitute a hypercomplex number system containing an invariant nilpotent subsystem, the group of transformations belonging to $R$ constituted by the aggregate of transformations $\Sigma_{i=1}^m x_i A_i$, for all values of the $x$'s rational with respect to $R$, is reducible with respect to $R$.*

If possible, let $\overline{G}$ be completely reducible with respect to $R$. Then, employing the notation of § 3, for any transformation of $\overline{G}$ (that is, for $x_1$, $x_2, \ldots, x_m$ rational with respect to $R$), we have

$$X = \Sigma_{i=1}^m x_i A_i = \omega \begin{vmatrix} \Sigma_{i=1}^m x_i A_{11}^{(i)}, & 0, & \ldots, & 0 \\ 0, & \Sigma_{i=1}^m x_i A_{22}^{(i)}, & \ldots, & 0 \\ \ldots & \ldots & \ldots & \ldots \\ 0, & 0, & \ldots, & \Sigma_{i=1}^m x_i A_{\kappa\kappa}^{(i)} \end{vmatrix} \omega^{-1}, \qquad (175)$$

where $\omega$ and the matrices $A_{uu}^{(i)}$ for $i = 1, 2, \ldots, m$ and $u = 1, 2, \ldots, \kappa$ belong to $R$, and the group $\overline{G}_u$ ($1 \leqq u \leqq \kappa$) of linear homogeneous transformations $\Sigma_{i=1}^m x_i A_{uu}^{(i)}$ for values of the $x$'s rational with respect to $R$ is irreducible with respect to $R$.

---

* *Math. Ann.*, Vol. LII (1889). In this paper Maschke does not refer explicitly to a domain of rationality. But, in the proof of the theorem given therein, p. 366, he shows that, if a non-diagonal constituent is zero throughout in the matrix of a group of linear homogeneous transformations, the group is reducible with respect to the domain of rationality in which the coefficients lie.

Let now $X^{(1)}$ denote the matrix or linear homogeneous transformation $\omega^{-1} X \omega$ whose constituent in the $u$-th row and $v$-th column is $\beta_{uv}(x)$ ($u, v = 1, 2, \ldots, n$); and let $\overline{\omega}$ denote the linear substitution defined by the equations

$$\xi'_i = \xi_i + b_{i\nu'} \xi_\lambda, \quad \xi'_\lambda = b_{\lambda\nu'} \xi_\lambda, \quad \xi'_{\nu'+j} = \xi_{\nu'+j},$$
$$(i = 1, 2, \ldots, \lambda - 1, \lambda + 1, \ldots, \nu' - 1;\ j = 0, 1, \ldots, n - \nu').$$

Then $\overline{\omega}$ belongs to $R$; and each of the group of transformations $\overline{\omega}^{-1} X^{(1)} \overline{\omega}$ (for all transformations $X$ of $G$) belongs to $R$. Let $\mu \geqq \nu'$. Then, since $\Sigma_{i=1}^{\nu'-1} b_{i\nu'} \beta_{\mu i}(x) = 0$, it follows that the constituent in the $\mu$-th row and $\lambda$-th column of the matrix of $\overline{\omega}^{-1} X^{(1)} \overline{\omega}$ is zero throughout. Therefore, by Maschke's theorem, this group is reducible with respect to $R$; and thus the group of transformations $X^{(1)} = \omega^{-1} X \omega$ is reducible with respect to $R$.

Let $1 \leq u \leq x$; and let $m_u (1 \leq m_u \leq m)$ be the number of the linear homogeneous transformations $A_{uu}^{(1)}, A_{uu}^{(2)}, \ldots, A_{uu}^{(m)}$ that are linearly independent. Without loss of generality, we may assume that $A_{uu}^{(1)}, A_{uu}^{(2)}, \ldots, A_{uu}^{(m_u)}$ are linearly independent; and then

$$A_{uu}^{(m_u+h)} = c_{h1}^{(u)} A_{uu}^{(1)} + c_{h1}^{(u)} A_{uu}^{(2)} + \ldots + c_{hm_u}^{(u)} A_{uu}^{(m_u)}, \quad (h=1, 2, \ldots, m-m_u). \quad (176)$$

where the $c$'s belong to $R$. Wherefore, by (152),

$$\begin{aligned} A_{uu}^{(i)} A_{uu}^{(j)} &= \Sigma_{k=1}^{m} \gamma_{ijk} A_{uu}^{(k)} \\ &= \Sigma_{k=1}^{m_u} \gamma_{ijk} A_{uu}^{(k)} + \Sigma_{h=1}^{m-m_u} \gamma_{i,j,m_u+h} \Sigma_{k=1}^{m_u} c_{hk}^{(u)} A_{uu}^{(k)} \\ &= \Sigma_{k=1}^{m_u} \bar{\gamma}_{ijk}^{(u)} A_{uu}^{(k)}, \qquad (i, j = 1, 2, \ldots, m_u), \end{aligned} \quad (177)$$

provided

$$\bar{\gamma}_{ijk}^{(u)} = \gamma_{ijk} + \Sigma_{h=1}^{m-m_u} \gamma_{i,j,m_u+h} c_{hk}^{(u)}, \qquad (i, j, k = 1, 2, \ldots, m_u). \quad (178)$$

Whence it follows that the linear homogeneous transformations $A_{uu}^{(1)}, A_{uu}^{(2)}, \ldots,$ $A_{uu}^{(m_u)}$ belonging to $R$ constitute a number system whose constants $\bar{\gamma}_{ijk}^{(u)}$ $(i, j, k = 1, 2, \ldots, m_u)$ of multiplication are rational with respect to $R$. From (176) it follows that the group $\bar{G}_u$ is constituted by the aggregate of numbers $\Sigma_{i=1}^{m_u} x_i A_{uu}^{(i)}$ of this number system for values of the $x$'s rational with respect to $R$. For this number system the determinant

$$\nabla^{(u)} = \begin{vmatrix} \bar{S}^{(u)} A_{uu}^{(1)} A_{uu}^{(1)}, & \ldots, & \bar{S}^{(u)} A_{uu}^{(1)} A_{uu}^{(m_u)} \\ \ldots & \ldots & \ldots \\ \bar{S}^{(u)} A_{uu}^{(m_u)} A_{uu}^{(1)}, & \ldots, & \bar{S}^{(u)} A_{uu}^{(m_u)} A_{uu}^{(m_u)} \end{vmatrix} \neq 0; \quad (179)$$

that is, this number system contains no invariant nilpotent subsystem. For otherwise, by the theorem just proved, it follows that $G_u$ is reducible with respect to $R$. Therefore, there is no number $\Sigma_{i=1}^{m_u} b_i A_{uu}^{(i)}$ of this number system whose product with every number of this system is nilpotent. *A fortiori*, there is no transformation $\Sigma_{i=1}^{m} b_i A_{uu}^{(i)}$ of $\bar{G}_u$ whose product with $\Sigma_{i=1}^{m} x_i A_{uu}^{(i)}$ is nilpotent for every system of values of the $x$'s.

Let now

$$N = \Sigma_{u=1}^{m_0} c_u N_u = \Sigma_{u=1}^{m_0} \Sigma_{i=1}^{m} c_u \alpha_{ui} A_i = \Sigma_{i=1}^{m} b_i A_i. \quad (180)$$

Then, by (175),

$$N = \omega \begin{vmatrix} \Sigma_{i=1}^{m} b_i A_{11}^{(i)}, & 0, & \ldots, & 0 \\ 0, & \Sigma_{i=1}^{m} b_i A_{22}^{(i)}, & \ldots, & 0 \\ \ldots & \ldots & \ldots & \ldots \\ 0, & 0, & \ldots, & \Sigma_{i=1}^{m} b_i A_{\kappa\kappa}^{(i)} \end{vmatrix} \omega^{-1}. \quad (181)$$

Since, by (163),

$$\begin{aligned} XN &= \Sigma_{i=1}^{m} x_i A_i \cdot \Sigma_{u=1}^{m_0} c_u N_u \\ &= \Sigma_{i=1}^{m} \Sigma_{u=1}^{m_0} x_i c_u A_i N_u \\ &= \Sigma_{i=1}^{m} \Sigma_{u=1}^{m_0} \Sigma_{v=1}^{m_0} x_i c_u \gamma_{iuv}^{(1)} N_v \end{aligned} \quad (182)$$

is expressible linearly in terms of the units of the nilpotent subsystem $(N_1, N_2, \ldots, N_{m_0})$, it is nilpotent; and thus $\omega^{-1}X\omega \cdot \omega^{-1}N\omega$ is nilpotent for any system of scalars $x_1, x_2, \ldots, x_m$. Whence, for each integer $u$ from 1 to $\varkappa$, it follows that some power of the product of $\Sigma_{i=1}^m x_i A_{uu}^{(i)}$ and $\Sigma_{i=1}^m b_i A_{uu}^{(i)}$ is zero for any system of scalars $x_1, x_2, \ldots, x_m$; and thus, either the product of $\Sigma_{i=1}^m x_i A_{uu}^{(i)}$ and $\Sigma_{i=1}^m b_i A_{uu}^{(i)}$ is nilpotent for every system of scalars $x_1, x_2, \ldots, x_m$, which as shown above is impossible, or $\Sigma_{i=1}^m b_i A_{uu}^{(i)}=0$. But it can not happen that $\Sigma_{i=1}^m b_i A_{uu}^{(i)}=0$ for $u=1, 2, \ldots, \varkappa$; for then, by (181), $N=0$, which is contrary to supposition. Wherefore, the hypothesis is absurd, that $\overline{G}$ is completely reducible with respect to $R$.

From the theorem on p. 362 and what has just been proved, we obtain the following theorem:

THEOREM V. *If $A_1, A_2, \ldots, A_m$ are $m$ linearly independent linear homogeneous transformations in $n$ variables that belong to an arbitrarily given domain of rationality $R$ and constitute a hypercomplex number system, so that*

$$A_i A_j=\Sigma_{k=1}^m \gamma_{ijk} A_k, \qquad (i, j=1, 2, \ldots, m),$$

*the group $\overline{G}$ of transformations*

$$X=\Sigma_{i=1}^m x_i A_i$$

*for all values of the $x$'s rational with respect to $R$, is completely reducible with respect to $R$ if, and only if,*

$$\begin{vmatrix} SA_1A_1, & SA_1A_2, & \ldots, & SA_1A_m \\ SA_2A_1, & SA_2A_2, & \ldots, & SA_2A_m \\ \ldots & \ldots & \ldots & \ldots \\ SA_mA_1, & SA_mA_2, & \ldots, & SA_mA_m \end{vmatrix} \neq 0;$$

*and, therefore, if and only if,*

$$\begin{vmatrix} S_1A_1A_1, & S_1A_1A_2, & \ldots, & S_1A_1A_m \\ S_1A_2A_1, & S_1A_2A_2, & \ldots, & S_1A_2A_m \\ \ldots & \ldots & \ldots & \ldots \\ S_1A_mA_1, & S_1A_mA_2, & \ldots, & S_1A_mA_m \end{vmatrix} = \begin{vmatrix} S_2A_1A_1, & S_2A_1A_2, & \ldots, & S_2A_1A_m \\ S_2A_2A_1, & S_2A_2A_2, & \ldots, & S_2A_2A_m \\ \ldots & \ldots & \ldots & \ldots \\ S_2A_mA_1, & S_2A_mA_2, & \ldots, & S_2A_mA_m \end{vmatrix} \neq 0.$$

## § 5.

As above, let the $m(m>1)$ linearly independent matrices $A_1, A_2, \ldots, A_m$ constitute a hypercomplex number system, so that

$$A_i A_j=\Sigma_{k=1}^m \gamma_{ijk} A_k, \qquad (i=1, 2, \ldots, m); \qquad (183)$$

and, as in § 4, let $A_1, A_2, \ldots, A_m$ belong to $R$. Then the constants $\gamma_{ijk}$ $(i, j, k=1, 2, \ldots, m)$ of multiplication of the number system $(A_1, A_2, \ldots, A_m)$ belong to $R$. I shall now assume that this number system contains but one idempotent number, the modulus 1.

In the *Transactions of the American Mathematical Society*, Vol. XXXIV (1904), pp. 529, 540, and 547, I have shown that in any number system, as $(A_1, A_2, \ldots, A_m)$, that contains a modulus but no other idempotent number, we may introduce new units—namely, the modulus,

$$1 = a_1^{(0)} A_1 + a_2^{(0)} A_2 + \ldots + a_m^{(0)} A_m, \tag{184}$$

and $m-1$ nilpotent numbers,

$$N_i = \tau_{1i} A_1 + \tau_{2i} A_2 + \ldots + \tau_{mi} A_m, \qquad (i = 1, 2, \ldots, m-1), \tag{185}$$

constituting an invariant nilpotent subsystem by a transformation rational with respect to the domain of rationality of the constants of multiplication of the system; and that there is a linear function,

$$N = c_1 N_1 + c_2 N_2 + \ldots + c_{m-1} N_{m-1} \neq 0, \tag{186}$$

with coefficients rational with respect to this domain, such that

$$N_i N = 0 = N N_i, \qquad (i = 1, 2, \ldots, m). \tag{187}$$

For such a transformation of the number system $(A_1, A_2, \ldots, A_m)$, the coefficients $a_i^{(0)}$, $\tau_{ij}$, $c_j$ for $i = 1, 2, \ldots, m$ and $j = 1, 2, \ldots, m-1$, being rational with respect to the domain of rationality of the constants of multiplication of this system, are rational with respect to $R$. And, therefore, since the $A$'s belong to $R$, it follows that $N$ and the new units $1, N_1, N_2, \ldots, N_{m-1}$ belong to $R$. Whence it follows that the group $\overline{G}$ of transformations belonging to $R$ that are numbers of the system $(A_1, A_2, \ldots, A_m)$ is constituted by the aggregate of transformations

$$Y = y_0 + \Sigma_{i=1}^{m-1} y_i N_i \tag{188}$$

for $y_0, y_1, \ldots, y_{m-1}$ rational with respect to $R$. Let

$$N_i N_j = \Sigma_{k=1}^{m-1} \gamma_{ijk}^{(0)} N_k, \qquad (i, j = 1, 2, \ldots, m-1). \tag{189}$$

Then $\gamma_{ijk}$, for $i, j, k = 1, 2, \ldots, m-1$, is rational with respect to $R$.

Since $N$ belongs to $R$, and since $N$ is nilpotent, and thus every root of its characteristic equation is zero, a linear substitution $\overline{\omega}$ belonging to $R$ can be found such that

$$N = \overline{\omega} \begin{vmatrix} 0, & b_{12}, & \ldots, & b_{1n} \\ 0, & 0, & \ldots, & b_{2n} \\ \cdots & \cdots & \cdots & \cdots \\ 0, & 0, & \ldots, & 0 \end{vmatrix} \overline{\omega}^{-1}, \tag{190}$$

where the $b$'s are rational with respect to $R$. Let

$$Y = y_0 + \Sigma_{i=1}^{m-1} y_i N_i, \tag{191}$$

where the $y$'s are rational with respect to $R$, be any transformation of $\bar{G}$; and let

$$\bar{\omega}^{-1} Y \bar{\omega} = y_0 + \Sigma_{i=1}^{m-1} y_i \bar{\omega}^{-1} N_i \bar{\omega} = \begin{vmatrix} \beta_{11}(y), & \beta_{12}(y), & \ldots, & \beta_{1n}(y) \\ \beta_{21}(y), & \beta_{22}(y), & \ldots, & \beta_{2n}(y) \\ \ldots & \ldots & \ldots & \ldots \\ \beta_{n1}(y), & \beta_{n2}(y), & \ldots, & \beta_{nn}(y) \end{vmatrix}. \qquad (192)$$

Then $\beta_{\mu\nu}(y)$, for $\mu, \nu = 1, 2, \ldots, n$, is rational with respect to $R$. Since, by (187),

$$\begin{aligned} \bar{\omega}^{-1} Y \bar{\omega} \cdot \bar{\omega}^{-1} N \omega &= \bar{\omega}^{-1} Y N \bar{\omega} \\ &= \bar{\omega}^{-1} (y_0 + \Sigma_{i=1}^{m-1} y_i N_i) N \bar{\omega} \\ &= y_0 \bar{\omega}^{-1} N \bar{\omega}, \end{aligned} \qquad (193)$$

it follows that the constituent in the $\mu$-th row and $\nu$-th column of $\bar{\omega}^{-1} Y \bar{\omega} \cdot \bar{\omega}^{-1} N \bar{\omega}$ is zero, provided $\nu \leqq \mu$; for this is true of $\bar{\omega}^{-1} N \bar{\omega}$. Therefore,

$$\beta_{\mu 1}(y) b_{1\nu} + \ldots + \beta_{\mu\lambda}(y) b_{\lambda\nu} + \ldots + \beta_{\mu, \nu-1}(y) b_{\nu-1, \nu} = 0, \quad (\mu = 2, 3, \ldots, n; \ \nu = 2, 3, \ldots, \mu). \qquad (194)$$

Not every constituent of $\bar{\omega}^{-1} N \bar{\omega}$ is zero, since otherwise $N = 0$. We may, therefore, then assume that $b_{\lambda\nu'} \neq 0$ for $1 < \nu' \leqq n$ and $\lambda < \nu'$. Therefore, not every constituent is zero in the linear relation we derive from (194) by putting $\nu = \nu'$; and thus the same rational relation with respect to $R$ obtains between the first $\nu' - 1$ constituents of the $\mu$-th row ($\mu \geqq \nu'$) of every transformation of the group of linear homogeneous transformations $\bar{\omega}^{-1} Y \bar{\omega}$ belonging to $R$. But then, by Maschke's theorem, this group is reducible with respect to $R$.* Wherefore, $\bar{G}$ is reducible with respect to $R$. That is to say, *if $A_1, A_2, \ldots, A_m$ are $m$ ($m > 1$) linearly independent transformations which belong to $R$ and constitute a hypercomplex number system containing a modulus but no other idempotent number, the group constituted by the aggregate of transformations $\Sigma_{i=1}^{m} x_i A_i$ for all values of the $x$'s rational with respect to $R$, is reducible with respect to $R$.*

We have now shown that it is possible to find a linear substitution $\omega$ belonging to $R$ such that, for any transformation $Y = y_0 + \Sigma_{i=1}^{m-1} y_i N_i$ of $\bar{G}$, that is, for $y_0, y_1, \ldots, y_{m-1}$ any system of scalars rational with respect to $R$,

$$Y = y_0 + \Sigma_{i=1}^{m-1} y_i N_i = \omega \begin{vmatrix} Y_{11}, & Y_{12}, & \ldots, & Y_{1\kappa} \\ 0, & Y_{22}, & \ldots, & Y_{2\kappa} \\ \ldots & \ldots & \ldots & \ldots \\ 0, & 0, & \ldots, & Y_{\kappa\kappa} \end{vmatrix} \omega^{-1}, \qquad (195)$$

where $Y_{uv}$ for $u = 1, 2, \ldots, \kappa$ and $v = u, u+1, \ldots, \kappa$ are matrices whose con-

* See note, p. 365.

stituents lie in $R$, and, for $1 \leqq u \leqq \varkappa$, the aggregate of transformations $Y_{uu}$ belonging to $R$ constitute a group $\overline{G}_u$ irreducible with respect to $R$.

Let

$$N_i = \omega \begin{vmatrix} N_{11}^{(i)}, & N_{12}^{(i)}, & \ldots, & N_{1\varkappa}^{(i)} \\ 0, & N_{22}^{(i)}, & \ldots, & N_{2\varkappa}^{(i)} \\ \cdots & \cdots & \cdots & \cdots \\ 0, & 0, & \ldots, & N_{\varkappa\varkappa}^{(i)} \end{vmatrix} \omega^{-1}, \qquad (i=1, 2, \ldots, m-1), \qquad (196)$$

where $N_{uv}^{(i)}$ $(u=1, 2, \ldots, \varkappa;\ v=u, u+1, \ldots, \varkappa)$ is rational with respect to $R$; and, for $1 \leqq u \leqq \varkappa$, let $1_u$ denote the identical transformation of the group $\overline{G}_u$ of transformations $Y_{uu}$. Then

$$Y_{uu} = y_0 1_u + \Sigma_{i=1}^{m-1} y_i N_{uu}^{(i)}, \qquad (u=1, 2, \ldots, \varkappa), \qquad (197)$$

$$Y_{uv} = \Sigma_{i=1}^{m-1} y_i N_{uv}^{(i)}, \qquad (u=1, 2, \ldots, \varkappa;\ v=u+1, u+2, \ldots, \varkappa). \qquad (198)$$

From (189) it follows that

$$\omega^{-1} N_i \omega \cdot \omega^{-1} N_j \omega = \omega^{-1} N_i N_j \omega = \Sigma_{k=1}^{m-1} \gamma_{ijk}^{(0)} \omega^{-1} N_k \omega,$$
$$(i, j=1, 2, \ldots, m-1); \qquad (199)$$

and, therefore,

$$N_{uu}^{(i)} N_{uu}^{(j)} = \Sigma_{k=1}^{m-1} \gamma_{ijk}^{(0)} N_{uu}^{(k)}, \qquad (i, j=1, 2, \ldots, m-1;\ u=1, 2, \ldots, \varkappa). \qquad (200)$$

Let $1 \leqq u \leqq \varkappa$; and let $m_u$ $(0 \leqq m_u \leqq m-1)$ be the number of linear homogeneous transformations $N_{uu}^{(1)}, N_{uu}^{(2)}, \ldots, N_{uu}^{(m-1)}$ that are linearly independent. We shall now show that $m_u=0$. For, if possible let $m_u>0$, in which case we may assume with loss of generality that $N_{uu}^{(1)}, N_{uu}^{(2)}, \ldots, N_{uu}^{(m_u)}$ are linearly independent; and thus

$$N_{uu}^{(m_u+h)} = c_{h1}^{(u)} N_{uu}^{(1)} + c_{h2}^{(u)} N_{uu}^{(2)} + \ldots + c_{hm_u}^{(u)} N_{uu}^{(m_u)}, \quad (h=1, 2, \ldots, m-m_u), \quad (201)$$

where the $c$'s are rational with respect to $R$. From (197) it then follows that the group $\overline{G}_u$ of transformations belonging to $R$ is constituted by the aggregate of numbers $y_0 + \Sigma_{i=1}^{m_u} y_i N_{uu}^{(i)}$ for values of the $y$'s rational with respect to $R$; and, from (200), that $N_{uu}^{(1)}, N_{uu}^{(2)}, \ldots, N_{uu}^{(m_u)}$ constitute a hypercomplex number system.* This system is nilpotent. For $\Sigma_{i=1}^{m_u} y_i N_i$ and, therefore, $\Sigma_{i=1}^{m_u} y_i \omega^{-1} N_i \omega$, is nilpotent, for every system of values of the $y$'s. Whence, by (196), it follows that $\Sigma_{i=1}^{m_u} y_i N_{uu}^{(i)}$ is nilpotent for every system of values of the $y$'s. The linear homogeneous transformations $1_u, N_{uu}^{(1)}, N_{uu}^{(2)}, \ldots, N_{uu}^{(m_u)}$ are linearly independent and constitute a number system with but one idempotent number, the modulus $1_u$.† Therefore, by the theorem proved p. 369, the group $\overline{G}_u$ of transformations belonging to $R$ constituted by the aggregate of transformations $y_0 + \Sigma_{i=1}^{m_u} y_i N_{uu}^{(i)}$,

---

* Cf. p. 366.

† See note, p. 363; also p. 372.

for values of the $y$'s rational with respect to $R$, is reducible with respect to $R$, which is contrary to supposition. Therefore, $m_u=0$, and thus

$$N_{uu}^{(i)}=0, \qquad (i=1,2,\ldots,m-1); \tag{202}$$

whence follows

$$Y_{uu}=y_0 1_u, \qquad (u=1,2,\ldots,\varkappa). \tag{203}$$

The order of the matrix $1_u$ is equal to unity; otherwise, $\overline{G}_u$ is reducible with respect to $R$. Since this is true for each value of $u$ from 1 to $\varkappa$, it follows that each of the matrices $Y_{uv}$ $(u=1,2,\ldots,\varkappa;\ v=u,u+1,\ldots,\varkappa)$ is of order one. Whence it follows that $\varkappa=n$; and, by (203),

$$Y_{uu}=y_0, \qquad (u=1,2,\ldots,n). \tag{204}$$

Again, by (196) and (202),

$$N_i=\omega\begin{vmatrix} 0, & b_{12}^{(i)}, & \ldots, & b_{1n}^{(i)} \\ 0, & 0, & \ldots, & b_{2n}^{(i)} \\ \ldots & \ldots & \ldots & \ldots \\ 0, & 0, & \ldots, & 0 \end{vmatrix}\omega^{-1}, \qquad (i=1,2,\ldots,m-1), \tag{205}$$

where the $b$'s are rational with respect to $R$. Wherefore, by (195),

$$Y=y_0+\Sigma_{i=1}^{m-1}y_iN_i=\omega\begin{vmatrix} y_0, & \Sigma_{i=1}^{m-1}y_ib_{12}^{(i)}, & \ldots, & \Sigma_{i=1}^{m-1}y_ib_{1n}^{(i)} \\ 0, & y_0, & \ldots, & \Sigma_{i=1}^{m-1}y_ib_{2n}^{(i)} \\ \ldots & \ldots & \ldots & \ldots \\ 0, & 0, & \ldots, & y_0 \end{vmatrix}\omega^{-1}. \tag{206}$$

The group $\overline{G}$ can not be completely reducible with respect to $R$. For it would then be possible to choose $\omega$ so that

$$Y_{uv}=0, \qquad (u=1,2,\ldots,\varkappa;\ v=u+1,u+2,\ldots,\varkappa);$$

and we should then have

$$Y=y_0+\Sigma_{i=1}^{m-1}y_iN_i=\omega\begin{vmatrix} Y_{11}, & 0, & \ldots, & 0 \\ 0, & Y_{22}, & \ldots, & 0 \\ \ldots & \ldots & \ldots & \ldots \\ 0, & 0, & \ldots, & Y_{\varkappa\varkappa} \end{vmatrix}\omega^{-1}.$$

But it has been shown that $Y_{uu}=y_0 1_u$ for $u=1,2,\ldots,\varkappa$; and we should then have $Y=y_0$, which is contrary to supposition, since $m>1$.

We have, therefore, established the following theorem:

Theorem VI. *If $A_1, A_2, \ldots, A_m$ are $m$ $(m>1)$ linearly independent linear homogeneous transformations belonging to an arbitrarily given domain of rationality $R$ and constituting a hypercomplex number system with but one idempotent number, the modulus* 1, *there is then a system of just $m-1$ linearly independent linear functions,*

$$N_i=\tau_{i1}A_1+\tau_{i2}A_2+\ldots+\tau_{im}A_m, \qquad (i=1,2\ldots,m-1)$$

*of the A's with coefficients rational with respect to R, constituting a maximal invariant nilpotent subsystem of* $(A_1, A_2, \ldots, A_m)$*; and there is a number of the subsystem,*

$$N = c_1 N_1 + c_2 N_2 + \ldots + c_{m-1} N_{m-1},$$

*where the c's are rational with respect to R, such that*

$$NN_i = 0 = N_i N, \qquad (i = 1, 2, \ldots, m-1).$$

*Moreover, a linear substitution* $\Omega$ *belonging to R can be found such that*

$$N_i = \Omega \begin{vmatrix} 0, & b_{12}^{(i)}, & \ldots, & b_{1n}^{(i)} \\ 0, & 0, & \ldots, & b_{2n}^{(i)} \\ \ldots & \ldots & \ldots & \ldots \\ 0, & 0, & \ldots, & 0 \end{vmatrix} \Omega^{-1},$$

*where the b's are rational with respect to R. Whence it follows that the group* $\overline{G}$ *of transformations belonging to R constituted by the aggregate of transformations* $\Sigma_{i=1}^{m} x_i A_i$ *for all values of the x's rational with respect to R, is reducible, but not completely reducible, with respect to R; and is similar to the group of transformations belonging to R constituted by the aggregate of transformations defined by the equations*

$$\xi'_\mu = y_0 \xi_\mu + \Sigma_{\nu=\mu+1}^{n} \Sigma_{i=1}^{m-1} y_i b_{\mu\nu}^{(i)} \xi_\nu, \qquad (\mu = 1, 2, \ldots, n),$$

*for all values of the parameters* $y_0, y_1, \ldots, y_{m-1}$ *rational with respect to R.*

A special case of this theorem is that for which $A_m = 1$, and $A_1, A_2, \ldots, A_{m-1}$ constitute a nilpotent subsystem of $(A_1, A_2, \ldots, A_m)$. For this system contains no idempotent number except the modulus 1, as may be shown as follows: Let

$$A = \alpha_m + \Sigma_{i=1}^{m-1} \alpha_i A_i = \Sigma_{i=1}^{m} \alpha_i A_i \neq 0$$

be idempotent. Then

$$A^2 = \alpha_m^2 + 2\alpha_m \Sigma_{i=1}^{m-1} \alpha_i A_i + (\Sigma_{i=1}^{m-1} \alpha_i A_i)^2 = \alpha_m + \Sigma_{i=1}^{m-1} \alpha_i A_i.$$

Therefore, $\alpha_m^2 = \alpha_m$, since $A_1, A_2, \ldots, A_{m-1}$ constitute a subsystem of $(A_1, A_2, \ldots, A_m)$. Since this subsystem is nilpotent, $\Sigma_{i=1}^{m-1} \alpha_i A_i$ is nilpotent. But there is no linear relation between the powers of a nilpotent number. Whence it follows that

$$(2\alpha_m - 1) \Sigma_{m=1}^{m-1} \alpha_i A_i = 0;$$

therefore, since $2\alpha_m - 1 \neq 0$,

$$\alpha_1 = \alpha_2 = \ldots \alpha_{m-1} = 0,$$

and thus $A = \alpha_m A_m = \alpha_m 1$. Wherefore, $\alpha_m = 1$, and thus $A = A_m = 1$; otherwise, $A = 0$.

# *On Sextic Surfaces Having a Nodal Curve of Order 8.*

BY C. H. SISAM.

1. In an article in this Journal (Vol. XXXVII, p. 445), I have discussed sextic surfaces whose plane sections are of genus 1. In this paper, non-ruled * sextics whose plane sections are of genus 2 are considered.

2. It is known † that such a surface is rational and that it is generated by a pencil (faisceau) of conics. The developable of the planes of these conics is of class 3. For, the section of the surface by a generic plane $\pi$ is an hyperelliptic sextic curve. The section by $\pi$ of the system of planes of the generating conics is the system of lines joining corresponding points of the $g_2'$ on the sextic. The involutorial (6, 6) correspondence of the lines of any pencil in $\pi$ such that corresponding lines pass through pairs of points of $g_2'$ has twelve coincidences. Six of these arise from coincidences of $g_2'$. The other six (in pairs) are due to lines in the pencil which join corresponding points of $g_2'$.

3. The system of conics on the given sextic surface is thus the locus of the conic of intersection of corresponding surfaces of the systems ‡

$$L_1k_1^3+L_2k_1^2k_2+L_3k_1k_2^2+L_4k_2^3=0, \tag{1}$$

$$L_1L_5k_1^2+Qk_1k_2+L_4L_6k_2^2=0, \tag{2}$$

where $L_i=0$ is the equation of a plane, $Q=0$ is the equation of a quadric and $k_1$, $k_2$ are parameters.

4. *Six Conics of the System are Composite.* The condition that a conic of the system is composite is that its plane (1) touches the corresponding quadric (2). If we write the equation of a quadric (2) in plane coordinates and impose the condition that the corresponding plane (1) satisfies the resulting equation, we obtain an equation of degree 12 in $k_1$ and $k_2$. It has an

* For the discussion of ruled sextics whose plane sections are of genus 2, see Snyder, in this Journal, Vol. XXVII, pp. 77–102, 173–188, where other references will be found.

† Castelnuovo–Enriques, *Math. Annalen*, Vol. XLVIII, p. 308.

‡ Compare an article by the author in this Journal, Vol. XXX, p. 99.

extraneous factor $k_1^3k_2^3$. The six values of $k_1/k_2$ which make the residual factor zero determine the composite conics of the system.

5. The given system of conics is also defined by (1) and any one of the systems of surfaces

$$(L_2L_5-Q)k_1^2+(L_3L_5-L_4L_6)k_1k_2+L_4L_5k_2^2=0, \qquad (3)$$

$$L_1L_6k_1^2+(L_2L_6-L_1L_5)k_1k_2+(L_3L_6-Q)k_2^2=0, \qquad (4)$$

$$(QL_6+L_1L_5^2-L_2L_5L_6)k_1+(QL_5+L_4L_6^2-L_3L_5L_6)k_2=0. \qquad (5)$$

If the coordinates of a point on the surface are substituted in (2), (3) and (4), the resulting equations in $k_1/k_2$ have a common root. By eliminating $k_1$ and $k_2$ from these equations we obtain at once the equation of the surface:

$$\begin{vmatrix} L_1L_5 & Q & L_4L_6 \\ L_2L_5-Q & L_3L_5-L_4L_6 & L_4L_5 \\ L_1L_6 & L_2L_6-L_1L_5 & L_3L_6-Q \end{vmatrix}=0. \qquad (6)$$

6. For points on the double curve, equations (2), (3) and (4) have both roots equal. Hence, the double curve lies on all the quartic surfaces determined by equating to zero the first minors of (6).

7. At a triple point of the surface, equations (2), (3) and (4) are satisfied identically. There are two such points, defined by

$$L_5=L_6=Q=0. \qquad (7)$$

8. For points on the double curve, (5) is satisfied identically; so that the double curve of (6) lies on all the cubic surfaces of the pencil (5). The triple points (7) are double points on these surfaces. The residual basis curve of the pencil is the line $L_5=0L_6=0$ joining the triple points. The octavic component of the basis curve of (5) has the points (7) as triple points. It is of genus 3 since* its projection on a plane from a triple point is a quintic curve with a triple point and no double points.

9. The complete intersection of a cubic of the pencil (5) with the given surface (6) consists of the double octic and a generating conic. It follows at once that if $C_n$ is a curve of order $n$ on the given surface, $\alpha$ the number of its intersections with the double curve and $\beta$ the number of its intersections with a generic generating conic, then

$$\alpha+\beta=3n,$$

since $\alpha+\beta$ is the number of its intersections with a generic cubic surface (5).

* This result also follows from the vanishing of the numerical genus. Cf. Castelnuovo-Enriques, *loc. cit.*, p. 312.

In particular, the right lines on the surface (Art. 4) are trisecants of the double curve. The generating conics intersect it in six points.

10. Denote the equation (6) of the given surface by $\Phi=0$ and those of two cubics of the system (5) by $\phi_1=0$ and $\phi_2=0$. The surfaces of the linear system $\infty^3$

$$\lambda_1\Phi+\lambda_2\phi_1^2+\lambda_3\phi_1\phi_2+\lambda_4\phi_2^2=0 \tag{8}$$

all have the double curve of $\Phi=0$ as double curve. Every sextic surface on which this curve is double belongs to the system (8). For, its residual intersection with $F=0$ breaks up into two conics of the system (1), (2) since the conic of the system through a generic point of the residual intersection has thirteen points in common with each surface (Art. 9). The surface in question thus belongs to the pencil of sextics defined by $\Phi=0$ and a pair of cubics (5) and hence to the system (8).

It follows that, through a generic point of space, there passes a single conic which intersects the double octavic in six points. Every such conic lies on a cubic of the pencil (5). Its plane intersects the cubic on which it lies in a right line which intersects the line $L_5=0L_6=0$ joining the triple points. The locus of this residual line is the ruled quartic

$$L_1L_5^3-L_2L_5^2L_6+L_3L_5L_6^2-L_4L_6^3=0.$$

Every plane tangent to this ruled quartic contains a conic of this doubly infinite system of conics.

The locus of the composite conics (Art. 4) on the system of surfaces (8) is the scroll of trisecants of the double octavic. This ruled surface has the octavic as sixfold curve. Its complete intersection with a surface (8) is the octavic and the six composite conics on the surface. It is of order 18.

11. Since the double octavic has two triple points and is of genus 3, the quartic surfaces which contain it form a linear system $\infty^8$ (cf. Art. 6). The residual octavic of intersection of such a quartic with $\Phi=0$ intersects the generating conics in two points (Art. 9), and the double octavic in twenty-two points. It is of genus 3.

12. The quartic surfaces of the system of Art. 11 which contain one component of each of three composite generating conics of $F=0$ form a bundle of quartics since the three lines are trisecants of the double octavic. The residual intersection of a quartic of the bundle with $\Phi=0$ is a quintic which intersects the generating conics in two points and the double octavic in thirteen points. It intersects each of the basis right lines of the bundle in two points.

For, a generic plane through such a basis line intersects $\Phi=0$ in a quintic and the given quartic surface in a cubic. Ten intersections of this quintic and cubic lie on the double octavic and one on each of the other two basis lines of the bundle so that only three are points of intersection of the given space quintic with the given plane.

The given quintic thus intersects any other quartic of the bundle in a single point not on the basis curve. Hence, any two quartics of the bundle determine uniquely a point on the surface. Moreover, a generic point on the surface is uniquely determined in this way, so that a system of parametric equations of the surface are determined by solving the equation $\Phi=0$ as simultaneous with

$$\frac{\psi_1}{\xi_1'}=\frac{\psi_2}{\xi_2'}=\frac{\psi_3}{\xi_3'}, \tag{9}$$

where $\psi_1=0$, $\psi_2=0$, $\psi_3=0$ are three suitably chosen quartics of the bundle.

13. Let

$$x_i=f_i'(\xi_1', \xi_2', \xi_3') \qquad i=1, 2, 3, 4 \tag{10}$$

be the parametric equations of the given surface obtained in this way. The curves $\Sigma u_i f_i'(S')=0$ which correspond to the plane sections of $\Phi=0$ are quintics (since to a generic line in the parametric plane corresponds a quintic on $\Phi=0$) of genus equal to that of the corresponding plane sections (Art. 1). By a suitable birational transformation of the plane, they may be transformed into a system of quartic curves having a common double point.

14. Let such a birational transformation be effected and let

$$x_i=f_i(\xi_1, \xi_2, \xi_3) \qquad i=1, 2, 3, 4 \tag{11}$$

be the resulting parametric equations of the surface. Denote the fundamental double point of the quartics $\Sigma u_i f_i(\xi)=0$ by $P_0$. Since a right line has just six points in common with the surface, these quartics have $4^2-2^2-6=6$ simple points in common. Denote these points by $P_1, P_2, \ldots, P_6$, respectively.

Let $P_0P_1P_2$ be chosen as vertices of the triangle of reference in the $\xi$-plane so that

$$\begin{aligned} x_i=f_i(\xi)\equiv a_{0i}\xi_1^2\xi_3^2+a_{1i}\xi_1\xi_2\xi_3^2+a_{2i}\xi_2^2\xi_3^2+a_{3i}\xi_1^3\xi_3+a_{4i}\xi_1^2\xi_2\xi_3+a_{5i}\xi_1\xi_2^2\xi_3 \\ +a_{6i}\xi_2^3\xi_3+a_{7i}\xi_1^3\xi_2+a_{8i}\xi_1^2\xi_2^2+a_{9i}\xi_1\xi_2^3, \qquad i=1, 2, 3, 4 \end{aligned} \tag{12}$$

where the coefficients are chosen so that $P_3P_4P_5P_6$ also lie on the curves $\Sigma u_i f_i=0$.

15. To the pencil of lines through $P_0$ corresponds the system of conics which generate the surface. Thus, to the line $k_2\xi_1=k_1\xi_2$ corresponds the conic

$$x_i=(a_{0i}k_1^2+a_{1i}k_1k_2+a_{2i}k_2^2)\xi_3^2+(a_{3i}k_1^3+a_{4i}k_1^2k_2+a_{5i}k_1k_2^2+a_{6i}k_2^3)\xi_3\xi_2$$
$$+(a_{7i}k_1^3k_2+a_{8i}k_1^2k_2^2+a_{9i}k_1k_2^3)\xi_2^2. \qquad i=1, 2, 3, 4 \qquad (13)$$

The six composite conics (Art. 4) of the system (13) are determined by the lines joining $P_0$ to $P_i (i=1, 2, \ldots, 6)$. One component of such a conic corresponds to the line $P_0P_i$, the other, to the fundamental point* $P_i$. There are, in general (cf. Art. 34), no other right lines on the surface.

16. There are, on the surface, thirty-two conics which do not belong to the system (13). One of these corresponds to $P_0$ (cf. Clebsch, *loc. cit.*) in such a way that to each direction through $P_0$ corresponds a point on the conic. Fifteen correspond to the lines joining the simple fundamental points in pairs; fifteen, to the conics through $P_0$ and four simple fundamental points. One other corresponds to the cubic which has a node at $P_0$ and passes through $P_1, P_2, \ldots, P_6$.

Each conic of the generating system (13) intersects each directrix conic in a single point. In particular, just one component of each composite conic intersects a given directrix conic. Denote the components of one of these composite conics by the symbols 1 and 1′, of another by 2 and 2′, etc , in such a way that the six lines 1, 2, 3, 4, 5, 6 all intersect the conic corresponding to $P_0$. If we put, in all possible ways, primes on an even number of the symbols 1, 2, 3, 4, 5, 6, we determine $2^5-1$ other sets of six lines on the surface such that all the lines of the set intersect a directrix conic. Let each such set of six symbols be used to indicate the directrix conic which intersects the corresponding lines. These thirty-two symbols individualize the thirty-two directrix conics.

Two directrix conics whose symbols have $2k$ corresponding component symbols unlike intersect in $k-1$ points as may be seen by counting the intersections, not at fundamental points, of the corresponding curves in the $\xi$-plane. Each directrix conic is intersected twice by just one other such conic, the one whose component symbols are all different from those of the given conic. Of two such pairs of conics, as, for example, 1 2 3 4 5 6, 1′ 2′ 3′ 4′ 5′ 6′ and 1′ 2′ 3 4 5 6, 1 2 3′ 4′ 5′ 6′, each conic of each pair intersects a single conic of the other pair. Given two non-intersecting conics, there are eight other conics which do not intersect either of them. Six directrix conics can be chosen (in thirty-two

* Clebsch, *Math. Annalen*, Vol. I, p. 266.

ways) so that no two intersect. Any five of six such conics intersect a right line lying on the surface.

Four directrix conics can be chosen in such a way that each intersects all the others just once. There are $\frac{32\cdot 15\cdot 6\cdot 1}{1\cdot 2\cdot 3\cdot 4}=120$ tetrads of this type.

17. The surface is generated by thirty-two linear systems of cubic curves. These systems of curves correspond to the six pencils of lines with vertices at $P_i (i=1, 2, \ldots, 6)$, the twenty pencils of conics through $P_0$ and three $P_i$, the six pencils of cubics which have a double point at $P_0$ and pass through five $P_i$. These thirty-two systems are characterized, in the notation of the preceding article, by putting primes on an odd number of the symbols, 1, 2, 3, 4, 5, 6. Two cubics which belong to systems having $2k$ component symbols unlike have $k$ points in common. In particular, cubics belonging to the same system $(k=0)$ do not intersect. Each directrix conic intersects the cubics of any system which has $2k+1$ symbols unlike the symbol of the conic in $k$ points. The generating conics intersect the cubics of each system just once.

There are twelve cubics on the surface each of which intersects each generating conic in two points. Six of these correspond to the six conics determined by five fundamental points $P_1, P_2, \ldots, P_6$. The others correspond to the cubics which pass through $P_0$, have a node at one of the points $P_1, \ldots, P_6$ and pass simply through the other five points. Each of these twelve cubics intersects one right line on the surface twice, the line coplanar with it not at all, and each of the remaining right lines on the surface once.

18. There are thirty-two linear systems each $\infty^2$ of rational quartics on the surface. One such system corresponds to the right lines in the $\xi$-plane, the others to the systems of curves which can be transformed into the right lines by birational transformations that transform the curves $\Sigma u_i f_i=0$ (Art. 13) into quartics. All the quartics of each system intersect six lines on the surface which intersect a directrix conic (Art. 16).

There are sixty pencils of rational quartics which intersect the generating conics twice. Any two non-intersecting lines on the surface determine such a system in such a way that the quartics of the system intersect each of the given lines twice.

There is a single linear system $\infty^2$ of quartics of genus 1 on the surface. It is determined by the system of cubics which pass through $P_0, P_1, \ldots, P_6$.

19. *The genus of a non-multiple curve of order $n$ on the given surface can not exceed the greatest integer in* $\frac{(n-2)^2+9}{12}$. *This limit is attained for all values of $n$.*

Let $C_n$ be the given curve of order $n$ and let $C_m$, the corresponding curve in the $\xi$-plane, be of order $m$ and have a point of multiplicity $\alpha_i$ at the fundamental point $P_i (i=0, 1, \ldots, 6)$. Then

$$4m-2\alpha_0-(\alpha_1+\alpha_2+\alpha_3+\alpha_4+\alpha_5+\alpha_6)=n,$$

since to each intersection of $C_n$ with a generic plane $\Sigma u_i x_i=0$ corresponds an intersection, not at a fundamental point of $C_m$ with $\Sigma u_i f_i=0$.

The curves $C_m$ and $C_n$ are in (1, 1) correspondence. Hence, the genus of $C_n$ is not greater than

$$\frac{(m-1)(m-2)}{2}-\sum_{i=0}^{6}\frac{\alpha_i(\alpha_i-1)}{2}.$$

This function has its greatest value when

$$\alpha_0=\frac{2n-1}{6} \quad \alpha_1=\alpha_2=\alpha_3=\alpha_4=\alpha_5=\alpha_6=\frac{n+1}{6} \text{ whence } m=\frac{4n+1}{6}.$$

The value of the function for these values of the variables is $\frac{(n-2)^2+9}{12}$ so that the genus can not exceed the greatest integer in this expression. By giving suitable integer values to $\alpha_0, \alpha_1, \ldots, \alpha_6$, this limit can be attained for all values of $n$.

20. Six generating conics (13) touch a given plane $(u)$. These correspond to the tangents from $P_0$ to the curve $\Sigma u_i f_i=0$ which corresponds to the section of the surface by the given plane. If we let

$$a_i=u_1 a_{i1}+u_2 a_{i2}+u_3 a_{i3}+u_4 a_{i4}$$

we find, as the equation of these six tangents:

$$(a_3\xi_1^3+a_4\xi_1^2\xi_2+a_5\xi_1\xi_2^2+a_6\xi_2^3)^2-4\xi_1\xi_2(a_7\xi_1^2+a_8\xi_1\xi_2+a_9\xi_2^2)(a_0\xi_1^2+a_1\xi_1\xi_2+a_2\xi_2^2)=0. \qquad (14)$$

The moduli of the section of the surface by the given plane are those of the corresponding quartic, that is, are the absolute invariants of the binary sextic (14). In particular, the condition that the given plane is tangent is that (14) has a double root. The equation of the surface in plane coordinates is thus found by equating to zero the discriminant of (14). The surface is of class 20.

21. The tangent plane at the point on the surface having the parameters

$(\xi_1, \xi_2, \xi_3)$ determines a quartic in the $\xi$-plane with a double point at $(\xi)$. The equation of this tangent plane is

$$\begin{vmatrix} x_1 & x_2 & x_3 & x_4 \\ \frac{\partial f_1}{\partial \xi_1} & \frac{\partial f_2}{\partial \xi_1} & \frac{\partial f_3}{\partial \xi_1} & \frac{\partial f_4}{\partial \xi_1} \\ \frac{\partial f_1}{\partial \xi_2} & \frac{\partial f_2}{\partial \xi_2} & \frac{\partial f_3}{\partial \xi_2} & \frac{\partial f_4}{\partial \xi_2} \\ \frac{\partial f_1}{\partial \xi_3} & \frac{\partial f_2}{\partial \xi_3} & \frac{\partial f_3}{\partial \xi_3} & \frac{\partial f_4}{\partial \xi_3} \end{vmatrix} = 0. \tag{15}$$

Hence, the parametric equations of the surface in plane coordinates are

$$u_i = J_i(\xi_1, \xi_2, \xi_3), \tag{16}$$

where $J_i$ is the cofactor of $x_i$ in (15).

The curves $\Sigma x_i J_i(\xi) = 0$ which correspond to the tangent cones from the points $(x)$ of space to the surface are of order 9. The point $P_0$ is fivefold and the points $P_1, P_2, \ldots, P_6$ are double points on these curves. Since the surface is of class 20 (Art. 20) there are twelve other points common to these curves. These twelve points correspond to pinch-points on the double curve. The tangent cone from a generic point to the surface is of genus 12.

22. The parabolic curve is of order 32. It touches each right line on the surface twice, intersects each generating conic in eight points and each directrix conic in twelve points. It has a double point at each pinch-point of the double curve. Its genus is 57. The developable of the stationary planes is of class 48 and order 80. These planes are determined by imposing the condition that (14) has a triple root.

23. The double developable is of class 111. It has as components the twelve pencils of planes (Art. 4) which contain a right line and a quintic of genus 1, and the cubic developable (Art. 2) of the planes which intersect the surface in a generating conic and a quartic of genus 1. The section by a generic plane of the residual developable is a non-composite rational sextic.

24. There are two hundred and thirty-eight triple tangent planes. The section of the surface by such a plane is composite since it has eleven double points. There are: (a) six planes which contain two lines and a quartic of genus 1; (b) one hundred and forty-four which contain a line and a rational quartic; (c) twenty-four which contain a generating conic and a rational quartic; (d) thirty-two which contain a directrix conic and a rational quartic; (e) thirty-two which contain two rational cubics. The two cubics in a plane of

type (e) belong to two systems of cubics on the surface (Art. 17) such that the curves of one system intersect those of the other in three points. In each pencil of cubics on the surface there are two cubics which lie in a plane.

25. The double curve intersects the generating conics in six points, the directrix conics in five points, and the right lines on the surface in three points, (Arts. 9 and 16). Hence, the curve $\sigma=0$ corresponding to the double curve has a fivefold point at $P_0$, threefold points at $P_1, P_2, \ldots, P_6$ and is of order 11. Each curve $\Sigma u_i f_i=0$ which passes through a given point of $\sigma=0$ passes through a second fixed point. To each triple point of the surface correspond three points of $\sigma$ such that each curve $\Sigma u_i f_i=0$ through one of these points passes through the other two. These points are double points on $\sigma=0$. The curve $\sigma=0$ is of genus 11. Its points determine, by means of equations (16), the tangent planes to the surface along the double curve. The developable of these planes is of class 26.

26. The quartic curves in the $\xi$-plane which have a double point at $P_0$ and pass through $P_1, P_2, \ldots, P_6$ form a linear system $\infty^5$. If $f_i(\xi_1, \xi_2, \xi_3)=0$ $(i=1, 2, \ldots, 6)$ are six linearly independent curves of this system, the sextic surface

$$x_i=f_i(\xi_1, \xi_2, \xi_3) \qquad i=1, 2, \ldots, 6 \tag{17}$$

defined by them belongs to a space of five dimensions $S_5$ and has the surface (10) as its projection from a generic line on an $S_3$. A generic hyperplane section of the surface is of genus 2, since it is in (1, 1) correspondence with a quartic curve of genus 2 in the $\xi$-plane.

27. No sextic surface whose hyperplane sections are of genus 2 can belong to a space of more than five dimensions. For, the sextic curve of section by a generic hyperplane belongs to an $S_4$ (at most). The $S_5$ determined by such an $S_4$ and a generic point on the surface contains the surface. If such a surface belongs to an $S_5$ (*i. e.*, does not lie in an $S_4$), it is either a cone projecting from a point a sextic curve of genus 2, or a rational surface of the type of equations (17). For, suppose first that the surface is ruled. Let $S_4$ be a four-space which contains two of its generators. If the generators of the surface intersect $S_4$ in a fixed point, the surface is a cone. If not, the locus of the intersection of the generators with $S_4$ is a plane quartic curve of genus 2. Hence, such a surface can be defined in infinitely many ways by a (1, 1) correspondence between two plane quartics. But the ruled surface defined by such a correspondence is of order 8 unless the quartics intersect. If the quartics intersect, their planes define a space of (at most) four dimensions which contains the ruled surface. If the surface is unruled, its projection from a

generic line on an $S_3$ is an unruled sextic surface with generic plane section of genus 2. The given surface is therefore rational (Art. 2) and its equations can be birationally reduced to the form (17).

28. To the cubic curves in the $\xi$-plane which pass through $P_0, P_1, \ldots, P_6$ correspond on the surface (17) a linear system $\infty^2$ of quartic curves (cf. Art. 18). Each of these quartics defines an $S_3$ in which it lies. Since through any two given points of the surface a quartic of the system passes, each bisecant line to the surface lies in such an $S_3$. In a generic bundle of quartics of the system in the $\xi$-plane which corresponds to the hyperplane sections of the surface, there is a unique pencil which breaks up into a fixed cubic through $P_0, P_1, \ldots, P_6$ and a pencil of lines. Hence, a generic point of $S_5$ lies in the $S_3$ of a single quartic of genus 1 on the surface. The apparent double points of the surface from the given point are the two apparent double points of the quartic lying in the $S_3$ which passes through the given point.

29. Through a generic point $P$ on the surface, there pass a pencil of quartic curves of genus 1 on the surface. These quartic curves have a second fixed point $P'$ in common determined by the ninth intersection of the corresponding cubics in the $\xi$-plane. The line $PP'$ is thus common to the $S_3$ defined by all these quartics. It intersects the plane of each of the generating conics on the surface. For, let $C_2$ be such a conic. Then $C_2$ and the pencil of quartics through $P$ are the sections of the surface by a pencil of hyperplanes, since the corresponding curves in the $\xi$-plane constitute a pencil of quartics with a double point at $P_0$ and passing through $P_1, P_2, \ldots, P_6$ (Art. 26). The basis $S_3$ of this pencil of $S_4$ contains the plane of the conic and the line $PP'$. Hence the line and the plane intersect. The system $\infty^2$ of lines $PP'$ joining all the pairs of corresponding points of the given surface thus constitute a rectilinear generation of the cubic three-spread $V_3^3$ defined by the planes of the generating conics of the surface.*

30. The given surface (17) lies on a web of hyperquadrics in $S_5$. For, the hyperquadrics which pass through eight generic fixed points of one quartic of the system of Art. 28, through six such points of another and through three generic fixed points of a directrix conic, contain these curves. They, therefore, contain the surface, since they have five points in common with each generating

* The lines joining corresponding pairs of points on the surface (11) in $S_3$ form a (3, 1) congruence. For, these lines are the projections from a fixed generic line $l$, of the lines on $V_3^3$ in $S_5$. Three such lines in $S_5$ intersect a generic plane through $l$ since $V_3^3$ is of order 3, and one lies in a generic $S_4$ through $l$ since the section of $V_3^3$ by $S_4$ is a ruled cubic with a single rectilinear directrix. Hence, three lines of the congruence by $S_3$ pass through a given point and one lies in a given plane.

conic. Since we have imposed, at most, seventeen conditions on the twenty-one homogeneous parameters in the equation of an hyperquadric in $S_5$, there exists a web of such hypersurfaces which contain the given surface.

The hyperquadrics of this web do not all contain $V_3^3$. For, the hyperquadrics which contain three skew planes of $V_3^3$, and hence contain the three-spread, constitute a bundle.* The given surface is thus the complete intersection of $V_3^3$ with an hyperquadric. Conversely, the intersection of a given hyperquadric in $S_6$ by a given cubic three-spread generated by planes is, in general, a surface of the type of equations (17), since the projection of the intersection from a generic line onto a generic $S_3$ is a surface of order 6 generated by conics whose planes envelope a rational developable of class 3.

31. Let the points of a non-singular hyperquadric $F_2=0$ be put in (1, 1) correspondence with the lines of $S_3$. To the section of $F_2=0$ by $V_3^3$ corresponds a (3, 3) cóngruence, since each plane contained in $F_2=0$ intersects $V_3^3$ in three points.

Conversely, let there be given a (3, 3) congruence which does not belong to a linear complex. The section of the congruence by a generic linear complex is a ruled sextic which does not belong to a linear congruence and hence is of genus not greater than 2. If the genus of such a section is equal to 2, the complex can not be generated by pencils of lines, since all the generating pencils would have a fixed line in common (Art. 27), and the congruence would belong to a special linear complex. The congruence is therefore rational and is generated in one, and only one, way by reguli. It belongs to a bundle of quadratic complexes, any two of which have for residual intersection a linear congruence which has in common with the given congruence a ruled quartic surface of genus 1 (Art. 30, footnote). A generic line in space belongs to one and only one such residual linear congruence of intersection of two quadratic complexes of the bundle (Art. 28).

32. To find the equation of the system of quadrics determined by the reguli which generate the congruence, let $p_{ij}(i=1, 2, 3, 4)$ be the current co-

---

* The parametric equations of $V_3^3$, referred to a suitable coordinate system, may be written in the form

$$x_1=k_1l_1,\quad x_2=k_1l_2,\quad x_3=k_1l_3,\quad x_4=k_2l_1,\quad x_5=k_2l_2,\quad x_6=k_2l_3,$$

where $k_i$ and $l_i$ are parameters. The hyperquadrics which contain $V_3^3$ belong to the bundle

$$\lambda_1(x_1x_5-x_2x_4)+\lambda_2(x_1x_6-x_3x_4)+\lambda_3(x_2x_6-x_3x_5)=0.$$

Each hyperquadric of this bundle has a double line. The intersection of any two hyperquadrics of the bundle consists of $V_3^3$ and an $S_3$ which intersects the surface (17) in a quartic curve of genus 1.

ordinates in $S_5$ and let the equation of $F_2=0$ be $p_{12}p_{34}+p_{13}p_{42}+p_{14}p_{23}=0$ so that the coordinates of a point on $F_2=0$ are also the coordinates

$$p_{ij}=y_ix_j-x_iy_j \qquad i,\ j=1,\ 2,\ 3,\ 4,$$

of a line in $S_3$.

The equations of a generating plane of $V_3^2$ can be written in the form (cf. Art. 30, footnote),

$$\Sigma\phi_{ij}^{(1)}p_{ij}=0, \quad \Sigma\phi_{ij}^{(2)}p_{ij}=0, \quad \Sigma\phi_{ij}^{(3)}p_{ij}=0, \tag{18}$$

where $\phi_{ij}^{(h)}=-\phi_{ji}^{(h)}=a_{ij}^{(h)}k_1+b_{ij}^{(h)}k_2$, $i,\ j=1,2,3,4$, $h=1,\ 2,\ 3$, $a_{ij}^{(h)}$ and $b_{ij}^{(h)}$ are constants and $k_1$, $k_2$ are parameters which define the planes of $V_3^2$.

If we put $p_{ij}=y_ix_j-y_jx_i$, equations (18) take the form

$$\Sigma\phi_{ij}^{(1)}y_ix_j=0, \quad \Sigma\phi_{ij}^{(2)}y_ix_j=0, \quad \Sigma\phi_{ij}^{(3)}y_ix_j=0.$$

The condition that these equations have solutions in $(y)$ other than

$$\frac{y_1}{x_1}=\frac{y_2}{x_2}=\frac{y_3}{x_3}=\frac{y_4}{x_4}$$

is that all the third order determinants in the matrix

$$\left\|\begin{matrix} \phi_{12}^{(1)}x_2+\phi_{13}^{(1)}x_3+\phi_{14}^{(1)}x_4 & \phi_{21}^{(1)}x_1+\phi_{23}^{(1)}x_3+\phi_{24}^{(1)}x_4 & \phi_{31}^{(1)}x_1+\phi_{32}^{(1)}x_2+\phi_{34}^{(1)}x_4 & \phi_{41}^{(1)}x_1+\phi_{42}^{(1)}x_2+\phi_{43}^{(1)}x_3 \\ \phi_{12}^{(2)}x_2+\phi_{13}^{(2)}x_3+\phi_{14}^{(2)}x_4 & \phi_{21}^{(2)}x_1+\phi_{23}^{(2)}x_3+\phi_{24}^{(2)}x_4 & \phi_{31}^{(2)}x_1+\phi_{32}^{(2)}x_2+\phi_{34}^{(2)}x_4 & \phi_{41}^{(2)}x_1+\phi_{42}^{(2)}x_2+\phi_{43}^{(2)}x_3 \\ \phi_{12}^{(3)}x_2+\phi_{13}^{(3)}x_3+\phi_{14}^{(3)}x_4 & \phi_{21}^{(3)}x_1+\phi_{23}^{(3)}x_3+\phi_{24}^{(3)}x_4 & \phi_{31}^{(3)}x_1+\phi_{32}^{(3)}x_2+\phi_{34}^{(3)}x_4 & \phi_{41}^{(3)}x_1+\phi_{42}^{(3)}x_2+\phi_{43}^{(3)}x_3 \end{matrix}\right\|$$

vanish. These determinants have in common the factor $\Delta=\Sigma\Phi_{ij}x_i,\ x_j$ where

$$\Phi_{11}=\begin{vmatrix} \phi_{14}^{(1)}\phi_{21}^{(1)}\phi_{31}^{(1)} \\ \phi_{14}^{(2)}\phi_{21}^{(2)}\phi_{31}^{(2)} \\ \phi_{14}^{(3)}\phi_{21}^{(3)}\phi_{31}^{(3)} \end{vmatrix}, \quad 2\Phi_{12}=\begin{vmatrix} \phi_{12}^{(1)}\phi_{24}^{(1)}\phi_{31}^{(1)} \\ \phi_{12}^{(2)}\phi_{24}^{(2)}\phi_{31}^{(2)} \\ \phi_{12}^{(3)}\phi_{24}^{(3)}\phi_{31}^{(3)} \end{vmatrix}+\begin{vmatrix} \phi_{12}^{(1)}\phi_{14}^{(1)}\phi_{32}^{(1)} \\ \phi_{12}^{(2)}\phi_{14}^{(2)}\phi_{32}^{(2)} \\ \phi_{12}^{(3)}\phi_{14}^{(3)}\phi_{32}^{(3)} \end{vmatrix},$$

etc. Hence, $\Delta=0$ is the equation required. Six of these surfaces break up into pairs of planes.

The opposite reguli on this system of surfaces generate another congruence of the same type. It corresponds to the section of the hyperquadric $p_{12}p_{34}+p_{13}p_{42}+p_{14}p_{23}=0$ by the $V_3^2$

$$p_{mn}=a_{ij}^{(1)}k_1l_1+a_{ij}^{(2)}k_1l_2+a_{ij}^{(3)}k_1l_3+b_{ij}^{(1)}k_2l_1+b_{ij}^{(2)}k_2l_2+b_{ij}^{(3)}k_2l_3,$$

where $i$, $j$, $m$, $n$ are the four numbers 1, 2, 3, 4 in some order and $k_1$, $k_2$, $l_1$, $l_2$, $l_3$ are parameters.

33. If the surface (17) in $S_5$ has a double point, the curves in the $\xi$-plane which correspond to the sections of the surface by the hyperplanes through the double point constitute a web of curves of genus 1. If these curves are proper quartics with a double point additional to $P_0$, two of the fundamental points $P_1, P_2, \ldots, P_6$ are consecutive at this point; if they degenerate into a web of cubics and a fixed line through $P_0$, two simple fundamental points lie on the fixed line. Conversely, if two simple fundamental points are consecutive, or are collinear with* $P_0$, the surface (17) has a double point. Sixteen of the directrix conics pass through the double point. These conics are consecutive in pairs. The projection of the surface on $S_3$ has a double point distinct from the double curve.

Not more than six distinct double points can exist on the surface (17). Six such points exist when $P_1, P_2, \ldots, P_6$, are consecutive in pairs, in such a way that consecutive points are collinear with $P_0$. The six double points lie in pairs on three generating conics, each of which reduces to a right line counted twice. Three generating conics touch a generic hyperplane (cf. Art. 20).

34. If three simple fundamental points are collinear the surface has a directrix line. Six directrix conics degenerate into this line and a component of a generating conic. Such a surface arises as the intersection of a $V_3^3$ with an hyperquadric which contains a rectilinear generator of $V_3^3$ (Art. 30). The residual section of the surface by an hyperplane which contains the directrix line is a rational quintic which intersects the directrix line in three points.

The surface can have at most four directrix right lines without having a double point. The surface has four such lines when the points $P_1, P_2, \ldots, P_6$ are the vertices of a complete quadrilateral. The four directrix lines do not intersect. There are eight non-composite directrix conics on the surface. None of these intersects the directrix lines.

35. If the fundamental points $P_1, P_2, \ldots, P_6$ lie on a conic $C'$, then to $C'$ corresponds a conic $C$ which intersects each generating conic in two points. Any two generating conics form, with $C$, the section of the surface by an hyperplane. Hence, any two planes of $V_3^3$ lie in an $S_4$ and intersect so that $V_3^3$ is a three-spread cone which projects a ruled cubic surface (belonging to an $S_4$) from a fixed point. Conversely, if $V_3^3$ is conical, any two generating conics of the given surface lie in an $S_4$ whose residual intersection with the surface is a conic which intersects all the generating conics in two points so that $P_1, P_2, \ldots, P_6$ lie on a conic. None of the directrix conics intersects $C$.

* These two configurations are birationally equivalent.

If $C'$ also passes through $P_0$, all the generating conics pass through the triple point of $V_3^3$ which is, in this case, a triple point on the given surface. Conversely, if the surface (17) has a triple point, all the generating conics pass through the triple point, and this point is triple on $V_3^3$. The $S_4$ determined by the planes of any two generating conics contains, as residual intersection with the surface, a directrix conic which passes through the triple point and intersects each generating conic on the surface in a second point. By a birational transformation in the $\xi$-plane, the conic $C'$ may be transformed into a right joining four simple fundamental points.

The given surface can have two triple points. This happens when $P_1$ and $P_2$ are each consecutive to $P_0$ and $P_3P_6P_5P_6$ are collinear. The planes of $V_3^3$ join the points of a cubic curve to a fixed right line $l$ which is triple on $l$. The triple points of the surface are the points of intersection of $l$ with an hyperquadric on which the surface lies (Art. 30). They may be consecutive.

THE UNIVERSITY OF ILLINOIS, *March* 29, 1915.

# *A Theorem Connected with Irrational Numbers.*

By William Duncan MacMillan.

Let us imagine in that portion of the plane in which the coordinates are both positive that the points whose coordinates are both integers are marked with a dot (lattice points), and let us imagine a straight line drawn through the origin making an angle $\phi$ with the $x$-axis. Let $\beta = \tan\phi$, $0 < \phi < \pi/2$. Consider the lattice points in the vertical line whose $x$-coordinates is $q_k$. One of these points is closer to the line $\phi$ than any other point in the same vertical line. Let the $y$-coordinate of this point be $p_k$, and let $y_k$ be the ordinate of the line $\phi$ corresponding to the abscissa $q_k$, so that $y_k = \beta q_k$. We will call $p_k - y_k = p_k - \beta q_k = d_k$ the distance of the point $q_k$, $p_k$ from the line $\phi$, though of course it is not the perpendicular distance. From the definition it is clear that $d_k$ is positive if the lattice point lies above the line $\phi$, and negative if it lies below the line $\phi$.

The theorem which is proved in the following pages relates to the limiting value of the geometric mean of the $|d_k|$. If $G_n$ denotes the geometric mean of the first $n$ quantities $|d_k|$, and if $\beta$ is an irrational number which satisfies a certain condition, then the limit of $G_n$, as $n$ approaches infinity, is $1/(2e)$, where $e$ is the logarithmic base. If $\beta$ is rational the limit of $G_n$ is, of course, zero. For values of $\beta$ other than the two classes already mentioned $G_n$ can not be said to have a limit as it oscillates between $1/(2e)$ and $l_\beta$, where $0 \overline{<} l_\beta < 1/(2e)$.

As the method of proof is based upon the properties of simple continued fractions, we shall set forth explicitly those formulas which will prove most useful. Let us suppose $\beta$ is developed as a simple continued fraction

$$\beta = a_0 + \frac{1}{a_1 +}\,\frac{1}{a_2 +}\,\cdots,$$

and let $p_n/q_n$ be the $n$-th principal convergent, and $p_n^{(s)}/q_n^{(s)}$ be an intermediate convergent. Then from the theory of continued fractions we will have

$$p_{n+1} = a_{n+1}p_n + p_{n-1}, \qquad \varepsilon_n = p_n - \beta q_n,$$

$$q_{n+1} = a_{n+1}q_n + q_{n-1}, \qquad |\varepsilon_{n-1}| > |\varepsilon_n| > |\varepsilon_{n+1}|,$$

$$\varepsilon_{n+1} = a_{n+1}\varepsilon_n + \varepsilon_{n-1},$$

$$p_n^{(s)} = (s+1)p_n + p_{n-1}, \quad q_n^{(s)} = (s+1)q_n + q_{n-1}, \quad \varepsilon_n^{(s)} = (s+1)\varepsilon_n + \varepsilon_{n-1}.$$

Since the principal convergent points $q_n$, $p_n$ are the closest approximations to $\beta$ it follows that these points lie closest to the line $\phi$. Since the $\varepsilon_n$ are alternately positive and negative, these principal convergent points lie alternately above and below the line $\phi$. If the convergent points $q_{n-1}$, $p_{n-1}$ and $q_{n+1}$, $p_{n+1}$ be joined by a straight line, the intermediate convergent points $q_n^{(s)}$, $p_n^{(s)}$ will lie on this straight line, and between this straight line and the line $\phi$ there are no lattice points. Finally,

$$\frac{1}{q_{n+1}} > |\varepsilon_n| > \frac{a_{n+2}}{q_{n+2}},$$

or, simpler,

$$\frac{1}{a_{n+1}\,q_n} > |\varepsilon_n| > \frac{1}{(a_{n+1}+2)\,q_n}.$$

Let us join the origin and the convergent point $q_n$, $p_n$ by a straight line and denote the distances of the lattice points from this straight line by $\delta_k$. In order to avoid confusion we shall call this straight line the rational line. It is clear that the distances $\delta_k$ in some order have the values

$$+\frac{1}{q_n},\ -\frac{1}{q_n},\ +\frac{2}{q_n},\ -\frac{2}{q_n},\ \ldots,\ \pm\frac{(q_n-1)/2}{q_n} \text{ or } \pm 1/2,$$

according as $q_n$ is odd or even, for there are $q_n-1$ such distances (excluding the end points). If we suppose two of them are equal, say $\delta_k$ and $\delta_j$, then we have $\delta_j-\delta_k=0$; that is,

$$(p_j-p_k)-\frac{p_n}{q_n}(q_j-q_k)=0,$$

whence

$$\frac{p_j-p_k}{q_j-q_k}=\frac{p_n}{q_n},$$

which is impossible since $|p_j-p_k|<p_n$, $|q_j-q_k|<q_n$, and $p_n$ and $q_n$ are relatively prime. Furthermore, from symmetry we have $\delta_k=-\delta_{n-k}$.

In computing the geometric mean of the quantities $|\delta_k|$ for the rational line we have

$$\prod_{k=1}^{q_n-1}|\delta_k|=\frac{1}{q_n}\cdot\frac{2}{q_n}\cdot\frac{3}{q_n}\cdot\,\ldots\,\cdot\frac{m}{q_n}\cdot\frac{m}{q_n}\cdot\,\ldots\,\cdot\frac{2}{q_n}\cdot\frac{1}{q_n} \quad \text{if } q_n=2m+1,$$

$$\prod_{k=1}^{q_n-1}|\delta_k|=\frac{1}{q_n}\cdot\frac{2}{q_n}\cdot\,\ldots\,\cdot\frac{m}{q_n}\cdot\frac{1}{2}\cdot\frac{m}{q_n}\cdot\,\ldots\,\cdot\frac{2}{q_n}\cdot\frac{1}{q_n} \quad \text{if } q_n=2m+2.$$

In general, the factors will not occur in the order above written, but all of the above factors occur with none others. It will be observed that the distance for the end point $\delta_n$ has been omitted since it is zero. It is seen, therefore, that we are computing the geometric mean out to the $n$-th convergent point, but that this point itself is excluded. We have then

$$\left[\prod_{k=1}^{q_n-1} |\delta_k|\right]^{\frac{1}{q_n-1}} = \left[\frac{(m!)^2}{(2m+1)^{2m}}\right]^{\frac{1}{2m}} = \left[\frac{m!}{(2m+1)^m}\right]^{\frac{1}{m}} \text{ if } q_n=2m+1,$$

$$\left[\prod_{k=1}^{q_n-1} |\delta_k|\right]^{\frac{1}{q_n-1}} = \left[\frac{1}{2}\frac{(m!)^2}{(2m+2)^{2m}}\right]^{\frac{1}{2m+1}} = \frac{1}{2}\left[\frac{m!}{(m+1)^m}\right]^{\frac{2}{2m+1}} \text{ if } q_n=2m+2.$$

In order to show that these expressions have definite limiting values as $n$, and consequently $m$, increases indefinitely we will use the theorem of Cauchy*

$$\underset{n=\infty}{L}\ [f(n)]^{\frac{1}{n}} = \underset{n=\infty}{L}\left[\frac{f(n+1)}{f(n)}\right].$$

If $q_n=2m+1$, we have $f(m)=\dfrac{m!}{(2m+1)^m}$. Consequently,

$$\frac{f(m+1)}{f(m)} = \frac{m+1}{2m+3}\left(\frac{2m+1}{2m+3}\right)^m = \frac{m+1}{2m+3}\left(1-\frac{1}{m+3/2}\right)^m,$$

and the limit as $m$ increases indefinitely is $1/(2e)$, where $e=2\cdot71828\ldots.$, the Naperian base. In precisely the same fashion it is found that $\frac{1}{2}\left[\frac{m!}{(m+1)^m}\right]^{\frac{2}{2m+1}}$ also has the limit $1/(2e)$. It follows then that whether $q_n$ is even or odd

$$\underset{n=\infty}{L}\left[\prod_{k=1}^{q_n-1} |\delta_k|\right]^{\frac{1}{q_n-1}} = \frac{1}{2e}.$$

Consider now the geometric mean for the irrational line. Let

$$G_s = \left[\prod_{k=1}^{s} |d_k|\right]^{\frac{1}{s}},$$

where $d_k = p_k - \beta q_k$ are the distances from the irrational line corresponding to the $\delta_k$ for the rational line.

If every $|d_k|$ is replaced by $|\delta_k| + \dfrac{1}{q_n}$, the resulting geometric mean will be too great, since

$$|d_k| < |\delta_k| + |\varepsilon_n| < |\delta_k| + \frac{1}{q_n};$$

and if every $|d_k|$ is replaced by $|\delta_k| - \dfrac{1}{q_n}$, except for the two points for which this expression is zero, the resulting geometric mean will be too small. The

* "Analyse Algebrique." See also Chrystal's "Algebra," II, p. 84.

two exceptional points for which $|\delta_k| = \dfrac{1}{q_n}$ are the points $(q_{n-1}, p_{n-1})$ and the symmetrical point $((a_n-1)q_{n-1}+q_{n-2}, (a_n-1)p_{n-1}+p_{n-2})$. These two points are at equal distances from the ends of the rational line and lie on opposite sides of it. When the rational line is rotated into coincidence with the irrational line, the distance of one of these points is increased while the other is decreased. It is readily verified that it is the distance of the point $q_{n-1}, p_{n-1}$ which is decreased. From the geometry it follows that the amount of this decrease is $\dfrac{q_{n-1}}{q_n}|\varepsilon_n| < \dfrac{1}{2q_n}$. Hence $|d_{q_{n-1}}| > \dfrac{1}{2q_n}$, and this is the only distance which is less than $1/q_n$. It results, then, that

$$\left[\frac{\frac{1}{2}(m-1)!}{(2m+1)^m}\right]^{\frac{1}{m}} < G_{2m} < \left[\frac{(m+1)!}{(2m+1)^m}\right]^{\frac{1}{m}}, \text{ if } q_n=2m+1,$$

and

$$\frac{1}{2}\left[\frac{m}{4}\cdot\frac{(m-1)!^2}{(m+1)^{2m+1}}\right]^{\frac{1}{2m+1}} < G_{2m+1} < \frac{1}{2}\left[(m+2)\frac{(m+1)!^2}{(m+1)^{2m+1}}\right]^{\frac{1}{2m+1}}, \text{ if } q_n=2m+2.$$

The limit of all of these functions of $m$ is $1/(2e)$. It follows, therefore, that *the limit of the geometric mean of the quantities $|d_k|$ taken out to, but not including, a principal convergent point is $1/(2e)$ for every irrational number.*

We wish, however, to include the distance of the convergent point itself. For this point $|d_{q_n}| = |\varepsilon_n|$ and

$$\frac{1}{(a_{n+1}+2)q_n} < |\varepsilon_n| < \frac{1}{a_{n+1}q_n}.$$

Since the limit of $\left(\dfrac{1}{q_n}\right)^{\frac{1}{q_n}}$ for large values of $q_n$ is unity, we have

$$\underset{n=\infty}{L}|\varepsilon_n|^{\frac{1}{q_n}} = \underset{n=\infty}{L}\left(\frac{1}{a_{n+1}}\right)^{\frac{1}{q_n}}.$$

The value of $q_n$ depends upon $a_1, \ldots, a_n$, but not upon $a_{n+1}$. Since $a_{n+1}$ is an integer, $\left(\dfrac{1}{a_{n+1}}\right)^{\frac{1}{q_n}}$ lies between zero and unity, and if it has a limit as $n$ increases indefinitely, say $l_\beta$, then it is clear that

$$\underset{n=\infty}{L} G_{q_n} = (l_\beta)/(2e),$$

and, in particular, if $l_\beta=1$ the limit, not only of $G_{q_n-1}$ but also of $G_{q_n}$, is $1/(2e)$. The limit $l_\beta=1$ will be assured if

$$a_{n+1} \leqq M\, q_n(q_n+1) \ldots . (q_n+s)$$

or, its equivalent,

$$a_{n+1} \leqq M q_n^{s+1}$$

for every $n$ greater than a fixed $n_1$, where $M$ is a chosen positive quantity and $s$ is any assigned positive integer. This condition is satisfied by every real algebraic number.* It is also satisfied by certain transcendental numbers, for example, the irrational number $e$, for which the values of the $a_n$ are known. Since $q_n > a_1 \cdot a_2 \cdot \ldots . \cdot a_n$ it is seen that $q_n$ increases very rapidly even for moderate values of the $a_j$; and, therefore, that the condition is satisfied by a very wide class of irrationals. In order that the condition be not satisfied it is necessary that ultimately the $a_n$ shall increase with exceeding rapidity.

If the quantity $\left(\frac{1}{a_{n+1}}\right)^{\frac{1}{q_n}}$ does not have any fixed limit as $n$ increases indefinitely but oscillates between zero and unity, then $G_{q_n}$ oscillates between zero and $1/(2e)$ although $G_{q_n-1}$ has the fixed limit $1/(2e)$.

In what follows we shall assume that the condition $\underset{n=\infty}{L} \left(\frac{1}{a_{n+1}}\right)^{\frac{1}{q_n}} = 1$ is satisfied, and free ourselves of the restriction that we proceed to the limit along a particular set of points. Let us take $b_n < a_{n+1}$ and consider the geometric mean out to the point $b_n q_n$. The distance of this point from the irrational line is $b_n |\epsilon_n|$. Since $a_{n+1}\epsilon_n = \epsilon_{n+1} - \epsilon_{n-1}$ and $|\epsilon_{n-1}| < \frac{1}{q_n}$ we have $b_n |\epsilon_n| < \frac{b_n}{a_{n+1}} \cdot \frac{1}{q_n} < \frac{1}{q_n}$. We draw now the rational line from the origin to the point $b_n q_n$, $b_n p_n$. For this rational line the $|\delta_k|$ in some order have the values $0, \frac{1}{q_n}, \frac{2}{q_n}, \ldots, \frac{m}{q_n}$ or $\frac{1}{2}$ according as $q_n$ is odd or even. The points for which $|\delta_k| = 0$ are the integral multiples of $q_n$; and the points for which $|\delta_k| = 1/q_n$ are the points whose abscissæ are $sq_n \pm q_{n-1}$, $s = 1, \ldots, b_n$. When the rational line is rotated into coincidence with the irrational line, the $|\delta_k|$ for the points $sq_n - q_{n-1}$ are increased, while for the points $sq_n + q_{n-1}$ they are decreased. It is, therefore, the points $sq_n + q_{n-1}$, the intermediate convergent points, which engage our attention. We have

$$\text{for } k = sq_n, \qquad |d_k| = 0 + s|\epsilon_n|, \qquad s \leqq b_n,$$

$$\text{for } k = sq_n + q_{n-1}, \qquad |d_k| = \frac{1}{q_n} - \left(s + \frac{q_{n-1}}{q_n}\right)|\epsilon_n|, \qquad s \leqq b_n - 1.$$

---

* Liouville *Journal de Mathématiques*, XVI (1851).

Since $|\epsilon_n| < \frac{1}{q_{n+1}}$ we have also

$$\left(s+\frac{q_{n-1}}{q_n}\right)|\epsilon_n| < \left(s+\frac{q_{n-1}}{q_n}\right)\frac{1}{a_{n+1}q_n+q_{n-1}} = \frac{1}{q_n}\frac{s+\frac{q_{n-1}}{q_n}}{a_{n+1}+\frac{q_{n-1}}{q_n}},$$

so that

$$\frac{1}{q_n}-\left(s+\frac{q_{n-1}}{q_n}\right)|\epsilon_n| > \frac{1}{q_n}\frac{a_{n+1}-s}{a_{n+1}+1}, \quad s \leqq b_n-1.$$

When the rational line is rotated into coincidence with the irrational line, no $|\delta_k|$ is altered by as much as $\frac{1}{q_n}$. If then all of the $|\delta_k| > \frac{1}{q_n}$ be diminished by $1/q_n$ we will obtain a geometric mean which is less than $G_{b_n q_n}$, and if all of the $|\delta_k|$ be increased by $\frac{1}{q_n}$, without exception, we will obtain a geometric mean which is greater than $G_{b_n q_n}$. Carrying out these operations first for $q_n$ odd and second for $q_n$ even, we find

$$\left[\left(\frac{|\epsilon_n|}{a_{n+1}+1}\right)^{b_n}\frac{(m-1)\,!^{2b_n}}{q_n^{b_n(q_n-1)}}\right]^{\frac{1}{b_n q_n}} < G_{b_n q_n} < \left[\frac{(m+1)\,!^{2b_n}}{q_n^{b_n q_n}}\right]^{\frac{1}{b_n q_n}} \text{ if } q_n = 2m+1,$$

$$\left[\left(\frac{|\epsilon_n|}{a_{n+1}+1}\right)^{b_n}\frac{m^{b_n}(m-1)\,!^{2b_n}}{q_n^{b_n(q_n-1)}}\right]^{\frac{1}{b_n q_n}} < G_{b_n q_n} < \left[\frac{(m+2)\,(m+1)\,!^{2b_n}}{q_n^{b_n q_n}}\right]^{\frac{1}{b_n q_n}} \text{ if } q_n = 2m+2.$$

These two inequalities can be replaced by the single one, whether $q_n$ be even or odd,

$$\left(\frac{1}{a_{n+1}+2}\right)^{\frac{2}{q_n}}\left[\frac{(m-1)\,!^2}{q_n^{q_n}}\right]^{\frac{1}{q_n}} < G_{b_n q_n} < \left[\frac{(m+2)\,!^2}{q_n^{q_n}}\right]^{\frac{1}{q_n}}.$$

Since by hypothesis the limit of $\left(\frac{1}{a_{n+1}+2}\right)^{\frac{1}{q_n}} = 1$, both extremes of this inequality have the limiting value $1/(2e)$. It follows that the limit of $G_{b_n q_n}$ is $1/(2e)$.

The general point $q$ can be written

$$q = b_n q_n + b_{n-1} q_{n-1} + \ldots\ldots + b_1 q_1 + b_0,$$

where $b_k \leqq a_{k+1}$. This point, therefore, lies between $b_n q_n$ and $b_{n+1} q_n$, which is an interval of length $q_n = a_n q_{n-1} + q_{n-2}$. If now we attempt to continue with the interval $b_{n-1} q_{n-1}$, a difficulty is encountered at the intermediate convergent point $b_n q_n + q_{n-1}$. For values of $b_n$ in the neighborhood of $a_{n+1}$ the distance of this lattice point is of the order $|\epsilon_n|$, while the root which we take is of the order $\frac{1}{q_{n-1}}$. Our hypothesis on the $a_{n+1}$ would not cover the situation at this point. If, therefore, $b_{n-1} \neq 0$, it will be convenient to include at one step the interval $q_n+q_{n-1}$ which immediately precedes this point; that is, the interval from $(b_n-1)q_n$ to $b_n q_n+q_{n-1}$, and we will suppose the straight line joining these two points to have been drawn.

Since $p_n+p_{n-1}$ is prime to $q_n+q_{n-1}$, the values of the $|\delta_k|$ for this rational line are integral multiples of $\frac{1}{q_n+q_{n-1}}$, none vanishing except for the end points. Counting $k$ from the beginning of this rational line we have $|\delta_k|=\frac{1}{q_n+q_{n-1}}$ for $k=q_n$, and $|\delta_k|=\frac{2}{q_n+q_{n-1}}$ for $k=q_n-q_{n-1}$. The $|\delta_k|$ for the symmetrical points have the same values.

Now $|b_n\epsilon_n+\epsilon_{n-1}| \leq |\epsilon_n+\epsilon_{n-1}| < \frac{2}{q_n+q_{n-1}}$, so that when the rational line is moved into coincidence with the irrational line, no $|\delta_k|$ is altered by as much as $\frac{2}{q_n+q_{n-1}}$. We have, therefore, to examine the points for which $|\delta_k|$ equals $\frac{1}{q_n+q_{n-1}}$ or $\frac{2}{q_n+q_{n-1}}$. For the points for which $|\delta_k|=\frac{1}{q_n+q_{n-1}}$, $k=q_n$ and $q_{n-1}$. Hence $|d_k|=b_n|\epsilon_n|$ and $|(b_n-1)\epsilon_n+\epsilon_{n-1}|$ respectively, both of which are greater than $|\epsilon_n|$. For the points for which $|\delta_k|=\frac{2}{q_n+q_{n-1}}$, we have $k=q_n-q_{n-1}$ and $2q_{n-1}$. At these points $|d_k|$ is equal to $|b_n\epsilon_n-\epsilon_{n-1}|$ and $|(b_n-1)\epsilon_n+2\epsilon_{n-1}|$ and both of these quantities are greater than $\frac{1}{q_n+q_{n-1}}$. If, for convenience, of notation we denote $q_n+q_{n-1}$ by $Q_n$ and the geometric mean of the quantities $|d_k|$ between the points $(b_n-1)q_n+1$ and $b_n q_n+q_{n-1}$, both inclusive, by $G_{Q_n}$, then whether $Q_n$ is even or odd we will have

$$Q_n=2m+1 \quad \text{or} \quad Q_n=2m+2,$$

$$\left[\frac{(m-2)\,!^2}{Q_n{}^{Q_n}\,a_{n+1}^2}\right]^{\frac{1}{Q_n}} < G_{Q_n} < \left[\frac{(m+3)\,!^2}{Q_n{}^{Q_n}}\right]^{\frac{1}{Q_n}},$$

and these expressions also have the limit $1/(2e)$.

No further difficulties are encountered in proceeding from the point $b_n q_n+q_{n-1}$ to the point $b_n q_n+b_{n-1}q_{n-1}$ since the minimum $|d_k|$ occur at the intermediate convergent points and have the values $|b_n\epsilon_n+s\epsilon_{n-1}|>|\epsilon_{n-1}|$ and $|d_k|$ does not differ from $|\delta_k|$ by as much as $\frac{1}{q_{n-1}}$.

Let us suppose that, proceeding in this manner, we have arrived at the point

$$b_n q_n + b_{n-1} q_{n-1} + \ldots + b_{n-r} q_{n-r},$$

and we endeavor to continue to the point

$$b_n q_n + \ldots + b_{n-r} q_{n-r} + b_{n-r-1} q_{n-r-1}.$$

If $b_{n-r-1}$ is zero the interval is of length less than $q_{n-r-1}$ and we proceed to an interval which is a multiple of $q_{n-r-2}$, and in such a case no difficulties arise at the subintermediate convergent points. But if $b_{n-r} \neq 0$, then

$$b_n q_n + \ldots + b_{n-r} q_{n-r} + q_{n-r-1}$$

is a subintermediate convergent point in an interval of length $q_{n-r}$. To arrive at this point we start from the point $b_n q_n + \ldots + (b_{n-r} - 1) q_{n-r}$ and include in one step the interval $(q_{n-r} + q_{n-r-1})$, which brings us to the point $b_n q_n + \ldots + b_{n-r} q_{n-r} + q_{n-r-1}$.

For notation let us take

$$P_{n-r} = p_{n-r} + p_{n-r-1}, \quad Q_{n-r} = q_{n-r} + q_{n-r-1}, \quad d_k^{(n-r)} = d_{b_n q_n + \ldots (b_n - 1) q_{n-r} + k}.$$

Then, since $P_{n-r}$ is prime to $Q_{n-r}$ the $|\delta_k^{(n-r)}|$ have values which are integral multiples of $\frac{1}{Q_{n-r}}$. If $k$ equals $q_{n-r-1}$ or $q_{n-r}$ then $|\delta_k^{(n-r)}| = \frac{1}{Q_{n-r}}$; and if $k$ equals $2q_{n-r-1}$ or $q_{n-r} - q_{n-r-1}$, then $|\delta_k^{(n-r)}| = \frac{2}{Q_{n-r}}$. The difference between the rational line and the irrational line for this interval at no point exceeds $|b_n \epsilon_n + \ldots + b_{n-r} \epsilon_{n-r} + \epsilon_{n-r-1}| \leq |\epsilon_{n-r} + \epsilon_{n-r-1}| < \frac{2}{Q_{n-r}}$. Hence, when the rational line is moved into coincidence with the irrational line no $|\delta_k^{(n-r)}|$ is altered by as much as $\frac{2}{Q_{n-r}}$. We need, therefore, to examine those points for which $|\delta_k^{(n-r)}| < \frac{3}{Q_{n-r}}$. For the points for which $|\delta_k^{(n-r)}| = \frac{1}{Q_{n-r}}$ we have $|d_k^{(n-r)}| = |b_n \epsilon_n + \ldots + (b_{n-r} - 1) \epsilon_{n-r} + \epsilon_{n-r-1}|$ and $|b_n \epsilon_n + \ldots + b_{n-r} \epsilon_{n-r}|$ respectively, and both of these quantities are less than $|\epsilon_{n-r}|$. For the points for which $|\delta_k^{(n-r)}| = \frac{2}{Q_{n-r}}$, we have

$$|b_n \epsilon_n + \ldots + b_{n-r} \epsilon_{n-r} + 2\epsilon_{n-r-1}|$$

and

$$|b_n \epsilon_n + \ldots + b_{n-r} \epsilon_{n-r} - \epsilon_{n-r-1}|,$$

each of which is greater than $\frac{1}{Q_{n-r}}$.

If, therefore, $G_{Q_{n-r}}$ denotes the geometric mean of the quantities $|d_k^{(n-r)}|$, $k=1, \ldots, Q_{n-r}$, and if $Q_{n-r}=2m+1$ or $2m+2$ we will have

$$\left[\frac{(m-2)!^2}{Q_{n-r}^{Q_{n-r}-3}}\,|\varepsilon_{n-r}|^3\right]^{\frac{1}{Q_{n-r}}} < G_{Q_{n-r}} < \left[\frac{(m+3)!}{Q_{n-r}^{Q_{n-r}}}\right]^{\frac{1}{Q_{n-r}}}.$$

For a fixed $r$ the two extremes of this inequality have the limit $1/(2e)$ as $n$ increases indefinitely.

We can now proceed from the point $b_n q_n + \ldots + b_{n-r} q_{n-r} + q_{n-r-1}$ to the point $b_n q_n + \ldots + b_{n-r-1} q_{n-r-1}$ without further difficulty. Draw the rational line connecting these two points. The distance between this line and the irrational line is nowhere greater than $\dfrac{1}{q_{n-r-1}}$. The values of the $|d|$ for the points for which $|\delta|=0$ or 1 are in every case greater than $|\varepsilon_{n-r-1}|$. Consequently, inequalities similar to the above can be constructed, and these inequalities have the same limits as before.

We have been compelled to break up the interval $q=b_n q_n + \ldots + b_{n-r} q_{n-r} + \ldots + b_0$ into the intervals $q=c_n q_n + Q_n + \ldots + c_{n-r} q_{n-r} + Q_{n-r} + \ldots + c_0$, where $c_n=b_n-1$, $c_j=b_j-2$, $j=2, \ldots, n-1$, $c_1=b_1-1$, $c_0=b_0$. Let us suppose that the geometric mean of the $|d_k|$ which belong to the interval $q_k$ are denoted by $G_k$ and for the interval $Q_k$ by $G_{k,1}$. Then the geometric mean $G_q$ for the entire interval $q$ will be

$$G_q = \left[G_0^{c_0} G_1^{c_1 q_1} \prod_{k=2}^{n} G_k^{c_k q_k} G_{k,1}^{Q_k}\right]^{\frac{1}{q}},$$

where the quantities $G_k$ and $G_{k1}$ approach the limit $1/(2e)$ as $k$ increases indefinitely.

The purpose of our investigation is now attained by the use of the following:

Lemma: *If $j_1, j_2, \ldots,$ is any set of positive integers such that $j_1 \leq J_1$, $j_2 \leq J_2, \ldots,$ where $J_1, J_2, \ldots,$ is a given set of integers, and if $x_1, x_2, \ldots,$ is a set of positive quantities which approach $x$ as a limit, then*

$$y_n = \left[\prod_{k=1}^{n} x_k^{j_k}\right]^{\frac{1}{j_1+\ldots+j_n}}$$

*also approaches $x$ as a limit as $n$ increases indefinitely, provided there does not exist an $n=N$ such that if $m>N$ every $j_m=0$.*

The proof of this lemma is simple and will be omitted. Since all of its conditions are satisfied by the expression for $G_q$, we have proved the following:

Theorem: *If $\beta$ is a given positive number, and $p_n/n$ is a rational fraction such that $|p_n - n\beta| \leq \frac{1}{2}$, and if $G_n = \left[\prod_{k=1}^{n} |p_k - k\beta|\right]^{\frac{1}{n}}$, then the limit of $G_n$ as $n$*

*increases indefinitely is zero if $\beta$ is rational, and is $1/(2e)$ if $\beta$ is an irrational number which, when expressed as a simple continued fraction*

$$\beta = a_0 + \frac{1}{a_1 +}\frac{1}{a_2 +}\frac{1}{a_3 +}\cdots,$$

*of which $p_n/q_n$ is the $n$-th convergent, satisfies the condition that the limit of* $\left(\frac{1}{a_{n+1}}\right)^{\frac{1}{q_n}} = 1.$

*For every irrational number, $\beta$, the limit of $G_{q_n-1}$ as $n$ increases indefinitely is $1/(2e)$, where $e$ is the logarithmic base.*

If $E(k\beta)$ is the greatest integer contained in $k\beta$ and if $f_{k\beta} = k\beta - E(k\beta)$, it is readily seen that limit of $[\prod_{k=1}^{n} f_{k\beta}]^{1/n}$ is $\frac{1}{e}$, provided $\beta$ satisfies the condition already stated.

In a manner, quite analogous to the preceding argument, it can be shown that the limit of the arithmetic mean of the quantities $d_k$ (signs considered) is zero whether $\beta$ is rational or irrational, provided that when $\beta$ is rational and $d_k = \pm\frac{1}{2}$ is ambiguous such terms be taken alternately positive and negative. Likewise the limit the arithmetic mean of $|d_k|$ (signs discarded) is $\frac{1}{4}$ if $\beta$ is any irrational number or any rational number which in its lowest terms has an even denominator; and is equal to $\frac{1}{4}\left(1 - \frac{1}{q^2}\right)$ if $\beta$ is a rational number which, when expressed in its lowest terms, has an odd denominator $q$. The proof for the theorems with respect to the arithmetic means will not be necessary inasmuch as substantially equivalent theorems have been given by Sierpinski.*

* See *Jahrbuch über die Fortschritte der Mathematik* (1909), p. 221; Hardy and Littlewood, *Proceedings of the Cambridge Congress of Mathematicians* (1912), Vol. I.

# *On the Representation of Arbitrary Functions by Definite Integrals.*

BY WALTER B. FORD.

1. *Introduction.* Suppose there is given an arbitrary function $f(x)$ defined throughout the real interval $(-\pi, \pi)$ and let the Fourier series for $f(\alpha)$ be formed, it being understood that $\alpha$ represents any special value of $x$ within the indicated interval. The sum of the first $n+1$ terms of the series may be put in the form

$$S_n(\alpha) = \int_{-\pi}^{\pi} f(x)\psi(n, x-\alpha)\,dx, \text{ where } \psi(n, x-\alpha) = \frac{\sin\frac{2n+1}{2}(x-\alpha)}{2\pi\sin\frac{x-\alpha}{2}}, \quad (1)$$

and it is a familiar fact that we then have, at least after suitable conditions have been placed upon $f(x)$ in the neighborhood of the point $x=\alpha$ *

$$\lim_{n=+\infty} S_n = \frac{f(\alpha-0)+f(\alpha+0)}{2}. \quad (2)$$

The facts just mentioned readily suggest a general problem as follows: Having given an arbitrary function $f(x)$ defined throughout any (finite) interval $(a, b)$, consider the integral (analogous to (1))

$$I_n(\alpha) = \int_a^b f(x)\phi(n, x-\alpha)\,dx, \quad (3)$$

wherein $\phi(n, x-\alpha)$ now represents an undetermined function of $n$ and $x-\alpha$, and let it be asked what conditions upon $\phi(n, x-\alpha)$ suffice to insure the following relation analogous to (2):

$$\lim_{n=+\infty} I_n(\alpha) = \frac{f(\alpha-0)+f(\alpha+0)}{2}, \quad (4)$$

it being assumed as before that $f(x)$ satisfies suitable conditions in the neighborhood of the point $x=\alpha$.

The question thus raised was first considered and in part answered by Du Bois Reymond as early as 1875,† while in more recent years it has formed

* For example, whenever $f(x)$ has limited total fluctuation in the neighborhood in question.

† *Crelle*, Vol. LXXIX.

the subject of numerous and extensive researches of which those of Dini,* Hobson † and Lebesgue ‡ should be especially mentioned. For the most part, however, little attention appears to have been given to the actual determination of functions $\phi$ satisfying the conditions in question, the sole desideratum being the determination of the conditions themselves. In this connection it is proposed in the present paper to point out a noteworthy class of possible functions $\phi$ with especial emphasis on the corresponding integrals (3) to which they give rise. In particular, we shall show (§ 3) that to every convergent improper integral of the form

$$\int_0^\infty p(x)\,dx=k\neq 0$$

there corresponds an integral (3) having the property (4) provided merely that $p(x)$ is an even function of $x$, *i. e.*, $p(-x)=p(x)$, and $|xp(x)|$ remains less than a constant when considered for all values of $x$ lying outside an arbitrarily small interval surrounding the origin.

2. We proceed at once to state and prove the following theorem:§

THEOREM I. *Let $F(x)$ be any single valued function of the real variable $x$ defined for all finite values of $x$ and satisfying the following three conditions:*

(a) $\lim\limits_{x=+\infty} F(x)$ *exists and* $=k\neq 0$,

(b) $F(-x)=-F(x)$,

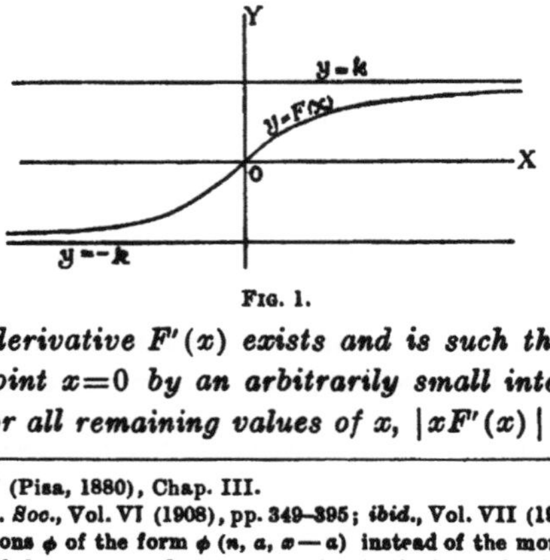


FIG. 1.

(c) *The derivative $F'(x)$ exists and is such that if we exclude the point $x=0$ by an arbitrarily small interval, we shall have for all remaining values of $x$, $|xF'(x)|<A=$ a constant.*

---

* "Serie di Fourier" (Pisa, 1880), Chap. III.

† *Proc. London Math. Soc.*, Vol. VI (1908), pp. 349–395; *ibid.*, Vol. VII (1909), pp. 338–358. Both Dini and Hobson consider functions $\phi$ of the form $\phi(n, a, x-a)$ instead of the more limited form $\phi(n, x-a)$ mentioned above. We shall, however, confine our attention to the latter in this paper.

‡ *Annales de Toulouse* (3), Vol. I (1909), pp. 25–128.

§ The author's attention has recently been called by Dr. W. W. Küstermann to the fact that Pringsheim, in his lectures for 1910–11, developed a theorem closely allied to the one here stated, though it does not appear to have been given in published form.

*Then, if $f(x)$ be an arbitrary function of the real variable $x$ defined throughout the interval $(a, b)$, we shall have for any special value $\alpha (a<\alpha<b)$*

$$\lim_{n=+\infty} \frac{1}{2k}\int_a^b f(x)\frac{d}{dx}F[n(x-\alpha)]dx = \frac{f(\alpha-0)+f(\alpha+0)}{2},$$

*provided merely that $f(x)$ satisfies suitable conditions (analogous to those under which the Fourier series for $f(\alpha)$ converges) in the neighborhood of the point $x=\alpha$.**

The proof of this theorem follows directly from the fact that if $f(x)$ satisfies the indicated conditions there exists the general relation

$$\lim_{n=+\infty} \int_a^b f(x)\phi(n, x-\alpha)dx = \frac{f(\alpha-0)+f(\alpha+0)}{2}$$

in case $\phi(n, t)$ is any function of the independent variables $n$ and $t$ satisfying the following three conditions, $\varepsilon$ always denoting an arbitrarily small positive quantity: †

$$\text{(I)} \quad \lim_{n=+\infty} \int_0^t \phi(n, t)dt = \begin{cases} -\frac{1}{2} \text{ when } -\varepsilon<t<0, \\ +\frac{1}{2} \text{ when } 0<t<\varepsilon, \end{cases}$$

$$\text{(II)} \quad \left|\int_0^t \phi(n, t)dt\right| < c_1 \text{ when } -\varepsilon<t<\varepsilon, \quad c_1 \text{ being a constant (independent of } n),$$

$$\text{(III)} \quad |\phi(n, t)| < c_2 \text{ when } \begin{Bmatrix} a-\alpha<t<-\varepsilon \text{ or} \\ \varepsilon<t<b-\alpha \end{Bmatrix}, \quad c_2 \text{ being a constant (independent of } n).$$

In fact, if $F(x)$ satisfies conditions (a), (b), (c), we have only to take

$$\phi(n, t) = \frac{1}{2k}\frac{d}{dt}F(nt) \tag{5}$$

in order to obtain a function $\phi(n, t)$ satisfying (I), (II) and (III), for we then have

$$\int_0^t \phi(n, t)dt = \frac{1}{2k}\Big[F(nt)\Big]_{t=0}^{t=t}, \tag{6}$$

and since by (b) we have $F(0)=0$, the right member of (6) reduces to $\frac{1}{2k}F(nt)$, and by (a) and (b) this is immediately seen to satisfy (I). Again, the assumption of the theorem that $F(x)$ is defined for all finite values of $x$

* For the sake of definiteness, let us suppose that $f(x)$ has limited total fluctuation in the neighborhood in question, cf. Hobson's "Theory of Functions of a Real Variable," § 457. Let us also suppose throughout (in order to confine the attention to the simplest considerations and thus avoid exceptional aspects of the problem into which we do not care to enter in this paper) that $|f(x)|$ is finite and integrable in the Riemann sense over the interval $(a, b)$.

† See for example Dini, *l. c.*, pp. 119–121; also, Ford, "Studies on Divergent Series and Summability" (1916), pp. 104–115.

and at the same time satisfies (a) and (b) shows at once, when employed in (6), that (II) is satisfied. Finally, (III) becomes satisfied as a result of (c) since (replacing $nt$ by $x$) the second member of (5) may be put into the form

$$\frac{1}{2k}\frac{d}{dt}F(nt)=\frac{n}{2k}F'(x)=\frac{x}{2kt}F'(x),$$

and we are concerned in (III) only with values of $t$ that do *not* include the value $t=0$.

Theorem I evidently enables one at once to construct an infinity of definite integrals (3) having the property (4). The simplest example of a function $F(x)$ such as described is perhaps that afforded by the branch of the transcendental curve $y=\text{arc}\tan x$ passing through the origin, the graph in this case having the general features indicated in the above figure, with $k=1$. Theorem I then gives the relation

$$\lim_{n=+\infty}\frac{n}{\pi}\int_a^b\frac{f(x)\,dx}{1+n^2(x-a)^2}=\frac{f(a-0)+f(a+0)}{2}.\text{*}$$

An example of an algebraic function $F(x)$ such as described in the theorem is easily supplied. Thus, consider the equation

$$(y^2-1)x+y=0. \tag{7}$$

The function $y$ of $x$ thus determined has a branch which passes through the origin and becomes asymptotic to the line $y=1$ when $x=+\infty$ and asymptotic to the line $y=-1$ when $x=-\infty$. The equation for this branch, as determined by solving (7) for $y$, is

$$y=\frac{-1+\sqrt{1+4x^2}}{2x}$$

and we readily see that this is a function of $x$ which satisfies all the conditions of the $F(x)$ of the theorem, the value of $k$ being 1. The resulting relation (3), as found by applying the theorem, becomes

$$\lim_{n=\infty}\frac{1}{4n^2}\int_a^b f(x)\left[\frac{\sqrt{1+4n^2(x-a)^2}-1}{(x-a)^2\sqrt{1+4n^2(x-a)^2}}\right]dx=\frac{f(a-0)+f(a+0)}{2}.$$

Without multiplying examples, we pass at once to certain applications of the theorem in the study of improper definite integrals.

3. As an application of the result in § 2 we may readily establish the following:

THEOREM II. *Given any convergent improper definite integral of the form*

$$\int_0^\infty p(x)\,dx=k\neq 0,$$

* This particular result has frequently been noted; cf. for example, Dini, *l. c.*, p. 36.

*wherein* (a) $p(x)$ *is an even function of* $x$, *i. e.*, $p(-x)=p(x)$ *and* (b) *the expression* $|xp(x)|$ *remains less than a constant when considered for all values of* $x$ *lying outside an arbitrarily small interval surrounding the point* $x=0$.

*Then, if* $f(x)$ *be an arbitrary function of the real variable* $x$ *defined throughout the interval* (a, b), *we shall have for any special value* $\alpha(a<\alpha<b)$

$$\lim_{n=+\infty}\frac{n}{2k}\int_a^b f(x)p[n(x-\alpha)]dx=\frac{f(\alpha-0)+f(\alpha+0)}{2}.$$

*provided merely that* $f(x)$ *satisfies suitable conditions* in the neighborhood of the point* $x=\alpha$.

In fact, we see at once that under the existing conditions for $p(x)$ we have but to place

$$F(x)=\int_0^x p(x)dx$$

in order to obtain a function $F(x)$ which satisfies the (a), (b) and (c) of Theorem I, and with this $F(x)$ substituted in the conclusion of that theorem we arrive at the conclusion stated in Theorem II.

In order to illustrate the consequences of Theorem II, several well-known definite integrals of the type under consideration are given in the left column below, and to the right occur the corresponding integrals (3) having the property (4) to which they give rise: †

$$\int_0^\infty e^{-x^2}dx=\frac{\sqrt{\pi}}{2}; \qquad \frac{n}{\sqrt{\pi}}\int_a^b f(x)e^{-n^2(x-\alpha)^2}dx,$$

$$\int_0^\infty \frac{\sin x}{x}dx=\frac{\pi}{2}; \qquad \frac{1}{\pi}\int_a^b f(x)\frac{\sin n(x-\alpha)}{x-\alpha}dx,$$

$$\int_0^\infty \frac{dx}{1-x^4}=\frac{\pi}{4}; \qquad \frac{2n}{\pi}\int_0^b \frac{f(x)dx}{1-n^4(x-\alpha)^4},$$

$$\int_0^\infty \frac{x^{2q}dx}{1-x^{2r}}=\frac{\pi}{2r}\cot\frac{2q+1}{2r}\pi; \qquad rn^{2q+1}\tan\frac{2q+1}{2r}\pi\int_a^b f(x)\frac{(x-\alpha)^{2r}dx}{1-n^{2r}(x-\alpha)^{2r}}.$$

(*q*, *r* *positive integers with* $r>q\neq 0$)

* See earlier footnote.

† It may be observed that the first integral in the right column is closely related to the one employed by Weierstrass in his well-known researches upon the representation of arbitrary functions by series of polynomials (cf. for example, Lebesgue, *l. c.*, pp. 25–28). In fact, the Weierstrass integral results from the one in question by merely placing $a=-\infty$, $b=+\infty$, thus involving an extension of the results embodied in the present paper to cover an infinite interval instead of the finite interval (a, b). Such extensions can doubtless be made, at least under suitable restrictions, in both Theorems I and II, but we shall leave the subject aside in this preliminary paper.

It may also be observed that the second integral in the right column is closely related to (1), though essentially different from the fact that the function $\psi$ in (1) is in reality dependent upon the

4. We proceed to deduce two noteworthy theorems analogous to those of §§ 2 and 3, but making less restrictive conditions upon the functions $F(x)$ and $p(x)$ there involved. The theorems in question will be found to bear an analogy to the well-known fact that in the study of the Fourier series for $f(a)$ if we demand merely that in the neighborhood *at the right* of the point $x=a$ the function $f(x)$ shall satisfy suitable conditions (in the sense heretofore specified) then, instead of (2) we may write

$$\lim_{n=+\infty}\int_a^{\pi} f(x)\,\psi(n, x-a)\,dx = \frac{1}{2}f(a+0),$$

the expression here appearing on the left being the same as (1) except for the substitution of $a$ for the lower limit of integration. Likewise, it is well known that if we demand merely that in the neighborhood *at the left* of the point $x=a$ the function $f(x)$ shall satisfy suitable conditions, we may write

$$\lim_{n=+\infty}\int_{-\pi}^{a} f(x)\,\psi(n, x-a)\,dx = \frac{1}{2}f(a-0).$$

To the facts just noted from the study of Fourier series there correspond in the more general studies of Du Bois Reymond, Dini and others, as found in the researches cited above, the following:

(A) If conditions (I), (II), (III) of § 2 are satisfied only for the *positive* values of $t$ there specified—*i. e.*, if (I) is satisfied when $0<t<\epsilon$, (II) when $0<t<\epsilon$ and (III) when $\epsilon<t<b-a$—then, whenever $f(x)$ satisfies suitable conditions at the right of the point $x=a$, we may write

$$\lim_{n=+\infty}\int_a^b f(x)\,\phi(n, x-a)\,dx = \frac{1}{2}f(a+0). \tag{8}$$

(B) If conditions (I), (II), (III) of § 2 are satisfied only for the *negative* values of $t$ there specified—*i. e.*, if (I) is satisfied when $-\epsilon<t<0$, (II) when $-\epsilon<t<0$ and (III) when $a-a<t<-\epsilon$—then, whenever $f(x)$ satisfies suitable conditions at the left of the point $x=a$, we may write

$$\lim_{n=\infty}\int_a^a f(x)\,\phi(n, x-a)\,dx = \frac{1}{2}f(a-0). \tag{9}$$

Relations (8) and (9) together with their accompanying remarks lead precisely as in § 2 to the following two theorems, in the first of which it is to be noted that we are in no wise concerned with the behavior of $F(x)$ for

limits of integration $-\pi$, $\pi$, while the functions $\phi$ of such integrals (3) as occur in Theorems I and II are independent of the limits $a$, $b$.

The third and fourth integrals of the left are selected as simple illustrations in which the integrand is algebraic. They are taken from Williamson's "Integral Calculus," pp. 140 and 142.

negative values of $x$, while in the second we are in no wise concerned with $F(x)$ for positive values of $x$, these facts being indicated by the dotting of certain portions of the accompanying diagrams:

THEOREM III. *Let $F(x)$ be any single valued function of the real variable $x$ which, when considered for positive values only (0 included) of $x$ satisfies the following three conditions:*

(a) $\lim\limits_{x=+\infty} F(x)$ *exists and* $=k \neq 0$,

(b) $F(0)=0$,

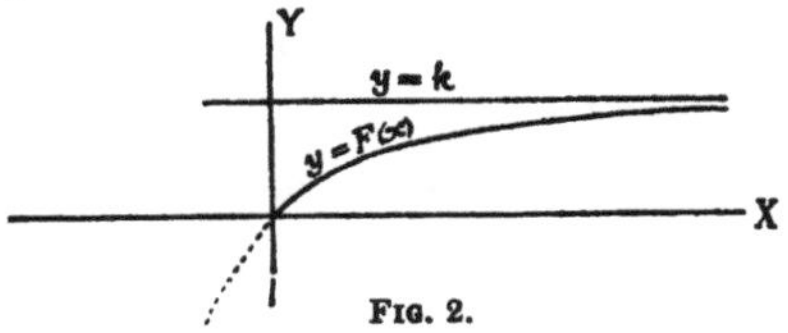


FIG. 2.

(c) *The derivative $F'(x)$ exists and is such that if we exclude the point $x=0$ by an arbitrarily small interval, we shall have for all remaining values of $x$, $|xF'_x(x)|<A=a$ constant.*

*Then, if $f(x)$ be an arbitrary function of the real variable $x$ defined throughout the interval $(a, b)$, we shall have for any special value $\alpha (a<\alpha<b)$*

$$\lim_{n=+\infty} \frac{1}{k} \int_\alpha^b f(x) \frac{d}{dx} F[n(x-\alpha)] = f(\alpha+0)$$

*provided merely that $f(x)$ satisfies suitable conditions at the right of the point $x=\alpha$.*

THEOREM IV. *Let $F(x)$ be any single valued function of the real variable $x$ which, when considered for negative values only (0 included) of $x$ satisfies the following three conditions:*

(a) $\lim\limits_{x=-\infty} F(x)$ *exists and* $=-k \neq 0$,

(b) $F(0)=0$,

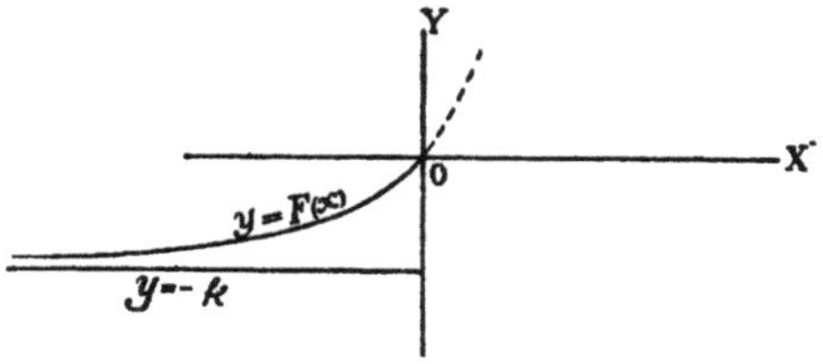


FIG. 3.

(c) *The derivative $F'(x)$ exists and is such that if we exclude the point $x=0$ by an arbitrarily small interval, we shall have for all remaining values of $x$, $|xF'(x)|<A=a$ constant.*

*Then, if $f(x)$ be an arbitrary function of the real variable $x$ defined throughout the interval $(a, b)$, we shall have for any special value $\alpha (a<\alpha<b)$*

$$\lim_{n=+\infty} \frac{1}{k} \int_a^\alpha f(x) \frac{d}{dx} F[n(x-\alpha)]dx = f(\alpha-0)$$

*provided merely that $f(x)$ satisfies suitable conditions at the left of the point $x=0$.*

By means of Theorems III and IV we may evidently arrive at other theorems analogous to that of § 3. These theorems are as follows, having been condensed, however, into one through the use of the ( ) to denote an alternative reading throughout:

THEOREM V. *Given any convergent improper integral of the form*

$$\int_0^\infty p(x)dx = k \neq 0 \quad \left(\int_{-\infty}^0 p(x)dx = k \neq 0\right),$$

*wherein $|xp(x)|$ remains less than a constant when considered for all positive (negative) values of $x$ lying outside an arbitrarily small interval to the right (left) of the point $x=0$.*

*Then, if $f(x)$ be an arbitrary function of the real variable $x$ defined throughout the interval $(a, b)$, we shall have for any special value $\alpha (a<\alpha<b)$*

$$\lim_{n=+\infty} \frac{n}{k} \int_\alpha^b f(x)p[n(x-\alpha)]dx = f(\alpha+0),$$

$$\left(\lim_{n=+\infty} \frac{n}{k} \int_a^\alpha f(x)p[n(x-\alpha)]dx = f(\alpha-0)\right)$$

*provided merely that $f(x)$ satisfies suitable conditions in the neighborhood to the right (left) of the point $x=\alpha$.**

Owing to the relative absence of restriction upon the $F(x)$ and $p(x)$ of Theorems III, IV, V, it is evident that illustrative examples are here supplied with even greater ease than was the case for Theorems I and II. The two following simple illustrations of Theorem V, wherein the same scheme as was indicated at the end of § 3 is carried out, will suffice:

$$\int_0^\infty x^m e^{-x} dx = \Gamma(m+1) \quad (m>-1); \qquad \lim_{n=+\infty} \frac{n^{m+1}}{\Gamma(m+1)} \int_\alpha^b f(x)(x-\alpha)^m e^{-n(x-\alpha)} dx = f(\alpha+0),$$

$$\int_{-\infty}^0 \frac{dx}{(x-1)^2} = 1; \qquad \lim_{n=+\infty} n \int_a^\alpha \frac{f(x)dx}{[n(x-\alpha)-1]^2} = f(\alpha-0).$$

* It may be observed at this point that Theorem I is a special consequence of Theorems III and IV obtained by applying III to the $F(x)$ of I when $x>0$ and IV to the $F(x)$ of I when $x<0$, then adding the two results and dividing by 2. Similarly, Theorem II is a special consequence of Theorem V.

5. The following general supplementary remarks may be noted in conclusion:

(1) Since the sum $s_n$ of the first $n$ terms of any convergent series approaches a limit as $n=+\infty$, we may frequently determine by means of the preceding theorems a certain integral (3) having the property (4) corresponding to a given convergent series. For example, consider the infinite geometric progression whose first term is $a$ and whose common ratio is $r<1$. Then $s_n=\frac{a(1-r^n)}{1-r}$ and $\lim\limits_{n=+\infty} s_n=\frac{a}{1-r}$ so that the idea is readily suggested of taking as a possible function $F(x)$ the following:

$$F(x)=\frac{a(1-r^x)}{1-r}$$

and this in reality is at once seen to satisfy conditions (a), (b), (c) of Theorem III, the value of $k$ being $\frac{a}{1-r}$. The theorem then gives

$$\lim_{n=+\infty} -n r^{-n\alpha}\log r \int_a^b f(x) r^{nx}dx=f(\alpha+0).$$

(2) The forms of representation discussed in the preceding §§ do not, strictly speaking, represent the given arbitrary function in terms of definite integrals, but rather in terms of the limits of such integrals as the parameter $n$ increases to $+\infty$. However, it may be observed that the first member of (4) may always be expressed as an infinite series, viz.:

$$I_0(\alpha)+\sum_{n=0}^{\infty}[I_{n+1}(\alpha)-I_n(\alpha)]$$

and thus to every integral (3) such as obtained by any one of the preceding theorems there corresponds an actual representation of the arbitrary function $f(x)$ in series of definite integrals.

(3) An inspection of the proof of Theorem I shows directly that the theorem holds equally true when, instead of using the integral form given in the statement of the theorem, we employ the more general form

$$\int_a^b f(x)\frac{d}{dx}F[q(n)r(x-\alpha)]\,dx, \tag{10}$$

wherein $q(n)$ represents any function of $n$ such that $\lim\limits_{n=+\infty} q(n)=+\infty$, while $r(x-\alpha)$ is such that $r(t)$ is positive for all positive values of $t$ and negative for all negative values of $t$, and possesses a derivative $r'(t)$ at all points $t$. The possible integrals (3) having the property (4) and derivable by Theorem I thus become exceedingly varied. Similar generalizations are naturally possible in the statements of Theorems II–V. The nature of the results thus reached is indicated in the following table wherein to each theorem number in the first column there is given in the second column the form of expression (corresponding to (10)) to be used in the generalized theorem, it being understood always that $q(n)$ represents any function having the properties above indicated and that accents denote the derivative with respect to $x$:

| | | |
|---|---|---|
| (II) | $\frac{q(n)r'(x-a)}{2k}\int_a^b f(x)p[q(n)r(x-a)]dx;$ | $r(t)$ *having the properties above indicated.* |
| (III) | $\frac{1}{k}\int_a^b f(x)F'[q(n)r(x-a)]dx;$ | $r(t)$ *having for positive values of* $t$ *the properties above indicated, and being such that* $r(0)=0$. |
| (IV) | $\frac{1}{k}\int_a^a f(x)F'[q(n)r(x-a)]dx;$ | $r(t)$ *having for negative values of* $t$ *the properties above indicated, and being such that* $r(0)=0$. |
| (V) | $\frac{q(n)r'(x-a)}{k}\int_a^b f(x)p[q(n)r(x-a)]dx;$ $r(t)$ *as in* (III), $\left(\frac{q(n)r'(x-a)}{k}\int_a^a f(x)p[q(n)r(x-a)]dx;\ r(t) \text{ as in (IV)}\right)$. | |

ANN ARBOR, *March*, 1915.

# *Metric Properties of Nets of Plane Curves.*

By H. R. Kingston.

## § 1. *Introduction.*

In a memoir entitled "One-Parameter Families and Nets of Plane Curves,"* Wilczynski has discussed the projective differential properties of nets of plane curves, by means of a completely integrable system of three partial differential equations of the second order. In the present paper the foundation is laid for a study of the metric differential properties of such nets. In order to accomplish this, it becomes necessary to consider, besides the coefficients of the partial differential equations used by Wilczynski, the coefficients $E$, $F$ and $G$ of the square of the element of arc when this is expressed in terms of the parameters of the net curves as coordinates. All of these quantities taken together, which must moreover satisfy certain relations, determine a net uniquely except for a motion combined with a reflection. The application of the general theory to the particular cases of orthogonal and isothermal nets gives rise to interesting results.

## § 2. *The Differential Equations and the Integrability Conditions.*

In order to investigate metrical properties it is convenient to use Cartesian coordinates. Let

$$x=x(u, v), \quad y=y(u, v)$$

be the equations of the net referred to a fixed Cartesian system, which shall hereafter be called the *fundamental Cartesian system.* This form of representation is contained in Wilczynski's form as a special case, if the homogeneous coordinates $y^{(1)}$, $y^{(2)}$, $y^{(3)}$, used by him, be regarded as homogeneous Cartesian coordinates, and if we put

$$y^{(1)}=x(u, v), \quad y^{(2)}=y(u, v), \quad y^{(3)}=1. \tag{1}$$

* *Transactions of the Am. Math. Soc.*, Vol. XII (1911), pp. 473–510. This paper will hereafter be referred to as "One-Parameter Families."

In consequence of this, any other point whose homogeneous coordinates are $\rho^{(1)}, \rho^{(2)}, \rho^{(3)}$, will have $\rho^{(1)}:\rho^{(3)}$ and $\rho^{(2)}:\rho^{(3)}$ as its Cartesian coordinates.

The differential equations of the net have the general form *

$$y_{uu}=ay_u+by_v+cy, \quad y_{uv}=a'y_u+b'y_v+c'y, \quad y_{vv}=a''y_u+b''y_v+c''y, \tag{2}$$

and $y^{(1)}, y^{(2)}, y^{(3)}$ must form a fundamental system of solutions of this system. Since $y^{(3)}=1$ must satisfy (2), we find, for our special form of representation,

$$c=c'=c''=0.$$

Equations (2) will therefore assume the form

$$\theta_{uu}=a\theta_u+b\theta_v, \quad \theta_{uv}=a'\theta_u+b'\theta_v, \quad \theta_{vv}=a''\theta_u+b''\theta_v, \tag{3}$$

and these equations must be satisfied by both $\theta=x(u, v)$ and $\theta=y(u, v)$.

The integrability conditions of (3) may be obtained from those of (2) by merely substituting in the latter $c=c'=c''=0$. They are †

$$\left.\begin{aligned} ba''+a_v&=a'b'+a'_u, & ab'+bb''+b_v&=a'b+b'^2+b'_u, \\ a'^2+b'a''+a'_v&=aa''+a'b''+a''_u, & a'b'+b'_v&=a''b+b''_u. \end{aligned}\right\} \tag{4}$$

Of course, it is evident that the equations (3) and (4) may be obtained also from the general equations of the metric theory of surfaces by proper specialization.

### § 3. *Discussion of the Triangle Determined by any Point of the Net and its Fundamental Covariant Points.*

The homogeneous coordinates of the covariant points $P_\rho$ and $P_\sigma$ are

$$\rho^{(k)}=y_u^{(k)}-b'y^{(k)}, \quad \sigma^{(k)}=y_v^{(k)}-a'y^{(k)}, \qquad (k=1, 2, 3). \tag{5}$$

The point $P_\rho$ is defined geometrically as the intersection of the tangents to two consecutive curves $v=$const., constructed at the points where these curves are met by a fixed curve $u=$const. Likewise, $P_\sigma$ is the point of intersection of the tangents to two consecutive curves $u=$const., constructed at the points where these curves are met by a fixed curve $v=$const. As $u$ and $v$ vary, each of these points, $P_\rho$ and $P_\sigma$, describes in general a new net of plane curves which is called a Laplacian transform of the original net. These nets are called, respectively, the first and the minus first Laplacian transforms of the

* "One-Parameter Families," equations (4). † "One-Parameter Families," equations (5).

original net, the reason for the terminology being that the first Laplacian transform of the net determined by $P_\rho$ is the original net.*

In view of (1), equations (5) give

$$\left.\begin{aligned} \rho^{(1)}&=x_u-b'x, & \sigma^{(1)}&=x_v-a'x, \\ \rho^{(2)}&=y_u-b'y, & \sigma^{(2)}&=y_v-a'y, \\ \rho^{(3)}&=-b', & \sigma^{(3)}&=-a'. \end{aligned}\right\} \tag{6}$$

Hence, in accordance with the remark made in § 2, the Cartesian coordinates of $P_\rho$ and $P_\sigma$ are given by the equations

$$\left.\begin{aligned} x_\rho&=x-x_u/b', & x_\sigma&=x-x_v/a', \\ y_\rho&=y-y_u/b', & y_\sigma&=y-y_v/a', \end{aligned}\right\} \tag{7}$$

which might also be obtained by specializing the corresponding equations in the theory of surfaces.

We proceed to find the angle between the two curves of the net passing through the point $P_y$. The slopes of the lines $P_yP_\rho$ and $P_yP_\sigma$ in rectangular Cartesian coordinates are

$$(y_\rho-y)/(x_\rho-x), \quad (y_\sigma-y)/(x_\sigma-x),$$

respectively. In virtue of (7) these slopes become $y_u:x_u$ and $y_v:x_v$, respectively, so that

$$\tan P_\rho P_y P_\sigma=(x_uy_v-x_vy_u)/(x_ux_v+y_uy_v). \tag{8}$$

We shall discuss later (in § 5) a more precise definition for this angle.

We find further

$$P_yP_\rho=\sqrt{(x_\rho-x)^2+(y_\rho-y)^2}\dagger=|\sqrt{x_u^2+y_u^2}/b'|,$$
$$P_yP_\sigma=\sqrt{(x_\sigma-x)^2+(y_\sigma-y)^2}=|\sqrt{x_v^2+y_v^2}/a'|.$$

Let us introduce the following permanent abbreviations:

$$x_u^2+y_u^2=E, \quad x_ux_v+y_uy_v=F, \quad x_v^2+y_v^2=G, \quad \sqrt{EG-F^2}=H. \tag{9}$$

Then

$$x_ux_v-y_uy_v=\varepsilon H, \tag{10}$$

where $\varepsilon=\pm1$. The significance of the double sign will be explained later (in § 5). Evidently all of these notations correspond to those of the classical theory of surfaces.

The case $H=0$ may be excluded from consideration. For, if $H=0$, we shall have $x_ux_v-y_uy_v=0$. Since the left member of this equation is precisely the Jacobian of $x$ and $y$, it follows that $x$ and $y$ are functions of the same combination of $u$ and $v$, so that our net degenerates into a single curve.

* "One-Parameter Families," § 4.

† In this paper all square roots are to be considered positive.

If we put $\theta_y$ for the angle $P_\rho P_y P_\sigma$, and make use of the notations (9) and (10), we have

$$\tan\theta_y=\varepsilon H/F, \quad P_yP_\rho=|\sqrt{E}/b'|, \quad P_yP_\sigma=|\sqrt{G}/a'|, \tag{11}$$

and further,

$$\begin{aligned}P_\rho P_\sigma&=|1/a'b'|\sqrt{(a'x_u-b'x_v)^2+(a'y_u-b'y_v)^2}\\&=|1/a'b'|\sqrt{a'^2E-2a'b'F+b'^2G}.\end{aligned}$$

From the first of equations (11) we see that the net will be orthogonal if, and only if,

$$F=0. \tag{12}$$

The slope of the line $P_\rho P_\sigma$ is $(y_\sigma-y_\rho):(x_\sigma-x_\rho)$ which reduces, in view of (7), to $(a'y_u-b'y_v):(a'x_u-b'x_v)$. The slope of the line $P_yP_\rho$ is $y_u:x_u$. Hence, we readily obtain the formula

$$\tan P_yP_\rho P_\sigma=\varepsilon b'H/(a'E-b'F).$$

Similarly, we find

$$\tan P_yP_\sigma P_\rho=\varepsilon a'H/(b'G-a'F).$$

The angle $P_yP_\rho P_\sigma$ will be a right angle if $G:F=a':b'$. The angle $P_yP_\sigma P_\rho$ will be a right angle if $E:F=b':a'$.

Again, the angles $P_yP_\rho P_\sigma$ and $P_yP_\sigma P_\rho$ will be equal if $E:G=b'^2:a'^2$.

Further, the triangle $P_yP_\rho P_\sigma$ will be equilateral if

$$E:2F:G=b'^2:a'b':a'^2.$$

## § 4. *The Angle formed by the two Families of the Laplacian Transformed Nets.*

Let $\theta_\sigma$ be the angle made by the two curves of the first Laplacian transformed net, which meet at $P_\sigma$. Then, corresponding to equation (8), we shall have

$$\tan\theta_\sigma=\left(\frac{\partial x_\sigma}{\partial u}\cdot\frac{\partial y_\sigma}{\partial v}-\frac{\partial x_\sigma}{\partial v}\cdot\frac{\partial y_\sigma}{\partial u}\right)\Big/\left(\frac{\partial x_\sigma}{\partial u}\cdot\frac{\partial x_\sigma}{\partial v}+\frac{\partial y_\sigma}{\partial u}\cdot\frac{\partial y_\sigma}{\partial v}\right).$$

By using equations (7) and (3) this reduces to

$$\tan\theta_\sigma=\frac{-\varepsilon a'a''H}{a'a''F+(a'b''-a'_v-a'^2)G}.$$

Hence, the first Laplacian transformed net will be orthogonal if, and only if,

$$a'a''F+(a'b''-a'_v-a'^2)G=0. \tag{13}$$

Similarly, if $\theta_\rho$ is the angle formed by the curves of the minus first Laplacian transformed net, which meet at $P_\rho$, we have

$$\tan\theta_\rho=\frac{-\varepsilon bb'H}{bb'F+(ab'-b'_u-b'^2)E}.$$

Hence, the minus first Laplacian transformed net will be orthogonal if, and only if,

$$bb'F+(ab'-b'_u-b'^2)E=0. \tag{14}$$

## § 5. *Introduction of a Local System of Cartesian Coordinates.*

In order to explore in detail the properties of a net in the vicinity of one of its points, it is desirable to introduce a system of Cartesian coordinates having this point as origin and the bisectors of the angles between the two net curves as axes. Let the positive direction of the tangent to a $u$-curve, *i. e.*, to a curve $v=$const., of the original net at the point $P_y$ be chosen as the direction in which $u$ increases. Similarly, let the positive direction of the tangent to a $v$-curve be that direction in which $v$ increases. Then, since a positive element of arc is, in general, given by $ds=\sqrt{Edu^2+2Fdudv+Gdv^2}$, the direction cosines of the positive $u$-tangent, referred to the fundamental Cartesian system, are $x_u:\sqrt{E}$ and $y_u:\sqrt{E}$, and the direction cosines of the positive $v$-tangent are $x_v:\sqrt{G}$ and $y_v:\sqrt{G}$.

Further, let $\theta_y$ be the angle through which the positive $u$-tangent must be rotated to coincide with the positive $v$-tangent, this angle to be counted from 0 to $+\pi$ when it is positive, and from 0 to $-\pi$ when it is negative. Thus $\theta_y$ is subject to the inequalities $-\pi\leqq\theta_y\leqq\pi$.

If $X$ and $Y$ are the axes of the fundamental Cartesian system we have, therefore,

$$\theta_y\equiv(X,v)-(X,u) \quad (\text{mod. } 2\pi)$$

where $(X, v)$ represents the angle through which the positive $X$-axis must be rotated to coincide with the direction of the positive $v$-tangent, etc. From this we at once obtain for the general case

$$\sin\theta_y=\epsilon H/\sqrt{EG}. \tag{15}$$

Thus, $\epsilon$ is positive or negative according as $\theta_y$ is positive or negative, and thus the significance of the double sign in (10) is explained. Similarly, we find

$$\cos\theta_y=F/\sqrt{EG}, \tag{16}$$

and these formulæ are consistent with the equation (11) for tan $\theta_y$. Moreover, they may be regarded as special cases of the corresponding formulæ in the theory of surfaces.

We shall now choose as the positive $\xi$-axis of our local Cartesian system, the bisector of the angle $\theta_y$; further, we shall choose the positive $\eta$-axis in such a manner as to make this system congruent to the fundamental Cartesian system.

In accordance with the convention we have made regarding the angle $\theta_y$, we find at once, by means of (16),

$$\cos\frac{\theta_y}{2}=+\sqrt{\frac{\sqrt{EG}+F}{2\sqrt{EG}}},\quad \sin\frac{\theta_y}{2}=\varepsilon\sqrt{\frac{\sqrt{EG}-F}{2\sqrt{EG}}}. \tag{17}$$

If we denote by $\theta$ the angle between the positive $X$-axis and the positive $\xi$-axis, we have

$$\theta=(\mathrm{X},u)+\theta_y/2.$$

Hence,

$$\cos\theta=\frac{x_u}{\sqrt{E}}\alpha-\frac{y_u}{\sqrt{E}}\cdot\varepsilon\beta, \tag{18}$$

$$\sin\theta=\frac{y_u}{\sqrt{E}}\alpha+\frac{x_u}{\sqrt{E}}\cdot\varepsilon\beta, \tag{19}$$

where we have made the abbreviations

$$\sqrt{\frac{\sqrt{EG}+F}{2\sqrt{EG}}}=\alpha,\quad \sqrt{\frac{\sqrt{EG}-F}{2\sqrt{EG}}}=\beta.$$

The transformation from the fundamental Cartesian system to the local system has the form

$$X=\xi\cos\theta-\eta\sin\theta+x,\quad Y=\xi\sin\theta+\eta\cos\theta+y, \tag{20}$$

where $\cos\theta$ and $\sin\theta$ are given by (18) and (19).

## § 6. *Transformation of Coordinates from the Covariant Triangle of Reference $P_yP_\rho P_\sigma$ to the Local Rectangular System.*

If $x_1, x_2, x_3$ are the homogeneous coordinates of any point referred to the triangle $P_yP_\rho P_\sigma$, and if $z_1, z_2, z_3$ are the homogeneous Cartesian coordinates of the same point referred to the fundamental Cartesian system, we have, by choosing our unit point in the proper way,*

$$\bar{\omega}z_i=x_1y^{(i)}+x_2\rho^{(i)}+x_3\sigma^{(i)},\qquad (i=1,2,3),$$

where $\bar{\omega}$ is a factor of proportionality. If we substitute in these equations from (1) and (6), and solve the resulting equations for $x_1$, $x_2$, and $x_3$ we get

$$\begin{aligned}\omega'x_1&=[z_1(b'y_v-a'y_u)-z_2(b'x_v-a'x_u)+z_3\\&\qquad\{\varepsilon H-x(b'y_v-a'y_u)+y(b'x_v-a'x_u)\}]/\varepsilon H,\\ \omega'x_2&=[z_1y_v-z_2x_v-z_3(xy_v-x_vy)]/\varepsilon H,\\ \omega'x_3&=[-z_1y_u+z_2x_u+z_3(xy_u-yx_u)]/\varepsilon H.\end{aligned}$$

If we divide the members of these equations by $z_3$, write $X$ for $z_1/z_3$ and $Y$ for $z_2/z_3$, and include the common factor $\varepsilon H/z_3$ in the proportionality factor

* E. J. Wilczynski, "Differential Geometry of Curves and Ruled Surfaces," p. 61.

$\omega''$, we obtain as the required transformation from the triangle $P_yP_\rho P_\sigma$ to the fundamental Cartesian system, the equations

$$\omega'' x_1 = X(b'y_v - a'y_u) - Y(b'x_v - a'x_u) + [\epsilon H - x(b'y_v - a'y_u) + y(b'x_v - a'x_u)],$$
$$\omega'' x_2 = Xy_v - Yx_v - (xy_v - yx_v),$$
$$\omega'' x_3 = -Xy_u + Yx_u + (xy_u - yx_u).$$

If in these equations we substitute the expressions for $X$ and $Y$ given by (20) and make use of equations (18) and (19), we obtain, as the transformation carrying us over directly from the triangle of reference $P_yP_\rho P_\sigma$ to the local rectangular system, the equations

$$\left.\begin{aligned} \omega x_1 &= \xi[\epsilon b'H\alpha + \epsilon(a'E - b'F)\beta] + \eta[(a'E - b'F)\alpha - b'H\beta] + \epsilon H\sqrt{E}, \\ \omega x_2 &= \xi[\epsilon H\alpha - \epsilon F\beta] + \eta[-F\alpha - \epsilon H\beta], \\ \omega x_3 &= \xi[\epsilon E\beta] + \eta[E\alpha]. \end{aligned}\right\} \tag{21}$$

We now proceed to find the position of the covariant points $P_\rho$ and $P_\sigma$ with reference to this local rectangular system. The coordinates of $P_\rho$ referred to the triangle of reference $P_yP_\rho P_\sigma$ are $x_1 = x_3 = 0$, $x_2 = 1$. If we substitute these values in (21) and solve for $\xi$ and $\eta$ we find the required coordinates of $P_\rho$ referred to the local rectangular system to be

$$\xi_\rho = -\alpha\sqrt{E}/b', \quad \eta_\rho = \epsilon\beta\sqrt{E}/b'. \tag{22}$$

Similarly, for the point $P_\sigma$ we find

$$\xi_\sigma = -\alpha\sqrt{G}/a', \quad \eta_\sigma = -\epsilon\beta\sqrt{G}/a'. \tag{23}$$

If we make use of equations (22) and (23) we may obtain from (21) the following set of transformation equations, whose coefficients are expressed in terms of $\xi_\rho$, $\eta_\rho$, $\xi_\sigma$, $\eta_\sigma$, $a'$, $b'$ and $H$:

$$\left.\begin{aligned} \omega x_1 &= a'b'(\eta_\rho - \eta_\sigma)\xi - a'b'(\xi_\rho - \xi_\sigma)\eta + \epsilon H, \\ \omega x_2 &= -a'\eta_\sigma\xi + a'\xi_\sigma\eta, \\ \omega x_3 &= b'\eta_\rho\xi - b'\xi_\rho\eta. \end{aligned}\right\} \tag{24}$$

## § 7. *The Osculating Conics of the Curves of the Net.*

Referred to the triangle $P_yP_\rho P_\sigma$, the equation of the conic osculating the curve $v = \text{const.}$ at the point $P_y$ is

$$\mathfrak{B}^2 x_2^2 + 4\mathfrak{B}\mathfrak{B}' x_2 x_3 - 2\mathfrak{B} x_1 x_3 + \phi x_3^2 = 0.^*$$

When we substitute in this equation the values of $x_1$, $x_2$, $x_3$ given by (24), we get the equation of the osculating conic referred to the local rectangular system, in the form

$$d_1\xi^2 + d_2\eta^2 + 2d_3\xi\eta + 2d_4\eta + 2d_5\xi = 0, \tag{25}$$

* "One-Parameter Families," equation (83).

where

$$\begin{aligned}
d_1&=a'^2\mathfrak{B}^2\eta_\sigma^2-(4a'b'\mathfrak{B}\mathfrak{B}'-2a'b'^2\mathfrak{B})\eta_\rho\eta_\sigma-(2a'b'^2\mathfrak{B}-b'^2\phi)\eta_\rho^2,\\
d_2&=a'^2\mathfrak{B}^2\xi_\sigma^2-(4a'b'\mathfrak{B}\mathfrak{B}'-2a'b'^2\mathfrak{B})\xi_\rho\xi_\sigma-(2a'b'^2\mathfrak{B}-b'^2\phi)\xi_\rho^2,\\
d_3&=-a'^2\mathfrak{B}^2\xi_\sigma\eta_\sigma+(2a'b'^2\mathfrak{B}-b'^2\phi)\xi_\rho\eta_\rho,\\
d_4&=\varepsilon b'H\mathfrak{B}\xi_\rho,\\
d_5&=-\varepsilon b'H\mathfrak{B}\eta_\rho.
\end{aligned}$$

Then

$$\begin{aligned}
d_1d_2-d_3^2\ &=H^2\mathfrak{B}^2[\phi-2a'\mathfrak{B}-(2\mathfrak{B}'-b')^2],\\
d_3d_4-d_2d_5&=H^2\mathfrak{B}^2[-a'\mathfrak{B}\xi_\sigma+b'(2\mathfrak{B}'-b')\xi_\rho],\\
d_3d_5-d_1d_4&=H^2\mathfrak{B}^2[-a'\mathfrak{B}\eta_\sigma+b'(2\mathfrak{B}'-b')\eta_\rho].
\end{aligned}$$

If $\xi_c$ and $\eta_c$ be the coordinates of the center of this osculating conic, we find

$$\left.\begin{aligned}
\xi_c&=[-a'\mathfrak{B}\xi_\sigma+b'(2\mathfrak{B}'-b')\xi_\rho]/[\phi-2a'\mathfrak{B}-(2\mathfrak{B}'-b')^2],\\
\eta_c&=[-a'\mathfrak{B}\eta_\sigma+b'(2\mathfrak{B}'-b')\eta_\rho]/[\phi-2a'\mathfrak{B}-(2\mathfrak{B}'-b')^2].
\end{aligned}\right\}\qquad(26)$$

It follows at once from equations (26) that *the conic osculating the curve $v=const.$ will be a parabola if*

$$\phi-2a'\mathfrak{B}-(2\mathfrak{B}'-b')^2=0.$$

If the origin be moved to the center [supposing $\phi-2a'\mathfrak{B}-(2\mathfrak{B}'-b')^2\neq 0$], and the axes be rotated to coincide with the principal axes, equation (25) will assume the form

$$\bar{d}_1\xi^2+\bar{d}_2\eta^2+\bar{d}_6=0,$$

where

$$\begin{aligned}
\bar{d}_1&=\tfrac{1}{2}[\mathfrak{B}^2G-2\mathfrak{B}(2\mathfrak{B}'-b')F-(2a'\mathfrak{B}-\phi)E+\sqrt{r_1}],\\
\bar{d}_2&=\tfrac{1}{2}[\mathfrak{B}^2G-2\mathfrak{B}(2\mathfrak{B}'-b')F-(2a'\mathfrak{B}-\phi)E-\sqrt{r_1}],\\
\bar{d}_6&=-H^2\mathfrak{B}^2/[\phi-2a'\mathfrak{B}-(2\mathfrak{B}'-b')^2],
\end{aligned}$$

in which

$$\begin{aligned}
r_1&=(d_1-d_2)^2+4d_3^2\\
&=\mathfrak{B}^4G^2+4\mathfrak{B}^2(2\mathfrak{B}'-b')^2EG+(2a'\mathfrak{B}-\phi)^2E^2-4\mathfrak{B}^3(2\mathfrak{B}'-b')FG\\
&\quad-2\mathfrak{B}^2(2a'\mathfrak{B}-\phi)EG+4\mathfrak{B}(2\mathfrak{B}'-b')(2a'\mathfrak{B}-\phi)EF+4H^2\mathfrak{B}^2(2a'\mathfrak{B}-\phi).
\end{aligned}$$

From the first form of $r_1$ we see that $\sqrt{r_1}$ is always real.

The conic will be an ellipse or an hyperbola according as $\phi-2a'\mathfrak{B}-(2\mathfrak{B}'-b')^2$ is positive or negative. Further, by a well-known procedure we find the semi-axes $\alpha_1$ and $\beta_1$ to be given by

$$\alpha_1^2,\ \beta_1^2=\left|\frac{2H^2\mathfrak{B}^2[\phi-2a'\mathfrak{B}-(2\mathfrak{B}'-b')^2]^{-1}}{\mathfrak{B}^2G-2\mathfrak{B}(2\mathfrak{B}'-b')F-(2a'\mathfrak{B}-\phi)E\pm\sqrt{r_1}}\right|.$$

The eccentricity $e$ is given by

$$e^2=2\sqrt{r_1}/|\mathfrak{B}^2G-2\mathfrak{B}(2\mathfrak{B}'-b')F-(2a'\mathfrak{B}-\phi)E+\sqrt{r_1}|.$$

Referred to the triangle $P_yP_\rho P_\sigma$, the equation of the conic osculating the curve $u=\text{const.}$ at the point $P_y$ is

$$\psi x_2^2+4\mathfrak{A}'\mathfrak{A}''x_2x_3+\mathfrak{A}''^2x_3^2-2\mathfrak{A}''x_1x_2=0.^*$$

When the substitutions (24) are made, this equation assumes the form

$$d_1'\xi^2+d_2'\eta^2+2d_3'\xi\eta+2d_4'\eta+2d_5'\xi=0, \tag{27}$$

where

$$\begin{aligned}
d_1'&=b'^2\mathfrak{A}''^2\eta_\rho^2-(4a'b'\mathfrak{A}'\mathfrak{A}''-2a'^2b'\mathfrak{A}'')\eta_\rho\eta_\sigma-(2a'^2b'\mathfrak{A}''-a'^2\psi)\eta_\sigma^2,\\
d_2'&=b'^2\mathfrak{A}''^2\xi_\rho^2-(4a'b'\mathfrak{A}'\mathfrak{A}''-2a'^2b'\mathfrak{A}'')\xi_\rho\xi_\sigma-(2a'^2b'\mathfrak{A}''-a'^2\psi)\xi_\sigma^2,\\
d_3'&=-b'^2\mathfrak{A}''^2\xi_\rho\eta_\rho+(2a'^2b'\mathfrak{A}''-a'^2\psi)\xi_\sigma\eta_\sigma,\\
d_4'&=-\varepsilon a'H\mathfrak{A}''\xi_\sigma,\\
d_5'&=\varepsilon a'H\mathfrak{A}''\eta_\sigma.
\end{aligned}$$

The coordinates of the center of this conic are given by the equations

$$\begin{aligned}
\xi_{c'}&=[-b'\mathfrak{A}''\xi_\rho+a'(2\mathfrak{A}'-a')\xi_\sigma]/[\psi-2b'\mathfrak{A}''-(2\mathfrak{A}'-a')^2],\\
\eta_{c'}&=[-b'\mathfrak{A}''\eta_\rho+a'(2\mathfrak{A}'-a')\eta_\sigma]/[\psi-2b\mathfrak{A}''-(2\mathfrak{A}'-a')^2].
\end{aligned}$$

Hence *the conic osculating the curve $u=\text{const.}$, will be a parabola if*

$$\psi-2b\mathfrak{A}''-(2\mathfrak{A}'-a')^2=0.$$

If the origin be moved to the center (supposing the conic not a parabola) and the principal axes be taken as axes of coordinates, equation (27) will take the form

$$\bar{d}_1'\xi^2+\bar{d}_2'\eta^2+\bar{d}_6'=0,$$

where

$$\begin{aligned}
\bar{d}_1'&=\tfrac{1}{2}[\mathfrak{A}''^2E-2\mathfrak{A}''(2\mathfrak{A}'-a')F-(2b'\mathfrak{A}''-\psi)G+\sqrt{r_1'}],\\
\bar{d}_2'&=\tfrac{1}{2}[\mathfrak{A}''^2E-2\mathfrak{A}''(2\mathfrak{A}'-a')F-(2b'\mathfrak{A}''-\psi)G-\sqrt{r_1'}],\\
\bar{d}_6'&=-H^2\mathfrak{A}''^2/[\psi-2b'\mathfrak{A}''-(2\mathfrak{A}'-a')^2],
\end{aligned}$$

in which

$$\begin{aligned}
r_1'=\mathfrak{A}''^4E^2&+4\mathfrak{A}''^2(2\mathfrak{A}'-a')^2EG+(2b'\mathfrak{A}''-\psi)^2G^2-4\mathfrak{A}''^3(2\mathfrak{A}'-a')EF\\
&-2\mathfrak{A}''^2(2b'\mathfrak{A}''-\psi)EG+4\mathfrak{A}''(2\mathfrak{A}'-a')(2b'\mathfrak{A}''-\psi)FG+4H^2\mathfrak{A}''^2(2b'\mathfrak{A}''-\psi),
\end{aligned}$$

and, as before, $\sqrt{r_1'}$ is always real.

The conic will be an ellipse or an hyperbola according as $\psi-2b'\mathfrak{A}''-(2\mathfrak{A}'-a')^2$ is positive or negative.

The principal semi-axes $\alpha_1'$ and $\beta_1'$ and eccentricity $e$ are given by

$$\alpha_1'^2,\ \beta_1'^2=\left|\frac{2H^2\mathfrak{A}''^2[\psi-2b'\mathfrak{A}''-(2\mathfrak{A}'-a')^2]^{-1}}{\mathfrak{A}''^2E-2\mathfrak{A}''(2\mathfrak{A}'-a')F-(2b'\mathfrak{A}''-\psi)G\pm\sqrt{r_1'}}\right|,$$

$$e^2=2\sqrt{r_1'}/|\mathfrak{A}''^2E-2\mathfrak{A}''(2\mathfrak{A}'-a')F-(2b'\mathfrak{A}''-\psi)G+\sqrt{r_1'}|.$$

---

* "One-Parameter Families," equation (87).

The formulæ of this section will find their application whenever it is desired to study the metric properties of the osculating conics of any given net, or else if it be proposed to characterize particular nets by metric properties of their osculating conics.

## § 8. *The Fundamental Theorem.*

From the definitions of $E$, $F$ and $G$ we obtain, by using (3),

$$\left.\begin{aligned} E_u&=2aE+2bF, & G_u&=2a'F+2b'G, & F_u&=a'E+(a+b')F+bG,\\ E_v&=2a'E+2b'F, & G_v&=2a''F+2b''G, & F_v&=a''E+(a'+b'')F+b'G. \end{aligned}\right\} \quad (28)$$

If we differentiate these equations again, we find that the relations $E_{uv}=E_{vu}$, $F_{uv}=F_{vu}$ and $G_{uv}=G_{vu}$ are satisfied as a consequence of the integrability conditions (4). Consequently, *if we regard $a$, $b$, ...., $b''$ as given functions of $u$ and $v$ which satisfy the integrability conditions* (4), *equations* (28) *form a completely integrable system of partial differential equations for $E$, $F$ and $G$.*

If $a$, ...., $b''$ are analytic in the vicinity of $u=u_0$, $v=v_0$, equations (28) may therefore be solved for $E$, $F$ and $G$ as convergent power series with arbitrary initial values. Further, if $E_k$, $F_k$, $G_k$, $(k=1, 2, 3)$, be any three linearly independent systems of analytic solutions, the most general analytic solution may be written as a homogeneous linear combination of these three with constant coefficients; thus

$$E=\Sigma c_k E_k, \quad F=\Sigma c_k F_k, \quad G=\Sigma c_k G_k,$$

where $c_1$, $c_2$, $c_3$ are arbitrary constants.

Suppose now that $a$, ...., $b''$ are given as functions of $u$ and $v$ satisfying the integrability conditions (4), and suppose further that $E$, $F$ and $G$ are given satisfying (28). To what extent will these nine functions determine the net? To answer this question we must find the most general transformation on $x$ and $y$ that will leave these nine functions unaltered.

Let us, in the first place, ignore $E$, $F$ and $G$. The most general pair of functions which satisfies equations (3) is obtained from the particular solution $x(u, v)$, $y(u, v)$, with which we started, by a transformation of the form

$$\bar{x}=a_1x+b_1y+c_1, \quad \bar{y}=a_2x+b_2y+c_2, \tag{29}$$

that is, by the most general affine transformation. Consequently, *those properties of the net which are expressed by conditions independent of $E$, $F$ and $G$ are of an affine character.*

If we denote by $\bar{E}$, $\bar{F}$, $\bar{G}$ the fundamental quantities of the net determined by $\bar{x}$, $\bar{y}$, we find

$$\bar{E}=(a_1^2+a_2^2)x_u^2+(b_1^2+b_2^2)y_u^2+2(a_1b_1+a_2b_2)x_uy_u,$$
$$\bar{F}=(a_1^2+a_2^2)x_ux_v+(b_1^2+b_2^2)y_uy_v,$$
$$\bar{G}=(a_1^2+a_2^2)x_v^2+(b_1^2+b_2^2)y_v^2+2(a_1b_1+a_2b_2)x_vy_v,$$

from which we see that $E$, $F$ and $G$ will remain unaltered by the transformation (29) if, and only if, the conditions $a_1^2+a_2^2=1$, $b_1^2+b_2^2=1$ and $a_1b_1+a_2b_2=0$ are fulfilled. These equations give $a_2=\pm b_1$ and $b_2=\mp a_1$, where $a_1^2+b_1^2=1$, so that the most general affine transformation (29) which leaves $E$, $F$ and $G$ unaltered, consists of a motion and a reflection. Hence a net is determined, except for a motion and a reflection, by the nine fundamental quantities, $E$, $F$, $G$, $a$, ...., $b''$, where the coefficients $a$, ...., $b''$ are subject to the integrability conditions (4), and $E$, $F$ and $G$ must satisfy the conditions (28).

We may formulate our final theorem as follows: *Let*

$$\theta_{uu}=a\theta_u+b\theta_v,\quad \theta_{uv}=a'\theta_u+b'\theta_v,\quad \theta_{vv}=a''\theta_u+b''\theta_v, \tag{3}$$

*be a completely integrable system of partial differential equations, so that its coefficients satisfy the integrability conditions* (4). *If two linearly independent non-constant solutions, $x(u,v)$, $y(u,v)$, of* (3) *be interpreted as the cartesian coordinates of a point, the equations*

$$X=x(u,v),\quad Y=y(u,v),$$

*determine a net $N$. The most general net obtained in this way from* (3) *is an affine transformation of $N$.*

*If the curves $u=const.$ and $v=const.$ of the net $N$ be regarded as curvilinear coordinates, the square of the element of arc assumes the form $ds^2=Edu^2+2Fdudv+Gdv^2$. If the coefficients of this quadratic form are given in any way subject to the conditions* (28), *the net is determined uniquely except for its position in the plane and a reflection in any line in the plane.*

*Finally, it is evident that any net may be studied in this way, since the nine quantities $a$, ...., $b''$, $E$, $F$, $G$ are easily determined when the finite equations of the net are given.*

The latter half of this theorem is, of course, an immediate consequence of the fundamental theorem of the theory of surfaces.

## § 9. *Orthogonal Nets.*

Let us assume that the net under consideration is orthogonal; that is, $F=0$, by (12). Then equations (28) become

$$\left.\begin{array}{lll} \text{a)}\ E_u=2aE, & \text{c)}\ 0=a'E+bG, & \text{e)}\ G_u=2b'G, \\ \text{b)}\ E_v=2a'E, & \text{d)}\ 0=a''E+b'G, & \text{f)}\ G_v=2b''G. \end{array}\right\} \tag{30}$$

If we differentiate $E_u$ with respect to $v$, and $E_v$ with respect to $u$, and equate the resulting values, we obtain the equation

$$a_v = a'_u. \tag{31}$$

Then the first equation of (4) gives

$$ba'' = a'b'. \tag{32}$$

From (31) we conclude that

$$a = \phi_u, \qquad a' = \phi_v, \tag{33}$$

where $\phi$ is an arbitrary function of $u$ and $v$.

If we differentiate $G_u$ and $G_v$, as expressed by (30), with regard to $v$ and $u$, respectively, we find

$$b'_v = b''_u, \tag{34}$$

whence, by the fourth of equations (4), we have

$$a'b' = a''b,$$

as before. Equation (34) tells us that

$$b' = \psi_u, \qquad b'' = \psi_v, \tag{35}$$

where $\psi$ is an arbitrary function of $u$ and $v$.

If we differentiate c) and d) of equations (30) with respect to $u$ and $v$, making use of a), b), e) and f), and equate the resulting values of the ratio $-G:E$, we obtain the equalities

$$\frac{a'_u + 2aa'}{b_u + 2bb'} = \frac{a'_v + 2a'^2}{b_v + 2bb''} = \frac{a''_u + 2aa''}{b'_u + 2b'^2} = \frac{a''_v + 2a'a''}{b'_v + 2b'b''} = \frac{a'}{b} = \frac{a''}{b'}.$$

If we make use of equations (33) and (35), these equalities become

$$\frac{\phi_{uv} + 2\phi_u\phi_v}{b_u + 2b\psi_u} = \frac{\phi_{vv} + 2\phi_v^2}{b_v + 2b\psi_v} = \frac{a''_u + 2\phi_u a''}{\psi_{uu} + 2\psi_u^2} = \frac{a''_v + 2\phi_v a''}{\psi_{uv} + 2\psi_u\psi_v} = \frac{\phi_v}{b} = \frac{a''}{\psi_u}. \tag{36}$$

From the equalities

$$(\phi_{uv} + 2\phi_u\phi_v)/(b_u + 2b\psi_u) = \phi_v/b \text{ and } (\phi_{vv} + 2\phi_v^2)/(b_v + 2b\psi_v) = \phi_v/b$$

of (36), we readily find

$$\log b = \log \phi_v + 2(\phi - \psi) + c_1(v) \text{ and } \log b = \log \phi_v + 2(\phi - \psi) + c_2(u),$$

where $c_1(v)$ is an arbitrary function of $v$ alone and $c_2(u)$ is an arbitrary function of $u$ alone. From these two values of $\log b$ we get $c_1(v) = c_2(u) = K_1$, where $K_1$ is an arbitrary constant. Hence, we have

$$b = k_1\phi_v e^{2(\phi-\psi)}.$$

In the same way, by using those equalities of (36) which involve $a''$, we find

$$a''=k_2\psi_u e^{-2(\phi-\psi)}.$$

Here $k_1$ and $k_2$ are arbitrary constants.

But we have found that $a''b$ must be equal to $\phi_v\psi_u$, so that $k_1k_2$ must be equal to unity. Hence, we have

$$b=k\phi_v e^{2(\phi-\psi)}, \quad a''=\frac{1}{k}\psi_u e^{-2(\phi-\psi)}, \tag{37}$$

where $k$ is an arbitrary constant which does not vanish if we assume that neither family of the net is composed of straight lines.* We shall suppose here that $k\neq 0$, and shall later consider the case where $k$ vanishes.

We have now made use of all our conditions except the second and third of (4), which must still be satisfied. If we substitute in the second equation of (4) the values we have found in (33), (35) and (37), we get

$$(\psi_u^2+\psi_{uu}-\phi_u\psi_u)e^{2\psi}-k(\phi_v^2+\phi_{vv}-\phi_v\psi_v)e^{2\phi}=0. \tag{38}$$

Similarly, if we substitute in the third of equations (4), we obtain precisely the same result. We have then found the following result:

Let $\phi$ and $\psi$ be any two functions of $u$ and $v$ satisfying the condition (38), and let

$$\left.\begin{aligned} &a=\phi_u, \quad a'=\phi_v, \quad a''=\frac{1}{k}\psi_u e^{-2(\phi-\psi)}, \\ &b=k\phi_v e^{2(\phi-\psi)}, \quad b'=\psi_u, \quad b''=\psi_v. \end{aligned}\right\} \tag{39}$$

Then any orthogonal net may be regarded as a solution of a system of partial differential equations of the form (3) with the coefficients (39). Of course, not all nets which satisfy such a system of form (3) are orthogonal; they are all affine to an orthogonal net. However, we may find $E$ and $G$ by quadratures from (30). We get

$$E=ce^{2\int(adu+a'dv)}=ce^{2\phi}, \quad G=c'e^{2\int(b'du+b''dv)}=c'e^{2\psi}, \tag{40}$$

where $c$ and $c'$ are constants satisfying the condition $c+kc'=0$, imposed by c) and d) of (30). If we adjoin the conditions that $E$ and $G$ shall have these values and that $F$ shall be equal to zero, the corresponding orthogonal net is determined uniquely except for its position in the plane and for a reflection. But this result may be simplified a little. The constant $k$, appearing here, has

* "One-Parameter Families," § 4, p. 490.

no geometrical significance for we may, by a linear transformation of the independent variables, make this constant assume any value we please. To show this, let us make the transformation

$$\bar{u}=\alpha u, \quad \bar{v}=\beta v,$$

where $\alpha$ and $\beta$ are arbitrary constants. Then $\phi_u=\phi_{\bar{u}}\cdot\alpha$, $\phi_{uu}=\phi_{\bar{u}\bar{u}}\cdot\alpha^2$, etc., and (38) becomes

$$\alpha^2(\psi_{\bar{u}}^2+\psi_{\bar{u}\bar{u}}-\phi_{\bar{u}}\psi_{\bar{u}})e^{2\psi}-\beta^2 k(\phi_{\bar{v}}^2+\phi_{\bar{v}\bar{v}}-\phi_{\bar{v}}\psi_{\bar{v}})e^{2\phi}=0.$$

We may now put $\alpha^2:\beta^2$ equal to any constant we please and thus make our new $k$ have any arbitrary value. In particular, put the ratio $\alpha^2:\beta^2$ equal to $-k$, and our new $k$ becomes equal to $-1$. The constants $c$ and $c'$ of (40) then become equal, the coefficients take the form of (39) with $-1$ in the place of $k$ and with $u$ and $v$ barred,* and our theorem becomes as follows:

*Let $\phi$ and $\psi$ be any two functions of $u$ and $v$ satisfying the condition*

$$(\psi_u^2+\psi_{uu}-\phi_u\psi_u)e^{2\psi}+(\phi_v^2+\phi_{vv}-\phi_v\psi_v)e^{2\phi}=0,$$

*and let*

$$\left.\begin{array}{lll} a=\phi_u, & a'=\phi_v, & a''=-\psi_u e^{-2(\phi-\psi)}, \\ b=-\phi_v e^{2(\phi-\psi)}, & b'=\psi_u, & b''=\psi_v. \end{array}\right\} \qquad (41)$$

*Moreover, let*

$$E=ce^{2\phi}, \quad F=0, \quad G=ce^{2\psi},$$

*where $c$ is an arbitrary constant. Then there exists an orthogonal net, uniquely determined by these quantities except for a motion and a reflection, and any orthogonal net, neither of whose families is composed of straight lines, may be obtained in this way.*

Let us now consider the case when $k=0$ in (39). If we proceed in a manner similar to that followed above, we readily obtain

$$\left.\begin{array}{lll} a=\phi_u, & a'=0, & a''=\bar{k}\psi_u e^{2(\psi-\phi)}, \\ b=0, & b'=\psi_u, & b''=\psi_v, \end{array}\right\} \qquad (42)$$

where $\phi$ is any function of $u$ alone, and $\psi$ any function of $u$ and $v$, such that $\psi_u^2+\psi_{uu}-\phi_u\psi_u=0$, and where $\bar{k}$ is an arbitrary constant.

If we now make the transformation

$$\bar{u}=U(u), \quad \bar{v}=V(v),$$

* "One-Parameter Families," equations (17).

where $U$ and $V$ are defined by the equations

$$U''-\phi_u U'=0, \quad V=v,$$

we readily obtain the following result:

*Let $\psi$ be any function of the form*

$$\psi=\log(V_0 u+V_1),$$

*where $V_0$ and $V_1$ are arbitrary functions of $v$ alone, and let*

$$a=0, \quad a'=0, \quad a''=-\psi_u e^{2\psi},$$
$$b=0, \quad b'=\psi_u, \quad b''=\psi_v.$$

*Moreover, let*

$$E=c, \quad F=0, \quad G=ce^{2\psi},$$

*where $c$ is an arbitrary constant. Then there exists an orthogonal net, of which the family of curves $v$=const. are straight lines,* uniquely determined by these quantities except for a motion and a reflection.*

If in (39) we had taken $k$ equal to infinity, that is, if we had taken $a''$ equal to zero, we should have obtained a similar theorem regarding the existence of an orthogonal net of which the curves $u$=const. are straight lines.

Again, let us suppose that the arbitrary constant $\bar{k}$ in (42) vanishes. We then find the following result:

*A system of partial differential equations of the form* (3), *in which all the coefficients vanish, when considered along with the equations*

$$E=c, \quad F=0, \quad G=c',$$

*where $c$ and $c'$ are arbitrary constants, determines uniquely, except for a motion and a reflection, an orthogonal net both of whose families of curves are composed of straight lines. Moreover, any such net can be represented in this way.*

This system of partial differential equations with vanishing coefficients, is readily integrated, and gives as one fundamental set of solutions, the equations

$$x=u, \quad y=v,$$

which represent the two families of straight lines parallel to the axes of coordinates.

---

* "One-Parameter Families," p. 489.

Let us consider again equation (38), and let us suppose first that $k\neq 0$. Then if

$$\psi_u^2+\psi_{uu}-\phi_u\psi_u=0, \tag{43}$$

it follows that

$$\phi_v^2+\phi_{vv}-\phi_v\psi_v=0 \tag{44}$$

and conversely. By comparing these equations with (13) and (14), and remembering that in the present case $F$ is equal to zero, we see that the equations (43) and (44) are respectively the conditions that the minus first and the first Laplacian transformed nets are orthogonal.

Again, if in (37) the constant $k$ should be zero, the family of curves $v=$const. is composed of straight lines, from which it follows that the minus first Laplacian transformed net degenerates.* Further, for this case we found that $a'$ vanishes, from which it follows by (13) that the first Laplacian transformed net is orthogonal unless it should be degenerate.

A similar conclusion is true if in (37) the constant $k$ should become infinite. We have then found the

THEOREM: *If a given net is orthogonal, and if either of its Laplacian transformed nets is orthogonal or is degenerate on account of the fact that one of the families of the original net is composed of straight lines, then the other Laplacian transformed nets will also be orthogonal or degenerate.*

Let us suppose we have the conditions

$$\psi_u^2+\psi_{uu}-\phi_u\psi_u=0, \quad \phi_v^2+\phi_{vv}-\phi_v\psi_v=0,$$

fulfilled. Then, by integration, we find the equations

$$\psi_u=V_0e^{\phi-\psi}, \quad \phi_v=U_0e^{\psi-\phi}, \quad \phi_v\psi_u=U_0V_0,$$

where $U_0$ and $V_0$ are, respectively, functions of $u$ alone and of $v$ alone. By a transformation of variables we may reduce these equations to

$$\psi_u=(V_0/U_0)e^{\phi-\psi}, \quad \phi_v=(U_0/V_0)e^{\psi-\phi}, \quad \phi_v\psi_u=1,$$

if we assume that neither $\phi_v$ nor $\psi_u$ is identically zero. Putting $(U_0/V_0)e^{\psi-\phi}=\chi$,

$$\phi=\int_{v_0}^{v}\chi dv+U_1, \quad \psi=\int_{u_0}^{u}\frac{du}{\chi}+V_1, \tag{45}$$

which gives rise to the following integral equation for $\chi$,

$$\log U_0-\log V_0+V_1-U_1+\int_{u_0}^{u}\frac{du}{\chi}-\int_{v_0}^{v}\chi dv=\log\chi. \tag{46}$$

---

* "One-Parameter Families," p. 490.

Hence, we may state the following result:

If $U_0$, $U_1$ are arbitrary functions of $u$, $V_0$, $V_1$ arbitrary functions of $v$, if $u_0$, $v_0$ are arbitrary constants, and if $\chi$ is any function of $u$ and $v$ which satisfies the integral equation (46), then the equations (41) and (45) together with

$$E=ce^{2\phi}, \quad F=0, \quad G=ce^{2\psi},$$

will determine (except for a motion and reflection) an orthogonal net whose Laplacian transforms are also orthogonal. The integral equation (46) may be replaced by the partial differential equation

$$\frac{\partial^2 \log \chi}{\partial u \partial v}+\frac{\partial \chi}{\partial u}-\frac{\partial \chi^{-1}}{\partial v}=0,$$

obtained from it by differentiation.

## § 10. *Isothermal Nets.*

Let us now suppose that the net is not only orthogonal but also isothermal, that is, such that

$$F=0, \quad E:G=U:V,$$

where $U$ is a function of $u$ alone, and $V$ a function of $v$ alone. Then

$$\frac{\partial^2 \log E/G}{\partial u \partial v}=0. \tag{47}$$

From (30) we obtain readily

$$\frac{\partial^2 \log E}{\partial u \partial v}=2a_v, \tag{48}$$

$$\frac{\partial^2 \log G}{\partial u \partial v}=2b'_v. \tag{49}$$

Uniting the results of equations (47), (48) and (49) we find

$$\frac{\partial^2 \log E/G}{\partial_u \partial_v}=2(a_v-b'_v)=0.$$

Hence,

$$a_v=b'_v. \tag{50}$$

Similarly,

$$a'_u=b''_u. \tag{51}$$

But since the isothermal net is a special case of the orthogonal net, the results of § 9 must hold here, and we obtain from equations (33), (50) and (51)

$$a=\phi_u, \quad a'=\phi_v, \quad b'=\phi_u+s, \quad b''=\phi_v+t,$$

where $s$ is an arbitrary function of $u$ alone, $t$ an arbitrary function of $v$ alone, and $\phi$ is a function of $u$ and $v$. Hence the function $\psi$ of § 9 is replaced here by $\phi+\int s du+\int t dv$. Putting this value of $\phi$ in (37) we get

$$a''=\frac{1}{k}(\phi_u+s)e^{2\int(sdu+tdv)}, \quad b=k\phi_v e^{-2\int(sdu+tdv)},$$

and (38) reduces to

$$(\phi_{uu}+\phi_u s+s_u+s^2)e^{2\int sdu}-k(\phi_{vv}-\phi_v t)e^{-2\int tdv}=0. \tag{52}$$

These expressions may be simplified a little by replacing $\int sdu$ by $\bar{s}$ and $\int tdv$ by $\bar{t}$, after which the bars may be dropped. Thus we see that the coefficients of the system of partial differential equations of an isothermal net may be written in the form

$$\left.\begin{aligned} &a=\phi_u, \quad a'=\phi_v, \quad a''=\frac{1}{k}(\phi_u+s_u)e^{2(s+t)}, \\ &b=k\phi_v e^{-2(s+t)}, \quad b'=\phi_u+s_u, \quad b''=\phi_v+t_v, \end{aligned}\right\} \tag{53}$$

where $\phi$ is a function of $u$ and $v$, $s$ a function of $u$, $t$ a function of $v$, and $k$ is an arbitrary constant (which we assume here to be different from zero), and where these quantities satisfy the relation

$$(\phi_{uu}+\phi_u s_u+s_{uu}+s_u^2)e^{2s}-k(\phi_{vv}-\phi_v t_v)e^{-2t}=0. \tag{54}$$

In order to make (53) more symmetrical, let us put $\phi+s=\bar{\phi}$, and then omit the bar. We shall find

$$\left.\begin{aligned} &a=\phi_u-s_u, \quad a'=\phi_v, \quad a''=\frac{1}{k}\phi_u e^{2(s+t)}, \\ &b=k\phi_v e^{-2(s+t)}, \quad b'=\phi_u, \quad b''=\phi_v+t_v, \end{aligned}\right\} \tag{55}$$

and the condition (54) becomes

$$(\phi_{uu}+\phi_u s_u)e^{2s}-k(\phi_{vv}-\phi_v t_v)e^{-2t}=0. \tag{56}$$

Let us now transform the independent variables by putting

$$\bar{u}=U(u), \quad \bar{v}=V(v).$$

Then (56) becomes

$$[U'^2\phi_{\bar{u}\bar{u}}+\phi_{\bar{u}}(\eta+U's_{\bar{u}})U']e^{2s}-k[V'^2\phi_{\bar{v}\bar{v}}+\phi_{\bar{v}}(\zeta-V't_{\bar{v}})V']e^{-2t}=0, \tag{57}$$

where

$$\eta=U''/U', \quad \zeta=V''/V'.$$

If $U$ and $V$ are so determined that

$$\eta+U's_{\bar{u}}=0, \quad \zeta-V't_{\bar{v}}=0, \quad \text{i. e.,} \quad U'=\alpha e^{-s}, \quad V'=\beta e^{t},$$

where $\alpha$ and $\beta$ are arbitrary constants, (57) reduces to

$$\alpha^2\phi_{\bar{u}\bar{u}}-k\beta^2\phi_{\bar{v}\bar{v}}=0.$$

If we put $\alpha^2=-k\beta^2$, this becomes

$$\phi_{\bar{u}\bar{u}}+\phi_{\bar{v}\bar{v}}=0,$$

and this is the fundamental condition which $\phi$ must satisfy. The coefficients (55) now assume the very simple form*

$$\begin{aligned}\bar{a}&=\phi_{\bar{u}}, & \bar{a}'&=\phi_{\bar{v}}, & \bar{a}''&=-\phi_{\bar{u}},\\ \bar{b}&=-\phi_{\bar{v}}, & \bar{b}'&=\phi_{\bar{u}}, & \bar{b}''&=\phi_{\bar{v}}.\end{aligned}$$

We then obtain from (30), by quadratures,

$$E=ce^{2\phi},\quad G=ce^{2\phi},$$

where $c$ is an arbitrary constant. Thus we have found the following

THEOREM: *If $\phi$ is any function of $u$ and $v$ satisfying the condition*

$$\phi_{uu}+\phi_{vv}=0, \tag{58}$$

*then any net which satisfies the differential equation* (3) *with the coefficients*

$$\left.\begin{aligned}a&=\phi_u, & a'&=\phi_v, & a''&=-\phi_u,\\ b&=-\phi_v, & b'&=\phi_u, & b''&=\phi_v,\end{aligned}\right\} \tag{59}$$

*will be affine to an isothermal net. If we have besides*

$$E=ce^{2\phi},\quad F=0,\quad G=ce^{2\phi}, \tag{60}$$

*where $c$ is any constant, the net will be isothermal, and moreover the parameters will be isometric parameters.*

To complete the proof of this theorem it suffices to remark that among the integral nets of a system of form (3) with the coefficients (59) there always exists an isothermal net; namely, one for which the $E$, $F$ and $G$ have the values (60). All other integral nets of the system are, of course, affine to this particular one.

It is of interest to note here that since the function $\phi$ satisfies equation (58), which is Laplace's equation, this function must be either the real or the imaginary part of some monogenic function of a complex variable.

Again, if we put the coefficients (59) in the first and last of equations (3), we get

$$\theta_{uu}=\phi_u\theta_u-\phi_v\theta_v,\quad \theta_{vv}=-\phi_u\theta_u+\phi_v\theta_v,$$

whence

$$\theta_{uu}+\theta_{vv}=0,$$

which says that each of the cartesian Coordinates of the generating point $P_y$, of our net, is the real or the imaginary part of some analytic function of a com-

* "One-Parameter Families," § 3, equation 17.

plex variable, a very familiar theorem. The whole theorem might, of course, be proved on this basis.

Further, the Laplace-Darboux invariants, $H$ and $K$, are given by the equations

$$H=C'+A'B'-A'_u, \quad K=C'+A'B'-B'_v.^*$$

If in these expressions we substitute the coefficients (59), and remember that $C'=0$, we find

$$H=\phi_u\phi_v-\phi_{uv}, \quad K=\phi_u\phi_v-\phi_{uv},$$

and, therefore, the invariants $H$ and $K$ are equal. A very elegant geometrical interpretation of the meaning of the equality of these two invariants has been given by Wilczynski in a memoir entitled "*Flächen mit unbestimmten Directrixkurven.*" This interpretation is embodied in the following

THEOREM: "*Let us suppose that $P$ is an arbitrary point of a net of plane curves with equal Laplace-Darboux invariants, and let $P_1$ and $P_{-1}$ be the corresponding points of the first and of the minus first Laplacian transformed nets. Then there exists a curve of the first net which touches the line $P_1P$ at $P_1$, and a curve of the second net which touches the line $P_{-1}P$ at $P_{-1}$. Let $M_1$ and $N_{-1}$ be the conics which osculate these two curves at $P_1$ and $P_{-1}$, respectively. Then either the conic of the pencil determined by $M_1$ and $N_{-1}$, which passes through $P$, intersects the segment $P_1P_{-1}$ harmonically, or this conic degenerates into a pair of straight lines, of which one constituent part is identical with $P_1P_{-1}$, or else the conics $M_1$ and $N_{-1}$ are coincident.*"

We now observe that this theorem is applicable to all isothermal nets, since any isothermal net may be represented by a system of form (3) with coefficients of form (59), and hence with equal Laplace-Darboux invariants.

We can easily prove a more specific theorem regarding isothermal nets than we have yet developed. Under conditions (59) the equations (28) become

$$\left.\begin{aligned} E_u&=2\phi_uE-2\phi_vF, & E_v&=2\phi_vE+2\phi_uF,\\ F_u&=\phi_v(E-G)+2\phi_uF, & F_v&=-\phi_u(E-G)+2\phi_vF,\\ G_u&=2\phi_vF+2\phi_uG, & G_v&=-2\phi_uF+2\phi_vG. \end{aligned}\right\} \qquad (61)$$

Suppose the net is orthogonal. Then, by the third and fourth of equations (61), either $E=G$ or $\phi_u=\phi_v=0$. In the latter case the coefficients (59) all vanish and the net consists of two families of straight lines. This case we exclude and therefore $E$ must be equal to $G$. We then have the result:

* "One-Parameter Families," equations (29).

*Any integral net of a system of form* (3) *with the coefficients* (59) *is isothermal if it is orthogonal, and the parameters u and v are isometric parameters.*

Again, suppose that $E=G$, $F=0$ for some particular regular point $u=u_0$, $v=v_0$, of the plane. We may express this hypothesis by saying that the net is isothermal *at that particular point.* Then $E$, $F$ and $G$ may be expanded by Taylor's theorem as power series whose coefficients have as factors the successive derivatives of $E$, $F$, and $G$ at the point $(u_0, v_0)$. When equations (61) are differentiated it is found that the successive derivatives of $F$ may all be put in the form, $p(E_0-G_0)+qF_0$, where $E_0$, $F_0$ and $G_0$ denote the values at $(u_0, v_0)$. Also, the successive derivatives of $E-G$ have this same form. Consequently, we find $F=0$, $E=G$ at all points of the plane. We then have the result:

*If an integral net of a system of form* (3), *with the coefficients* (59), *is isothermal at a single regular point, it is isothermal over the entire plane, and the parameters are isometric.*

§ 11. *Effect of the Laplace Transformation on an Isothermal Net.*

Let $E_1$, $F_1$ and $G_1$ be the fundamental quantities of the first order for the first Laplacian transformed net. Then

$$E_1=\left(\frac{\partial x_\sigma}{\partial u}\right)^2+\left(\frac{\partial y_\sigma}{\partial u}\right)^2,$$

with corresponding expressions for $F_1$ and $G_1$. If we make use of (7) and (3) we find

$$E_1=\frac{(a'_u-a'b')^2}{a'^4}\cdot G,$$

$$F_1=\frac{a'_u-a'b'}{a'^4}\left[-a'a''F+(a'_v-a'b''+a'^2)G\right],$$

$$G_1=\frac{1}{a'^4}\left[a'^2a''^2E-2a'a''(a'_v-a'b''+a'^2)F+(a'_v-a'b''+a'^2)^2G\right].$$

Let us suppose that the given net is isothermal with coefficients given by (59). Then $F=0$, $E=G$, and we find

$$E_1=\frac{(\phi_{uv}-\phi_u\phi_v)^2}{\phi_v^4}\cdot G,\quad F_1=\frac{\phi_{uv}-\phi_u\phi_v}{\phi_v^4}\cdot\phi_{vv}G,\quad G_1=\frac{\phi_{vv}^2+\phi_u^2\phi_v^2}{\phi_v^4}\cdot G.$$

The first Laplacian transformed net will be orthogonal also, if $F_1=0$, that is, if either $\phi_{uv}-\phi_u\phi_v=0$, or $\phi_{vv}=0$. The first case is not permissible, otherwise $E_1$ would vanish and the Laplacian net would degenerate, the curves $u=$const.

reducing to points, and the curves $v=$const. to one and the same curve. The second condition gives, by (58), $\phi_{uu}=0$. Hence we can only have

$$\phi=\alpha u+\beta v+\delta uv+\gamma,$$

where $\alpha$, $\beta$, $\gamma$, $\delta$ are constants. Thus, we find in this case

$$\frac{E_1}{G_1}=\frac{[\delta-(\alpha+\delta v)(\beta+\delta u)]^2}{(\alpha+\delta v)^2(\beta+\delta u)^2}.$$

If we take the second logarithmic derivative of $E_1/G_1$ with respect to $u$ and $v$, we get

$$\frac{\partial^2 \log E_1/G_1}{\partial u\partial v}=\frac{-2\delta^3}{[\delta-(\alpha+\delta v)(\beta+\delta u)]^2}.$$

Hence $\dfrac{\partial^2 \log E_1/G_1}{\partial u\partial v}$ will vanish if, and only if, $\delta=0$. It follows that the ratio $E_1:G_1$ is of the form, a function of $u$ alone divided by a function of $v$ alone, if, and only if, $\delta=0$, in which case $E_1=G_1$. Therefore, the first Laplacian transform will be isothermal together with the original net, if, and only if,

$$\phi=\alpha u+\beta v+\gamma. \tag{62}$$

Again, if we let $E_{-1}$, $F_{-1}$, $G_{-1}$ be the fundamental quantities of the first order for the minus first Laplacian transform, and proceed as before, we find that the minus first Laplacian transform will also be isothermal together with the original net, if, and only if, the condition (62) is satisfied.

When $\phi$ is of the form (62), the coefficients (59) become

$$\left.\begin{aligned} a&=\alpha, & a'&=\beta, & a''&=-\alpha,\\ b&=-\beta, & b'&=\alpha, & b''&=\beta, \end{aligned}\right\} \tag{63}$$

where $\alpha$ and $\beta$ are arbitrary constants. We may now make the following statement:

*The isothermality of a net and of one of its Laplacian transforms implies that of all of the others. Every net of this kind may be regarded as an integral net of a system of partial differential equations with constant coefficients of the form* (63).

This system of partial differential equations with constant coefficients is readily integrated. The system, written out explicitly, is

$$x_{uu}=\alpha x_u-\beta x_v,\quad x_{uv}=\beta x_u+\alpha x_v,\quad x_{vv}=-\alpha x_u+\beta x_v. \tag{64}$$

It may be satisfied by putting $x=e^{au+bv}$, where $a$ and $b$ are constants to be determined. We readily find the non-constant solutions

$$x=e^{(\beta-i\alpha)(iu+v)},\quad y=e^{(\beta+i\alpha)(-iu+v)}.$$

The corresponding net is imaginary for real values of $\alpha$ and $\beta$. However, these solutions may be put into the form

$$x=e^{\alpha u+\beta v}[\cos(\beta u-\alpha v)+i\sin(\beta u-\alpha v)],$$
$$y=e^{\alpha u+\beta v}[\cos(\beta u-\alpha v)-i\sin(\beta u-\alpha v)],$$

whence

$$\frac{x+y}{2}=e^{\alpha u+\beta v}\cos(\beta u-\alpha v),\quad \frac{x-y}{2}=e^{\alpha u+\beta v}\sin(\beta u-\alpha v).$$

If we make the imaginary affine transformation

$$\xi=(x+y)/2,\ \eta=(x-y)/2i,$$

we obtain

$$\xi=e^{\alpha u+\beta v}\cos(\beta u-\alpha v),\quad \eta=e^{\alpha u+\beta v}\sin(\beta u-\alpha v), \tag{65}$$

and these equations represent a net which satisfies (64) and which is real if $\alpha$ and $\beta$ are real. If we now put $\sqrt{\xi^2+\eta^2}=r$, $\eta/\xi=\tan\theta$, we obtain

$$r=e^{\alpha u+\beta v},\quad \theta=\beta u-\alpha v+2n\pi, \tag{66}$$

as the polar coordinates of the generating point of the net. For the curves $u=$const. and $v=$const. in (66) we find

$$r\frac{d\theta}{dr}=-\frac{\alpha}{\beta} \text{ and } r\frac{d\theta}{dr}=\frac{\beta}{\alpha},$$

respectively. But $r\dfrac{d\theta}{dr}$ is the tangent of the angle at which the radius vector cuts the curve. Hence, our net is composed of two families of equiangular spirals, which intersect their radii vectores at the angles whose tangents are $-\alpha/\beta$ and $\beta/\alpha$, respectively. Hence the two families are actually orthogonal to each other and therefore (by a theorem of § 10) form an isothermal net. Hence, we have the result:

*The isothermal net composed of two orthogonal families of equiangular spirals has the property that all of its Laplacian transformed nets are also isothermal.* Moreover, it can be readily shown that this is the only real isothermal net having this property.

In order to study the Laplacian transforms of the nets considered in the foregoing paragraph, we set up the Laplacian transforms for the general net (3). If we proceed in a way similar to that followed by Wilczynski,* using equations (7), we find for the first Laplacian transform the system of partial differential equations,

$$\sigma_{uu}=a_1\sigma_u+b_1\sigma_v,\quad \sigma_{uv}=a_1'\sigma_u+b_1'\sigma_v,\quad \sigma_{vv}=a_1''\sigma_u+b_1''\sigma_v,$$

* "One-Parameter Families," § 4.

where

$$a_1=\frac{1}{a'_u-a'b'}\left[b'a'_u+a'_u a+\frac{a''_u a'_u}{a''}-b'a'a-\frac{b'a'a''_u}{a''}-b'_u a'+a'_{uu}-\frac{2a'^2_u}{a'}\right],$$

$$a'_1=\frac{1}{a'_u-a'b'}\left[-a'b'_v-b'a'^2+a'_{uv}-\frac{a'_v a'_u}{a'}+a'a'_u\right],$$

$$a''_1=\frac{1}{a'_u-a'b'}\left[-b''_v a'+a'_{vv}-b'a''a'+\frac{b''a''_v a'}{a''}-b''a'_v-\frac{a''_v a'_v}{a''}+2a'a'_v-\frac{a''_v a'^2}{a''}\right],$$

$$b_1=\frac{a'b'-a'_u}{a''},\quad b'_1=\frac{a'b'-a'_u}{a'},\quad b''_1=b''-\frac{2a'_v}{a'}+\frac{a''_v}{a''}.$$

Similarly, we find for the minus first Laplacian transformed net, the equations

$$\rho_{uu}=a_{-1}\rho_u+b_{-1}\rho_v,\quad \rho_{uv}=a'_{-1}\rho_u+b'_{-1}\rho_v,\quad \rho_{vv}=a''_{-1}\rho_u+b''_{-1}\rho_v,$$

where

$$a_{-1}=a-\frac{2b'_u}{b'}+\frac{b_u}{b},\quad a'_{-1}=\frac{a'b'-b'_v}{b'},\quad a''_{-1}=\frac{a'b'-b'_v}{b},$$

$$b_{-1}=\frac{1}{b'_v-a'b'}\left[-a_u b'+b'_{uu}-a'bb'+\frac{ab_u b'}{b}-ab'_u-\frac{b_u b'_u}{b}+2b'b'_u-\frac{b_u b'^2}{b}\right],$$

$$b'_{-1}=\frac{1}{b'_v-a'b'}\left[-b'a'_u-a'b'^2+b'_{uv}-\frac{b'_u b'_v}{b'}+b'b'_v\right],$$

$$b''_{-1}=\frac{1}{b'_v-a'b'}\left[a'b'_v+b'_v b''+\frac{b_v b'_v}{b}-a'b'b''-\frac{a'b'b_v}{b}-a'_v b'+b'_{vv}-\frac{2b'^2_v}{b'}\right].$$

When the coefficients (63) are substituted in these two sets of equations, it is found that the coefficients of each transformed net are exactly the same as the corresponding coefficients of the original net. Hence the Laplacian transformed nets are affine to the original net. We may state our result as follows:

*The only real isothermal nets both of whose Laplacian transformed nets are isothermal, are those which are composed of two families of equiangular spirals, and then both Laplacian transforms are affine to the original net.*

# *On a Special Elliptic Ruled Surface of the Ninth Order.*

BY HARRY CLINTON GOSSARD.

The object of this paper is to discuss the following problem connected with a tetrahedron:

*Are there lines connected with a tetrahedron, such that if the vertices are reflected in these lines, the reflections will fall on the opposite faces? If so, what is the locus of these lines?*

It has been shown by G. T. Bennett* that for any given tetrahedron there are $\infty^1$ such lines, and that when the opposite edges are equal, the locus consists of three cylindroids. The existence of these lines will be established by direct geometric considerations, and their locus will be discussed together with other related questions.

## § 1. *Special Displacements of a Given Tetrahedron.*

Let $(\alpha x)(a\eta)=0$ be a collineation between a point $x$ and a point $x'$. To within a factor of proportionality, the coefficients of the $\eta$'s are the coordinates of the point $x'$, *i. e.*, $kx'_i=a_i(\alpha x)$, where $i=0, 1, 2, 3$, and where $k$ is the factor of proportionality. If we put $x'_i=x_i$, the eliminant of the four equations, giving the fixed point, is

$$\begin{vmatrix} a_0\alpha_0-k & a_0\alpha_1 & a_0\alpha_2 & a_0\alpha_3 \\ a_1\alpha_0 & a_1\alpha_1-k & a_1\alpha_2 & a_1\alpha_3 \\ a_2\alpha_0 & a_2\alpha_1 & a_2\alpha_2-k & a_2\alpha_3 \\ a_3\alpha_0 & a_3\alpha_1 & a_3\alpha_2 & a_3\alpha_3-k \end{vmatrix}=0,$$

which may be written as

$$k^4-I_1k^3+I_2k^2-I_3k+I_4=0,$$

where

$$I_1=(a\alpha)=a_0\alpha_0+a_1\alpha_1+a_2\alpha_2+a_3\alpha_3, \qquad I_2=(\alpha\beta\cdot ab),$$
$$I_3=(\alpha\beta\gamma\cdot abc), \qquad I_4=(\alpha\beta\gamma\delta\cdot abcd).$$

We are concerned with those collineations for which $(\alpha a)=0$.†

---

* *Proc. London Math. Soc.*, Series 2, Vol. X (1911), Parts 4 and 5.

† These are Study's "Normal Collineations."

Suppose the above collineation to be a displacement. The normal form of a displacement is

$$x' = e^{i\alpha} x,$$
$$\bar{x}' = e^{-i\alpha} \bar{x},$$
$$z' = z + p\alpha,$$

where $p$ is the pitch of the screw, $\alpha$ the angle turned through, and the barred variables are the conjugates of the unbarred variables.

Written homogeneously these equations become

$$x'_0 = e^{i\alpha} x_0,$$
$$x'_1 = e^{-i\alpha} x_1,$$
$$x'_2 = x_2 + p\alpha x_3,$$
$$x'_3 = x_3,$$

for which $I_1$ is $e^{i\alpha} + e^{-i\alpha} + 2$ or $I_1 = 2 + 2\cos\alpha$.

For a displacement $I_1$ will equal 0 when $\alpha = 180°$. Thus we see that

*There are displacements of $T$ which send each vertex onto the face opposite it, and these displacements consist of rotations about an axis, through an angle of* 180°, *and a translation.*

The general collineation has 15 independent constants. In the displacement herein considered, the plane at infinity (which we shall name $W$), the absolute in this plane and the size of $T$ are all left unaltered. This puts $3+5+1$ or 9 conditions on the displacement. It is one condition for each vertex to go onto the face opposite it. This leaves $15-13=2$ constants at our disposal and consequently $\infty^2$ axes of rotation. Taking the pitch of the screw motion to be 0, which is one more condition, leaves $\infty^1$ axes of rotation and reduces the above displacements to rotations of 180° only, about these $\infty^1$ axes, and therefore to reflections in these lines.

Throughout this paper $T$ is taken to be the reference tetrahedron, and the one formed by the reflection of the vertices in any one of these $\infty^1$ lines will be designated as $T'$. Then for a given $T$, there are $\infty^1$ $T'$'s. Or,

*For a given tetrahedron $T$, there exists $\infty^1$ lines such that reflections of $T$ in these lines give $\infty^1$ tetrahedrons $T'$, which are inscribed to $T$.*

## § 2. *The Axes of Rotation.*

Consider the four reflections of $T$, given by

$$R_0:\quad x'_0 = -x_0,\quad x'_1 = x_1,\quad x'_2 = x_2,\quad x'_3 = x_3;$$
$$R_1:\quad x'_0 = x_0,\quad x'_1 = -x_1,\quad x'_2 = x_2,\quad x'_3 = x_3;$$
$$R_2:\quad x'_0 = x_0,\quad x'_1 = x_1,\quad x'_2 = -x_2,\quad x'_3 = x_3;$$
$$R_3:\quad x'_0 = x_0,\quad x'_1 = x_1,\quad x'_2 = x_2,\quad x'_3 = -x_3.$$

$R_0$ sends a line $p$, whose coordinates are

$$p_{01},\ p_{02},\ p_{03},\ p_{12},\ p_{23},\ p_{31}$$

into a line whose coordinates are

$$-p_{01},\ -p_{02},\ -p_{03},\ p_{12},\ p_{23},\ p_{31},$$

which we shall call $pR_0$. $R_1$ sends $p$ into a line $pR_1$ whose coordinates are

$$-p_{01},\ p_{02},\ p_{03},\ -p_{12},\ p_{23},\ -p_{31}.$$

If $\pi$ be an axis with coordinates $\pi_{ij}$, meeting $pR_0$, then

$$-p_{01}\pi_{01}-p_{02}\pi_{02}-p_{03}\pi_{03}+p_{23}\pi_{23}+p_{31}\pi_{31}+p_{12}\pi_{12}=0.$$

If $\pi$ meets $pR_1$,

$$-p_{01}\pi_{01}+p_{02}\pi_{02}+p_{03}\pi_{03}+p_{23}\pi_{23}-p_{31}\pi_{31}-p_{12}\pi_{12}=0.$$

If $\pi$ meets both $pR_0$ and $pR_1$, then

$$p_{01}\pi_{01}=p_{23}\pi_{23}, \tag{1}$$

*i. e.*, $\pi$ belongs to a linear complex. Similarly, if $\pi$ meets $pR_0$ and $pR_2$,

$$p_{02}\pi_{02}=p_{31}\pi_{31}, \tag{2}$$

and if $\pi$ meets $pR_0$ and $pR_3$,

$$p_{03}\pi_{03}=p_{12}\pi_{12}. \tag{3}$$

Or, if $\pi$ meets $pR_0$, $pR_1$, $pR_2$ and $pR_3$, then it is common to three linear complexes. Consequently, there are $\infty^1$ lines $\pi$ meeting the four lines above, and these four lines $pR_i$ belong to a regulus. $p$ and $\pi$ play dual rôles and so $p$ must also meet the same four lines.

We now seek the equation of the quadric on which $pR_i$ lie. From (1), (2) and (3) we have for any given line $\pi$, equations of type

$$(x_0y_1-x_1y_0)\pi_{01}-(x_2y_3-x_3y_2)\pi_{23}=0.$$

Eliminating $y_1$, $y_2$ and $y_3$,

$$\rho\, y_0=x_0(x_0^2\pi_{01}\pi_{02}\pi_{03}+x_1^2\pi_{12}\pi_{13}\pi_{10}++)=0,$$

with three similar equations. These equations are satisfied by $x\equiv y$, and by

$$x_0^2\pi_{01}\pi_{02}\pi_{03}+x_1^2\pi_{12}\pi_{13}\pi_{10}+x_2^2\pi_{20}\pi_{21}\pi_{23}+x_3^2\pi_{30}\pi_{31}\pi_{32}=0, \tag{4}$$

which is the desired equation. It represents *a quadric referred to a self-polar tetrahedron of reference*, on which the four lines $pR_i$ must lie, *as well as the lines $p$ and $\pi$.*

The relation of $p$ and $\pi$ in (1), (2) and (3) is that the reflection of a vertex $e_i$ of $T$ in $p$ and $\pi$ is on $\varepsilon_i$, the face opposite $e_i$. This reflection can be made a physical one by sending the line $\pi$ to infinity in a direction perpendicular to $p$. This makes $\pi$ the polar line, as to the absolute in $W$, of the point of intersection of the line $p$ and the plane $W$. This makes the quadric (4) touch the plane $W$, *i. e.*, makes it a paraboloid, and because of the way $\pi$ is sent to infinity, makes this paraboloid orthic, *i. e.*, makes it apolar to the absolute.

Making a slight change in coordinates we shall write (4) as

$$\left(\frac{x^2}{y}\right) \equiv \frac{x_0^2}{y_0} + \frac{x_1^2}{y_1} + \frac{x_2^2}{y_2} + \frac{x_3^2}{y_3} = 0,$$

which is any quadric referred to a self-polar tetrahedron, adding the necessary requirements to make it an orthic paraboloid. To be a paraboloid it must touch the plane at infinity, whose equation is $(x) = 0$. This condition is

$$\frac{1}{y_1 y_2 y_3} + \frac{1}{y_0 y_2 y_3} + \frac{1}{y_0 y_1 y_3} + \frac{1}{y_0 y_1 y_2} = 0, \text{ or } y_0 + y_1 + y_2 + y_3 = 0.$$

To be orthic, *i. e.*, to be apolar to the absolute, the equation of which we shall write as $(A\xi)^2 = 0$, where $A_{ii}$ is the square of the area of the face $\varepsilon_i$ of $T$, the condition is

$$\left(\frac{A^2}{y}\right) \equiv \frac{A_{00}}{y_0} + \frac{A_{11}}{y_1} + \frac{A_{22}}{y_2} + \frac{A_{33}}{y_3} = 0.$$

Then all lines $p$ lie on the quadrics represented by $\left(\frac{x^2}{y}\right) = 0$, where these quadrics are subject to the relations $\left(\frac{A^2}{y}\right) = 0$ and $(y) = 0$, where $y$ is the point of $\pi$ where the axis of the quadric meets $W$, *i. e.*, is the point of contact of $W$ and the quadric. The quadric being orthic, its axis and principal generators form three mutually perpendicular lines, which meet $W$ in three points forming a triad, self-conjugate with regard to the absolute. As $\pi$ is the polar line of the point $p$ (*i. e.*, the point where the line $p$ meets the plane $W$), and the principal tangent plane meets $W$ in a line which is the polar line of the point $y$, the line $p$ must be one of the principal generators of the quadric. It is at once seen that both of the principal generators play the same rôle; and, consequently, both are lines $p$ of the kind herein discussed. For either as a line $p$, the corresponding $\pi$ is the polar of the meet of $p$ and $W$ as to the absolute, and these two corresponding lines $\pi$ and $\pi'$ meet on the point $y$. So we have this theorem:

*All the lines sought are the principal generators of the quadrics* $\left(\frac{x^2}{y}\right) = 0$, *where* $(y) = 0$ *and* $\left(\frac{A^2}{y}\right) = 0$.

A case in point is the rectangular hyperboloid $xy=z$. The polar plane of a point $x', y', z'$ is $xy'+x'y=z+z'$. This equation is satisfied by $x', -y', -z'$ and $-x', y', -z'$, the reflections of $x', y', z'$ in the lines $x=0$ and $y=0$, respectively. These two lines are the principal generators of $xy=z$.

## § 3. *The Plane at Infinity and the Surface Sought.*

The reflection of a vertex $e_i$ of $T$ in *any* line at infinity, will lie on the face $\varepsilon_i$, opposite $e_i$. For, to get a reflection of a point in two lines, one takes the fourth harmonic of this point and the two points where the two given lines are intersected by the line on the given point, meeting the two given lines. As one of the two lines is a line $\pi$ in the plane $W$, any other line in $W$, that is on the meet of $\pi$ and the intersection of $W$ and $\varepsilon_i$, would do for a line $p$. For, then the reflection of $e_i$ in $p$ and $\pi$ would be at their intersection and consequently on $\varepsilon_i$. Therefore, in a certain sense the *entire plane W is* to be considered *a part of the locus of lines* for which we are seeking.*

## § 4. *The Curve at Infinity.*

Let $\Omega$ denote the surface on which the lines lie. We seek the intersection of $\Omega$ and $W$. The cubic surface $\left(\frac{A^2}{y}\right)$ meets the plane $W$ in a cubic curve, the locus of all points $y$, where the axes of the quadrics $\left(\frac{x^2}{y}\right)=0$ meet the plane $W$. Also, as before stated, these points $y$ are the intersections of pairs of lines $\pi$ and $\pi'$ on which are the points, where the principal generators meet the plane $W$.

These line pairs $\pi$ and $\pi'$ are conjugate as to the absolute and form the degenerate conics of the net of apolar conics of some cubic. The cubic $\left(\frac{A^2}{y}\right)=0$, being on the double points of these degenerate conics, is the Jacobian of this net, and the Hessian of the cubic of which this is the apolar net.† The Cayleyan of this latter cubic would be enveloped by these line pairs $\pi$ and $\pi'$, and *the reciprocal of this Cayleyan as to the absolute passes through the point-pairs on* $\pi$ *and* $\pi'$. Therefore:

*The intersection of all lines p and the plane W is a cubic curve,* which we shall name $R$. This cubic $R$ meets the absolute in six points, and the tangents to this cubic at these points can be seen at once to be special lines of $\Omega$. The

* This plane $W$ will not be included as a part of $\Omega$ in this paper.

† See Clifford's Papers, "Polar Theory of the Plane Cubic."

$\pi$'s corresponding to these six tangents are the tangents to the absolute, where it meets the cubic $R$. Moreover, an examination of other possible special lines of $\Omega$ in the plane $W$ shows that there are none other than these six. Consequently,

*The intersection of the plane W and the surface $\Omega$ consists of a cubic curve and six special lines and is therefore of the ninth order.*

These six lines meet in fifteen points; they meet the cubic in six points other than the points of tangency; and the six lines meet lines $p$ at the six points of tangency. In all the intersection of the plane $W$ and the surface $\Omega$ has 27 double points, one less than the total maximum number for a curve of the ninth order. Therefore,

*The intersection of the plane W and the surface $\Omega$ is an elliptic curve of the ninth order, and it follows that $\Omega$ is an elliptic surface of the ninth order.*

## § 5. *Special Lines of $\Omega$.*

There are some lines of $\Omega$ that are, in a special way, connected with the tetrahedron $T$. G. T. Bennett* has mentioned the two following types:

(1) The two perpendicular normals to an edge at its middle point, each equallv inclined to the faces through that edge, *i. e.*, lying in the planes that bisect the dihedral angles on an edge.

(2) Any line which can be drawn, meeting two opposite edges and bisecting, internally or externally, at each of its extremities, the angle subtended by the opposite edge.

It is seen from simple geometric observation that these lines are proper lines of $\Omega$. The six pairs of perpendicular lines of type (1) are evidently pairs of principal generators of six of the quadrics of § 2. For each pair of opposite edges of $T$ there are seven lines of type (2). This is shown by the correspondence set up on the pairs of opposite edges. For, if $\overline{12}$ and $\overline{34}$ are opposite edges of $T$, then for two points 1 and 2 on one edge, we have one point $y$ on the other edge. This point $y$ definitely determines two points $x$ on the first edge, one on the internal and one on the external bisector of the angle $\underline{|1\,y\,2}$, *i. e.*, for one point $y$ there are four points on the edge $\overline{12}$. Similarly, for one point $x$ on the edge $\overline{12}$ are four points on the edge $\overline{34}$. Therefore, the correspondence between the two edges, thus established is a 4—4 correspondence. The lines $\overline{xy}$ are given by the coincidences of this correspondence which must be eight in number. However, the plane $W$ contains one of these

* See page 431, footnote.

lines, which is not to be included as part of $\Omega$, or there are seven lines of type (2). It is evident that no other lines of $\Omega$ meet an edge of $T$, and consequently,

*A given edge of $T$ meets nine lines of the surface $\Omega$—two of type* (1) *and seven of type* (2)*—or, what is the same thing, an edge of $T$ meets the surface $\Omega$ in nine points, which verifies the fact that $\Omega$ is of the ninth order.*

The equation of $\Omega$ can be written from this point of view, by expressing the coordinates of any line of $\Omega$ in terms of an elliptic parameter. To do this, it is only necessary to suppose the cubic curve $R$ to be expressed in terms of an elliptic parameter $u$. Then since for each point of $R$ there corresponds one line, in general, the lines can be expressed in terms of this parameter $u$. The six coordinates $p_{ij}$ of each line of $\Omega$ will each equal the product of nine terms like $\sigma(u-a)$, one for each of the lines that meets an edge of the tetrahedron of reference, *i. e.*,

$$p_{ij}=\sigma(u-a_1)\sigma(u-a_2)\ldots.\sigma(u-a_9).$$

Since two of the nine lines on an edge of $T$ are perpendicular, their parameters would differ only by a half period, *i. e.*, if one is $(u-\alpha)$, the other would be $(u-\alpha+\omega)$. Moreover, since seven of the nine lines on an edge of $T$, also meet the opposite edge of $T$, seven of the $\sigma$ factors would be the same for the two coordinates $p_{ij}$ and $p_{kl}$. The coordinates of any line of $\Omega$ would then be written:

$$\begin{aligned}
p_{01}&=\sigma(u-a_1)\sigma(u-a_2)\ldots.\sigma(u-a_7)\sigma(u-\alpha_{01})\sigma(u-\alpha_{01}+\omega),\\
p_{23}&=\sigma(u-a_1)\sigma(u-a_2)\ldots.\sigma(u-a_7)\sigma(u-\alpha_{23})\sigma(u-\alpha_{23}+\omega),\\
p_{02}&=\sigma(u-b_1)\sigma(u-b_2)\ldots.\sigma(u-b_7)\sigma(u-\alpha_{02})\sigma(u-\alpha_{02}+\omega),\\
p_{31}&=\sigma(u-b_1)\sigma(u-b_2)\ldots.\sigma(u-b_7)\sigma(u-\alpha_{31})\sigma(u-\alpha_{31}+\omega),\\
p_{03}&=\sigma(u-c_1)\sigma(u-c_2)\ldots.\sigma(u-c_7)\sigma(u-\alpha_{03})\sigma(u-\alpha_{03}+\omega),\\
p_{12}&=\sigma(u-c_1)\sigma(u-c_2)\ldots.\sigma(u-c_7)\sigma(u-\alpha_{12})\sigma(u-\alpha_{12}+\omega).
\end{aligned}$$

## § 6. *Introduction of Elliptic Functions.*

The cubic surface $\left(\dfrac{A^2}{y}\right)=0$ contains the edges of the reference tetrahedron $T$. By putting $(y)=0$, and thereby getting the cubic curve which is the intersection of the plane $W$ and this cubic surface, one gets a cubic curve which passes through the six points where the edges of $T$ meet $W$.

Let this cubic bear an elliptic parameter $u$, and for the above six points of this cubic let $u$ have the values $a$, $b$, $c$, $x$, $y$ and $z$, respectively. As these six values of the elliptic parameter are collinear by three's, we have

$$a+b+c=0, \quad b+x+y=0,$$
$$a+y+z=0, \quad c+x+z=0.$$

Adding these four equations and substituting 0 for $a+b+c$, the result is $x+y+z=0$, or $=\omega_1$, where $\omega_1$ is a half-period. Then,

$$x=\omega_1-y-z=\omega_1+a, \text{ and similarly, } y=\omega_1+b, \; z=\omega_1+c.$$

Consider $y_0=\lambda_0\sigma(u-a)\sigma(u-b)\sigma(u-c)$. This represents a line on the points whose parameters are $a$, $b$ and $c$ respectively, where $a+b+c=0$. The line on the points whose parameters are $a$, $y$ and $z$ respectively, is

$$y_1=\lambda_1\sigma(u-a)\sigma(u-b-\omega_1)\sigma(u-c-\omega_1),$$

and similar equations for the lines that are on $b$, $y$, $z$ and $c$, $x$, $z$. The function $\sigma(u-b-\omega_1)$ will be replaced by $\sigma_1(u-b)$, where $\sigma_1$ is the *allied sigma-function*. Then we have

$$y_0=\lambda_0\sigma(u-a)\sigma(u-b)\sigma(u-c),$$
$$y_1=\lambda_1\sigma(u-a)\sigma_1(u-b)\sigma_1(u-c), \text{ etc.}$$

It is necessary now to so determine $\lambda_i$ that $(y)=0$, and $\left(\frac{A^2}{y}\right)=0$. If $(y)=0$,

$$\lambda_0\sigma(u-a)\sigma(u-b)\sigma(u-c)+\lambda_1\sigma(u-a)\sigma_1(u-b)\sigma_1(u-c)$$
$$+\lambda_2\sigma_1(u-a)\sigma(u-b)\sigma_1(u-c)+\lambda_3(\ldots.)=0.$$

Dividing by the coefficient of $\lambda_0$,

$$\lambda_0+\lambda_1\frac{\sigma(u-a)\sigma_1(u-b)\sigma_1(u-c)}{\sigma(u-a)\sigma(u-b)\sigma(u-c)}+\lambda_2(\ldots.)+\lambda_3(\ldots.)=0. \tag{5}$$

To remove the infinities for $a$, $b$ and $c$, put $u=a$, $u=b$, $u=c$ in succession in (5), observing that $\sigma_1(0)=1$, this gives

$$\lambda_2\frac{\sigma_1(a-c)}{\sigma(a-c)}+\lambda_3\frac{\sigma_1(a-b)}{\sigma(a-b)}=0,$$
$$\lambda_1\frac{\sigma_1(b-c)}{\sigma(b-c)}+\lambda_3\frac{\sigma_1(a-b)}{\sigma(a-b)}=0,$$
$$\lambda_1\frac{\sigma_1(b-c)}{\sigma(b-c)}+\lambda_2\frac{\sigma_1(a-c)}{\sigma(a-c)}=0.$$

Solving for the $\lambda$'s, we have

$$\lambda_1=\frac{\sigma(b-c)}{\sigma_1(b-c)}, \quad \lambda_2=\frac{\sigma(a-c)}{\sigma_1(a-c)}, \quad \lambda_3=\frac{\sigma(a-b)}{\sigma_1(a-b)},$$

which substituted in (5) reduces that equation to

$$\lambda_0 + \frac{\sigma(b-c)\,\sigma_1(u-b)\,\sigma_1(u-c)}{\sigma_1(b-c)\,\sigma(u-b)\,\sigma(u-c)} + (\ldots.) + (\ldots.) = 0. \tag{6}$$

To determine $\lambda_0$, let $u=a+\omega_1$. Then $u-a=\omega_1$ and $\sigma_1(u-a)=\sigma_1\omega=0$. This eliminates the last two terms in (6), and leaves

$$\lambda_0 + \frac{\sigma(b-c)\,\sigma_1(u-b)\,\sigma_1(u-c)}{\sigma_1(b-c)\,\sigma(u-b)\,\sigma(u-c)} = 0, \tag{7}$$

whence

$$\lambda_0 = (e_1-e_2)\,(e_1-e_3)\,\frac{\sigma(b-c)\,\sigma(c-a)\,\sigma(a-b)}{\sigma_1(b-c)\,\sigma_1(c-a)\,\sigma_1(a-b)},$$

which gives as the equation corresponding to $(y)=0$, by substituting in (6),

$$(e_1-e_2)\,(e_1-e_3)\,\frac{\sigma(b-c)\,\sigma(c-a)\,\sigma(a-b)}{\sigma_1(b-c)\,\sigma_1(c-a)\,\sigma_1(a-b)} + \sum^{3}\frac{\sigma(b-c)\,\sigma_1(u-b)\,\sigma_1(u-c)}{\sigma_1(b-c)\,\sigma(u-b)\,\sigma(u-c)} = 0, \tag{8}$$

or,

$$\left.\begin{aligned} y_0 &= (e_1-e_2)\,(e_1-e_3)\,\frac{\sigma(b-c)\,\sigma(c-a)\,\sigma(a-b)}{\sigma_1(b-c)\,\sigma_1(c-a)\,\sigma_1(a-b)},\\ y_1 &= \frac{\sigma(b-c)\,\sigma_1(u-b)\,\sigma_1(u-c)}{\sigma_1(b-c)\,\sigma(u-b)\,\sigma(u-c)},\\ &\text{and similar expressions for } y_2 \text{ and } y_3. \end{aligned}\right\} \tag{9}$$

If we substitute $u+\omega_1$ for $u$ in equation (8), and for $\frac{\sigma_1(u-b)}{\sigma(u-b)}$ its corresponding value $\frac{\sigma(u-b)}{\sigma_1(u-b)}\sqrt{(e_1-e_2)\,(e_1-e_3)}$, and then divide the result by $(e_1-e_2)\,(e_1-e_3)$, we obtain the equation:

$$\frac{\sigma^2(b-c)\,\sigma^2(c-a)\,\sigma^2(a-b)}{\sigma_1^2(b-c)\,\sigma_1^2(c-a)\,\sigma_1^2(a-b)}\,(e_1-e_2)\,(e_1-e_3)\,\frac{1}{y_0} + \sum^{3}\frac{\sigma^2(b-c)}{\sigma_1^2(b-c)}\,\frac{1}{y_1} = 0,$$

which is manifestly of the form $\left(\frac{A^2}{y}\right)=0$, where

$$\left.\begin{aligned} A_{00} &= (e_1-e_2)\,(e_1-e_3)\,A_{11}\,A_{22}\,A_{33},\\ A_{ii} &= \frac{\sigma^2(b-c)}{\sigma_i^2(b-c)},\quad i=1,2,3. \end{aligned}\right\} \tag{10}$$

By means of these equations (9) and (10) we shall now get an equation of that part of the double curve on $\Omega$, which is the locus of the intersections of those generators of $\Omega$ which meet in perpendicular pairs.

## § 7. *The Double Curve.*

The plane section of $\Omega$, as stated in § 4, has twenty-seven double points, and so the surface $\Omega$ must have a curve of double points of the 27th degree. At least a part of this double curve will be the locus of the vertices of the quadrics of this paper. This part we shall now find. We first seek the locus of the tangent planes of the quadrics at their vertices. Since the principal generators meet the plane $W$ in points which, with the corresponding point $y$ where the axis of the quadric pierces $W$, form a self-conjugate triangle, the plane on these two generators will meet $W$ in a line that is the polar line of $y$. Consequently, the polar plane of $y$ as to, say the circumsphere of $T$, will be a plane parallel to the principal tangent plane of the quadric above.

The equation of this circumsphere is given in quadri-planar coordinates as:

$$\frac{e_{21}^2\bar{x}_1\bar{x}_2}{x_1'x_2'}+\frac{e_{20}^2\bar{x}_2\bar{x}_0}{x_2'x_0'}+\ldots\ldots+\frac{e_{23}^2\bar{x}_2\bar{x}_3}{x_2'x_3'}=0,^*$$

where $e_{ij}$ is an edge of $T$, and $x_i'$ is the altitude to the face $x_i=0$. In barycentric projective coordinates, this equation is:

$$e_{21}^2x_1x_2+\ldots\ldots+e_{23}^2x_2x_3=0, \text{ or } \overset{6}{\Sigma}e_{ij}^2x_ix_j=0,$$

where ($i=0, 1, 2, 3$, and $j=0, 1, 2, 3$).

The polar plane of $y$ as to this sphere is:

$$\overset{6}{\Sigma}e_{ij}^2(x_iy_j+x_jy_i)=0,$$

which for convenience we shall write as $(\xi x)=0$ where

$$\xi_i=(e_{ij}^2y_j+e_{ik}^2y_k+e_{im}^2y_m) \qquad (j\neq k\neq m).$$

To get the planes tangent to the quadrics $\left(\frac{x^2}{y}\right)=0$ at their vertices, it is only necessary to get the plane parallel to $(\xi x)=0$, which touches the quadric. Any plane parallel to $(\xi x)=0$, is given by

$$(\xi x)+\lambda(x)=0. \tag{11}$$

The condition for this plane to touch the quadric $\left(\frac{x^2}{y}\right)=0$ reduces at once to

$$\lambda^2(y)+2\lambda(\xi y)+(\xi^2y)=0, \tag{12}$$

where the $y$'s are subjected to the two relations $(y)=0$ and $\left(\frac{A^2}{y}\right)=0$.

Substituting the values for $\xi_i$ in $(\xi^2y)$ this term reduces to

$$\overset{4}{\Sigma}y_1y_2y_3[2(e_{12}^2e_{23}^2+e_{23}^2e_{31}^2+e_{31}^2e_{12}^2)-e_{12}^4-e_{13}^4-e_{23}^4].$$

* "Rogers's revision of Salmon's Geometry of Three Dimensions," p. 235.

As the coefficient of $y_1y_2y_3$ is sixteen times the square of the area of the face $x_0=0$ of $T$, *i. e.*, equals $16A_{00}$, this last quantity can be written as

$$16y_0y_1y_2y_3\left(\frac{A^2}{y}\right).$$

Therefore $(\xi^2y)=16y_0y_1y_2y_3\left(\frac{A^2}{y}\right)=0$, and (12) reduces to $2\lambda(\xi y)=0$. This last equation has the roots $\lambda=0, \infty$. If $\lambda=\infty$, (11) reduces to $(x)=0$, the plane $W$ already discussed. If $\lambda=0$, (11) becomes $(\xi x)=0$, the principal tangent plane of $\left(\frac{x^2}{y}\right)=0$, and as $y$ traces out the cubic curve $\left(\frac{A^2}{y}\right)=0$ in the plane $W$, $(\xi x)=0$ gives the locus of these planes.

The $y$'s, as given by equations (9), are cubics in $\sigma$, and as $(\xi x)=0$ is linear in $y$, when the $\sigma$'s are substituted the equation resulting will be a cubic in $\sigma$. Therefore *these double tangent planes of $\Omega$ envelope a cubic curve.*

An interesting property of these double tangent planes is derived very simply. The points of any plane cubic reciprocate, with regard to a sphere, into planes on a cone of class 3, with its vertex at the point which is the reciprocal of the plane on which the cubic lies. In this case the latter plane is the plane $W$ which reciprocates with regard to the sphere circumscribing $T$ into the centre of this sphere. Thus:

*All the planes $(\xi x)=0$ pass through the centre of the sphere circumscribing the reference tetrahedron $T$.*

The equation $\left(\frac{x^2}{y}\right)=0$, in planes, is $(\xi^2y)=0$. The pole of $\xi$, one of the planes enveloping $(\xi x)=0$, will be the vertex of the quadric of which $\xi$ is the principal tangent plane. Taking the coefficients of $\xi'$ in $(\xi\xi'y)=0$ as the coordinates of $x$, we have $x_i=\xi_iy_i$ as the locus of the vertices of all the quadrics $\left(\frac{x^2}{y}\right)=0$. This is a quadratic in $y$ and therefore a sextic in $\sigma$, *i. e.*:

*The locus of the intersections of those generators of $\Omega$ which meet in perpendicular pairs, is an elliptic space sextic.*

The six points where this sextic meets the plane $W$ are the points where the cubic $R$ meets the absolute. It is at once seen that the tangent to $R$ at any one of these points—which is a line of $\Omega$, as before stated—is perpendicular to that generator of $\Omega$ which pierces $W$ at that point.

As the mid-points of the edges of $T$ are points on this sextic, we have at once that three of the points where this sextic pierces any face of $T$, are the mid-points of the edges of $T$ which lie in that face.

The complete curve of double points being of degree 27, there remains a part of degree 27—6=21, which is the intersection of those génerators which are not perpendicular. This part meets each generator of $\Omega$ seven times. Also since the curve of double planes is of degree 3 less than the degree of the curve of double points, the complete curve of double planes would be of degree 24. Of this the cubic $(\xi x)=0$, has already been accounted for, there remaining a part of degree 21.

### § 8. *A Quadratic Transformation Connected with* $\Omega$.

The equation of the sextic curve of double points suggests a quadratic transformation. Let this equation be written as

$$x_i = \alpha_{ij} y_i y_j + \alpha_{ik} y_i y_k + \alpha_{il} y_i y_l , \qquad (13)$$

where $\alpha_{ij} = e_{ij}^2$ and $i, j, k, l = 0, 1, 2, 3$, $i \neq j \neq k \neq l$. If in $(\xi x)=0$, the relation of incidence between a point $x$ and a plane $\xi$, we substitute these values for $x_i$, the correspondence established by (13) takes the form

$$\sum^{3} (\alpha_{ij} y_i y_j \xi_i + \alpha_{ik} y_i y_k \xi_i + \alpha_{il} y_i y_l \xi_i) = 0. \qquad (14)$$

Let us regard this as a correspondence between two spaces $s_x$ and $s_y$. To a plane $\xi$ in $s_x$ corresponds a quadric in $s_y$. To a line in $s_x$ corresponds a quartic curve in $s_y$. To a point in $s_x$ corresponds four points in $s_y$. A plane in $s_y$ meeting a quartic curve in four points, the correspondent in $s_x$ must meet a line in four points; and, therefore, must be a quartic surface which can be shown to be a Steiner's Quartic Surface. To a line in $s_y$ corresponds a conic in $s_x$. A plane $\eta$ meets the three-fold system of quadrics in a three-fold system of conics, which map the plane $\eta$ onto the corresponding Steiner's Quartic Surface.

The Jacobian of the system of quadrics is a quartic surface whose equation is

$$J \equiv \begin{vmatrix} a_{01}y_1 + a_{02}y_2 + a_{03}y_3 , & a_{01}y_0 , & a_{02}y_0 , & a_{03}y_0 , \\ a_{01}y_1 , & a_{01}y_0 + a_{12}y_2 + a_{13}y_3 , & a_{12}y_1 , & a_{13}y_1 , \\ a_{02}y_2 , & a_{12}y_2 , & a_{02}y_0 + a_{12}y_1 + a_{23}y_3 , & a_{23}y_2 , \\ a_{03}y_3 , & a_{13}y_3 , & a_{23}y_3 , & a_{03}y_0 + a_{13}y_1 + a_{23}y_2 , \end{vmatrix} = 0.$$

Or,

$$\begin{aligned} J \equiv\, & \alpha_{01}\alpha_{02}\alpha_{12} y_0 y_1 y_2 (\alpha_{03} y_0 + \alpha_{13} y_1 + \alpha_{23} y_2) \\ & + \alpha_{01} a_{03} \alpha_{13} y_0 y_1 y_3 (\alpha_{02} y_0 + \alpha_{12} y_1 + \alpha_{23} y_3) \\ & + \alpha_{02}\alpha_{03}\alpha_{23} y_0 y_2 y_3 (\alpha_{01} y_0 + \alpha_{12} y_2 + \alpha_{13} y_3) \\ & + \alpha_{12}\alpha_{13}\alpha_{23} y_1 y_2 y_3 (\alpha_{01} y_0 + \alpha_{02} y_2 + \alpha_{03} y_3) = 0. \end{aligned} \qquad (15)$$

Or,

$$J \equiv \sum^{4} \alpha_{ij}\alpha_{ik}\alpha_{jk} y_i y_j y_k (\alpha_{il} y_i + \alpha_{jl} y_j + \alpha_{kl} y_k) = 0.$$

Since to a line in $s_x$ corresponds a quartic in $s_y$, the correspondent of $J$ will meet the line in the same number of points as the number in which the quartic curve will meet $J$, or in sixteen points. Then the correspondent of $J$ is a surface of order 16. A plane $\xi$ will touch this latter surface whenever $J$ touches the quadrics of the system, *i. e.*, when the discriminant of (15) vanishes. The equation of the correspondent of $J$ is given in planes, by bordering this discriminant, and is

$$\sum^{3} a_{01}^2 a_{23}^2 (\xi_0+\xi_1)^2 (\xi_2+\xi_3)^2 - 2\sum^{3} a_{01} a_{02} a_{13} a_{23} (\xi_0+\xi_1)(\xi_0+\xi_2)(\xi_1+\xi_3)(\xi_2+\xi_3) = 0,$$

or, which can be written

$$Q \equiv \sqrt{a_{01}a_{23}(\xi_0+\xi_1)(\xi_2+\xi_3)} + \sqrt{a_{02}a_{13}(\xi_0+\xi_2)(\xi_1+\xi_3)} + \sqrt{a_{03}a_{12}(\xi_0+\xi_3)(\xi_1+\xi_2)} = 0. \qquad (16)$$

As this is of the fourth degree in $\xi$ the correspondent of the Jacobian is a surface of class 4. Since the reflection of the plane $\xi_0+\xi_1=0$ in the centroid of $T$, whose coordinates are (1, 1, 1, 1), is the plane $\xi_2+\xi_3$, and similarly for $\xi_0+\xi_2$ and $\xi_0+\xi_3$, which reflect into the planes $\xi_1+\xi_3$ and $\xi_1+\xi_2$, $Q$ reflects into itself, and is therefore symmetrical about this centroid. Thus:

*The correspondent of the Jacobian of the system of quadrics is a surface of order* 16 *and class* 4, *which is symmetrical with regard to the centroid of T.*

The equation of the Steiner's Quartic Surface corresponding to the plane $W$ is found by forming the Hessian of (14), changing the resulting equations in planes into the point equation, and making this surface touch $W_x$. The result is

$$R \equiv (a_{12}+a_{13}-a_{23})(a_{01}a_{23}+a_{02}a_{03}) + (a_{12}-a_{13}+a_{23})(a_{02}a_{13}+a_{01}a_{03}) + (-a_{12}+a_{13}+a_{23})(a_{03}a_{12}+a_{01}a_{02}) = a_{01}^2 a_{23} + a_{02}^2 a_{13} + a_{03}^2 a_{12} + a_{12}a_{13}a_{23}, \qquad (17)$$

where

$$a_{ij} = a_{ij}(\xi_i+\xi_j).$$

The cubic $c_3$, *i. e.*, the intersection of $W$ and $\Omega$, maps into the sextic curve $x_i=\xi_i y_i$, the curve of double points of § 7. The quadric corresponding to the plane $W$, whose coordinates are (1, 1, 1, 1), is seen from (14) to be $(\xi y)=0$, *i. e.*, is the circumsphere of $T$. This is one of the quadrics of the system and consequently the absolute is one of the conics of the system which map the plane $W$ onto its corresponding surface $R$. Since to the plane $W$ in $s_y$ corresponds the quartic surface $R$ in $s_x$ and to $W$ in $s_x$ corresponds the quadric $(\xi y)=0$ in $s_y$, to the intersection of $W$ and $(\xi y)$, *i. e.*, to the absolute, will correspond the intersection of $W$ and $R$, *i. e.*, a rational quartic on $R$. As the absolute meets the cubic $c_3$ in six points, this quartic on $R$ will be on six points of the sextic $x_i = \xi_i y_i$. These two sets of six points are one and the same.

To the four lines $y_i = 0$, which are the lines of $W$ cut out by the planes of $T$, correspond a set of four lines on $R$ whose equations are

$$e_{01}^2 y_1 + e_{02}^2 y_2 + e_{03}^2 y_3 = 0,$$

and three similar ones.

## § 9. *Surfaces Connected with $\Omega$.*

If three points are chosen, whose Cartesian coordinates are $(a, b, c)$, $(d, e, f)$, $(r, s, t)$ and their polar planes, with regard to the quadric $xy = z$, are determined, they will meet in a fourth point, whose coordinates are $\left(\frac{B}{C}, -\frac{A}{C}, -\frac{D}{C}\right)$, where $A, B, C, D$ are the first minors of $x, y, z, 1$, respectively in

$$\begin{vmatrix} x & y & z & 1 \\ a & b & c & 1 \\ d & e & f & 1 \\ r & s & t & 1 \end{vmatrix} = 0.$$

If we put the $z$-coordinates of these four points, which we shall consider as the vertices of a tetrahedron $T$, equal to zero, the resulting coordinates represent the projections of the points in the plane, $z = 0$. This is the principal tangent plane of the above quadric. These four projected points lie on a circle, for if the coordinates are substituted in the determinant

$$\Delta = \begin{vmatrix} a^2 + b^2, & a & b & 1 \\ d^2 + e^2, & d & e & 1 \\ r^2 + s^2, & r & s & 1 \\ \left(\frac{A}{C}\right)^2 + \left(\frac{B}{C}\right)^2 & \frac{B}{C}, & -\frac{A}{C} & 1 \end{vmatrix},$$

which vanishes if the points are on a circle, $\Delta$ becomes

$$\Delta = (a^2+b^2)(e^2r^2 - d^2s^2) + (d^2+e^2)(a^2s_2 - b^2r^2) + (r^2+s^2)(b^2d^2 - a^2e^2) = 0.$$

$\therefore$ *The vertices of $T$ are on a circular cylinder which is perpendicular to the principal tangent planes of all the quadrics* $\left(\frac{x^2}{y}\right) = 0.$

The reflections of the vertices of $T$ in either of the principal generators of $xy = z$; *i. e.*, in either of the lines $x = 0$, or $y = 0$, will give a new tetrahedron which is inscribed to $T$, *i. e.*, of the kind called $T'$ in the first part of this paper. Let them be $T'$ and $T''$. It is easily seen, either analytically or geometrically, that the vertices of $T'$ and $T''$ are also on circular cylinders perpendicular to the plane $z=0$. It follows directly from the way in which $T'$ and $T''$ are determined that $T$, $T'$ and $T''$ are symmetrically placed with regard to the $z$-axis of the quadric.

Thus, $T$, $T'$ *and* $T''$ *are inscribed in circular cylinders of equal radii, all of which are perpendicular to the principal tangent plane of the quadric* $xy=z$, *and the axes of these three cylinders are on a circular cylinder whose axis is the* $z$-*axis of the quadric.*

For a point to lie on a cylinder is one condition, and as five conditions determine a cylinder, for it to be on four points leaves one degree of freedom. Then there are $\infty^1$ cylinders on four points. Beltrami has shown that axes of such cylinders meet the plane at infinity in the cubic curve $\left(\frac{A^2}{y}\right)=0$. The axes of the above cylinders will all meet $W$ on this same cubic.

The axes of the circular cylinders on $T$ form a ruled surface of the ninth order.* Each generator of this surface is then perpendicular to a double plane of $\Omega$, and conversely, each double plane of $\Omega$ is perpendicular to one generator of the above surface.

The axes of the paraboloids $\left(\frac{x^2}{y}\right)=0$ also form a ruled surface whose order *can not be more than nine*, for its generators pass through and are determined by the sextic curve of double points, and the cubic curve $\left(\frac{A^2}{y}\right)=0$ at infinity.

---

* J. B. Eck, "Über die Verteilung der Axen der Rotationsflächen zweiten Grades, welche durch gegebene Punkte gehen" (Dissertation).

# CONTENTS.



The American Journal of Mathematics will appear four times yearly.

The subscription price of the Journal is $5.00 a volume (foreign postage, 50 cents); single numbers, $1.50. A few complete sets of the Journal remain on sale.

It is requested that all editorial communications be addressed to the Editor of the American Journal of Mathematics, and all business or financial communications to The Johns Hopkins Press, Baltimore, Md., U. S. A.

The Lord Baltimore Press
BALTIMORE, MD., U. S. A.

Lightning Source UK Ltd.
Milton Keynes UK
UKHW012242110219
337137UK00006B/1006/P

9 780260 708731